ADVANCES IN REGENERATIVE MEDICINE

Edited by **Sabine Wislet-Gendebien**

Advances in Regenerative Medicine
Edited by Sabine Wislet-Gendebien

Published by InTech

Janeza Trdine 9, 51000 Rijeka, Croatia

Publishing Process Manager Romana Vukelic

Technical Editor Teodora Smiljanic

Cover Designer InTech Design Team

Additional hard copies can be obtained from orders@intechopen.com

Advances in Regenerative Medicine, Edited by Sabine Wislet-Gendebien
p. cm.
ISBN 978-953-307-732-1

For EU product safety concerns:
IN TECH d.o.o., Prolaz Marije Krucifikse Kozulić 3, 51000 Rijeka, Croatia,
info@intechopen.com or visit our website at intechopen.com.

Meet the editor

Dr. Sabine Wislet-Gendebien obtained her PhD from the Center of Cellular and Molecular Neurobiology at the University of Liège under the supervision of Dr. Bernard Rogister and Dr. Gustave Moonen. She completed her postdoctoral training at the Center for Research in Neurodegenerative Disorders directed by Professor St George-Hyslop, in Dr. Anurag Tandon's laboratory, at the university of Toronto. She is currently a Senior Scientist at the GIGA Neurosciences and an Assistant Professor in the Department of Biomedical and Preclinical Sciences, at the University of Liège. Dr Wislet-Gendebien research focuses on cell replacement therapy in neurological disorders, using adult bone marrow stromal cells.

Contents

Preface

In order to better introduce this book, it is important to define regenerative medicine. This field is built through a combination of multiple elements including living cells, matrix to support the living cells (i.e. a scaffold), and cell communicators (or signaling systems) to stimulate the cells, and their surrounding environment to grow and develop into new tissue or organ. Indeed, regenerative medicine is an emerging multidisciplinary field involving biology, medicine, and engineering that is likely to revolutionize the ways we improve the health and quality of life for millions of people worldwide by restoring, maintaining, or enhancing tissue and organ function.

Even if the origins of regenerative medicine can be found in Greek mythology, as attested by the story of Prometheus, the Greek god whose immortal liver was feasted on day after day by Zeus' eagle; many challenges persist in order to successfully regenerate lost cells, tissues or organs and rebuild all connections and functions. In this book, we will cover a few aspects of regenerative medicine highlighting major advances and remaining challenges in cellular therapy (including cell communicators) and tissue/organ engineering.

Cell replacement therapy

The types of cells that are used are dependent on the type of tissue that needs to be repaired. Several cells have been suggested as suitable for cellular therapies: i.e. embryonic stem cells (ES), induced pluripotent stem cells (iPS); somatic stem cells from fetal or adult tissues. The potential use of fetal tissue or differentiated embryonic stem cells from allogenic sources suffer limitations due to tissue availability, ethical issues or safety concerns. On the contrary, adult somatic stem cells can be used in autologus graft procedure, avoiding patient's immunosuppression. In this book, several chapters will discuss stem cell applications in regenerative medicine focusing on several organs or tissues like brain, heart, liver or retina.

Cell communicators

The circulatory system is involved in the transport of a wide variety of biological molecules and cells and can be considered as the body's basic communication system. Cell communicators act as a signaling system, which stimulates the cells into action.

In some cases those communicators could lead cells to integrate damage tissues and rebuild lost connections, however, some signals could also induce cellular stress responses conducting to cell death. Few of those aspects will be directly addressed in this book.

Tissue engineering or …where biology meets engineering.

All cells within tissues are separated and interlinked by a matrix or structure. The consistency of the matrix may vary from liquid, as in blood; to semi-solid, as in cartilage; to solid, as in bone. Tissue engineers either implant cells into a matrix or create the proper conditions for the living cells to build their own three dimensional matrix. Such a matrix provides the structure that supports the cells and creates the physiological environment for them to interact within the host tissue. The success or failure of an implant material in the body depends on a complex interaction between a synthetic 'foreign body' and the 'host tissue', which involves not only biological, but also mechanical, physical and chemical mediated factors. The latest advances in tissue engineering will be discussed in this book underlying many challenges that remain pending in this field.

Finally, I would like to conclude this preface by expressing my deepest gratitude to all the authors who contributed to the realization of this book.

Sabine Wislet-Gendebien, PhD
GIGA Neurosciences
University of Liège,
Belgium

Part 1

Stem Cells

1

Neural Crest Stem Cells from Adult Bone Marrow: A New Source for Cell Replacement Therapy?

Aneta Glejzer[1], Virginie Neirinckx[1], Bernard Rogister[1,2,3] and Sabine Wislet-Gendebien[1]
[1]GIGA-Neurosciences, University of Liège,
[2]GIGA-Development, Stem cells and Regeneative Medicine, University of Liège,
[3]Neurology Department, CHU,
Belgium

1. Introduction

Neurodegenerative disease is a generic term used for a wide range of acute and chronic conditions whose etiology is unknown such as Parkinson's disease, Huntington's disease, amyotrophic lateral sclerosis (ALS), Alzheimer's disease, but also now for other neurological diseases whose etiology is better known but which are also concerned by a chronic lost of neurons and glial cells such as multiple sclerosis (MS), stroke, and spinal cord injury. Although the adult brain contains small numbers of stem cells in restricted areas, the central nervous system exhibits limited capacity of regenerating lost tissue. Therefore, cell replacement therapies of lesioned brain have provided the basis for the development of potentially powerful new therapeutic strategies for a broad spectrum of human neurological diseases. However, the paucity of suitable cell types for cell replacement therapy in patients suffering from neurological disorders has hampered the development of this promising therapeutic approach.

Stem cells are classically defined as cells that have the ability to renew themselves continuously and possess pluripotent or multipotent ability to differentiate into many cell types. Besides the germ stem cells devoted to give rise to ovocytes or spermatozoïdes, those cells can be classified in three subgroups: embryonic stem cells (ES), induced pluripotent stem cells (iPS) and somatic stem cells (Figure 1). ES cells are derived from the inner mass of blastocyst and are considered as pluripotent stem cells as these cells can give rise to various mature cells from the three germ layers. iPS cells are also pluripotent stem cells, however, those cells derived from adult somatic cells such as skin fibroblasts are genetically modified by introduction of four embryogenesis-related genes (Takahashi et al., 2007; Park et al., 2008). Finally, tissue-specific stem cells known as somatic or adult stem cells are more restricted stem cells (multipotent stem cells) and are isolated from various fetal or adult tissues (i.e. hematopoietic stem cells, bone marrow mesenchymal stem cells, adipose tissue-derived stem cells, amniotic fluid stem cells, neural stem cells, etc.; Reviewed by Kim and de Vellis, 2009).

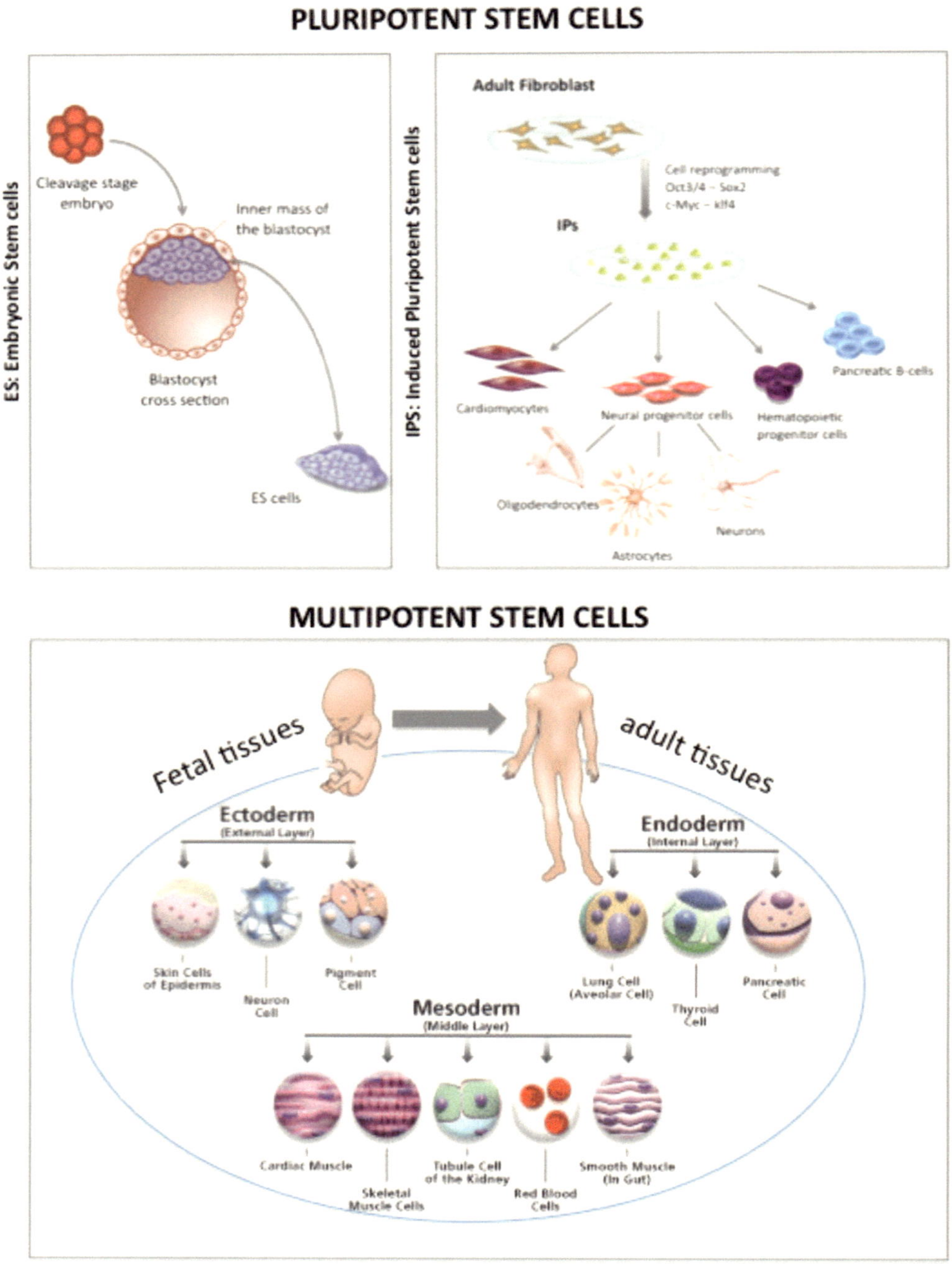

Fig. 1. Stem cell type and origin. Besides germ stem cells, three group of stem cells can be defined according to their differentiating abilities: A. pluripotent embryonic stem cells (ES), B. induced pluripotent stem cells (iPS) and C. multipotent fetal or adult somatic stem cells (Figure adapted from Sigma-Aldrich).

In recent years, neurons and glial cells have been successfully generated from stem cells such as embryonic stem cells (Patani et al., 2010), iPS (Swistowski et al., 2010), mesenchymal stem cells (MSC) (Wislet-Gendebien et al., 2005), and adult neural stem cells (reviewed by Ming et Song, 2011), and extensive efforts by investigators to develop stem cell-based brain transplantation therapies have been carried out. Over the last decade, convincing evidence has emerged of the capability of various stem cell populations to induce regeneration in animal models of Parkinson's disease (PD), Huntington's disease, Alzheimer's disease (AD), multiple sclerosis or cerebral ischemia (Reviewed by Gögel et al., 2011). Some of the studies have already been carried out to clinical trials. In example, in the case of Parkinson's disease, transplantation of fetal ventral mesencephalon tissue directly into the brains of PD patients has been done in a few centers with varying results (Kordower et al., 2008; Li et al., 2008 ; Mendez et al., 2008) and it appeared that using fetal ventral mesencephalon tissue raised numerous problems from ethical issues to heterogeneity and relative scarcity of tissue (reviewed by Wakeman et al., 2011) suggesting that other stem cells (like adult somatic stem cells) may be more suitable for such a therapy. Likewise, ES cells have also been grafted in patients with injured spinal cord, as USA Federal Regulators have cleared the way for the first human trials of human ES-cell research, authorizing researchers to test whether those cells are safe or not (Schwarz et al., 2010). It is still to early to know the effect of ES cells on patient recovery; however, several concerns have been previously raised on animal models as ES cells induced teratocarcimas and some exploratory clinical trials are confirming the animal studies (reviewed by Solter, 2006).
In this chapter, we will review our results concerning identification and characterization of neural crest stem cells (NCSC) in adult bone marrow as a potential source for cellular therapy in neurological disorders. We will also discuss what are the main questions that remain pending concerning the use of those cells in cellular therapy protocols for neurological disorders.

2. Somatic stem cells isolated from adult bone marrow

The post-natal bone marrow has traditionally been seen as an organ composed of two main systems rooted in distinct lineages—the hematopoietic tissue and the associated supporting stroma. The evidence pointing to a putative stem cell upstream of the diverse lineages and cell phenotypes comprising the bone marrow stromal system has made marrow the only known organ in which two separate and distinct stem cells and dependent tissue systems not only coexist but functionally cooperate, defining hematopoietic stem cells (HSC) and mesenchymal stem cells (MSC) (reviewed by Bianco et al., 2001).
MSC were first isolated from the bone marrow (BM-MSC) stem cell niche. More recently, extensive research has revealed that cells with morphological and functional characteristics similar to BM-MSC can be identified in a large number of organs or tissues including adipose tissue and peripheral blood. Despite having different origins, these MSC populations maintain cell biological properties typically associated with stem cells. These include continuous cell cycle progression for self-renewal and the potential to differentiate into highly specialized cell types of the mesodermal phenotype including chondroblast, osteoblast, and adipocyte lineages. Interestingly, BM-MSC have also been reported to be inducible via the ectodermal or endodermal germline, demonstrating the expression of neuron-like factors insulin production or hepatic lineage-associated genes respectively. In addition to these general stem cell properties, the International Society for Cellular Therapy proposed a more specific panel of markers for the characterization of

MSC. Due to the failure to identify a certain unique MSC cell-surface molecule, a set of minimal criteria for MSC was recommended, which includes the capability of adherence to plastic surfaces and the expression of the cell surface markers CD44, CD73, CD90, and CD105 with a concomitant absence of CD14, CD19, CD34, CD45, and HLA-DR expression (Reviewed by Hilfiker et al., 2011).

Originally analyzed because of their critical role in the formation of the hematopoietic microenvironment (HME), bone marrow stromal cells became interesting because of their surprising ability to differentiate into mature neural cell types. More recently, a third stem cell group has been identified as originating from the neural crest, which could explain the capacity of stromal stem cells to differentiate into functional neurons.

2.1 Neural phenotypic plasticity of adult bone marrow stromal cells

Several years ago, we demonstrated that a fraction of bone marrow stromal cells were able to differentiate into functional neurons. Those specific cells were characterized as nestin-positive mesenchymal stem cells (Wislet-Gendebien, 2003-2005). Electrophysiological analyses using the whole-cell patch-clamp technique revealed that adult rat bone marrow stromal cells (Wislet-Gendebien et al., 2005a and 2005b) were able to differentiate into excitable neuron-like cells when they were co-cultivated with mouse cerebellar granule neurons. First, we demonstrated that those cells express several neuronal markers (NeuN and Beta-III-tubulin ; Figure 2), an axonal marker (neurofilament protein recognized by the

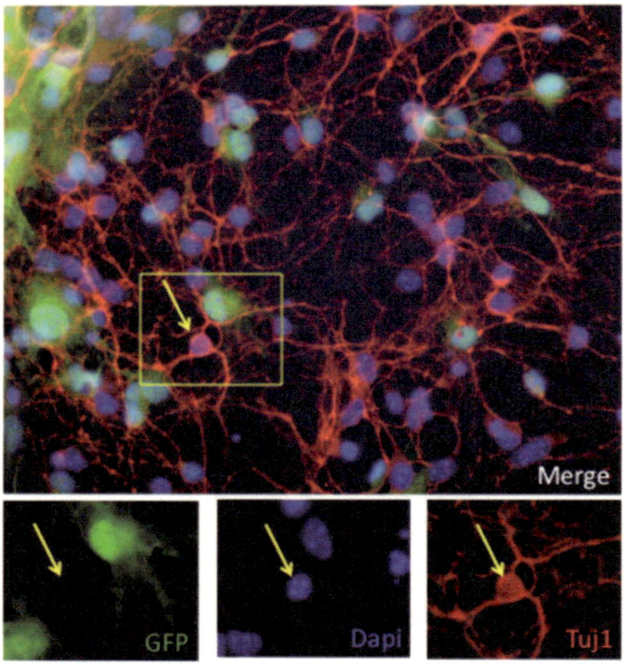

Fig. 2. Neuronal marker expressed by bone marrow stromal cells. Bone marrow stromal cells were co-cultivated for 5 days with GFP-positive cerebellar granule neurons (green). Immunofluorescence labeling showed that beta-III tubulin recognize by Tuj1 antibodies (red) was expressed by about 20% of bone marrow stromal cells (GFP-negative or non-green cells) (Wislet-Gendebien et al., 2005).

monoclonal antibody, SMI31) and a dendritic marker (MAP2ab). Electrophysiological recordings of these nestin-positive bone marrow-derived neuron-like cells (BMDN) were performed and three maturation stages were observed (Table 1). At 4–6 days of co-culture, BMDN showed some neurotransmitter responsiveness (GABA, glycine, serotonin and glutamate) and voltage-gated K^+ currents inhibited by TEA (tetraethylammonium). However, those cells did not express functional sodium voltage-gated channels and have a low membrane potential (*V*rest) (-37.6° ± 3mV, n = 61). During the second week of co-culture, BMDN started to display Na^+ currents reversely inhibitsed by TTX (tetrodotoxin) and became able to fire single spike of action potential. In those older co-cultures, the *V*rest reaches a more negative value, which was closer to the value usually measured in neurons (7–9 days, -50.3 ± 2mV, n = 76 and 10–15 days, -56.7 ± 2.3mV, n = 97).
As only nestin-positive bone marrow stromal cells were able to differentiate into functional neurons, we performed several proteomic and transcriptomic comparisons that pointed out several characteristics like ErbB3 and Sox10 over-expression in nestin-positive MSC, suggesting that these cells could actually be neural-crest derived cells (reviewed by Wislet-Gendebien et al., 2008). Few months later, Nogoshi et al. (2008) confirmed the presence of neural crest derived cells in adult bone marrow.

Maturation of BMDN	5 Days in vitro	8 Days in vitro	12 Days in vitro
Neurotransmitter sensitivities	GABA, Glycin, Glutamate	GABA, Glycin, Glutamate	GABA, Glycin, Glutamate
Potassic voltage-gated channels	+++	+++	+++
Sodic voltage-gated channels	_	+++	+++
Action potentials	_	+++	+++
Trains of action potentials	_	_	_
Synaptic activities	_	_	_
Membrane potential (mV)	-37 ± 3	-50,3 ± 2	- 57,7 ± 2,3

Table 1. Maturation steps of bone marrow derived neuron-like cells

2.2 Characterization of neural crest stem cells from adult bone marrow

2.2.1 Neural crest stem cell origin

In early vertebrate development, the neural crest is specified in the embryonic ectoderm at the boundary of the neural plate and the ectoderm. Once specified, the neural crest cells undergo a process of epithelium to mesenchyme transition (EMT) that will confer them the ability to migrate. The EMT involves different molecular and cellular machineries and implies deep changes in cell morphology and in the type of cell surface adhesion and recognition molecules. When the EMT is complete, they delaminate from the neural folds/neural tube and migrate along characteristic pathways to differentiate into a wide variety of derivates (**Figure 3**; reviewed by Kalcheim, 2000).

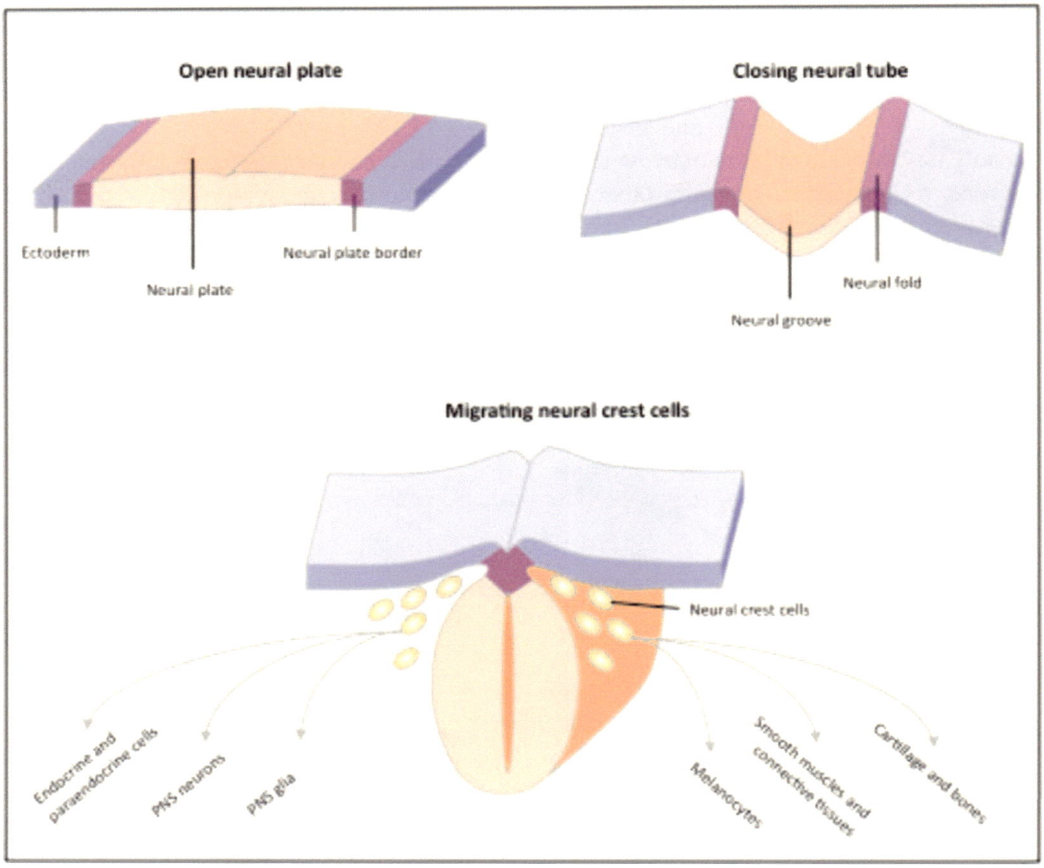

Fig. 3. Neurulation and neural crest migration. As neurulation proceeds, the neural plate rolls up and the neural plate border becomes the neural folds. Near the time of neural tube closure (depending on the species), the neural crest cells go through an epithelial to mesenchymal transition (EMT) and delaminate from the neural folds or dorsal neural tube and migrate along defined pathways.

In 2000, Jiang et al. developed a two-component genetic system based on Cre/lox recombination to label indelibly the entire mouse neural crest population at the time of its formation, and to detect it at any time thereafter. Briefly, the fate of neural crest cells was mapped *in vivo* by mating *ROSA26* Cre reporter (*R26R*) mice, which express β-galactosidase upon Cre-mediated recombination, with mice expressing Cre recombinase under the control of the *Wnt1* promoter. In *Wnt1-Cre/R26R* double transgenic mice, virtually all neural crest stem cells express β-galactosidase. Using this transgenic model, Sieber-Blum and Grim (2004) demonstrated the presence of pluripotent neural crest stem cells in adult follicle hairs, Wong et al. (2006) demonstrated the presence of neural crest cells in the mouse adult skin and Nagoshi et al. (2008) confirmed the presence of NCSC in adult bone marrow (Table 2).

Place	Marker	Animal	Genotype	Reference
Gut	P75NTR	Rat	Wild type	Kruger et al., 2002
DRG		Rat	Wild type	Li et al., 2007
DRG, Whisker pad, bone marrow	EGFP	Mouse	P0 Wnt1-CRE/CAG-EGFP	Nagoshi et al., 2008
Skin		Mouse	Wild type	Toma et al., 2005
Skin	LacZ	Mouse	Wnt1-CRE/ROSA-LacZ	Sieber-Blum et al., 2004
Skin	EYFP	Mouse	Dct-Cre/ROSA-EYFP	Wong et al., 2006
Cornea	EGFP	Mouse	P0 Wnt1-CRE/CAG-EGFP	Yoshida et al., 2006
Carotide body	EGFP	Mouse	GFAP promotor-EGFP	Pardal et al., 2007

Table 2. Presence of neural crest derived cells in adult tissues.

2.2.2 Self-renewal ability and multipotency of adult bone marrow NCSC

To consider NCSC from adult bone marrow as a potential source for cellular therapy protocol, a better characterization of those cells was mandatory. In our study, we first address the self-renewal ability, as first characteristic of stemness. Indeed, we demonstrated that NCSC were able to grow as spheres, which is one of the main hallmarks of immature neural cells and proliferate from a single cell culture (clonal culture). We then addressed the multipotency and verify if those NCSC clones were able to differentiate into multiple mature cell types. Indeed, we observed that NCSC were able to differentiate into adipocytes, melanocytes, smooth muscles, osteocytes, neurons and astrocytes (Figure 4, Glejzer et al., 2011).

2.2.3 Maintenance and proliferation of adult bone marrow NCSC

Before using NCSC from adult bone marrow, we have to face some limiting factors like the fact that NCSC are a minority population (less than 1%) in adult bone marrow. As Wnt1 and BMP2 factors were described to help for maintenance and proliferation of NCSC isolated from embryo (Sommer, 2006), we tested those two factors, on adult NCSC isolated from adult bone marrow. Interestingly, we demonstrated that Wnt1 and BMP2 were able to increase the number of NCSC present in bone marrow stromal cell culture, up to four times within 2 passages (Glejzer et al., 2011) reaching 20 % of NCSC.

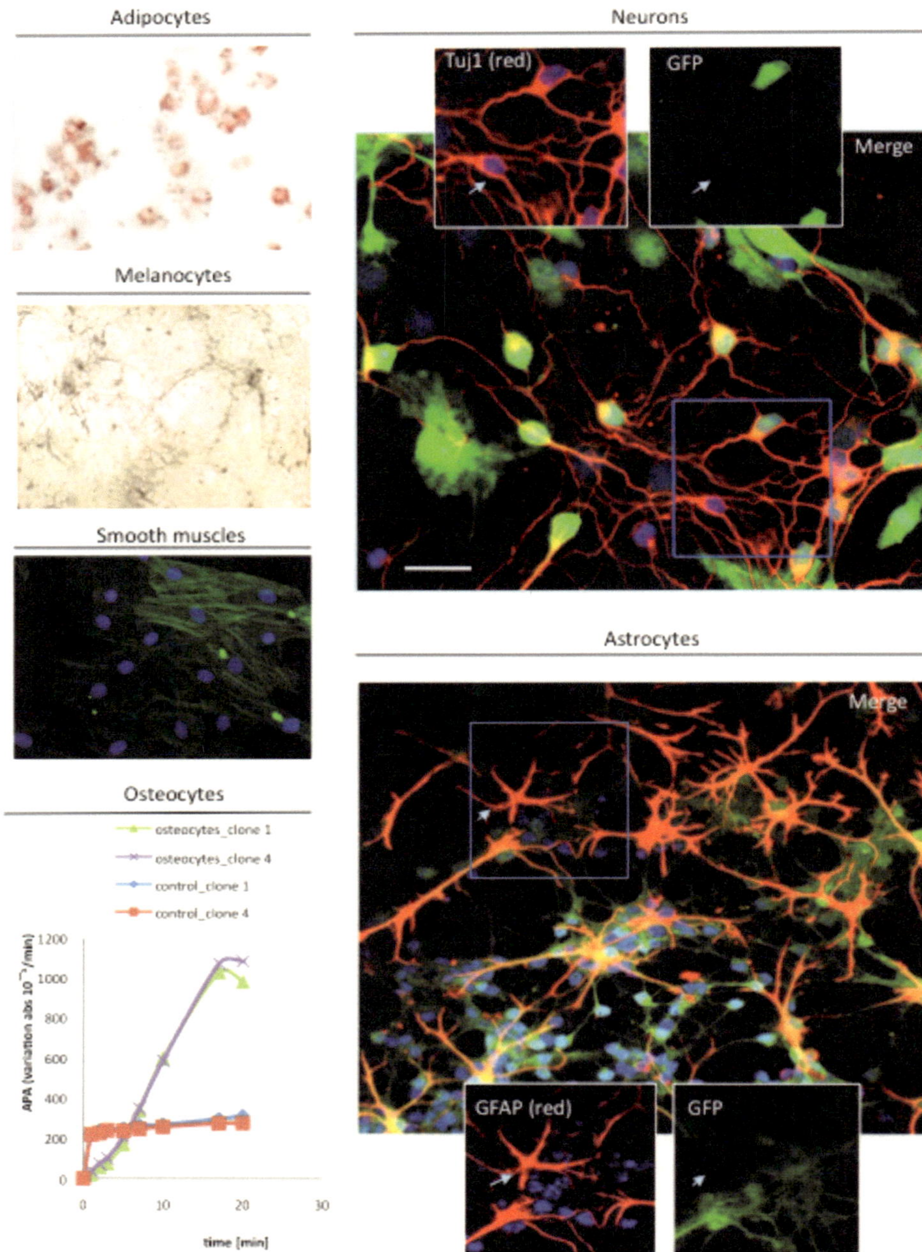

Fig. 4. Multipotency of adult bone marrow NCSC. NCSC clones were subjected to differentiating protocols and were shown to be able to differentiate into adipocytes (Oil Red O labeling), melanocytes (L-DOPA labeling), smooth muscles (SMA-labeling) and Osteocytes (alkaline phosphatase activity). Moreover, when co-cultured with cerebellar granule neurons, we were able to differentiate NCSC clones into neurons (betaIII-tubulin labeling by Tuj1 monoclonal antibody) or astrocytes (GFAP labeling).

3. In vivo characterization of neural crest stem cells and/or bone marrow stromal cells in neurological disorder mice models

3.1 Spinal stroke

Among others, the spinal cord is the collection of fibers that runs from or to the brain through the spine, carrying signals from or to the brain to or from the rest of the body. Those signals control a person's muscles and enable the person to feel various sensations. The main consequence of injuries to the spinal cord is the interference with those signals. Those injuries are characterized as "complete" or "incomplete": if the injured person loses all sensation and all ability to control the muscles below the point of the injury, the injury is said "complete"; in the case of an "incomplete" injury, the victim retains some ability to feel sensations or control movement below the injured area.

Main goals in spinal cord repair include reconnecting brain and lower spinal cord, building new circuits, re-myelination of demyelinated axons, providing trophic support, and bridging the gap of the lesion (Reviewed by Enzmann et al., 2006). Overcoming myelin-associated and/or glial-scar-associated growth inhibition are experimental approaches that have been most successfully studied in *in vivo* experiments. Further issues concern gray matter reconstitution and protecting neurons and glia from secondary death (Reviewed by Enzmann et al., 2006).

In this purpose, neural crest stem cells isolated from the bulge of hair follicle have been grafted in rat model of spinal cord lesion (reviewed by Sieber-Blum 2010). Those cells survived, integrated and intermingled with host neurites in the lesioned spinal cord. NCSC were non-migratory and did not proliferate or form tumors. Significant subsets of grafted cells expressed the neuron-specific beta-III tubulin, the GABAergic marker glutamate decarboxylase 67 (GAD67), the oligodendrocyte markers RIP or myelin basic protein (MBP) (Sieber-Blum et al., 2006). More interestingly, functional improvement was shown by two independent approaches, spinal somatosensory evoked potentials (SpSEP) and the Semmes-Weinstein touch test (Hu et al., 2010). The strength of NSCS was fully characterized as they can exert a combination of pertinent functions in the contused spinal cord, including cell replacement, neuroprotection, angiogenesis and modulation of scar formation. However, those results have never been confirmed with human NCSC, which should be the next promising step.

Similar studies were previously performed with bone marrow stromal cells. Indeed, several researches reported the anti-proliferative, anti-inflammatory and anti-apoptotic features of bone marrow stromal cells (reviewed by Uccelli et al., 2011). Indeed, Zeng et al. (2011) demonstrated that BMSC seeded in a three dimensions gelatin sponge scaffold and transplanted in a transected rat spinal cord resulted in attenuation of inflammation, promotion of angiogenesis and reduction of cavity formation. Those BMSC were isolated from 10 weeks old rats and passaged 3 to 6 times. Likewise, Xu et al. (2010) demonstrated that a co-culture of Schwann cell with BMSC had greater effects on injured spinal cord recovery than untreated BMSC. Indeed, analyses of chemokine and cytokine expression revealed that BMSC/Schwann cell co-cultures produced far less MCP-1 and IL-6 than BMSC or Schwann cells cultured alone. Transplanted BMSC may thus improve recovery in spinal cord injured mice through immunosuppressive effects that can be enhanced by a Schwann cell co-culturing step. These results indicate that the temporary presence of BMSC in the injured cord is sufficient to alter the cascade of pathological events that normally occurs after spinal cord injury and therefore, generating a microenvironment which favours an improved recovery. In this study, BMSC were isolated from adult mice and used after 4 passages.

3.2 Multiple sclerosis

Multiple sclerosis (MS) is a common neurological disease and a major cause of disability, particularly affecting young adults. It is characterized by patches of damage occurring throughout the brain and spinal cord with loss of myelin sheaths accompanied by loss of cells that make myelin (oligodendrocytes) (reviewed by Scolding, 2011). In addition, we now know that there is damage to neurons and their axons too, and that this occurs both within these discrete patches and in tissue between them. The cause of MS remains unknown, but an autoimmune reaction against oligodendrocytes and myelin is generally assumed to play a major role and early acute MS lesions almost invariably show prominent inflammation. Efforts to develop cell therapy of nervous system lesion in MS have long been directed towards directly implanting cells capable of replacing lost oligodendrocytes and regenerating myelin sheaths.

To our knowledge, no experiment has been performed to characterize the effect of neural crest stem cells on the improvement of Multiple Sclerosis disease; however, several data can be collected concerning the positive effect of Schwann cells (derived from NCSC) and of bone marrow stromal cells.

As previously described in injured spinal cord, bone marrow stromal cells have been characterized on their anti-proliferative, anti-inflammatory and anti-apoptotic features. These properties have been exploited in the effective treatment of experimental autoimmune encephalomyelitis (EAE), an animal model of multiple sclerosis where the inhibition of the autoimmune response resulted in a significant neuroprotection (reviewed by Uccelli et al., 2011). Based on recent experimental data, a number of clinical trials have been designed for the intravenous (IV) and/or intrathecal (ITH) administration of BMSCs in MS patients (Grigoriadis et al., 2011).

3.3 Parkinson disease

Parkinson's disease (PD) is a chronic, progressive neurodegenerative disorder characterized by a continuous and selective loss of dopaminergic neurons in the *substantia nigra pars compacta* with a subsequent reduction of dopamine release mainly in the striatum. This ongoing loss of nigral dopaminergic neurons leads to clinical diagnosis mainly due to occurrence of motor symptoms such as rigidity, tremor and bradykinesia, which result from a reduction of about 70% of striatal dopamine (reviewed by Meyer et al., 2010).

Levy et al. (2008) analyzed the effect of differentiated human BMSC onto dopaminergic precursor on hemi-Parkinsonian rats, after transplantation into striatum. This graft resulted in improvement of rat behavioral deficits quantified by apomorphine-induced rotational behavior. The transplanted induced-neuronal cells proved to be of superior benefit compared with the transplantation of naive BMSC. Immunohistochemical analysis of grafted brains revealed that abundant induced cells survived the grafting procedure and some of these cells displayed dopaminergic traits.

Similarly, Zhang et al. (2008) isolated and characterized MSCs from **Parkinson**'s **disease** (PD) patients and compared them with MSCs derived from normal adult **bone marrow**. These authors show that PD-derived MSCs are similar to normal MSCs in phenotype, morphology, and differentiation capacity. Moreover, PD-derived MSCs are able of differentiating into neurons in a specific medium with up to 30% having the characteristics of dopamine cells. At last, PD-derived MSCs could inhibit T-lymphocyte proliferation induced by mitogens. These findings indicate that MSCs derived from PD patients' **bone marrow** could be a promising cell type for cellular therapy and somatic gene therapy applications.

3.4 Huntington disease

Huntington disease (HD) is an autosomal dominant genetic disorder caused by the expansion of polyglutamine encoded by CAG repeats in Exon 1 of the *IT15* gene encoding for Huntingtin (Htt). The polyglutamine repeat length determines the age of onset and the overall level of function, but not the severity of the disease (Vassos et al., 2007). Although the exact mechanism underlying HD disease progression remains uncertain, the hallmark of this disease is a gross atrophy of the striatum and cortex and a decrease of GABAergic neurons (DiFiglia et al., 1997).

One strategy for HD therapy is to enhance neurogenesis, which has been studied by the administration of Stem/progenitor cells, including BMSC. Several studies (reviewed by Snyder et al., 2010) showed that BMSC promote repair of the CNS by creating a more favorable environment for neuroprotection and regeneration through the secretion of various cytokines and chemokines. Moreover, Snyder et al. (2010) demonstrated that BMSC injected into the dentate gyrus of HD mice model increased neurogenesis and decreased atrophy of the striatum.

3.5 Alzheimer disease

Alzheimer's disease (AD) is the most common form of dementia, affecting more than 18 million people worldwide. With increased life expectancy, this number is expected to rise in the future. AD is characterized by progressive memory deficits, cognitive impairment, and personality changes associated with the degeneration of multiple neuronal types and pathologically by the presence of neuritic or amyloid plaques and neurofibrillary tangles (Reviewed by Selko, 2001). Amyloid β-peptide (Aβ) appears to play a key pathogenic role in AD, and studies have connected Aβ plaques with the formation of intercellular tau tangles, another neurotoxic feature of AD (Reviewed by Mattson, 2004). Currently, no treatment is available to cure or prevent the neuronal cell death that results in inevitable decline in AD patients.

The innate immune system is the vital first line of defense against a wide range of pathogens and tissue injuries, triggering inflammation through activation of microglia and macrophages. Many studies have shown that microglia are attracted to and surround senile plaques both in human AD samples and in rodent transgenic models that develop AD-related disease (Simard et al., 2006). In this context, Lee and al. (2010) demonstrated that treated APP/PS1 mice (mouse model of AD) with BM-MSCs promoted microglial activation, rescued cognitive impairment, and reduced Aβ and tau pathology in the mouse brain.

4. Conclusions

The NCSC is one of the most intriguing cells in the field of regenerative medicine, because it is easily harvested from various accessible peripheral tissues, which could make autologous transplantation possible. Autologous transplantation would avoid immunological complications as well as the ethical concerns associated with the use of embryonic stem cells. Of the various NCSC, research on skin-derived NCSC is the most advanced mainly due to their easy isolation process. One of the critical questions for the application of NCSC to regenerative medicine is whether cells that are differentiated from NCSCs are functional. Some evidence supports this (reviewed by Nagoshi et al., 2009), however, lots of questions remained pending. By example, a very important question is the differentiation abilities of NCSC isolated from various tissues: are they similar or different?

On the other hand, even if bone marrow stromal cells did not show a strong ability to replace lost neurons in neurodegenerative disorders such as Parkinson or Huntington disease, their impact on inflammation modulation or stimulation of endogenous cells were quite remarkable. This impact is also illustrated by a high number of ongoing clinical trials with these cells (Reviewed by Sensebé et Bourin, 2011). However, the main challenges remain the standardization of cell culture and isolation, to meet the international rules. Indeed, more than ever, it has been demonstrated that bone marrow stromal cells are constituted of an heterogenous population containing multiple stem/progenitor cell types including mesenchymal stem cells and neural crest stem cells, among other. Most of the studies describing the effects of BMSC on inflammation modulation or stimulation of endogenous cells were performed on low passages (< 4), which mainly contain MSC and less than 10 % of NCSC. So we could stipulate that most of these effects were probably due to MSC. However, in a perspective of cell therapy, a strong characterization of the role of each cell type in neuronal recovery seemed mandatory to establish strong and safe protocols.

5. Acknowledgment

Our work presented in this review was supported by grants from the Fonds National de la Recherche Scientifique (FNRS) of Belgium, by a grant of the Action de Recherche Concertée de la Communauté Française de Belgique, and by the Belgian League against Multiple Sclerosis associated with Leon Fredericq Foundation. AG is a Marie Curie Host Fellow for Early Stage Research Training, EURON 020589 within the 6th FP of the EU, Marie Curie Actions - Human Resources and Mobility.

6. References

Bianco P, Riminucci M, Gronthos S, Robey PG (Bone marrow stromal stem cells: nature, biology, and potential applications. Stem Cells 19:180-192.2001).

DiFiglia M, Sapp E, Chase KO, Davies SW, Bates GP, Vonsattel JP, Aronin N (Aggregation of huntingtin in neuronal intranuclear inclusions and dystrophic neurites in brain. Science 277:1990-1993.1997).

Enzmann GU, Benton RL, Talbott JF, Cao Q, Whittemore SR (Functional considerations of stem cell transplantation therapy for spinal cord repair. J Neurotrauma 23:479-495.2006).

Glejzer A, Laudet E, Leprince P, Hennuy B, Poulet C, Shakhova O, Sommer L, Rogister B, Wislet-Gendebien S (Wnt1 and BMP2: two factors recruiting multipotent neural crest progenitors isolated from adult bone marrow. Cell Mol Life Sci 68:2101-2114.2011).

Gogel S, Gubernator M, Minger SL (Progress and prospects: stem cells and neurological diseases. Gene Ther 18:1-6.2011).

Grigoriadis N, Lourbopoulos A, Lagoudaki R, Frischer JM, Polyzoidou E, Touloumi O, Simeonidou C, Deretzi G, Kountouras J, Spandou E, Kotta K, Karkavelas G, Tascos N, Lassmann H (Variable behavior and complications of autologous bone marrow mesenchymal stem cells transplanted in experimental autoimmune encephalomyelitis. Exp Neurol 230:78-89.2011).

Hilfiker A, Kasper C, Hass R, Haverich A (Mesenchymal stem cells and progenitor cells in connective tissue engineering and regenerative medicine: is there a future for transplantation? Langenbecks Arch Surg 396:489-497.2011).

Hu SL, Luo HS, Li JT, Xia YZ, Li L, Zhang LJ, Meng H, Cui GY, Chen Z, Wu N, Lin JK, Zhu G, Feng H (Functional recovery in acute traumatic spinal cord injury after transplantation of human umbilical cord mesenchymal stem cells. Crit Care Med 38:2181-2189.2010).

Jiang X, Rowitch DH, Soriano P, McMahon AP, Sucov HM (Fate of the mammalian cardiac neural crest. Development 127:1607-1616.2000).

Kalcheim C (Mechanisms of early neural crest development: from cell specification to migration. Int Rev Cytol 200:143-196.2000).

Kim SU, de Vellis J (Stem cell-based cell therapy in neurological diseases: a review. J Neurosci Res 87:2183-2200.2009).

Kordower JH, Chu Y, Hauser RA, Freeman TB, Olanow CW (Lewy body-like pathology in long-term embryonic nigral transplants in Parkinson's disease. Nat Med 14:504-506.2008).

Kruger GM, Mosher JT, Bixby S, Joseph N, Iwashita T, Morrison SJ (Neural crest stem cells persist in the adult gut but undergo changes in self-renewal, neuronal subtype potential, and factor responsiveness. Neuron 35:657-669.2002).

Levy YS, Bahat-Stroomza M, Barzilay R, Burshtein A, Bulvik S, Barhum Y, Panet H, Melamed E, Offen D (Regenerative effect of neural-induced human mesenchymal stromal cells in rat models of Parkinson's disease. Cytotherapy 10:340-352.2008).

Li HY, Say EH, Zhou XF (Isolation and characterization of neural crest progenitors from adult dorsal root ganglia. Stem Cells 25:2053-2065.2007).

Li JY, Englund E, Holton JL, Soulet D, Hagell P, Lees AJ, Lashley T, Quinn NP, Rehncrona S, Bjorklund A, Widner H, Revesz T, Lindvall O, Brundin P (Lewy bodies in grafted neurons in subjects with Parkinson's disease suggest host-to-graft disease propagation. Nat Med 14:501-503.2008).

Mattson MP (Pathways towards and away from Alzheimer's disease. Nature 430:631-639.2004).

Mendez I, Vinuela A, Astradsson A, Mukhida K, Hallett P, Robertson H, Tierney T, Holness R, Dagher A, Trojanowski JQ, Isacson O (Dopamine neurons implanted into people with Parkinson's disease survive without pathology for 14 years. Nat Med 14:507-509.2008).

Meyer M, Jensen P, Rasmussen JZ ([Stem cell therapy for neurodegenerative disorders]. Ugeskr Laeger 172:2604-2607.2010).

Ming GL, Song H (Adult neurogenesis in the Mammalian brain: significant answers and significant questions. Neuron 70:687-702.2011).

Nagoshi N, Shibata S, Kubota Y, Nakamura M, Nagai Y, Satoh E, Morikawa S, Okada Y, Mabuchi Y, Katoh H, Okada S, Fukuda K, Suda T, Matsuzaki Y, Toyama Y, Okano H (Ontogeny and multipotency of neural crest-derived stem cells in mouse bone marrow, dorsal root ganglia, and whisker pad. Cell Stem Cell 2:392-403.2008).

Nagoshi N, Shibata S, Nakamura M, Matsuzaki Y, Toyama Y, Okano H (Neural crest-derived stem cells display a wide variety of characteristics. J Cell Biochem 107:1046-1052.2009).

Pardal R, Ortega-Saenz P, Duran R, Lopez-Barneo J (Glia-like stem cells sustain physiologic neurogenesis in the adult mammalian carotid body. Cell 131:364-377.2007).

Park IH, Zhao R, West JA, Yabuuchi A, Huo H, Ince TA, Lerou PH, Lensch MW, Daley GQ (Reprogramming of human somatic cells to pluripotency with defined factors. Nature 451:141-146.2008).

Patani R, Hollins AJ, Wishart TM, Puddifoot CA, Alvarez S, de Lera AR, Wyllie DJ, Compston DA, Pedersen RA, Gillingwater TH, Hardingham GE, Allen ND, Chandran S (Retinoid-independent motor neurogenesis from human embryonic stem cells reveals a medial columnar ground state. Nat Commun 2:214.2011).

Schwarz SC, Schwarz J (Translation of stem cell therapy for neurological diseases. Transl Res 156:155-160.2010).

Scolding N (Adult stem cells and multiple sclerosis. Cell Prolif 44 Suppl 1:35-38.2011).

Sensebe L, Bourin P, Tarte K (Good manufacturing practices production of mesenchymal stem/stromal cells. Hum Gene Ther 22:19-26.2011).

Sieber-Blum M (Epidermal neural crest stem cells and their use in mouse models of spinal cord injury. Brain Res Bull 83:189-193.2010).

Sieber-Blum M, Grim M, Hu YF, Szeder V (Pluripotent neural crest stem cells in the adult hair follicle. Dev Dyn 231:258-269.2004).

Simard AR, Soulet D, Gowing G, Julien JP, Rivest S (Bone marrow-derived microglia play a critical role in restricting senile plaque formation in Alzheimer's disease. Neuron 49:489-502.2006).

Snyder BR, Chiu AM, Prockop DJ, Chan AW (Human multipotent stromal cells (MSCs) increase neurogenesis and decrease atrophy of the striatum in a transgenic mouse model for Huntington's disease. PLoS One 5:e9347.2010).

Solter D (From teratocarcinomas to embryonic stem cells and beyond: a history of embryonic stem cell research. Nat Rev Genet 7:319-327.2006).

Sommer L (Growth factors regulating neural crest cell fate decisions. Adv Exp Med Biol 589:197-205.2006).

Swistowski A, Peng J, Liu Q, Mali P, Rao MS, Cheng L, Zeng X (Efficient generation of functional dopaminergic neurons from human induced pluripotent stem cells under defined conditions. Stem Cells 28:1893-1904.2010).

Takahashi K, Okita K, Nakagawa M, Yamanaka S (Induction of pluripotent stem cells from fibroblast cultures. Nat Protoc 2:3081-3089.2007).

Toma JG, McKenzie IA, Bagli D, Miller FD (Isolation and characterization of multipotent skin-derived precursors from human skin. Stem Cells 23:727-737.2005).

Uccelli A, Benvenuto F, Laroni A, Giunti D (Neuroprotective features of mesenchymal stem cells. Best Pract Res Clin Haematol 24:59-64.2011).

Vassos E, Panas M, Kladi A, Vassilopoulos D (Effect of CAG repeat length on psychiatric disorders in Huntington's disease. J Psychiatr Res 42:544-549.2008).

Wakeman DR, Dodiya HB, Kordower JH (Cell transplantation and gene therapy in Parkinson's disease. Mt Sinai J Med 78:126-158.2011).

Wautier F, Wislet-Gendebien S, Chanas G, Rogister B, Leprince P (Regulation of nestin expression by thrombin and cell density in cultures of bone mesenchymal stem cells and radial glial cells. BMC Neurosci 8:104.2007).

Wislet-Gendebien S, Hans G, Leprince P, Rigo JM, Moonen G, Rogister B (Plasticity of cultured mesenchymal stem cells: switch from nestin-positive to excitable neuron-like phenotype. Stem Cells 23:392-402.2005).

Wislet-Gendebien S, Leprince P, Moonen G, Rogister B (Regulation of neural markers nestin and GFAP expression by cultivated bone marrow stromal cells. J Cell Sci 116:3295-3302.2003).

Wislet-Gendebien S, Wautier F, Leprince P, Rogister B (Astrocytic and neuronal fate of mesenchymal stem cells expressing nestin. Brain Res Bull 68:95-102.2005).

Wong CE, Paratore C, Dours-Zimmermann MT, Rochat A, Pietri T, Suter U, Zimmermann DR, Dufour S, Thiery JP, Meijer D, Beermann F, Barrandon Y, Sommer L (Neural crest-derived cells with stem cell features can be traced back to multiple lineages in the adult skin. J Cell Biol 175:1005-1015.2006).

Xu X, Geremia N, Bao F, Pniak A, Rossoni M, Brown A (Schwann cell co-culture improves the therapeutic effect of bone marrow stromal cells on recovery in spinal cord-injured mice. Cell Transplant.2010).

Yoshida S, Shimmura S, Nagoshi N, Fukuda K, Matsuzaki Y, Okano H, Tsubota K (Isolation of multipotent neural crest-derived stem cells from the adult mouse cornea. Stem Cells 24:2714-2722.2006).

Zeng X, Zeng YS, Ma YH, Lu LY, Du BL, Zhang W, Li Y, Chan WY (Bone Marrow Mesenchymal Stem Cells in a Three Dimensional Gelatin Sponge Scaffold Attenuate Inflammation, Promote Angiogenesis and

Reduce Cavity Formation in Experimental Spinal Cord Injury. Cell Transplant.2011).

Zhang Z, Wang X, Wang S (Isolation and characterization of mesenchymal stem cells derived from bone marrow of patients with Parkinson's disease. In Vitro Cell Dev Biol Anim 44:169-177.2008).

2

Regenerative Medicine for Cerebral Infarction

Masahiro Kameda and Isao Date
Department of Neurological Surgery, Okayama University Graduate School of Medicine, Dentistry and Pharmaceutical Sciences
Japan

1. Introduction

Tissue plasminogen activator (t-PA) is the gold standard drug for cerebral infarction in the acute phase (Adams et al., 2007), but it cannot be administered to all cerebral infarction patients. Some patients who survive the acute phase of cerebral infarction suffer from permanent hemiparesis in the chronic phase, which highlights the need for regenerative medicine to play a more important role in treating such individuals. There are two primary approaches to the use of regenerative medicine for patients with cerebral infarction: exogenous stem cell therapies, and enhancement of endogenous stem cells.

Stem cell transplantation is one of the most widely employed strategies using exogenous stem cells. Many studies with experimental animals have shown that stem cell transplantation enhances functional recovery after cerebral infarction (Kameda et al., 2007; Takahashi et al., 2008). Based on the results of animal experiments, several clinical trials for patients with cerebral infarction are currently ongoing, using stem cell transplantation techniques, typically with mesenchymal stem cells (Detante et al.). However, these clinical trials have been started despite a lack of results showing that transplanted stem cells can reliably replace infarct areas. The principal purpose of cell transplantation in these cases is cell replacement, or replacement and restoration of infarct areas. Nevertheless, only a few percent of the transplanted cells typically survive during the chronic phase of cerebral infarction (Lindvall & Kokaia, 2006). Even more problematic is the fact that few of these transplanted stem cells differentiate into neurons with immunohistological and electrophysiological properties (Anderova et al., 2006). Based on these reports, some scientists maintain that functional improvements can be achieved without cell-replacement, that the effects of trophic factors secreted by the transplanted cells are sufficient (Cabrer et al., 2010; Shimada & Spees, 2011).

Another approach that regenerative medicine can take is enhancement of endogenous stem cells, based on methods that are less invasive than the use of exogenous stem cells. Deep brain stimulation (DBS) for Parkinson's disease patients is an example of a standard therapy now used in clinical situations. A previous report using animal subjects has shown that DBS can enhance the neurogenesis of endogenous stem cells (Toda et al., 2008). Based on this report, we evaluated the effectiveness of electrical stimulation on animals with cerebral infarction. Recently, we showed that electrical stimulation of the cerebral cortex during the acute phase of cerebral infarction exerted anti-apoptotic, angiogenic and anti-inflammatory effects through the PI3K-Akt signaling pathway (Baba et al., 2009). Moreover, we showed that striatal electrical stimulation during the chronic phase of cerebral infarction was effective due to enhancement of endogenous stem cells in response to glial cell-line derived neurotrophic factor (GDNF) and vascular endothelial growth factor (VEGF) upregulation (Morimoto et al., 2010). Electrical stimulation had therapeutic benefit in cerebral infarction cases not only during the acute phase, but also during the chronic phase, which suggests that electrical stimulation has considerable therapeutic potential.

This review summarizes the current consensus concerning regenerative medicine for cerebral infarction, focusing on stem cell transplantation and electrical stimulation techniques, and briefly describes strategies for applying these methods in a clinical setting.

2. Approaches to regenerative medicine for cerebral infarction

2.1 Stem cell transplantation using exogenous stem cells

2.1.1 Donor cell sources

Stem cell transplantation is one of the most established strategies based on the use of exogenous stem cells. Currently, many different types of stem cell can be cultured and transplanted, including induced pluripotent stem cells (iPS cells) (Takahashi & Yamanaka, 2006), embryonic stem cells (ES cells) (Wang et al., 2011), neural stem cells (NSCs) (Kameda et al., 2007; Muraoka et al., 2006) , mesenchymal stem cells (MSCs) (Kurozumi et al., 2004; Wang et al., 2010) , and hematopoietic stem cells (Shyu et al., 2006), and some of these have already been used in clinical trials. Every type of stem cell has unique advantages and disadvantages. Deciding what type of stem cell to transplant requires careful consideration of availability, immune system response, ethical concerns, and the possibility of tumor genesis.

Concerning the availability of stem cells for transplantation, iPS and ES cells are the most promising candidates, but the possibility of tumor formation must be addressed. Many researchers have searched for methods that prevent tumor formation. Regarding iPS cells, for example, Maekawa et al. showed that using maternal transcription factor Glis1 instead of oncogenic c-Myc enhanced the generation of iPS cells when expressed together with key transcription factors Oct3/4, Sox2, and Klf4 (Maekawa et al., 2011).

The majority of animal experiments have demonstrated the neuroprotective effect of transplantation using allografts of adult NSCs or MSCs in the acute phase of ischemia (Kameda et al., 2007; Takahashi et al., 2008), but the effectiveness of stem cell transplantation during the subacute or chronic phase of ischemia was not seen (de Vasconcelos Dos Santos., 2011). To avoid the problem of rejection by the immune system, and ethical issues, autologous stem cell transplantation using adult NSCs and MSCs is attractive, but considerable time is needed to expand these cells so that sufficient quantities are available. Muraoka et al. established an autologous NSC transplantation model using adult rats, in which NSCs were removed from the subventricular zone of adult Fischer 344 rats using stereotactic methods (Muraoka et al., 2008). The NSCs were expanded, which required approximately three weeks, and microinjected into normal hippocampus in the autologous brain. At the present time, MSCs are considered to be a more useful donor cell source in clinical settings than adult NSCs. In Japan, the first clinical trial of autologous MSC transplantation was performed for a patient in the chronic phase of cerebral infarction (Honmou et al., 2011). Detante et al. have started a clinical Phase II trial using autologous MSC transplantation for patients with cerebral infarction, with inclusion criteria that patients must have ischemic stroke confirmed by MRI within the previous 14 days. Based on the previous results from animal experiments as well as the current clinical situation, autologous stem cell transplantation is considered to be of significant benefit to patients who were not administered rt-PA, provided it is performed in the acute phase three hours after onset.

2.1.2 Delivery methods

There are several stem cell delivery methods, such as intraparenchymal transplantation, intravenous administration and intraarterial administration. With any of these methods, the number of surviving transplanted donor cells is thought to affect the extent of recovery from cerebral infarction. Intraparenchymal transplantation can be performed so that stem cells are delivered in the ischemic penumbra, and it is thus the best method for placing the largest number of stem cells in the desired area of the brain, but this is the most invasive method. Intravenous administration, on the other hand, is the least invasive, but most stem cells end up in the liver and lung (Wang et al., 2010), leaving a much smaller number of stem cells surviving in the area of the ischemic penumbra compared with the intraparenchymal transplantation method. This is why the quantity of stem cells administrated intravenously in animal experiments is roughly an order of magnitude larger than that used in other delivery methods (Li et al., 2008; Lundberg et al., 2011). Lappalainen et al. detected an accumulation of graft cells which were intraarterially administrated in the ischemic brain, using SPECT/CT, but such cells were not observed when administrated intravenously (Lappalainen et al., 2008). At the present time, relatively few papers have explored intraarterial methods of administering stem cells for cerebral infarction in animal models. However, an endovascular technique, superselective intraarterial administration to the penumbra via a micro-catheter, can be performed in a clinical situation, and this method is expected to be less invasive than intraparenchymal transplantation.

2.1.3 Graft survival

Although the therapeutic effect of stem cell transplantation depends upon the rate of stem cell survival, research to date has reported that only approximately 5% survives after transplantation (Lindvall et al., 2004). Cytoprotection can enhance the percentage of graft survival. In particular, GDNF has been shown to be an effective neurotropic factor against ischemic injury. Its neuroprotective effect mainly derives from activation of the phosphatidylinositide-3-kinase/Akt (PI3K/Akt) and mitogen-activated protein kinase/ERK (MAPK/ERK) pathways (Nicole et al., 2001; Treanor et al., 1996). During transplantation, stem cells are subject to hypoxic-ischemic injury. Wang et al. showed that graft survival was enhanced by pretreatment with GDNF for three days before NSC transplantation (Wang et al., 2011). Because brain tissue architecture is disrupted in the ischemic brain, the use of biodegradable scaffolds may help transplanted stem cells regenerate and/or restore damaged brain structures and functions, by affecting cell differentiation, morphology, adhesion, or gene expression (Kleinman & Martin, 2005; Steindler, 2002). Jin et al. showed that transplantation of human neural precursor cells (NPCs) in Matrigel scaffolding at the time of transplantation partially improved therapeutic outcome compared to that of NPCs without Matrigel scaffolding, and that the use of NPC/Matrigel cultures dramatically improved the therapeutic effect (Jin et al., 2010). These reports indicate that, when using transplantation methods employing pre-treatment with GDNF, or Matrigel scaffolding, cytoprotective effects that enhance the survival rate of stem cells require time to develop *in vitro* before transplantation and that such development might be related to the enhanced cytoprotective effects observed after transplantation.

2.1.4 Functional recovery mechanisms: cell-replacement versus paracrine effects

In experiments with animals, many researchers have confirmed that stem cell transplantation provides neuroprotective effects immediately after transplantation, based on behavioral analyses and histological analyses that show reductions in infarct volume (Kameda et al., 2007; Kurozumi et al., 2004; Takahashi et al., 2008). Histological analyses also showed that the neuroprotective effects were due to enhanced angiogenesis (Onda et al., 2008), anti-apoptotic effects (Kameda et al., 2007; Kurozumi et al., 2004), and so on. A recent paper showed that mononuclear bone marrow cells played a role in a rapidly developed neuroprotective effect by increasing cerebral blood flow six hours after transplantation, followed by evidence of angiogenesis (Fujita et al., 2010). Yilmaz et al. described remarkable induction of genes for nerve guidance survival (e.g., cytokine receptor-like factor 1, glypican 1, Dickkopf homolog2, osteopontin), as well as increased expression of neurogenerative, nerve guidance, and angiogenic factors (bFGF, bone morphogenetic protein, angiopoietins, neural growth factor), after transplantation with bone marrow stromal cells (Yilmaz et al., 2010). Angiogenesis and anti-apoptotic effects are preferable for neuroprotection, and if the goal were limited to functional recovery, these neuroprotective mechanisms might be sufficient. However, in a strict sense, stem cell transplantation is expected to provide for cell replacement, since stem cells have two outstanding capacities, namely, self-renewal and pluripotency, which means that they can produce neurons, astrocytes and oligodendrocytes (Gage, 2000; Okano, 2002; Temple, 2001).

Cell-replacement therapy requires that transplanted stem cells survive in the damaged brain, differentiate into mature cells, then replace neurons of several phenotypes, and reconstruct new networks with host cells. Several approaches have been studied to enhance neuronal differentiation, and one approach is to transplant site-specific cells. If the site-specific characteristics of NSCs can be maintained during *in vitro* expansion, such cells may differentiate into site-specific neurons after transplantation. Another approach is to modify the cellular characteristics of the stem cells differentiation by transfecting a trophic factor gene (Kameda et al., 2007; Kurozumi et al., 2004). We have showed that, compared with unmodified stem cells, neuronal differentiation is enhanced by transplanting into the ischemic brain adult neural stem/progenitor cells that were modified to secrete GDNF. This enhancement of the differentiation is usually difficult to detect in the ischemic core, and is typically found only in the small ischemic boundary zone. Also, we are still not able to effect a complete replacement of the damaged infarct area using transplanted stem cells.

Liu et al. have shown that DCX-expressing immature neurons in the subventricular zone (SVZ) do not exhibit a Na+ current, and their resting membrane potential is approximately -25mV in the absence of ischemic insult, however, after ischemic insult, such neurons do exhibit a Na+ current, and the membrane potential is hyperpolarized to about -54mV, a voltage similar to that of mature neurons. They also showed that gene analysis indicates that immature DCX cells express immature markers for Sox 2 and nestin in the absence of ischemic insult, but tyrosine hydroxylase (TH) is expressed as a mature marker after ischemic insult (Liu et al., 2009). After ischemic insult, immature stem cells become able to express the same phenotypes as mature neurons.

Research published during the last three years, however, indicates that cell replacement via stem cells transplantation is not essential to functional recovery (Ramos-Cabrer et al., 2010; Shimada & Spees, 2011). Previously, there was a research trend that aimed specifically at developing cell replacement therapies, as many researchers sought better methods to improve graft survival rates and enhance neuronal differentiation. Nevertheless, the present survival rate for transplanted cells during the chronic phase of cerebral infarction remains in the single digits (Lindvall et al., 2004), and few transplanted stem cells differentiate into neurons with immunohistological and electrophysiological properties (Anderova et al., 2006). This has been interpreted by some researchers as indicating that functional improvements can be achieved without cell replacement, and that the effects of trophic factors secreted by the transplanted cells are sufficient. Ramos-Cabrer et al. recently have shown that post-stroke functional recovery after stem cell transplantation is due to paracrine mechanisms, not cell replacement. They found no evidence of surviving grafted stem cells six months after stem cell transplantation, and, compared with control animals, functional recovery was confirmed even during the chronic phase of cerebral infarction (Ramos-Cabrer et al., 2010). This report supports a new interpretation concerning the importance of paracrine mechanisms to functional recovery from cerebral infarction. Based on this new interpretation, clinical trials using stem cells to treat cerebral infarction are presently ongoing (Detante et al.).

2.1.5 Future directions

An increasing number of research papers on stem cell transplantation are focused on the neuroprotective effects provided by paracrine mechanisms (Shimada & Spees, 2011; Sun et al., 2010). Although cell replacement therapy is likely to remain an important target of stem cell transplantation, the focus on paracrine mechanisms will spur the development of clinical trials, for which long-term efficacy and safety are crucial evaluative factors. Thus, standardization of techniques will be more important compared with the procedures used in animal experiments. Furthermore, it has been reported that the quality of transplanted stem cells profoundly affects the functional outcome. Assmus et al. reported the results of the REPAIR-AMI (Reinfusion of Enriched Progenitor cells And Infarct Remodeling in Acute Myocardial Infarction) trial, and showed that contamination by red blood cells affected the functionality of isolated bone marrow-derived progenitor cells, and ultimately inhibited recovery from acute myocardial infarction (Assmus et al., 2010). An increasing number of clinical trials applying stem cell therapies for cerebral infarction will be started in the near future, and cell preparation will play a vital role for proper interpretation of the results.

2.2 Enhancement of endogenous stem cells

2.2.1 Long-term potentiation (LTP)

Another useful approach for regenerative medicine in cerebral infarction cases is to enhance endogenous stem cells. LTP is thought to be a cellular and molecular mechanism of hippocampal learning and memory (Bliss & Collingridge, 1993). LTP is observed as a long-lasting enhancement in the efficacy of synaptic transmissions, which requires NMDA receptor activation and increased Ca^{2+} influx. LTP can be induced by brief high-frequency stimulation. Chen et al. showed that high-frequency stimulation or tetanic stimulation induced the release of Wnt3a from hippocampal neurons (Chen et al., 2006). Wnt3a has been shown to be a major regulator of neurogenesis *in vivo* and *in vitro*, and blocking Wnt3a expression has been reported to cause a significant decline in neuronal generation (Davidson et al., 2007). Moreover, recent papers suggest that LTP *per se* enhances neurogenesis (Bruel-Jungerman et al., 2006; Chun et al., 2006, 2009). These findings, linking LTP stimulation to the activation of a large latent neural precursor pool in the dentate gyrus, could explain the ability of specific environmental stimuli to increase the rate of neurogenesis in the hippocampus over prolonged periods.

2.2.2 Exercise

It is now widely accepted that exercise (van Praag et al., 1999), enriched environments (Kempermann et al., 1997), and learning tasks (Gould et al., 1999), can enhance the neurogenesis of endogenous stem cells, and affect regulatory mechanisms that may be linked to LTP. Exercise has been shown to enhance neurogenesis in both intact and disease animal models. Tajiri et al. showed that exercise had neuroprotective effects on a Parkinson's disease model in rats, with enhanced neurogenesis and migration toward the lesioned striatum observed. Furthermore, brain-derived neurotrophic factor (BDNF) and glial cell line-derived neurotrophic factor (GDNF) showed increases in the striatum as a consequence of exercise (Tajiri et al., 2010). Li et al. showed that whisker stimulation, as a peripheral stimulation after focal barrel cortex ischemia, enhanced migration from the subventricular zone, and increased the neurogenesis of endogenous stem cells, due to increased vascular endothelial growth factor (VEGF) and stromal-derived factor-1 expression (SDF-1) in the penumbra. They also showed that local cerebral blood flow recovered to a greater degree due to whisker stimulation (Li et al., 2008). This indicates that increased peripheral stimulation and afferent signals to the ischemic cortex can activate endogenous neural stem cells, cause them to migrate to the injured region, and differentiate into mature neurons. Thus, the beneficial aspects of rehabilitation are recognized as being vital to long-term recovery from ischemic stroke.

2.2.3 Electrical stimulation and repetitive transcranial magnetic stimulation

Electrical stimulation and repetitive transcranial magnetic stimulation have been used in clinical situations for treatment of many central nervous system diseases, for example Parkinson's disease (PD) (Fasano et al., 2010), epilepsy (Fisher et al., 2010), depression (Horvath et al.,2010) , and chronic pain (Rasche et al., 2006). Deep brain stimulation (DBS) has become a standard clinical therapy for PD patients and it can ameliorate motor function in such individuals, although the mechanism remains poorly understood. Toda et al. showed that DBS of the anterior nucleus of the thalamus (AN) enhanced the presence of endogenous stem cells in the hippocampal dentate gyrus and enhanced neurogenesis, which were associated with enhanced behavioral performance. Moreover, they showed that DBS of the AN reversed steroid-induced reductions in neurogenesis (Toda et al., 2008), implying that DBS can modulate synaptic plasticity and hippocampal neurogenesis.

Based on these reports, we have tried to evaluate whether electrical stimulation has a neuroprotective effect that mitigates cerebral infarction damage by the enhancement of endogenous stem cells and neurogenesis. We began with epidural stimulation because we wanted to stimulate the brain using a method that is less invasive compared with the use of exogenous stem cells for stem cell transplantation. In one of our experiments, rats received continuous electrical stimulation above the cerebral cortex during the acute phase of cerebral infarction. This stimulation increased cerebral blood flow, enhanced behavioral recovery and reduced infarct volume. The neuroprotective effect was derived from anti-apoptotic, angiogenic and anti-inflammatory effects through the PI3K-Akt signaling pathway (Baba et al., 2009). As a next step, we hypothesized that electrical stimulation of the striatum could enhance the proliferation, migration, and neuronal differentiation of endogenous stem cells in the subventricular zone even during the chronic phase of the ischemic brain. Striatal electrical stimulation during the chronic phase of cerebral infarction was observed to enhance behavioral recovery and reduce infarct size. This neuroprotective effect was derived from stimulation of endogenous angiogenesis and neurogenesis with GDNF and VEGF upregulation (Morimoto et al., 2010). Machado et al. showed that chronic contralesional electrical stimulation of the lateral cerebellar nucleus improved motor recovery in rats following ischemic strokes, an effect derived from increased perilesional cortical excitability via chronic activation of the dentatothalamocortical pathway (Machado et al., 2009).

Transcranial direct current stimulation (tDCS) has been used in animal experiments and for chronic stroke patients. tDCS is thought to strengthen synaptic connections (Cheeran et al., 2008; Hummel et al., 2005; Nitsche et al., 2003, 2004), through a mechanism similar to that of LTP. Fritsch et al. showed that tDCS improved motor skill learning through enhanced synaptic plasticity that required brain-derived neurotrophic factor (BDNF) secretion and TrkB activation (Fritsch et al., 2010).

Vagus nerve stimulation (VNS) has been used in an animal model of ischemia, and patients given this therapy have demonstrated enhanced behavioral recovery and reduced infarction size (Ay et al., 2009). The potential mechanisms for the observed beneficial effects of VNS are thought to be the suppression of increased neuronal excitability, and the reduction of cytokine overproduction and inflammation. Collectively, electrical stimulation has been shown to have therapeutic benefit in cases of cerebral infarction not only in the acute phase but also in the chronic phase, which suggests that electrical stimulation has considerable therapeutic potential.

In addition to electrical stimulation, repetitive transcranial magnetic stimulation (rTMS) has been used in animal models of ischemia (Kaga et al., 2003), and in infarct patients with aphasia (Weiduschat et al., 2011), and has shown enhanced functional recovery. Compared with electrical stimulation, rTMS is more widely used because patients can avoid the surgery required for electrode implantation. It is presumed that the mechanism of its neuroprotective effect derives from increased glucose metabolism, inhibition of apoptosis in the ischemic hemisphere (Gao et al., 2010), and increased expression of c-fos, which is followed by upregulation of BDNF (Zhang et al., 2007).

2.2.4 Rehabilitation combined with electrical stimulation

Currently, after cerebral infarction, electrical stimulation (especially epidural electrical stimulation) is mainly performed together with rehabilitation, to enhance the functional recovery that normally occurs during rehabilitation. Northstar Neuroscience has performed clinical trials using epidural electrical stimulation for infarct patients with upper extremity hemiparesis (Northstar Neuroscience, formerly of Seattle, WA, U.S.A.). Unfortunately, the results of a Phase III randomized trial were unsuccessful, but the Phase II study showed that this therapeutic intervention is both safe and effective. Recently, preliminary study results from the same group showed that patients with non-fluent aphasia benefitted from speech-language therapy in combination with epidural electrical stimulation of the premotor cortex, identified by fMRI. Patients with moderate as well as severe aphasia showed functional improvements after epidural electrical stimulation (Cherney et al., 2010).

2.2.5 Parameters and stimulation patterns

As mentioned above, electrical stimulation and transcranial magnetic stimulation can be of significant functional benefit to individuals who have suffered a cerebral infarction. However, stimulation parameters can vary widely even for the same stimulation method. For example, in some of our research on epidural stimulation, we used continuous stimulation with 2Hz pulses of 1ms width at an intensity of 100μA. On the other hand, Moon et al. used intermittent stimulation with 50Hz pulses of 194ms width whose intensity was flexibly adjusted to evoke movement of a forelimb. They also compared the duration of stimulation, and observed that compared to continuous stimulation, intermittent stimulation enhanced functional recovery more effectively (Moon et al., 2009).

The neuroprotective and neurorestorative effects of electromagnetic stimulation depend on the stimulation parameters and pattern. Moon et al. mentioned that the pattern and intensity of stimulation should be modified on an individual basis depending on the extent of the infarct. The efficacy of the results is affected by a large number of parameters, such as the frequency, intensity, pulse width, and duration of the stimulation (i.e., whether it is continuous or intermittent), and the stimulation target area and electrode resistance. Since the best combination of stimulation parameters and pattern are unknown at the outset, researchers tend to stimulate the ischemic brain using different stimulation parameters and patterns, searching for the combination that best enhances the neuroprotective and neurorestorative effects.

Regarding the frequency of stimulation, the difference in effect between high-frequency stimulation and low-frequency stimulation is usually explained as a consequence of different cellular and molecular mechanisms, LTP versus long-term depression (LTD), because synaptic plasticity is one of the mechanisms responsible for enhanced functional recovery due to electrical stimulation. Based on numerous previous reports, brief high-frequency stimulation (100Hz or higher) can induce long-term potentiation (LTP). On the other hand, brief low-frequency stimulation (1 or 2Hz) can induce long-term depression (LTD), which impairs long-lasting enhancement of synaptic transmission. Thus, the mechanism of enhanced recovery observed in response to high-frequency stimulation is discussed in terms of LTP, whereas low-frequency stimulation fails to induce functional recovery.

The duration of stimulation also affects the outcome. Brief high-frequency stimulation induces LTP, but this does not mean that continuous high-frequency stimulation will do the same. In a preliminary study, we could not confirm reductions in infarction volume after continuous high-frequency stimulation, compared with application of low-frequency stimulation (Baba et al., 2009). This implies that inappropriate parameters and stimulation patterns may simply cause tissue damage, and provide no therapeutic effect.

The condition of the brain also affects the results. Most LTP experiments conducted in the field of electrophysiological research are performed using intact rats and mice. As described above, NMDA receptor activation and increased Ca^{2+} influx are required for the induction of LTP. Increased Ca^{2+} influx is also observed in ischemia and because the ischemic brain has already been exposed to increased Ca^{2+} influx, the response to high-frequency stimulation that can induce LTP in intact animals would be different in animals with cerebral infarction.

Although LTP is a possible mechanism whereby electrical stimulation enhances functional recovery, it is likely that other mechanisms are also involved. Electrical stimulation can induce increased regional cerebral blood flow, suppress inflammatory responses, induce anti-apoptotic responses, and enhance angiogenesis, which, individually and in combination, modulate the microenvironment of the infarct brain to enable functional recovery. This implies that electrical stimulation should be performed using appropriate parameters and stimulation patterns that are tailored for the condition of the brain.

2.2.6 Future directions

Due to a recent finding, that aged mice contain a larger pool of latent stem cells than can be activated (Walker et al., 2008), determining the best parameter settings and pattern of electromagnetic stimulation that will yield the best possible functional outcomes in patients with cerebral infarction is of paramount importance. To find ideal parameter values and patterns, the mechanism of electrical and magnetic stimulation must be elucidated in more detail. Given the failure of the Northstar Neuroscience Phase III trial in which epidural electric stimulation was used, rTMS will likely play a more important role for cerebral infarction patients in the future, because it is a less invasive technique. And, although improving cerebral infarction treatment procedures is of vital importance, primary stroke prevention is also essential. Simvastatin enhances hippocampal LTP in mice and causes a significant increase in Akt phosphorylation (Mans et al., 2010), and since LTP *per se* can enhance neurogenesis, as mentioned earlier, this medication should be effective for preventing primary stroke as well as hyperlipidemia. In short, the protection and enhancement of endogenous stem cells may be a key factor in the maintenance and prolongation of health.

2.3 Progress in ischemia analysis methods

Thus far, we have outlined basic therapeutic strategies for treating cerebral ischemia, based on the use of exogenous stem cells and the enhancement of endogenous stem cells. Improvement in therapeutic strategies and further elucidation of the ischemia mechanism in greater detail are both important.

To elucidate the mechanism of ischemia, and the actions of regenerative medicine, we need to analyze the neuronal activity in the ischemic brain by examining the response of single cells, as well as the response of large populations of neurons. For this purpose, the combination of MRI examinations (especially fMRI) and electrophysiological analysis are useful, because fMRI signals are thought to be proportional to the local average of neuronal activity.

The use of fMRI provides a major breakthrough not only in animal experiments but also in treatment of patients. As mentioned above, fMRI enables the enhancement of patient rehabilitation after stroke, via epidural electrical stimulation and detection of premotor cortex function. Using fMRI techniques, functional recovery after ischemia can be monitored, as originally functional areas are reactivated, with preservation of neurovascular coupling (Weber et al., 2008). fMRI also can reveal ipsilateral cortical fMRI responses after peripheral nerve damage, so that increased interneuron activity can be observed. Thus, fMRI enables analysis of modifications in fiber connections, such as callosal interhemispheric projections (Pelled et al., 2009). The use of voltage-sensitive dyes also provides a similar correlation with extracellular direct current potential recording, which enables the analysis of molecular mechanisms of ischemia from an electrophysiological point of view (Farkas et al., 2008). Voltage-sensitive dye techniques allow sensory-evoked depolarization after cerebral infarction to be analyzed in considerable detail (Siglera et al., 2009).

The comparison of results derived from MRI examinations with those obtained from electrophysiological analysis would also be useful. The ischemic penumbra is a major target when attempting to treat cerebral infarction. From an electrophysiological point of view, depolarization is induced in the ischemic core and brief depolarization is induced in peri-infarct areas. Unlike electrophysiological analysis, diffusion-weighted imaging (DWI) in MRI does not include peri-infarct areas, defined as areas where a brief depolarization is seen during an electrophysiological examination. Thus, electrophysiological analysis can detect wider areas of damage than those detectable using histological or MRI techniques (Breschi et al., 2010). Behavioral analysis and fMRI analysis are typically usually used when analyzing and evaluating functional recovery, but electrophysiological analysis is seldom performed. The pursuit of cross-sectional analysis in greater depth should help to clarify the mechanism of cerebral infarction.

3. Conclusion

We have reviewed the therapeutic effect of stem cell transplantation and techniques for the enhancement of endogenous stem cells. As described above, previous studies have shown that functional recovery after cerebral infarction can be enhanced by stem cell transplantation, and that electromagnetic stimulation can provide neuroprotective and/or neurorestorative effects in animal models of ischemia. These results will stimulate additional clinical studies, but the development of more effective and reliable therapies will require further analysis.

4. Acknowledgments

We deeply appreciate the valuable support provided by Prof. Perry Bartlett, Prof. Cliff Abraham, Dr. Tara Walker, and Dr. Takao Yasuhara. We also appreciate the effort and contributions of Drs. Tanefumi Baba, Naoki Tajiri, Feifei Wang, and Takamasa Morimoto.

5. References

Adams, Jr, HP.; Zoppo, Gd.; Alberts, MJ.; Bhatt, DL.; Brass, L.; Furlan, A.; Grubb, RL.; Higashida, RT.; Jauch, EC.; Kidwell, C.; Lyden, PD.; Morgenstern, LB.; Qureshi, AI.; Rosenwasser, RH.; Scott, PA.; & Wijdicks, EFM. (2007). Guidelines for the Early Management of Adults With Ischemic Stroke. *Stroke*, 38, pp.1655-1711.

Anderova, M.; Kubinova, S.; Jelitai, M.; Neprasova, H.; Glogarova, K.; Prajerova, I.; Urdzikova, L.; Chvatal, A.; & Sykova, E. (2006). Transplantation of embryonic neuroectodermal progenitor cells into the site of a photochemical lesion: immunohistochemical and electrophysiological analysis. *J Neurobiol.*, 66(10), pp. 1084-1100.

Assmus, B.; Tonn, T.; Seeger, FH.; Yoon, CH.; Leistner, D.; Klotsche, J.; Schächinger, V.; Seifried, E.; Zeiher, AM.; & Dimmeler, S. (2010). Red Blood Cell Contamination of the Final Cell Product Impairs the Efficacy of Autologous Bone Marrow Mononuclear Cell Therapy. *Journal of the American College of Cardiology*, Vol. 55, No. 13, ISSN 0735-1097/10/$36.00, doi:10.1016/j.jacc.2009.10.059

Ay, I.; Lu, J.; Ay, H.; Sorensen, AG. (2009). Vagus nerve stimulation reduces infarct size in rat focal cerebral ischemia. *Neuroscience Letters*, 459, pp. 147–151.

Baba, T.; Kameda, M.; Yasuhara, T.; Morimoto T.; Kondo, A.; Shingo, T.; Tajiri, N.; Wang, F.; Miyoshi, Y.; Borlongan, CV.; Matsumae, M.; Date, I. (2009). Electrical Stimulation of the Cerebral Cortex Exerts Antiapoptotic, Angiogenic, and Anti-Inflammatory Effects in Ischemic Stroke Rats Through Phosphoinositide 3-Kinase/Akt Signaling Pathway. *Stroke*, 40, pp. e598-e605.

Bliss, TV.; & Collingridge, GL. (1993). A synaptic model of memory: long-term potentiation in the hippocampus. *Nature*, 361, pp. 31-39.

Breschi, GL.; Librizzi, L.; Pastori, C.; Zucca, I.; Mastropietro, M.; Cattalini, A. & Curtis, Md. (2010). Functional and structural correlates of magnetic resonance patterns in a new in vitro model of cerebral ischemia by transient occlusion of the medial cerebral artery, *Neurobiology of Disease*, 39, pp. 181–191.

Bruel-Jungerman, E.; Davis, S.; Rampon, C.; & Laroche, S. (2006). Long-term potentiation enhances neurogenesis in the adult dentate gyrus. *The Journal of Neuroscience*, 26, pp. 5888-5893.

Cabrer, PR.; Justicia, C.; Wiedermann, D.; & Hoehn, M. (2010). Stem Cell Mediation of Functional Recovery after Stroke in the Rat. *PLos ONE*, Volume 5, Issue 9, e12779.

Cheeran, B.; Talelli, P.; Mori, F.; Koch, G.; Suppa, A.; Edwards, M.; Houlden, H.; Bhatia, K.; Greenwood, R.; & Rothwell, JC. (2008). A common polymorphism in the brain-derived neurotrophic factor gene (BDNF) modulates human cortical plasticity and the response to rTMS. *J. Physiol.*, 586, pp. 5717–5725.

Chen, J.; Park, CS.; & Tang, SJ. (2006). Activity-dependent synaptic Wnt release regulates hippocampal long term potentiation. *J Biol Chem.*, 281, pp. 11910-11916.

Cherney, LR.; Erickson, RK.; & Small, SL. (2010). Epidural cortical stimulation as adjunctive treatment for non-fluent aphasia: preliminary findings. *J Neurol Neurosurg Psychiatry*, 81, pp.1014e1021, doi:10.1136/jnnp.2009.184036.

Chun, SK.; Sun, W.; Park, JJ.; & Jung, MW. (2006). Enhanced proliferation of progenitor cells following long-term potentiation induction in the rat dentate gyrus. *Neurobiol Learn Mem*, 86, pp. 322-329.

Chun, SK.; Sun, W.; & Jung, MW. (2009). LTD induction suppresses LTP-induced hippocampal adult neurogenesis. *Neuroreport*, 23, 20(14), pp 1279-1283.

Davidson, KC.; Jamshidi, P.; Daly, R.; Hearn, MT.; Pera, MF.; & Dottori, M. (2007). Wnt3a regulates survival, expansion, and maintenance of neural progenitors derived from human embryonic stem cells. *Mol Cell Neurosci.*, 36, pp. 408-415.

Detante, O.; Intravenous Stem Cells After Ischemic Stroke (ISIS), Available from http://clinicaltrials.gov/ct2/show/NCT00875654?term=mesenchymal+stem+cell+infarction&rank=10 de Vasconcelos Dos Santos, A.; da Costa Reis, J.; Diaz Paredes, B.; Moraes, L.; Jasmin.; Giraldi-Guimaraes, A.; & Mendez-Otero, R. (2011). Therapeutic window for treatment of cortical ischemia with bone marrow-derived cells in rats. *Brain Research*, 1306, pp. 149-158.

Farkas, E.; Pratt, R.; Sengpiel, F.; & Obrenovitch, TP. (2008). Direct, live imaging of cortical spreading depression and anoxic depolarisation using a fluorescent, voltage-sensitive dye. *J Cereb Blood Flow Metab.*, 28(2), pp.251–262, doi:10.1038/sj.jcbfm.9600569

Fasano, A.; Romito,LM.; Daniele, A.; Piano, C.; Zinno, M.; Bentivoglio, AR.; & Albanese, A. (2010). Motor and cognitive outcome in patients with Parkinson's disease 8 years after subthalamic implants. *Brain*, 133(9), pp 2664-2676.

Fisher, R.; Salanova, V.; Witt, T.; Worth, R.; Henry, T.; Gross, R.; Oommen, K.; Osorio, I.; Nazzaro, J.; Labar, D.; Kaplitt, M.; Sperling, M.; Sandok, E.; Neal, J.; Handforth, A.; Stern, J.; DeSalles, A.; Chung, S.; Shetter, A.; Bergen, D.; Bakay, R.; Henderson, J.; French, J.; Baltuch, G.; Rosenfeld, W.; Youkilis, A.; Marks, W.; Garcia, P.; Barbaro, N.; Fountain, N.; Bazil, C.; Goodman, R.; McKhann, G.; Babu, Krishnamurthy, K.; Papavassiliou, S.; Epstein, C.; Pollard, J.; Tonder, L.; Grebin, J.; Coffey, R.; Graves, N.; & SANTE Study Group. (2010). Electrical stimulation of the anterior nucleus of thalamus for treatment of refractory epilepsy. *Epilepsia*, 51(5), pp. 899-908.

Fritsch, B.; Reis, J.; Martinowich, K.; Schambra, HM.; Ji, Y.; Cohen, LG.; & Lu, B. (2010). Direct Current Stimulation Promotes BDNF-Dependent Synaptic Plasticity: Potential Implications for Motor Learning. *Neuron*, 66, pp.198–204.

Fujita, Y.; Ihara, M.; Ushiki, T.; Hirai, H.; Kizaka-Kondoh, S.; Hiraoka, M.; Ito, H.; Takahashi, R. (2010). Early Protective Effect of Bone Marrow Mononuclear Cells Against Ischemic White Matter Damage Through Augmentation of Cerebral Blood Flow. *Stroke*, 41, pp.2938-2943.

Gage, FH. (2000). Mammalian neural stem cells. *Science*, 287, pp. 1433–1438.

Gao, F.; Wang, S.; Guo, Y.; Wang, J.; Lou, M.; Wu, J.; Ding, M.; Tian, M.; & Zhang, H. (2010). Protective effects of repetitive transcranial magnetic stimulation in a rat model of transient cerebral ischaemia: a microPET study. *Eur J Nucl Med Mol Imaging*, 37, pp. 954–961

Gould, E.; Beylin, A.; Tanapat, P.; Reeves, A.; & Shors, TJ. (1999). Learning enhances adult neurogenesis in the hippocampal formation. *Nat Neurosci.*, 2, 260-265.

Honmou, O.; Houkin, K.; Matsunaga, T.; Niitsu, Y.; Ishiai, S.; Onodera, R.; Waxman, SG.; & Kocsis, JD. (2011). Intravenous administration of auto serum-expanded autologous mesenchymal stem cells in stroke. *Brain*, 134(Pt 6), pp. 1790-1807.

Horvath, JC.; Mathews, J.; Demitrack, MA.; Pascual-Leone, A. (2010). The NeuroStar TMS device: conducting the FDA approved protocol for treatment of depression. *J Vis Exp.*, 45, pii: 2345, doi: 10.3791/2345.

Hummel, F.; Celnik, P.; Giraux, P.; Floel, A.; Wu, WH.; Gerloff, C.; & Cohen, LG. (2005). Effects of non-invasive cortical stimulation on skilled motor function in chronic stroke. *Brain*, 128, pp. 490–499.

Jin, K.; Mao, XO.; Xie, L.; Galvan, V.; Lai, B.; Wang, Y.; Gorostiza, O.; Wang, X.; & Greenberg, DA. (2010). Transplantation of Human Neural Precursor Cells in Matrigel Scaffolding Improves Outcome from Focal Cerebral Ischemia after Delayed Postischemic Treatment in Rats. *J Cereb Blood Flow Metab.*, 30(3), pp. 534–544, doi:10.1038/jcbfm.2009.219.

Kaga, A.; Fujiki, M.; Hori, S.; Nakano, T.; & Isono, M. (2003). Motor evoked potentials following transcranial magnetic stimulation after middle cerebral artery and/or basilar artery occlusions in rats. *Journal of Clinical Neuroscience*, 10(4), pp. 470–475

Kameda, M.; Shingo, T.; Takahashi, K.; Muraoka, K.; Kurozumi, K.; Yasuhara, T.; Maruo, T.; Tsuboi, T.; Uozumi, T.; Matsui, T.; Miyoshi, Y.; Hamada, H.; & Date, I. (2007). Adult neural stem and progenitor cells modified to secrete GDNF can protect, migrate and integrate after intracerebral transplantation in rats with transient forebrain ischemia. *European Journal of Neuroscience*, Vol. 26, pp. 1462–1478.

Kempermann, G.; Kuhn, H.G.; & Gage, FH. (1997). More hippocampal neurons in adult mice living in an enriched environment. *Nature* 386, pp. 493-495.

Kleinman, HK.; & Martin, GR. (2005). Matrigel: basement membrane matrix with biological activity. *Seminars in Cancer Biology*, 15:5, pp. 378-386.

Kurozumi, K.; Nakamura, K.; Tamiya, T.; Kawano, Y.; Kobune, M.; Hirai, S.; Uchida, H.; Sasaki, K.; Ito, Y.; Kato, K.; Honmou, O.; Houkin, K.; Date, I.; & Hamada, H. (2004). BDNF gene-modified mesenchymal stem cells promote functional recovery and reduce infarct size in the rat middle cerebral artery occlusion model. Mol Ther., 9(2), pp. 189-197.

Lappalainena, RS.; Narkilahti, S.; Huhtal, T.; Liimatainen, T.; Suuronen. T.; N¨arv¨anen, A.; Suuronena, R.; Hovatta, O.; & Jolkkonend, J. (2008). The SPECT imaging shows the accumulation of neural progenitor cells into internal organs after systemic administration in middle cerebral artery occlusion rats. *Neuroscience Letters*, 440, pp. 246–250.

Li, WL.; Yu, SP.; Ogle, ME.; Ding, XS.; & Wei, L. (2008). Enhanced Neurogenesis and Cell Migration following Focal Ischemia and Peripheral Stimulation in Mice. *Dev Neurobiol.*, 68(13), pp.1474–1486, doi:10.1002/dneu.20674.

Lindvall, O.; Kokaia, Z.; & Martinez-Serrano, A. (2004). Stem cell therapy for human Neurodegenerative disorders–how to make it work. *Nature Medicine*, 10, pp. S42–S50

Lindvall, O.; & Kokaia, Z. (2006). Stem cells for the treatment of neurological disorders. *Nature*, Vol 441, doi:10.1038/nature04960

Liu, XS.; Chopp, M.; Zhang, XG.: Zhang, RL.; Buller, B.; Hozeska-Solgot, A.; Gregg, SR.; & Zhang. ZG. (2009). Gene profiles and electrophysiology of doublecortin-expressing cells in the subventricular zone after ischemic stroke. *J Cereb Blood Flow Metab.*, 29(2), pp. 297–307., doi:10.1038/jcbfm.2008.119.

Lundberg J, Södersten E, Sundström E, Blanc KL, Andersson T, Hermanson O, Holmin S. (2011) Targeted Intra-arterial Transplantation of Stem Cells to the Injured CNS is More Effective than Intravenous Administration - Engraftment is Dependent on Cell Type and Adhesion Molecule Expression. *Cell Transplant.*, Epub ahead of print.

Maekawa M, Yamaguchi K, Nakamura T, Shibukawa R, Kodanaka I, Ichisaka T, Kawamura Y, Mochizuki H, Goshima N, Yamanaka S. (2011). Direct reprogramming of somatic cells is promoted by maternal transcription factor Glis1. *Nature*, 474(7350), pp. 225-229.

Mans, RA.; Chowdhury, N.; Cap, D.; McMahon, LL.; & Li, L. (2010). Simvastatin enhances hippocampal long-term potentiation in C57BL/6 mice. *Neuroscience*, 166, pp. 435–444.

Moon, SK.; Shin, YI.; Kim, HI.; Kim, H.; Lee, JO.; & Lee, MC. (2009). Effect of prolonged cortical stimulation differs with size of infarct after sensorimotor cortical lesions in rats. *Neuroscience Letters*, 460, pp. 152–155

Morimoto, T.; Yasuhara, T.; Kameda, M.; Baba, T.; Kuramoto, S.; Kondo, A.; Takahashi, K.; Tajiri, N.; Wang, F.; Meng, J.; Ji, YW.; Kadota, T.; Maruo, T.; Kinugasa, K.; Miyoshi, Y.; Shingo, T.; Borlongan, CV.; & Date, I. (2010). Striatal stimulation nurtures endogenous neurogenesis and angiogenesis in chronic-phase ischemic stroke rats. DOI: 10.3727/096368910X544915

Muraoka, K.; Shingo, T.; Yasuhara, T.; Kameda, M.; Yuan, W.; Hayase, H.; Matsui, T.; Miyoshi, Y.; & Date, I. (2006). The high integration and differentiation potential of autologous neural stem cell transplantation compared with allogeneic transplantation in adult rat hippocampus. *Experimental Neurology*, 199, pp. 311–327.

Muraoka, K.; Shingo, T.; Yasuhara, T.; Kameda, M.; Yuen, WJ.; Uozumi, T.; Matsui, T.; Miyoshi, Y.; & Date, I. (2008). Comparison of the therapeutic potential of adult and embryonic neural precursor cells in a rat model of Parkinson disease. *J Neurosurg*, 108, pp. 149–159.

Nicole, O.; Ali, C.; Docagne, F.; Plawinski, L.; MacKenzie, ET.; Vivien, D.; & Buisson, A. (2001). Neuroprotection mediated by glial cell line-derived neurotrophic factor: involvement of a reduction of NMDA-induced calcium influx by the mitogen-activated protein kinase pathway. *J Neurosci.*, 21(9), pp. 3024-3033.

Nitsche, MA.; Fricke, K.; Henschke, U.; Schlitterlau, A.; Liebetanz, D.; Lang, N.; Henning, S.; Tergau, F.; & Paulus, W. (2003). Pharmacological modulation of cortical excitability shifts induced by transcranial direct current stimulation in humans. *J. Physiol.*, 553, pp. 293–301.

Nitsche, MA.; Liebetanz, D.; Schlitterlau, A.; Henschke, U.; Fricke, K.; Frommann, K.; Lang, N.; Henning, S.; Paulus, W.; & Tergau, F. (2004). GABAergic modulation of DC stimulation-induced motor cortex excitability shifts in humans. *Eur. J. Neurosci.*, 19, pp. 2720–2726

Northstar Neuroscience.; Safety and Effectiveness of Cortical Stimulation in the Treatment of Stroke Patients With Upper Extremity Hemiparesis (EVEREST), available from http://clinicaltrials.gov/ct2/show/NCT00170716?term=epidural+stimulation&rank=5

Okano, H. (2002) Neural stem cells: progression of basic research and perspective for clinical application. *Keio J. Med.*, 51, pp. 115–128.

Onda, T.; Honmou, O.; Harada, K.; Houkin, K.; Hamada, H.; & Kocsis, JD. (2008). Therapeutic benefits by human mesenchymal stem cells (hMSCs) and Ang-1 gene-modified hMSCs after cerebral ischemia. *J Cereb Blood Flow Metab.*, 28(2), pp. 329-340.

Pelled, G.; Bergstrom, DA.; Tierney, PL.; Conroy, RS.; Chuanga, KH.; Yud, D.; Leopolde, DA.; Waltersb, JR.; & Koretskya, AP. (2009). Ipsilateral cortical fMRI responses after peripheral nerve damage in rats reflect increased interneuron activity. *PNAS*, vol. 106, no. 33, pp. 14114–14119.

Ramos-Cabrer, P.; Justicia, C.; Wiedermann, D.; & Hoehn, M. (2010). Stem Cell Mediation of Functional Recovery after Stroke in the Rat. *PLoS one*, 5(9), e12779

Rasche, D.; Rinaldi, PC.; Young, RF.; & Tronnier, VM. (2006). Deep brain stimulation for the treatment of various chronic pain syndromes. *Neurosurg Focus*, 21(6), E8.

Shimada, IS.; & Spees, JL. (2011). Stem and Progenitor Cells for Neurological Repair: Minor Issues, Major Hurdles, and Exciting Opportunities for Paracrine-Based Therapeutics. *Journal of Cellular Biochemistry*, 112, pp. 374–380.

Shyu, WC.; Lin, SZ.; Chiang, MF.; Su, CY.; & Li, H. (2006). Intracerebral peripheral blood stem cell (CD34+) implantation induces neuroplasticity by enhancing beta1 integrin-mediated angiogenesis in chronic stroke rats. *The Journal of Neuroscience*, 29,26(13), pp. 3444-3453.

Siglera, A.; Mohajerania, MH.; & Murphya, TH. (2009). Imaging rapid redistribution of sensory-evoked depolarization through existing cortical pathways after targeted stroke in mice. *PNAS* ,vol. 106, no. 28, pp. 11759–11764.

Steindler, DA. (2002). Neural stem cells, scaffolds, and chaperones. *Nat Biotechnol.*, 20(11), pp. 1091-1093.

Sun, C.; Zhang, H.; Li, J.; Huang, H.; Cheng, H.; Wang, Y.; Li, P.; & An, Y.(2010). Modulation of the major histocompatibility complex by neural stem cell-derived neurotrophic factors used for regenerative therapy in a rat model of stroke. *Journal of Translational Medicine*, 8, 77, http://www.translational-medicine.com/content/8/1/77

Takahashi, K.; Yamanaka, S. (2006). Induction of pluripotent stem cells from mouse embryonic and adult fibroblast cultures by defined factors. *Cell*, 25, 126(4), pp. 663-676.

Takahashi, K.; Yasuhara, T.; Shingo, T.; Muraoka, K.; Kameda, M.; Takeuchi, A.; Yano, A.; Kurozumi, K.; Agari, T.; Miyoshi, Y.; Kinugasa, K.; & Date, I. (2008). Embryonic neural stem cells transplanted in middle cerebral artery occlusion model of rats demonstrated potent therapeutic effects, compared to adult neural stem cells. *Brain Research*, 1234, pp. 172-82.

Tajiri, N.; Yasuhara, T.; Shingo ,T.; Kondo, A.; Yuan, W.; Kadota, T.; Wang, F.; Baba, T.; Tayra, JT.; Morimoto, T.; Jing, M.; Kikuchi, Y.; Kuramoto, S.; Agari, T.; Miyoshi, Y.; Fujino, H.; Obata, F.; Takeda, I.; Furuta, T.; & Date, I. (2010). Exercise exerts neuroprotective effects on Parkinson's disease model of rats. *Brain Research*, Volume 1310, pp. 200-207

Temple, S. (2001). The development of neural stem cells. *Nature*, 414, pp. 112–117.

Toda, H.; Hamani, C.; Fawcett, A.; Hutchison, W.; & Lozano, A. (2008). The regulation of adult rodent hippocampal neurogenesis by deep brain stimulation. *J Neurosurg.*, 108, pp. 132–138.

Treanor, JJ.; Goodman, L.; de Sauvage, F.; Stone, DM.; Poulsen, KT.; Beck, CD.; Gray, C.; Armanini, MP.; Pollock, RA.; Hefti, F.; Phillips, HS.; Goddard, A.; Moore, MW.; Buj-Bello, A.; Davies, AM.; Asai, N.; Takahashi, M.; Vandlen, R.; Henderson, CE.; & Rosenthal, A. (1996). Characterization of a multicomponent receptor for GDNF. *Nature*, 382(6586), pp. 80-83. van Praag, H.; Christie, BR.; Sejnowski, TJ.; & and Gage, FH. (1999). Running enhancesneurogenesis, learning, and long-term potentiation in mice. *Proc Natl Acad Sci U S A*, 96, pp. 13427-13431.

Walker, TL.; White, A.; Black, DM.; Wallace, RH.; Sah, P.; & Bartlett, PF. (2008). Latent Stem and Progenitor Cells in the Hippocampus Are Activated by Neural Excitation. *The Journal of Neuroscience*, 28(20), pp. 5240 –5247.

Wang F, Yasuhara T, Shingo T, Kameda M, Tajiri N, Yuan WJ, Kondo A, Kadota T, Baba T, Tayra JT, Kikuchi Y, Miyoshi Y, Date I. (2010). Intravenous administration of mesenchymal stem cells exerts therapeutic effects on parkinsonian model of rats: focusing on neuroprotective effects of stromal cell-derived factor-1alpha. *BMC Neuroscience*, 11, 52, doi:10.1186/1471-2202-11-52

Wang, F.; Kameda, M.; Yasuhara, T.; Tajiri, N.; Kikuchi, Y.; Liang, HB.; Tayra, JT.; Shinko, A.; Wakamori, T.; Agari, T.; Miyoshi, Y.; & Date, I. (2011). GDNF pretreatment enhances the survival of neural stem cell following transplantation in Parkinson's disease model of rats. *Accepted by Neuroscience research*

Weber, R.; Ramos-Cabrer, P.; Justicia, C.; Wiedermann, D.; Strecker, C.; Sprenger, C.; & Hoehn, M. (2008). Early Prediction of Functional Recovery after Experimental Stroke: Functional Magnetic Resonance Imaging, Electrophysiology, and Behavioral Testing in Rats. *J. Neurosci.*, 28(5), pp.1022–1029.

Weiduschat, N.; Thiel, A.; Rubi-Fessen, I.; Hartmann, A.; Kessler, J.; Merl, P.; Kracht, L.; Rommel, T.; & Heiss, WD. (2011). Effects of repetitive transcranial magnetic stimulation in aphasic stroke: a randomized controlled pilot study. *Stroke*, 42(2), pp. 409-15.

Yilmaz, G.; Alexander, JS.; Yilmaz, CE.; & Granger, DN. (2010). Induction of neuro-protective/regenerative genes in stem cells infiltrating post-ischemic brain tissue. *Experimental & Translational Stroke Medicine*, 2,11. http://www.etsmjournal.com/content/2/1/11

Zhang, X.; Mei, Y.; Liu, C.; & Yu, S. (2007). Effect of Transcranial Magnetic Stimulation on the Expression of c-Fos and Brain-derived Neurotrophic Factor of the Cerebral Cortex in Rats with Cerebral Infarct. *Journal of Huazhong University of Science and Technology*, 27 (4), pp. 415-418.

3

Human Umbilical Cord Blood Stem Cells Rescue Ischemic Tissues

Dong-Hyuk Park[1], Jeong-Hyun Lee[2], David J. Eve[3], Cesario V. Borlongan[3], Paul R. Sanberg[3,4], Yong-Gu Chung[1] and Tai-Hyoung Cho[1]

[1]*Department of Neurosurgery, Korea University Medical Center, Korea University College of Medicine, Seoul,*
[2]*Department of Anesthesiology and Pain Medicine, Samsung Medical Center, Sungkyunkwan University School of Medicine, Seoul,*
[3]*Center of Excellence for Aging and Brain Repair, Department of Neurosurgery and Brain Repair, University of South Florida College of Medicine, Tampa,*
[4]*Office of Research and Innovation, University of South Florida, Tampa,*
[1,2] *Korea*
[3,4]*USA*

1. Introduction

Stroke causes irreversible and permanent damage in the brain immediately adjacent to the region of reduced blood perfusion. The only Food and Drug Administration (FDA)-approved drug treatment for acute ischemic stroke is the thrombolytic, tissue-plasminogen activator (t-PA), which acts by dissolving the blood clot in the vessel, and restoring proper blood flow, but is only effective if administered within 3 hours of onset. However, only a few stroke patients are seen within 3 hours of the initial onset of an ischemic stroke and thus have the opportunity at full restoration of blood flow with rapid management by t-PA therapy. Meanwhile, the irreversible loss of myocytes after myocardial infarction (MI) also caused by impaired blood flow, followed by fibrosis of the myocardial scar, infarct expansion, concentric hypertrophy, and left ventricular dilation ultimately leads to progressive heart failure. Current clinical interventions to minimize the devastating effects of myocardial infarction are frequently not sufficient to prevent left ventricular remodeling and subsequent development of heart failure, emphasizing the need for more effective therapies.

These two disorders are related by common features as they are both prevalent in the elderly worldwide. The obstruction of blood vessels reduces the availability of blood to the tissues, causes the specific tissue-related symptoms, and eventually damages or destroys the tissues. Despite interventional and medical advances, the number of people suffering from these disorders is on the rise. Most importantly, there are currently no effective treatments for these ischemic disorders after permanent tissue injury.

Human umbilical cord blood (HUCB) has been used to treat children with various hematological malignant and nonmalignant diseases (Sirchia and Rebulla, 1999). More than six thousand HUCB transplantations have been performed worldwide (Laughlin et al., 2001;

Tse and Laughlin, 2005; Wagner et al., 1992), since HUCB was first used to successfully treat a 5-year-old child suffering from Fanconi anemia in 1988 (Gluckman et al., 1989). Recently, a number of reports reveal the many advantages of HUCB cells for cellular therapies, especially compared to bone marrow (BM) stem cells. HUCB cells are immature and elicit a lower incidence of graft rejection, graft versus host disease (GvHD), and post-transplant infections even though they primarily come from an allogeneic origin (Knutsen and Wall, 1999; Tse and Laughlin, 2005). The immaturity of cord blood stem cells has been implicated in the optimal effects of these cells for hematopoietic and somatic organ therapy. Over the past years, many researchers have investigated the therapeutic potential of progenitor cells found within HUCB for stroke and MI. We have demonstrated that systemic administration of HUCB cells provides significant benefits in the stroke and MI animal models. Interestingly, rather than the conventional cell replacement mechanism, we advance alternative pathways of graft-mediated brain repair involving trophic effects resulting from release of various growth factors that afford cell survival, angiogenesis, and anti-inflammation. Here is a critical literature review of the therapeutic benefits of HUCB stem cell transplantation for stroke and MI in pre-clinical experiments.

2. The nature of human umbilical cord blood cells

Hematopoietic progenitors in HUCB are the most primitive and are able to repopulate blood lineages for a long time (Broxmeyer et al., 1992; Nayar et al., 2002; Todaro et al., 2000). The number of myeloid progenitors in HUCB is similar to the number found in BM (Broxmeyer et al., 1992), however, HUCB cells have a greater colony-forming ability (Nakahata and Ogawa, 1982). Thus, they can be expanded in long-term cultures *in vitro* using different growth factors and have longer telomeres than other adult stem cells (Vaziri et al., 1994). Furthermore, it has been shown that HUCB grafts, compared to adult BM stem cells, are more efficient at restoring the host's hematopoietic progenitor cell reservoir (Frassoni et al., 2003).

The mononuclear fraction of HUCB, is primarily composed of lymphocytes and monocytes (Pranke et al., 2001). It has a comparable B-cell population and a lower absolute number of cluster of differentiation 3 positive ($CD3^+$) T-cells but a higher $CD4^+/CD8^+$ ratio compared to adult peripheral blood (Harris et al., 1992; Pranke et al., 2001). When comparing the characteristics of B-cell differentiation *in vitro* from $CD34^+$ cord blood cells with those of peripheral blood, B-cell precursors that are differentiated from cord blood, appear to be more immature (Hirose et al., 2001). The relative immunological immaturity of HUCB, compared to adult cell sources, is further classified by a higher proportion of immature T-cells and decreased numbers of mature memory T-cells (Harris et al., 1992). Moreover, cord blood lymphocytes express cytokine receptor profiles such as interleukin (IL)-2, IL-6, IL-7, tumor necrosis factor (TNF)-α, and interferon (IFN)-γ at lower levels compared to adult blood cells (Zola et al., 1995) and produce greater amounts of the anti-inflammatory cytokines such as IL-10 (Gluckman and Rocha, 2005, Rainsford and Reen, 2002). IL-10 down-regulates expression of CD86 on dendritic cells (DCs), which may prevent initiation of the T-cell mediated inflammatory reaction (Buelens et al., 1995). Increased levels of IL-10 may stimulate regulatory T-cells, which in turn inhibit antigen-specific immune responses (Asseman and Powrie, 1998). HUCB also contains more natural killer (NK) cells but lower numbers of $CD56^+$ cytotoxic T-cells (D'Arena et al., 1998). NK cells can inhibit T-cell proliferation and reduce TNF- α production (El Marsafy et al., 2001). The DCs, which are the sentinel cells of the body and initiate immune responses within the lymph nodes, in cord blood reflect lymphoid DCs that are most likely to be responsible for colonizing neonatal tissue, while they are more myeloid in adults (Willing et al., 2007). The lymphoid DCs induce T-helper cell 2 (anti-inflammatory) responses, which along with the naïve T-cells, may promote down-regulation of immune responses (Arpinati et al., 2000; Willing et al., 2007). The immaturity of the immunological properties of HUCB cells probably causes a prolonged immunodeficient status after HUCB transplantation (Garderet et al., 1998; Thomson et al., 2000). This may explain the low incidence of GvHD and viral transmission. Such a cellular constitution could also allow for less strict donor-recipient matching requirements, hence leading to a shorter waiting period for treatment (Newcomb et al., 2007). Rocha *et al.* found that GvHD incidence was significantly lower in children receiving HUCB transplants compared to BM recipients when the source was from an human leukocyte antigen (HLA)-identical sibling (Rocha et al., 2000).

The level of maturity of a cell is identified by the cell's combination of surface antigens. For example, the $CD34^+$ population, a marker designated for its role in early hematopoiesis, in HUCB can be defined as more primitive than those found in BM because a higher proportion of them are negative for CD38, a marker for pre-lymphoid cells (Cardoso et al., 1993; Conrad and Emerson, 1998). Not only does the mononuclear fraction contain roughly 1% $CD34^+$ cells but these cells appeared to be more immature than those found in BM (Newcomb et al., 2007). Another human antigen, CD133, has also been identified as a hematopoietic stem cell marker that may provide an alternative to CD34 for the selection and expansion of hematopoietic cells for transplantation (Kobari et al., 2001; Yin et al., 1997). It has been shown that about 80% of $CD34^+$ cells express CD133 and more than 97% of CD 133^+ cells are $CD133^+CD34^+$ in fresh cord blood (Hao et al., 2003). Although $CD133^+$ cells comprised 0.67% of the total mononuclear HUCB cells (Ma et al., 2002), expansion of CD 133^+ and $CD133^+CD34^+$ cells was significantly higher than those from the $CD34^+$ cells (Hao et al., 2003). These findings suggest that $CD133^+$ cells may be more primitive hematopoietic progenitor/stem cells than $CD34^+$ cells. Furthermore, $CD133^+$ cells have been identified in fetal brain and in this area they are considered to be neural stem cells (NSCs) (Tamaki et al., 2002; Uchida et al., 2000). Meanwhile, we identified *in vitro* two different subpopulations of mononuclear HUCB cells- adherent and floating (N. Chen et al., 2005). Whereas the adherent cell population mainly contained lymphocytes expressing hematopoietic antigens, there are a significant number of stem cell and neural cell antigens expressed on cells in the floating population. These results suggest that a nonhematopoietic subpopulation of cells exists within the mononuclear fraction of HUCB cells and they seem to have the potential to become neural-like cells. Thus, HUCB seems to include a primitive stem cell population that may give rise to both hematopoietic and neural cells (Garbuzova-Davis et al., 2006). A recent study suggests that the $CD133^+$ population of cells within cord blood can be induced to differentiate along specific lineages, but cell-cell contact is required (Park et al., 2009). However, as described below, other studies suggest that culture in the presence of specific factors is normally sufficient.

A nonhematopoietic stem cell, the mesenchymal stem cell (MSC), has also been found in UCB (Goodwin et al., 2001; Yang et al., 2004). It has been confirmed that MSCs and MSC-like progenitors can be isolated from amniotic fluid, placenta, and HUCB. Moreover, MSCs derived from HUCB show impressive plasticity and differentiate into cells of all three germ-line derivatives (Jeong et al., 2004; Lee et al., 2004; Yang et al., 2004). However, determining a definitive phenotype as well as the surface antigens of MSCs from HUCB has proven to be controversial, unlike that from BM. HUCB MSCs failed to produce macrophage, granulocyte-erythroid-macrophage-megakaryocyte, or granulocyte-macrophage hematopoietic colonies in methylcellulose, whereas supernatants from cultured HUCB MSCs promoted survival of NT2N neural cells and peripheral blood mononuclear cells when cultured under conditions designed to induce cell stress and limit protein synthesis (El-Badri et al., 2006). Furthermore, after incubation in neural differentiation medium, HUCB MSCs expressed the neural cell surface antigen A2B5, the neurofilament polypeptide NF 200, the oligodendrocyte precursor marker 04, intermediate filament proteins characteristic of neural differentiation (nestin and vimentin), as well as the astrocyte marker glial fibrillary acidic protein (GFAP) and the neural progenitor marker class III β-tubulin (El-Badri et al., 2006). We also demonstrated the immunomodulatory effects of HUCB MSCs after co-culture with murine splenocytes (El-Badri et al., 2006). These results indicate that HUCB MSCs possess multiple utilities that may contribute to their therapeutic potency in the treatment of neurological diseases.

3. HUCB in preclinical stroke research

The therapeutic window of current stroke treatment is narrow and restrictive. Many patients seen outside of this window suffer from irreversible deficits. This reality should be taken into consideration when developing other therapeutic options, especially cell-based therapies. HUCB is one of the most promising sources of multipotent stem cells that have shown affirmative effects in *in vivo* studies for the treatment of stroke. We reviewed the literature focusing on delivery techniques such as the route of administration, time, and dose of HUCB cells given and underlying mechanisms responsible for beneficial effects from the transplantation of HUCB cells.

3.1 Delivery routes, time, and dose of HUCB stem cells

Chen *et al.* showed that the intravenous administration of mononuclear HUCB cells at 24 hours or 7 days after middle cerebral artery occlusion (MCAO) in a rat model of stroke significantly improved neurological function (Chen et al., 2001). Upon histological examination of the brains, mononuclear HUCB cells were observed mainly in the cortex and striatum of the injured hemisphere in the ischemic boundary zone. Few cells were found in the contralateral hemisphere. It was determined that some of these mononuclear HUCB cells were immunoreactive for the endothelial cell marker blood clotting factor VIII (FVIII), GFAP, microtubule-associated protein 2 (MAP2), and neuronal nuclei (NeuN) by immunohistochemistry. Xiao *et al.* produced a line of cells isolated from HUCB that they called 'nonhematopoietic umbilical cord blood stem cells' by culturing of the mononuclear fraction and cell sorting for specific CD antigens (Xiao et al., 2005). A reduction in infarction volume was observed after intravenous transplantation of these cells into rats with ischemic brain injury. Histological analysis revealed that some of the transplanted cells were double labeled for human nuclei and NeuN, even though it was unlikely that they contributed to the recovery.

Meanwhile, we compared the effect of intravenous versus intrastriatal injection of mononuclear HUCB cells to assess which route produced the greatest behavioral recovery in rats with a permanent MCAO (Willing et al., 2003a). Behavioral improvement was similar for both cell delivery routes. Spontaneous activity was significantly less when cells were transplanted 24 hours after stroke compared with non-treated animals. However, 2 months after transplantation, significant improvements were found in the step test only after intravenous delivery of the mononuclear HUCB cells. Also, in the passive avoidance test, transplanted animals learned the task faster than non-transplanted rats. These results suggest that intravenous delivery of mononuclear HUCB cells may be more effective than direct striatal delivery in producing long-term functional benefits to the stroke animal. Next, we investigated the dose effect of mononuclear HUCB cells after MCAO (Vendrame et al., 2004). Twenty-four hours after MCAO, rats were intravenously infused with from 10^4 up to 3-5 x 10^7 mononuclear HUCB cells. At 4 weeks after infusion, there was a significant recovery in behavioral function when 10^6 or more mononuclear HUCB cells were delivered. Infarct volume analysis showed an inverse relationship between cell dose and damage volume, which reached significance at the higher doses of mononuclear HUCB cells. Moreover, transplanted cells were observed by immunofluorescence for human nuclei antigen expression, to be localized only in the injured brain hemisphere and spleen. Taken together, this suggests that there may be a dose relationship between introduced transplanted cells, behavioral improvement, and neuronal sparing using mononuclear HUCB cell infusion in the MCAO rat stroke model.

For translation from research to clinical application, it will be important to determine how soon after onset, HUCB cells should be transplanted for the best treatment of stroke. Our group determined the optimal time to administer these cells after a stroke. Using ischemic tissue extracts in an *in vitro* assay, we investigated the migration capability of mononuclear HUCB cells (Newman et al., 2005). This assay revealed increased migration of mononuclear HUCB cells towards both the hippocampal and striatal extracts that were harvested 24-72 hours after stroke. Cytokine-induced neutrophil chemoattractant-1 (CINC-1) and monocyte chemotactic protein-1 (MCP-1) were increased in the extracts at 48 hours after MCAO. This suggests that these substances probably participate in the cell migration. Further analysis showed that growth-regulated oncogene/CINC-1 (the rat equivalent of human IL-8) and MCP-1 were expressed in a time-dependent pattern similar to that of the migration assays. The chemokines which are present in the supernatant may play a major role in the specific mechanisms responsible for the *in vivo* migration of mononuclear HUCB cells after stroke induction (Newman et al., 2005). Recently, we demonstrated that, *in vivo*, the HUCB treatment window is not particularly wide. When we intravenously injected the HUCB cells at times ranging from 3 hours to 30 days after MCAO, maximal improvements were observed with treatment at 48 hours (Newcomb et al., 2006). According to a recent study by Riegelsberger *et al.*, neither the infarct volume nor cleaved caspase 3 activity was significantly affected by intravenous HUCB cell transplantation 24 hours after stroke onset (Riegelsberger et al., 2011). In addition, a recent study of MCAO in spontaneously hypertensive rats demonstrated a therapeutic window of upto 72 hours with intravenous injection of 1 x 10^6 HUCB mononuclear cells based on lesion size and behavioral impairment (Boltze et al., 2011a). The different rat species may explain the slightly longer therapeutic window compared with our studies. These results are promising in that by using mononuclear HUCB cell therapy, the current 3 hour time window of t-PA for the treatment of stroke may be extended to at least 24 hours after the stroke event. These findings suggest that relatively delayed systemic transplantation of HUCB cells failed to produce neuroprotective effects in a permanent stroke rat model. Taken together these results suggest that the grafts actually exist over a very short time frame to either appropriately control the immune system or provide secretory products to cells within the brain to confer therapeutic effects.

3.2 Probable mechanisms responsible for neural protection

Although several reports revealed that HUCB cells expressed neural phenotypes both *in vitro* and *in vivo* (Garbuzova-Davis et al., 2003; Ha et al., 2001; Sanchez-Ramos et al., 2001; Zigova et al., 2002), few cord blood cells survive in the ischemic region compared to the number of infused cells (Chen et al., 2006; Vendrame et al., 2004, 2005; Willing et al., 2003a, 2003b). In addition, two studies using mannitol to permeate the blood brain barrier showed that the passage of growth factors into the brain following MCAO and neonatal hypoxic ischemia was more important than cell migration from the circulation following intravenous injection (Borlongan et al., 2004; Yasuhara et al., 2010). These findings suggest that cell replacement is not the primary mechanism responsible for the functional improvements seen in these animal studies. Recent research in our team has found that HUCB may not only act as a cell replacement source, but also as a neurotrophic, neuroprotective and anti-inflammatory agent. We found that there was an increase in the number of $CD45^+/CD11b^+$ and $CD45^+/B220^+$ B-cells in the brain of rats with permanent MCAO, while the number of $CD45^+/CD11b^+$ and $CD45^+/B220^+$ B-cells in the brain significantly decreased after mononuclear HUCB cell transplantation (Vendrame et al., 2005). The reduction of $CD45^+/CD11b^+$ cells is of particular interest, given that chronic microgliosis is believed to mediate neuronal damage, not only in ischemic injury but also in other neurodegenerative diseases (Streit et al., 1999; Tan et al., 1999). HUCB treatment also decreases the proinflammatory cytokines such as TNF-α and IL-1β, in the brain following stroke as well as reducing the presence of both activated microglia and astrocytes (Vendrame et al., 2005). Therefore, the potential anti-inflammatory effects of HUCB therapy may protect against neuronal death while evidence suggests that the $CD34^+$ component of the transplant may facilitate revascularization (Taguchi et al., 2004). These cellular changes were accompanied by decreases in mRNA and protein expression of pro-inflammatory cytokines in the brain of stroke animals treated with mononuclear HUCB cells. In our next studies, we revealed that, following MCAO, rat spleen size was decreased concomitantly with their $CD8^+$ T-cell counts (Vendrame et al., 2006). Interestingly, the MCAO-induced reduction in spleen size correlated with the extent of ischemic damage, however, HUCB cell treatment prevented the reduction in spleen weight and splenic $CD8^+$ T-cell counts, as well as reducing the degree of brain injury. Additionally, splenocyte proliferation assays demonstrated that HUCB cell treatment opposed MCAO associated T-cell proliferation by increasing the production of IL-10 while decreasing IFN-γ.

As mentioned above, the HUCB cell preparation is a mixed population composed of immature T-cells, B-cells, monocytes/macrophages, and stem cells. Thus, it is important to determine whether the beneficial effects of HUCB injection can be attributed to a specific cell population. Recently, we found that the whole mononuclear fraction of cord blood, stem cell-, T-cell-, and B-cell-depleted preparations improved the function of the impaired left forelimb to a similar extent that exceeded the performance of MCAO-only animals (Womble et al., 2009). However, animals that received the monocyte/macrophage-depleted HUCB preparation performed more poorly than those receiving the other HUCB cells. Additionally, HUCB administration significantly reduced MCAO-induced hyperactivity, while depletion of stem cells, monocytes ($CD14^+$), and B cells prevented this recovery following stroke. These results suggest that stem cells, monocytes ($CD14^+$), and B cells are critical to HUCB-induced recovery following stroke. A study performed in Germany confirmed that the mixed mononuclear cell population was most effective at neuroprotection against ischemia caused by MCAO in a spontaneously hypertensive rats (Boltze et al., 2011b). They showed that $CD34^+$ cells contributed to this effect *in vitro* against oxygen-glucose deprivation (OGD), but did not appear to be important *in vivo*, since a CD34- cell population from HUCB was equally effective.

The expression of cytokines and chemokines produced by HUCB cells under various culturing conditions was also investigated (Newman et al., 2006). The heterogeneous cells from mononuclear fractions of HUCB produced a variety of cytokines when grown in various culture conditions. In particular, IL-8, MCP-1, and IL-1β that have been implicated as the first line of defense in the inflammatory reaction, are more extensively produced than other chemokines in the human body, especially the brain. These results suggest that these factors may be partially responsible for the functional improvements that were observed in the animal stroke models investigating the therapeutic use of HUCB cells. On the other hand, we recently showed that chemokines are also elevated in the ischemic brain area in an *in vivo* study (Jiang et al., 2008). MCP-1 and macrophage inflammatory protein (MIP)-1α seem to play a role in the infiltration of monocytes into the central nervous system under pathological conditions (Babcock et al., 2003; Glabinski et al., 1998). We found that MCP-1 and MIP-1α expression were significantly increased in the ischemic brain, and significantly promoted HUCB cell migration compared to the contralateral side (Jiang et al., 2008). This cell migration was neutralized with polyclonal antibodies against MCP-1 and MIP-1α. Chemokine receptors were also expressed on the surface of transplanted HUCB cells. These findings suggest that the increased chemokines within the ischemic lesion can bind surface receptors on the HUCB cells, and induce homing of systemically delivered HUCB cells into the CNS *in vivo*. Recently, Ou *et al.* showed the potential of intravenous transplantation of HUCB CD34+ cells transfected with the glial cell line-derived neurotrophic factor (GDNF) gene to exert therapeutic benefits in spontaneous hypertensive rats exposed to a transient MCAO (Ou et al., 2010). The stroke animals transplanted with GDNF-modified CD34+ cells showed a significant increase in GDNF protein levels in the infarcted hemisphere, reduced brain infarction volume, and enhanced functional recovery compared to those that received CD34+ cells only.

The immune system and inflammation are known to play a role in the development of stroke. Possible mechanisms of action for HUCB cells include the modification of this effect. Using an organotypic slice culture and OGD, a recent study showed that an intact immune system is not required for the beneficial effects of HUCB cells (Hall et al., 2009a). Cell survival was increased and the release of nitric oxide by resident microglia (an innate inflammatory response) was reduced back to normal levels following incubation with the cells. Further study showed that the secretion of IL-1β by microglia was impaired following hypoxia in the presence of HUCB cells and microglia viability was also decreased (Jiang et al., 2010). *In vivo*, HUCB cells were shown to impair the recruitment of pro-inflammatory microglia and hence the secretion of matrix metalloproteinase 9 and nitric oxide at the site of injury following MCAO (Leonardo et al., 2010). The microglia found at the site of injury tended to be of a ramified shape, which is believed to be a quiescent inactive form.

Protein and microarray analysis of neuronal co-cultures with HUCB cells following OGD revealed the upregulation of a number of specific genes which coincided with improved survival (Shahaduzzaman et al., 2010). Several of these genes e.g. peroxiredoxin were verified as being upregulated *in vivo* also, demonstrating that transplantation of HUCB cells promotes survival gene expression within the brain. OGD treatment of PC12 cells followed by co-culture with HUCB also demonstrated increased cell survival which was shown to be related to the appearance of antioxidants as well as neurotrophic and angiogenic factors (Arien-Zakay et al, 2009).

The majority of studies on stroke and potential therapies focus on the neurons. However, ischemia also affects the white matter and therefore therapies should focus on both the neurons and the oligodendrocytes that myelinate the axons found in the white matter. Two recent studies by the Willing and Pennypacker group demonstrate that infusion of HUCB cells 48 hours after MCAO was also able to protect oligodendrocytes by reducing caspase 3 activation (Hall et al., 2009b) and promoting the expression of genes associated with oligodendrocyte survival, proliferation and function (Rowe et al., 2010). A further study using HUCB-derived MSCs transduced to express hepatocyte growth factor revealed increased behavioral recovery and enhanced remyelination in an intracranial hemorrhage model compared to untransduced MSCs (Liu et al., 2010). In the latter study, the cells were transplanted one week after hemorrhage directly into the left ventricle. These studies demonstrate an additional means by which HUCBs could promote recovery from a stroke by aiding survival of oligodendrocyte survival and myelination of axons.

While the majority of studies suggest mechanisms other than cell replacement, a recent study in canines demonstrated survival of intra-arterially injected HUCB-derived MSCs in a thromboembolic brain ischemia model at 4 weeks (Chung et al., 2009). The cells were injected one day after lesion and were shown to decrease the lesion volume and differentiate into neurons and astrocytes in the neighborhood of von Willebrand factor-positive endothelial cells suggesting an association with new blood vessels.

4. HUCB in preclinical myocardial infarction research

As shown in stroke investigations, the mechanisms by which HUCB stem cells protect cardiomyocytes and improve cardiac function seem to be complex and multifactorial. Transplanted HUCB cells can differentiate into cardiomyocyte-like cells (Cheng et al., 2003; Wu et al., 2006), induce neovascularization in the necrotic area (Botta et al., 2004; Ma et al., 2005; Wu et al., 2007a), modulate the inflammatory reaction induced by ischemic cascades (Henning et al., 2006), and secrete growth factors including vascular endothelial growth factor (VEGF) (Hu et al., 2006; Tang et al., 2005). However, there are some *in vitro* studies that suggest that HUCB stem cells in culture do not differentiate into cardiomyocytes. Roura *et al.* used a number of different methods including culture of HUCB cells with 5-azacytidine, early cardiomyogenesis growth factor cocktails, Wnt signaling activators or direct neonatal rat cardiomyocyte contact and did not observe any transdifferentiation (Roura et al., 2010). By comparison, Avitabile *et al.* recently reported that $CD34^+$ HUCB cells adopted a cardiomyocyte-like phenotype following co-culture with mouse neonatal cardiomyocytes not due to transdifferentiation, but instead by cell fusion with the neonatal cardiomyocytes (Avitabile et al, 2011). These studies would seem to suggest that the *in vivo* influences which appear to cause transdifferentiation (Cheng et al., 2003; Wu et al., 2006), may not be adequately modeled *in vitro*.

The optimal timing and dose of HUCB stem cell administration for treating MI are also critical issues as shown in the animal stroke studies. In most of the previous investigations,

HUCB stem cells that were administered within 24 hours after MI exerted benefits on cardiac function (Botta et al., 2004; H. K. Chen et al., 2005; Henning et al., 2004; Hirata et al., 2005). By contrast, a wide range of HUCB cell numbers were examined in the MI animal models and functional benefits seem to be partly dose-dependent (Botta et al., 2004; H. K. Chen et al., 2005; Henning et al., 2004; Hirata et al., 2005; Kim et al., 2005; Ma et al., 2005). Meanwhile, as with stroke treatments, a variety of administration routes for HUCB cells, including intramyocardial injection, intracoronary injection and intravenous routes, were demonstrated as effective and safe. Henning *et al.* looked at a variety of doses and a variety of routes within 2 hours of ligation of the left coronary artery (Henning et al., 2007). They found that intramyocardial, intra-arterial and intravenous injection of cells, all produced similar significant reductions in infarct size when examined 1 month later, but with different optimum doses of cells (4×10^6, 4×10^6 and 16×10^6 respectively) and concluded that intramyocardial injection appeared to be best. Below we review further reports focusing on the delivery routes of HUCB stem cells to treat MI in animal models.

4.1 Intramyocardial injection

Direct transplantation of HUCB stem cells into the myocardium of the infarcted heart has been widely investigated in *in vivo* studies due to the advantage that it provides a direct route to the damaged area. Hirata *et al.* showed the therapeutic effects of HUCB $CD34^+$ cells on MI in rats (Hirata et al., 2005). $CD34^+$ cell transplantation significantly improved ventricular function as revealed by echocardiography and hemodynamic analysis. Immunohistochemistry studies for human CD34, CD45, and CD31 (Platelet endothelial cell adhesion molecule; PECAM-1) revealed that transplanted HUCB cells survived in the host rat myocardium. These findings suggest that transplanted HUCB cells survived, enhanced neovascularization and improved cardiac function after MI in the animal models. Interestingly, Higuchi *et al.* recently showed, using iron oxide and sodium iodide symporter reporter gene transfected $CD34^+$ HUCB cells, that transplanted cells did not survive for more than 7 days when 4×10^6 cells were injected directly into normal rat hearts (Higuchi et al, 2009). This suggests that the damaging effects of a myocardial infarction may promote the survival of transplanted cells Botta *et al.* transplanted different subpopulations of HUCB cells into the peri-ischemic area at 24 hours after left anterior descending coronary artery (LAD) ligation in the non-obese diabetic–severe combined immune deficient (NOD-SCID) mouse model (Botta et al., 2004). They showed that $CD34^+KDR^+$ (kinase insert domain receptor) cells or $CD34^+$ cells significantly improved the rate of rise of left ventricular pressure (dP/dt) and left ventricle (LV) end diastolic pressure in cardiac function after MI, whereas the same number of $CD34^+KDR^-$ cells did not have any benefit. In histological studies, a limited number of human nuclear antigen positive cells were observed within the cardiomyocytes of mice receiving $CD34^+KDR^+$ HUCB cells. These results suggest that $CD34^+KDR^+$ cells are probably an active subpopulation within $CD34^+$ cells. Their remarkable resistance to apoptosis and angiogenic ability may contribute to their beneficial effects on the damaged cardiac tissues.

Meanwhile, Henning *et al.* injected 1×10^6 HUCB mononuclear cells into the infarcted myocardium 1 hour after rats underwent LAD ligation (Henning et al., 2004). Compared to the vehicle group, the ejection fraction of the HUCB cell-treated group was not decreased significantly at 1 month, but progressively increased at 3 and 4 months. At 4 months, in the HUCB cell-treated group, anteroseptal wall thickening and increased dp/dt (max) were markedly greater than in the vehicle group. Infarct sizes in the HUCB cell treated group were smaller than the vehicle group at 3 and 4 months. These findings suggest that HUCB cell transplantation substantially reduces infarct size associated with improvement of left ventricular function in the rat MI model. Of interest, Kim *et al.* reported that HUCB cells improved cardiac function even when transplanted 4 weeks after ischemia (Kim et al., 2005). In this study, the distal LAD of Yorkshire pigs was occluded by endovascular coil embolization. Human unrestricted somatic stem cells (USSCs) isolated from HUCB following culturing of the mononuclear fraction under specific conditions (Kogler et al., 2004) or culture media were directly injected into the infarcted area at 4 weeks after ischemia. At 4 weeks after transplantation, single-photon emission computed tomography technetium 99m sestamibi radioisotope scans revealed improved regional perfusion and wall motion of the infarct region in the cell-treated group compared to the non-treated control group. Ejection fraction decreased in the control group while it increased in the transplant group. The grafted cells were detected 4 weeks after transplantation by both immunohistochemistry and *in situ* hybridization, and they improved regional and global function after MI. Transplanting 1×10^5 or 1×10^6 USSCs intramyocardially, 20 minutes after ligation in a rat model, was recently shown to lead to a dose-dependent generation of human cardiomyocytes and vasculogenesis which could be detected 4 weeks later. This was determined by the colocalization of cardiac troponin-I and human nuclear antigen as well as class I HLA staining and smooth muscle actin and von Willebrand factor and human nuclear antigen respectively (Iwasaki et al, 2009). These findings suggest that USSCs from HUCB also can be a good candidate for cell-based therapies to treat MI and suggest that they may function to promote new blood vessel growth and cell replacement. Wu *et al.* also reported a significant improvement of cardiac function in rats with MI that were treated with HUCB-derived stem cells compared to the control untreated group (Wu et al., 2007b). Four weeks after transplantation, histological studies revealed that some of the grafted HUCB-derived stem cells survived in the host infarcted myocardium and accumulated around arterioles and scattered in capillary networks. Some of the cells expressed cardiac troponin-T, von Willebrand factor, and smooth muscle actin, suggesting that transplanted HUCB-derived stem cells in the infarcted myocardium can differentiate into cardiomyocytic, endothelial, and smooth muscle cells to repair damaged myocardium. The capillary and arteriole density were also significantly elevated in the HUCB cell-treated group, while the number of apoptotic cells decreased significantly. Taken together, cardiogenesis and angiogenesis probably exert a favorable influence on the recovery of cardiac function even after chronic administration of HUCB cells in the animals with acute MI. Hu *et al.* also demonstrated that angiogenesis induced by transplanted cells plays a major role in cardiac remodeling and improvement of cardiac function in animal MI models (Hu et al., 2006). HUCB mononuclear cells were transplanted into the marginal area of the infarcted myocardium immediately after LAD ligation. One month after MI, the grafted cells survived in the host heart and the collagen density in the LV was significantly lower in the transplanted group, which coincided with higher mRNA expression of VEGF. These results suggest that angiogenesis promoted by transplanted HUCB mononuclear cells in the infarcted area play a critical role in cardiac repair in the MI animal models. Recently, Hu *et al.* investigated the therapeutic efficacy and feasibility of intramyocardial transplantation of 1×10^6 human umbilical cord blood-derived endothelial progenitor cells (hUCB-EPCs) in rats with acute MI (Hu et al., 2009, 2010). These cells were obtained by culturing adherent cells from the mononuclear fraction under specific media conditions. The cell therapy group had increased microvessel formation and a decreased degree of myocardial fibrosis compared to the control non-treated group. The increased microvessel formation was shown to come from both upregulation of endogenous rat cells (as shown by increased rat proliferating cell nuclear antigen and rat PECAM expression) and colocalization of human nuclear antigen and CD31 and anti-VIII staining. Moreover, the degree of myocardial fibrosis was less than that of the control group. The improved global heart function and decreased cardiac fibrosis in rats with acute MI implies the potential benefit of hUCB-EPC transplantation.

Meanwhile, Henning *et al.* revealed that the transplantation of HUCB cells attenuated the inflammatory and immune reaction induced by ischemic cascades (Henning et al., 2006), which likely contributed to the cardiac regeneration of the infarcted heart. Implantation of HUCB cells into infarcted myocardium of non-immunosuppressed rats within 2 or at 24 hours after LAD occlusion, limited the expression of pro-inflammatory cytokines such as TNF-α, MCP-1, MIP-1α, and IFN-γ in the infarcted myocardium and it may be associated with significant decreases in infarct volume. Further analysis was carried out by transplanting 4 x 10^6 cells and performing cytokine analysis and inflammatory cell recruitment at 2, 6, 12, 24 and 72 hours after LAD ligation (Henning et al., 2008). They observed that the infiltration of inflammatory cells and secretion of inflammatory cytokines was significantly reduced by HUCB cells within 2 hours compared to non-treatment after LAD occlusion and this was maintained through to 72 hours.

The number of cells required to be effective in clinical applications, is likely to mean that the cells would need to be expanded *in vitro* prior to delivery. Senegaglia *et al.* investigated whether *in vitro* expansion of $CD133^+$ endothelial progenitor cells derived from HUCB affected their ability to improve the left ventricular ejection fraction following MI in a rat model (Senegaglia et al., 2010). 200,000 expanded or purified (non-expanded) cells were transplanted intramyocardially 7 days after infarction and the left ventricular ejection fraction was found to be equally improved. This suggested that *in vitro* expansion did not alter the transplanted cell's function suggesting that it may be safe to expand these cells *in vitro*. Further studies would be necessary to see if any benefit is conferred by expansion (apart from the more readily available supply of cells) since the authors did report that the expanded cells expressed VEGF mRNA whereas non-expanded did not *in vitro*. However the expansion methods are currently not sufficiently reliable to be used routinely (see later).

Of interest, Ma *et al.* compared the effects of cell transplantation between BM stem cells and HUCB stem cells for injured myocardium. HUCB $CD133^+$ cells (5 x 10^5) were injected directly into the infarcted myocardium induced by cryoinjury of NOD/SCID mice (Ma et al., 2006a; 2006b). An equivalent number of $CD133^+$ BM cells were transplanted and a sham injection was performed in the respective control groups. Both BM $CD133^+$ and HUCB $CD133^+$ cells increased capillary density in the injured myocardium, whereas only BM $CD133^+$ cells improved myocardial contractility assessed by echocardiography *in vivo*. Armiñán *et al.* also suggest that mesenchymal stem cells derived from bone marrow maybe more effective than $CD34^+$ cells isolated from UCB following intramyocardial injection into a nude rat model of MI 7 days after infarction (Armiñán et al., 2010). Both cell types led to improved cardiac function, cardiac cell proliferation and neoangiogenesis, but the MSCs appeared to be more effective at reducing infarct size. However, it is important to note that

in their study Armiñán *et al.* used twice as many MSCs than $CD34^+$ cells (1.2×10^6 vs. 6×10^5) making it difficult to really make a comparison. Gaebel *et al.* demonstrated that MSCs derived from 3 different sources - human BM, HUCB and human adipose tissue - were differentially able to heal male SCID mice after LAD ligation (Gaebel et al, 2011). 400,000 cells were transplanted immediately after ligation and the hearts examined 6 weeks later. While each cell type had some benefit, the best results were seen with the BM-derived cells. However, they also showed that HUCB-derived MSCs had a lower proportion of $CD105^+$ (endoglin, a type 1 membrane glycoprotein) cells compared to the other sources, which decreased further with additional passaging. A purified preparation of $CD105^+$ HUCB cells was shown to possess a similar benefit as BM-derived MSCs. Endoglin is known to be involved in the development of the cardiovascular system and vascular remodeling, which may explain its apparent importance in transplanted cells (Gaebel et al, 2011).
Using a right ventricle overload model, autologous sheep UCB mononuclear cells were shown to improve the diastolic properties of the right ventricle (Yerebakan et al., 2009). The authors suggested that this occurred due to increased angiogenesis. This is one of the few examples of autologous UCB cell transplantation for myocardial impairments so far reported.

4.2 Systemic injection

A number of studies explored the usefulness and effectiveness of the systemic route including intravenous and intra-arterial administration, which could have clinical potential because the procedure is minimally invasive and a minimal burden to the recipients suffering from critical diseases, even though it raises the doubt of adequate homing to the lesion. However, as shown in animal stroke or MI models, there has been emerging evidence suggesting that a significant proportion of the HUCB cells which are systemically administered, will migrate into the infarcted brain or heart lesion and the animals show improvement in their behavioral deficits.
Ma *et al.* revealed the homing ability of HUCB mononuclear cells that were injected in the tail vein of NOD/SCID mice that had (MI+) or had not (MI–) received ligation of the LAD (Ma et al., 2005). Human DNA (hDNA) was detected in marrow, spleen, and liver of both MI+ and MI– mice from 24 hours to 3 weeks after cell injection. By contrast, hDNA was found in the hearts of 53% of MI+ mice but in none of the MI– mice. Infarct size was smaller and collagen deposition was reduced in the cell-treated group. Capillary density in the border zone of the MI was approximately 20% higher, and clusters of HUCB-derived cells were detected in the perivascular interstitium in cell-treated MI+ mice, though the vast majority of new vessels seemed to originate from the endogenous mouse cells. There was no evidence of cardiomyocyte differentiation of the intravenously grafted HUCB mononuclear cells. Interestingly, in infarcted myocardium, stromal cell-derived factor (SDF)-1 mRNA was expressed approximately 7-fold higher than in normal hearts. These results suggest that systemically administered HUCB cells can migrate into the faraway infarcted lesion where they play a role in neovascularization, and beneficially influence tissue-remodeling processes. Leor *et al.* administered HUCB-derived $CD133^+$ progenitor cells intravenously at 7 days after permanent coronary artery ligation in athymic nude rats (Leor et al., 2006). One month after transplantation, LV fractional shortening was much improved. Moreover, anterior wall thickness decreased significantly in control group but not in HUCB cell-treated group. The grafted HUCB cells were found to have migrated and survived in the infarcted myocardium on microscopic examination. Human originated-cells were detected near vessel

walls and LV cavity in the HUCB cell-treated rats. They were incorporated into endothelial cells in six out of the nine cell-treated animals. Scar tissue from cell-treated animals was significantly populated with autologous myofibroblasts as indicated by colocalization of HLA-DR and alpha-smooth muscle actin staining. These findings suggest that intravenous transplanted HUCB cells can migrate, survive, and be incorporated into the host myocardium and produce functional recovery in animal MI models.

A couple of studies focused on the mechanisms of how the systemically transplanted HUCB cells are recruited into the site of tissue damage. Henning *et al.* revealed that the greatest HUCB cell migration into the infarcted myocardium persisted at 2 and 24 hours after LAD occlusion (Henning et al., 2006). Increased SDF-1 in the infarcted myocardium attracts circulating stem cells through the CXCR4 receptor, leading to activation of surface integrins and subsequent recruitment into the vasculature (Ma et al., 2005). These findings suggest that the recruitment of transplanted cells to the injured site was likely due to the concentrations of chemoattractants within the ischemic and infarcted tissues.

Another study using an unspecified number of cells attempted to explore the optimal time for transplantation via the caudal vein (Xing et al., 2009). They injected HUCB cells 1, 5, 10 or 30 days after LAD ligation and they measured cardiac function 4 weeks later. Animals transplanted 5 and 10 days after LAD ligation showed improved cardiac function, with the best results occurring in the 10 day transplant animals. The best survival, more pronounced angiogenesis and VEGF expression was observed within these animals, though no evidence of cardiac regeneration was seen. The authors proposed that the angiogenic effects were the main cause of the improvement. It is worth noting that the authors characterized the cells prior to transplantation and showed that only 35-42% of the cells were $CD34^+$.

Lineage-depleted HUCB cells that express high levels of the cytosolic enzyme aldehyde dehydrogenase (ALDH) were found to possess a high long term repopulation potential whereas cells with low levels of ALDH had negligible long term repopulation potential (Sondergaard et al., 2010). Intravenous administration of these cells one day after LAD ligation in NOD/SCID mice showed specific localization to the injured area within 48 hours and engraftment was higher 4 weeks later using the high ALDH expressing population compared to the low expressing population (Sondergaard et al., 2010). However none of the transplanted cells were observed to became cardiomyocytes, but a few became endothelial cells and the surviving cells were associated with increased vascular density. It is worth noting that only a small number of cells were transplanted (~400,000 cells), which may not be sufficient to see the best effects when administered systemically. These studies show that there are a number of cell types in HUCB which differ in their CD markers and enzyme levels which could affect their ability to be effective.

4.3 Intracoronary delivery

Intracoronary injection can be classified as via the systemic route as it is a subtype of intra-arterial infusion. This method is somewhat more complicated than peripheral access, but can easily be accomplished within several minutes in a well-equipped coronary suite with well-trained physicians. Thus, this method is directly applicable to the real clinical situation. Moreover, intracoronary delivery has an advantage over the intravenous route since the stem cells can be delivered directly to the infarcted myocardium without passing through the systemic circulation. This aspect alleviates the possibility of systemic cell loss such as cell trapping within the lungs and spleen. However, there is also a disadvantage with intra-arterial infusion, since the transplanted cells can form clumps or aggregates if very large quantities of cells are infused, which can subsequently become an embolus (Qian et al., 2006).

While a number of studies have examined the feasibility and efficacy of intracoronary infusion of BM stem cells or peripheral blood-derived mononuclear cells for treating MI in animal and human models (Bartunek et al., 2005; Manginas et al., 2007; Qian et al., 2006), there are few published studies using HUCB cells. Moelker *et al.* failed to show improvement of LV function by intracoronary administration of HUCB cells after MI (Moelker et al., 2007). In this study, approximately 10^8 HUCB stem cells were injected via the intracoronary route 1 week after MI and reperfusion in a swine model. By this time point, end-diastolic volume and LV mass were larger, while ejection fraction was smaller. Regional wall thickening in the left circumflex coronary artery area became akinetic. There was no difference in global and regional LV function at 5 weeks between MI animals receiving HUCB stem cells or normal medium, while surprisingly the infarct size after HUCB stem cell treatment was significantly larger. Transplanted HUCB cells survived only in the infarct border zone at 5 weeks and did not express cardiomyocyte or endothelial markers. Of interest, histological examination revealed that HUCB stem cells transplanted through intracoronary infusion caused micro-infarctions by obstructing vessels. Although the feasibility and efficacy of intracoronary delivery of HUCB could not be determined because there has been few studies about this issue, these results raise safety concerns due to embolic infarction related with the high dose of transplanted HUCB cells required for the intracoronary route (Moelker et al., 2007; Qian et al., 2006).

4.4 Tissue engineering and gene modification

Recent investigations focusing on gene therapy or tissue engineering demonstrate that the combination of stem cell and gene modification may be a useful treatment modality for MI treatment (Mangi et al., 2003; Matsumoto et al., 2005). However, gene manipulation of stem cells outside the body and in the laboratory before transplantation raises safety problems as well as possible moral and ethical concerns when it is applied clinically.

Cortes-Morichetti *et al.* examined the feasibility of a collagen matrix seeded with HUCB stem cells and their engraftment onto infarcted ventricles (Cortes-Morichetti et al., 2007). Echocardiography and histological examination implied that the cell-loaded matrix and the cell implants appeared to be an effective alternative to prevent post-MI ventricular dilation and cardiac remodeling compared to either HUCB cells or collagen matrix alone-treated groups. These results suggest that stem cell implantation accompanied with tissue engineering techniques, probably augment the efficacy of simple cellular therapy and could emerge as a new therapeutic modality to prevent adverse remodeling and progressive heart failure.

Meanwhile, Chen *et al.* studied the efficacy of HUCB cells with genes inserted to overexpress angiogenic factors that were subsequently transplanted into infarcted ventricles (H. K. Chen et al., 2005). In this study, HUCB $CD34^+$ cells were transfected with adeno-associated virus (AAV) vectors encoding either human angiopoietin (Ang1; AAV-Ang1) or VEGF (AAV-VEGF) cDNA alone, or both combined (AAV-Ang1 plus VEGF). After LAD ligation in SCID mice, they administered the expanded $CD34^+$ cells transduced with AAV-Ang1, AAV-VEGF or AAV-Ang1 plus VEGF intramyocardially at the left anterior wall. Western blot analysis revealed that Ang1 and VEGF expressions were increased in the $CD34^+$ cells transduced with AAV-Ang1 and AAV-VEGF, respectively. Infarct size significantly decreased and capillary density remarkably increased after treatment with CD34+/AAV-Ang1 plus VEGF compared to the treatment with CD34+ cells only. Combination therapy with CD34+/AAV-Ang1, CD34+/AAV-VEGF, and CD34+/AAV-Ang1 plus VEGF, all expressed remarkably higher cardiac performance in echocardiography than the CD34+ cells alone 4 weeks after MI. These results suggest that combination therapy with HUCB CD34+ cells and overexpression of both Ang1 and VEGF genes decreases infarct size, improves cardiac dysfunction and elevates capillary density in acute MI in mice to a greater extent compared to HUCB hematopoietic stem cell only transplantation.

Pretreatment of HUCB cells with oxytocin prior to transplant has been shown to promote their activity. Oxytocin stimulates cell proliferation and may promote cardiomyogenesis (Kim et al., 2010). Kim *et al.* transplanted 10^6 pretreated cells intramyocardially after LAD ligation and found that cardiac fibrosis and macrophage (CD68+) infiltration was reduced by oxytocin pretreatment. Engraftment, connexion 43 expression and cardiac contractility were increased, but angiogenesis did not appear to be altered in animals administered pretreated cells (Kim et al., 2010). This suggests that oxytocin pretreatment may increase the proliferation and differentiation of the transplanted cells, but does not appear to influence blood vessel growth.

5. Conclusions

Ischemic insults to neural tissue and the myocardium involve complicated inflammatory cascades that eventually lead to a pronounced cell death adjacent to the obstructed vasculature. The appropriate delivery of HUCB cells most likely prevents the apoptotic cascade, regenerates damaged cells and tissues, and modulates the inflammatory response to ischemic injury. Another advantage in using HUCB cells for the treatment of ischemic damage is the potential restoration of vascularity since cord blood contains endothelial progenitor cells and hematopoietic stem cells which may be of use in neovascularization therapy.

However, there are still a number of obstacles that prevent the routine application of HUCB cell transplantation to treat human stroke or MI patients as follows. First, safety has not been effectively determined in terms of immunological rejection when allogeneic transplantation is considered. HUCB cells are likely to be used primarily for transplantation into allogeneic recipients, though as HUCB cell banking becomes more popular autologous transplants may become more common in the future. Although HUCB cells are less immunogenic, long term studies should be considered to evaluate the long term safety with or without immunosuppression. Jozwiak *et al.* and Yang *et al.* have reported on using HUCB cells for the safe treatment of a single case of global brain ischemia and in a number of neurological disorders (including stroke) respectively with short term follow-up (Jozwiak et al., 2010; Yang et al., 2010). Second, because HUCB collection can only be performed at single time-points (i.e. during birth), and the number of stem cells extracted from one donor is very limited, expansion of the stem cells is mandatory. Unfortunately, current methods of expanding HUCB cells do not preserve the quality of the hematopoietic progenitor cells through to the end product (Jaroscak et al., 2003; Robinson et al., 2005). Current expansion methods cannot make up for the cells lost in the storage process, let alone augment them to a suitable dosing regimen (Hows et al., 1992; Liu et al., 2006; Traycoff et al., 1995; Yao et al., 2006). However, in the last couple of years, there have been several studies looking at ways to maximize yield, storage and recovery, and expansion which require validation (Gonzalez et al., 2010; Koestenbauer et al., 2009; Lin et al., 2010; Song et al., 2010; Vanneaux et al., 2010; Zeisburger et al., 2010). Third, the cause of limited graft survival in the recipient remains to be determined even though the paradigm of underlying mechanisms on the functional improvement after stem cell transplantation has been changed from cell replacement to a paracrine bystander effect. The longer the grafts survive, probably the better the neurological or cardiac function recovers. When these hurdles are overcome in the not too distant future, HUCB cell therapy may become a promising treatment modality for ischemic diseases including stroke and MI with multiple therapeutic potentials and a longer effective time frame in a single transplant that no other currently available pharmacological agent could mimic.

6. Acknowledgements

This work was supported in part by a Korea University Grant (K1030571) and Basic Science Research Program through the National Research Foundation of Korea (NRF) funded by the Ministry of Education, Science and Technology (2011-0013525). PRS is a co-founder and CVB a consultant for Saneron CCEL Therapeutics, Inc, a company exploring the therapeutic applications of stem cells including HUCB. CVB and PRS hold patents on the use of HUCB for the treatment of neurodegenerative disorders.

7. References

Arien-Zakay, H., Lecht, S., Bercu, M. M., Tabakman, R., Kohen, R., Galski, H., Nagler, A. & Lazarovici, P. (2009). Neuroprotection by cord blood neural progenitors involves antioxidants, neurotrophic and angiogenic factors. *Exp Neurol* 216, 83-94.

Armiñán, A., Gandía, C., García-Verdugo, J. M., Lledó, E., Trigueros, C., Ruiz-Saurí, A., Miñana, M. D., Solves, P., Payá, R., Montero, J. A. & Sepúlveda, P. (2010). Mesenchymal stem cells provide better results than hematopoietic precursors for the treatment of myocardial infarction. *J Am Coll Cardiol* 55, 2244-2253.

Arpinati, M., Green, C. L., Heimfeld, S., Heuser, J. E. & Anasetti, C. (2000). Granulocyte-colony stimulating factor mobilizes T helper 2-inducing dendritic cells. *Blood* 95, 2484-2490.

Asseman, C. & Powrie, F. (1998). Interleukin 10 is a growth factor for a population of regulatory T cells. *Gut* 42, 157-158.

Avitabile, D., Crespi, A., Brioschi, C., Parente, V., Toietta, G., Devanna, P., Baruscotti, M., Truffa, S., Scavone, A., Rusconi, F., Biondi, A., D'Alessandra, Y., Vigna, E., Difrancesco, D., Pesce, M., Capogrossi, M. C. & Barbuti, A. (2011). Human cord blood CD34+ progenitor cells acquire functional cardiac properties through a cell fusion process. *Am J Physiol Heart Circ Physiol* 300, H1875-1884.

Babcock, A. A., Kuziel, W. A., Rivest, S. & Owens, T. (2003). Chemokine expression by glial cells directs leukocytes to sites of axonal injury in the CNS. *J Neurosci* 23, 7922-7930.

Bartunek, J., Vanderheyden, M., Vandekerckhove, B., Mansour, S., De Bruyne, B., De Bondt, P., Van Haute, I., Lootens, N., Heyndrickx, G. & Wijns, W. (2005). Intracoronary injection of CD133-positive enriched bone marrow progenitor cells promotes cardiac recovery after recent myocardial infarction: feasibility and safety. *Circulation* 112, I178-183.

Boltze, J., Schmidt, U. R., Reich, D. M., Kranz, A., Reymann, K. G., Strassburger, M., Lobsien, D., Wagner, D-C., Förschler, A. & Schäbitz, W-R. (2011a). Determination of the therapeutic time window for human umbilical cord blood mononuclear cell transplantation following experimental stroke in rats. *Cell Transplant* Epub ahead of print. DOI: 10.3727/096368911X589609.

Boltze, J., Reich, D. M., Hau, S., Reymann, K. G., Strassburger, M., Lobsien, D., Wagner, D-C., Kamprad, M. & Stahl, T. (2011b). Assessment of neuroprotective effects of human umbilical cord blood mononuclear cell subpopulations in vitro and in vivo. *Cell Transplant* Epub ahead of print. DOI: 10.3727/096368911X586783.

Borlongan, C. V., Hadman, M., Sanberg, C. D. & Sanberg, P. R. (2004). Central nervous system entry of peripherally injected umbilical cord blood cells is not required for neuroprotection in stroke. *Stroke* 35, 2385-2389.

Botta, R., Gao, E., Stassi, G., Bonci, D., Pelosi, E., Zwas, D., Patti, M., Colonna, L., Baiocchi, M., Coppola, S., Ma, X., Condorelli, G. & Peschle, C. (2004). Heart infarct in NOD-SCID mice: therapeutic vasculogenesis by transplantation of human CD34+ cells and low dose CD34+KDR+ cells. *FASEB J* 18, 1392-1394.

Broxmeyer, H. E., Hangoc, G., Cooper, S., Ribeiro, R. C., Graves, V., Yoder, M., Wagner, J., Vadhan-Raj, S., Benninger, L., Rubinstein, P. & Broun, E. R. (1992). Growth characteristics and expansion of human umbilical cord blood and estimation of its potential for transplantation in adults. *Proc Natl Acad Sci U S A* 89, 4109-4113.

Buelens, C., Willems, F., Delvaux, A., Pierard, G., Delville, J. P., Velu, T. & Goldman, M. (1995). Interleukin-10 differentially regulates B7-1 (CD80) and B7-2 (CD86) expression on human peripheral blood dendritic cells. *Eur J Immunol* 25, 2668-2672.

Cardoso, A. A., Li, M. L., Batard, P., Hatzfeld, A., Brown, E. L., Levesque, J. P., Sookdeo, H., Panterne, B., Sansilvestri, P., Clark, S. C. & Hatzfeld, J. (1993). Release from quiescence of CD34+ CD38- human umbilical cord blood cells reveals their potentiality to engraft adults. *Proc Natl Acad Sci U S A* 90, 8707-8711.

Chen, J., Sanberg, P. R., Li, Y., Wang, L., Lu, M., Willing, A. E., Sanchez-Ramos, J. & Chopp, M. (2001). Intravenous administration of human umbilical cord blood reduces behavioral deficits after stroke in rats. *Stroke* 32, 2682-2688.

Chen, H. K., Hung, H. F., Shyu, K. G., Wang, B. W., Sheu, J. R., Liang, Y. J., Chang, C. C. & Kuan, P. (2005). Combined cord blood stem cells and gene therapy enhances angiogenesis and improves cardiac performance in mouse after acute myocardial infarction. *Eur J Clin Invest* 35, 677-686.

Chen, N., Hudson, J. E., Walczak, P., Misiuta, I., Garbuzova-Davis, S., Jiang, L., Sanchez-Ramos, J., Sanberg, P. R., Zigova, T. & Willing, A. E. (2005). Human umbilical cord blood progenitors: the potential of these hematopoietic cells to become neural. *Stem Cells* 23, 1560-1570.

Chen, S. H., Chang, F. M., Tsai, Y. C., Huang, K. F., Lin, C. L. & Lin, M. T. (2006). Infusion of human umbilical cord blood cells protect against cerebral ischemia and damage during heatstroke in the rat. *Exp Neurol* 199, 67-76.

Cheng, F., Wang, H. W., Cuenca, A., Huang, M., Ghansah, T., Brayer, J., Kerr, W. G., Takeda, K., Akira, S., Schoenberger, S. P., Yu, H., Jove, R. & Sotomayor, E. M. (2003). A critical role for Stat3 signaling in immune tolerance. *Immunity* 19, 425-436.

Chung, D. J., Choi, C. B., Lee, S. H., Kang, E. H., Lee, J. H., Hwang, S. H., Han, H., Lee, J. H., Choe, B. Y., Lee, S.Y. & Kim, H. Y. (2009). Intraarterially delivered human umbilical cord blood-derived mesenchymal stem cells in canine cerebral ischemia. *J Neurosci Res* 87, 3554-3567.

Conrad, P. D. & Emerson, S. G. (1998). Ex vivo expansion of hematopoietic cells from umbilical cord blood for clinical transplantation. *J Leukoc Biol* 64, 147-155.

Cortes-Morichetti, M., Frati, G., Schussler, O., Duong Van Huyen, J. P., Lauret, E., Genovese, J. A., Carpentier, A. F. & Chachques, J. C. (2007). Association between a cell-seeded collagen matrix and cellular cardiomyoplasty for myocardial support and regeneration. *Tissue Eng* 13, 2681-2687.

D'Arena, G., Musto, P., Cascavilla, N., Di Giorgio, G., Fusilli, S., Zendoli, F. & Carotenuto, M. (1998). Flow cytometric characterization of human umbilical cord blood lymphocytes: immunophenotypic features. *Haematologica* 83, 197-203.

El-Badri, N. S., Hakki, A., Saporta, S., Liang, X., Madhusodanan, S., Willing, A. E., Sanberg, C. D. & Sanberg, P. R. (2006). Cord blood mesenchymal stem cells: Potential use in neurological disorders. *Stem Cells Dev* 15, 497-506.

El Marsafy, S., Dosquet, C., Coudert, M. C., Bensussan, A., Carosella, E. & Gluckman, E. (2001). Study of cord blood natural killer cell suppressor activity. *Eur J Haematol* 66, 215-220.

Frassoni, F., Podesta, M., Maccario, R., Giorgiani, G., Rossi, G., Zecca, M., Bacigalupo, A., Piaggio, G. & Locatelli, F. (2003). Cord blood transplantation provides better reconstitution of hematopoietic reservoir compared with bone marrow transplantation. *Blood* 102, 1138-1141.

Gaebel, R., Furlani, D., Sorg, H., Polchow, B., Frank, J., Bieback, K., Wang, W., Klopsch, C., Ong, L. L., Li, W., Ma, N. & Steinhoff, G. (2011). Cell origin of human mesenchymal stem cells determines a different healing performance in cardiac regeneration. *PLoS One* 6, e15652.

Garbuzova-Davis, S., Willing, A. E., Zigova, T., Saporta, S., Justen, E. B., Lane, J. C., Hudson, J. E., Chen, N., Davis, C. D. & Sanberg, P. R. (2003). Intravenous administration of human umbilical cord blood cells in a mouse model of amyotrophic lateral sclerosis: distribution, migration, and differentiation. *J Hematother Stem Cell Res* 12, 255-270.

Garbuzova-Davis, S., Willing, A. E., Saporta, S., Bickford, P. C., Gemma, C., Chen, N., Sanberg, C. D., Klasko, S. K., Borlongan, C. V. & Sanberg, P. R. (2006). Novel cell therapy approaches for brain repair. *Prog Brain Res* 157, 207-222.

Garderet, L., Dulphy, N., Douay, C., Chalumeau, N., Schaeffer, V., Zilber, M. T., Lim, A., Even, J., Mooney, N., Gelin, C., Gluckman, E., Charron, D. & Toubert, A. (1998). The umbilical cord blood alphabeta T-cell repertoire: characteristics of a polyclonal and naive but completely formed repertoire. *Blood* 91, 340-346.

Glabinski, A. R., Tuohy, V. K., & Ransohoff, R. M. (1998). Expression of chemokines RANTES, MIP-1alpha and GRO-alpha correlates with inflammation in acute experimental autoimmune encephalomyelitis. *Neuroimmunomodulation* 5, 166-171.

Gluckman, E., Broxmeyer, H. A., Auerbach, A. D., Friedman, H. S., Douglas, G. W., Devergie, A., Esperou, H., Thierry, D., Socie, G., Lehn, P., Cooper, S., English, D., Kurtzberg, J., Bard, J. & Boyse, E. A. (1989). Hematopoietic reconstitution in a patient with Fanconi's anemia by means of umbilical-cord blood from an HLA-identical sibling. *N Engl J Med* 321, 1174-1178.

Gluckman, E. & Rocha, V. (2005). History of the clinical use of umbilical cord blood hematopoietic cells. *Cytotherapy* 7, 219-227.

Gonzalez, R., Griparic, L., Umana, M., Burgee, K., Vargas, V., Nasrallah, R., Silva, F. & Patel, A. (2010). An efficient approach to isolation and characterization of pre- and postnatal umbilical cord lining stem cells for clinical applications. *Cell Transplant* 19, 1439-1449.

Goodwin, H. S., Bicknese, A. R., Chien, S. N., Bogucki, B. D., Quinn, C. O. & Wall, D. A. (2001). Multilineage differentiation activity by cells isolated from umbilical cord blood: expression of bone, fat, and neural markers. *Biol Blood Marrow Transplant* 7, 581-588.

Ha, Y., Choi, J. U., Yoon, D. H., Yeon, D. S., Lee, J. J., Kim, H. O. & Cho, Y. E. (2001). Neural phenotype expression of cultured human cord blood cells in vitro. *Neuroreport* 12, 3523-3527.

Hall, A. A., Leonardo, C. C., Collier, L. A., Rowe, D. D., Willing, A. E. & Pennypacker, K. R. (2009a). Delayed treatments for stroke influence neuronal death in rat organotypic slice cultures subjected to oxygen glucose deprivation. *Neuroscience* 164, 470-477.

Hall, A. A., Guyer, A. G., Leonardo, C. C., Ajmo, C. T., Jr., Collier, L. A., Willing, A. E. & Pennypacker, K. R. (2009b). Human umbilical cord blood cells directly suppress ischemic oligodendrocyte cell death. *J Neurosci Res* 87, 333-341.

Hao, S. G., Sun, G. L., Wu, W. L. & Wu, Y. L. (2003). [Studies on the dynamics of biological characteristics of CD133+ cells from human umbilical cord blood during short-term culture]. *Zhongguo Shi Yan Xue Ye Xue Za Zhi* 11, 569-575.

Harris, D. T., Schumacher, M. J., Locascio, J., Besencon, F. J., Olson, G. B., DeLuca, D., Shenker, L., Bard, J. & Boyse, E. A. (1992). Phenotypic and functional immaturity of human umbilical cord blood T lymphocytes. *Proc Natl Acad Sci U S A* 89, 10006-10010.

Henning, R. J., Abu-Ali, H., Balis, J. U., Morgan, M. B., Willing, A. E. & Sanberg, P. R. (2004). Human umbilical cord blood mononuclear cells for the treatment of acute myocardial infarction. *Cell Transplant* 13, 729-739.

Henning, R. J., Burgos, J. D., Ondrovic, L., Sanberg, P., Balis, J. & Morgan, M. B. (2006). Human umbilical cord blood progenitor cells are attracted to infarcted myocardium and significantly reduce myocardial infarction size. *Cell Transplant* 15, 647-658.

Henning, R. J., Burgos, J. D., Vasko, M., Alvarado, F., Sanberg, C. D., Sanberg, P. R. & Morgan, M. B. (2007). Human cord blood cells and myocardial infarction: effect of dose and route of administration on infarct size. *Cell Transplant* 16, 907-917.

Henning, R. J., Shariff, M., Eadula, U., Alvarado, F., Vasko, M., Sanberg, P. R., Sanberg, C. D. & Delostia, V. (2008). Human cord blood mononuclear cells decrease cytokines and inflammatory cells in acute myocardial infarction. *Stem Cells Dev* 17, 1207-1219.

Higuchi, T., Anton, M., Dumler, K., Seidl, S., Pelisek, J., Saraste, A., Welling, A., Hofmann, F., Oostendorp, R. A., Gansbacher, B., Nekolla, S. G., Bengel, F. M., Botnar, R. M. & Schwaiger, M. (2009). Combined reporter gene PET and iron oxide MRI for monitoring survival and localization of transplanted cells in the rat heart. *J Nucl Med* 50, 1088-1094

Hirata, Y., Sata, M., Motomura, N., Takanashi, M., Suematsu, Y., Ono, M. & Takamoto, S. (2005). Human umbilical cord blood cells improve cardiac function after myocardial infarction. *Biochem Biophys Res Commun* 327, 609-614.

Hirose, Y., Kiyoi, H., Itoh, K., Kato, K., Saito, H. & Naoe, T. (2001). B-cell precursors differentiated from cord blood CD34+ cells are more immature than those derived from granulocyte colony-stimulating factor-mobilized peripheral blood CD34+ cells. *Immunology* 104, 410-417.

Hows, J. M., Bradley, B. A., Marsh, J. C., Luft, T., Coutinho, L., Testa, N. G. & Dexter, T. M. (1992). Growth of human umbilical-cord blood in longterm haemopoietic cultures. *Lancet* 340, 73-76.

Hu, C. H., Wu, G. F., Wang, X. Q., Yang, Y. H., Du, Z. M., He, X. H. & Xiang, P. (2006). Transplanted human umbilical cord blood mononuclear cells improve left ventricular function through angiogenesis in myocardial infarction. *Chin Med J (Engl)* 119, 1499-1506.

Hu, C. H., Li, Z. M., Du, Z. M., Zhang, A. X., Yang, D. Y. & Wu, G. F. (2009). Human umbilical cord-derived endothelial progenitor cells promote growth cytokines-mediated neorevascularization in rat myocardial infarction. *Chin Med J (Engl)* 122, 548-555.

Hu, C. H., Li, Z. M., Du, Z. M., Zhang, A. X., Rana, J. S., Liu, D. H., Yang, D. Y. & Wu, G. F. (2010). Expanded human cord blood-derived endothelial progenitor cells salvage infarcted myocardium in rats with acute myocardial infarction. *Clin Exp Pharmacol Physiol* 37, 551-556.

Iwasaki, H., Kawamoto, A., Willwerth, C., Horii, M., Oyamada, A., Akimaru, H., Shibata, T., Hirai, H., Suehiro, S., Wnendt, S., Fodor, W. L. & Asahara, T. (2009). Therapeutic potential of unrestricted somatic stem cells isolated from placental cord blood for cardiac repair post myocardial infarction. *Arterioscler Thromb Vasc Biol* 29, 1830-1835.

Jaroscak, J., Goltry, K., Smith, A., Waters-Pick, B., Martin, P. L., Driscoll, T. A., Howrey, R., Chao, N., Douville, J., Burhop, S., Fu, P. & Kurtzberg, J. (2003). Augmentation of umbilical cord blood (UCB) transplantation with ex vivo-expanded UCB cells: results of a phase 1 trial using the AastromReplicell System. *Blood* 101, 5061-5067.

Jeong, J. A., Gang, E. J., Hong, S. H., Hwang, S. H., Kim, S. W., Yang, I. H., Ahn, C., Han, H. & Kim, H. (2004). Rapid neural differentiation of human cord blood-derived mesenchymal stem cells. *Neuroreport* 15, 1731-1734.

Jiang, L., Newman, M., Saporta, S., Chen, N., Sanberg, C. D., Sanberg, P. R. & Willing, A. E. (2008). MIP-1alpha and MCP-1 Induce Migration of Human Umbilical Cord Blood Cells in Models of Stroke. *Curr Neurovasc Res* 5, 118-124.

Jiang, L., Womble, T., Saporta, S., Chen, N., Sanberg, C. D., Sanberg, P. R. & Willing, A. E. (2010). Human umbilical cord blood cells decrease microglial survival in vitro. *Stem Cells Dev* 19, 221-228.

Jozwiak, S., Habich, A., Kotulska, K., Sarnowska, A., Kropiwnicki, T., Janowski, M., Jurkiewicz, E., Lukomska, B., Kmiec, T., Walecki, J., Roszkowski, M., Litwin, M., Oldak, T., Boruczkowski, D. & Domanska-Janik, K. (2010) Intracerebroventricular Transplantation of Cord Blood-derived Neural Progenitors in a Child with Severe Global Brain Ischemic Injury. *Cell Med* 1, 71-80.

Kim, B. O., Tian, H., Prasongsukarn, K., Wu, J., Angoulvant, D., Wnendt, S., Muhs, A., Spitkovsky, D. & Li, R. K. (2005). Cell transplantation improves ventricular function after a myocardial infarction: a preclinical study of human unrestricted somatic stem cells in a porcine model. *Circulation* 112, I96-104.

Kim, Y. S., Kwon, J. S., Hong, M. H., Kim, J., Song, C. H., Jeong, M. H., Cho, J. G., Park, J. C., Kang, J. C. & Ahn, Y. (2010). Promigratory activity of oxytocin on umbilical cord blood-derived mesenchymal stem cells. *Artif Organs* 34, 453-461.

Knutsen, A. P. & Wall, D. A. (1999). Kinetics of T-cell development of umbilical cord blood transplantation in severe T-cell immunodeficiency disorders. *J Allergy Clin Immunol* 103, 823-832.

Kobari, L., Giarratana, M. C., Pflumio, F., Izac, B., Coulombel, L. & Douay, L. (2001). CD133+ cell selection is an alternative to CD34+ cell selection for ex vivo expansion of hematopoietic stem cells. *J Hematother Stem Cell Res* 10, 273-281.

Koestenbauer, S., Zisch, A., Dohr, G. & Zech, N. H. (2009) Protocols for hematopoietic stem cell expansion from umbilical cord blood. *Cell Transplant* 18, 1059-1068.

Kogler, G., Sensken, S., Airey, J. A., Trapp, T., Muschen, M., Feldhahn, N., Liedtke, S., Sorg, R. V., Fischer, J., Rosenbaum, C., Greschat, S., Knipper, A., Bender, J., Degistirici, O., Gao, J., Caplan, A. I., Colletti, E. J., Almeida-Porada, G., Muller, H. W., Zanjani, E. & Wernet, P. (2004). A new human somatic stem cell from placental cord blood with intrinsic pluripotent differentiation potential. *J Exp Med* 200, 123-135.

Laughlin, M. J., Barker, J., Bambach, B., Koc, O. N., Rizzieri, D. A., Wagner, J. E., Gerson, S. L., Lazarus, H. M., Cairo, M., Stevens, C. E., Rubinstein, P. & Kurtzberg, J. (2001). Hematopoietic engraftment and survival in adult recipients of umbilical-cord blood from unrelated donors *N Engl J Med* 344, 1815-1822.

Lee, K. D., Kuo, T. K., Whang-Peng, J., Chung, Y. F., Lin, C. T., Chou, S. H., Chen, J. R., Chen, Y. P. & Lee, O. K. (2004). In vitro hepatic differentiation of human mesenchymal stem cells. *Hepatology* 40, 1275-1284.

Leonardo, C. C., Hall, A. A., Collier, L. A., Ajmo, C. T., Jr., Willing, A.E. & Pennypacker, K.R. (2010). Human umbilical cord blood cell therapy blocks the morphological change and recruitment of CD11b-expressing, isolectin-binding proinflammatory cells after middle cerebral artery occlusion. *J Neurosci Res* 88, 1213-1222.

Leor, J., Guetta, E., Feinberg, M. S., Galski, H., Bar, I., Holbova, R., Miller, L., Zarin, P., Castel, D., Barbash, I. M. & Nagler, A. (2006). Human umbilical cord blood-derived CD133+ cells enhance function and repair of the infarcted myocardium. *Stem Cells* 24, 772-780.

Lin, R. Z., Dreyzin, A., Aamodt, K., Dudley, A. C. & Melero-Martin, J. M. (2011). Functional endothelial progenitor cells from cryopreserved umbilical cord blood. *Cell Transplant* 20, 515-522.

Liu, A. M., Lu, G., Tsang, K. S., Li, G., Wu, Y., Huang, Z. S., Ng, H. K., Kung, H. F. & Poon, W. S. (2010). Umbilical cord-derived mesenchymal stem cells with forced expression of hepatocyte growth factor enhance remyelination and functional recovery in a rat intracerebral hemorrhage model. *Neurosurgery* 67, 357-365; discussion 365-366.

Liu, Y., Liu, T., Fan, X., Ma, X. & Cui, Z. (2006). Ex vivo expansion of hematopoietic stem cells derived from umbilical cord blood in rotating wall vessel. *J Biotechnol* 124, 592-601.

Ma, N., Stamm, C., Kaminski, A., Li, W., Kleine, H. D., Muller-Hilke, B., Zhang, L., Ladilov, Y., Egger, D. & Steinhoff, G. (2005). Human cord blood cells induce angiogenesis following myocardial infarction in NOD/scid-mice. *Cardiovasc Res* 66, 45-54.

Ma, N., Ladilov, Y., Kaminski, A., Piechaczek, C., Choi, Y. H., Li, W., Steinhoff, G. & Stamm, C. (2006a). Umbilical cord blood cell transplantation for myocardial regeneration. *Transplant Proc* 38, 771-773.

Ma, N., Ladilov, Y., Moebius, J. M., Ong, L., Piechaczek, C., David, A., Kaminski, A., Choi, Y. H., Li, W., Egger, D., Stamm, C. & Steinhoff, G. (2006b). Intramyocardial delivery of human CD133+ cells in a SCID mouse cryoinjury model: Bone marrow vs. cord blood-derived cells. *Cardiovasc Res* 71, 158-169.

Ma, Y., Zou, P., Xiao, J. & Huang, S. (2002). [The expression and functional characteristics of AC133 antigen in cord blood hematopoietic cells]. *Zhonghua Nei Ke Za Zhi* 41, 798-800.

Mangi, A. A., Noiseux, N., Kong, D., He, H., Rezvani, M., Ingwall, J. S. & Dzau, V. J. (2003). Mesenchymal stem cells modified with Akt prevent remodeling and restore performance of infarcted hearts. *Nat Med* 9, 1195-1201.

Manginas, A., Goussetis, E., Koutelou, M., Karatasakis, G., Peristeri, I., Theodorakos, A., Leontiadis, E., Plessas, N., Theodosaki, M., Graphakos, S. & Cokkinos, D. V. (2007). Pilot study to evaluate the safety and feasibility of intracoronary CD133(+) and CD133(-) CD34(+) cell therapy in patients with nonviable anterior myocardial infarction. *Catheter Cardiovasc Interv* 69, 773-781.

Matsumoto, R., Omura, T., Yoshiyama, M., Hayashi, T., Inamoto, S., Koh, K. R., Ohta, K., Izumi, Y., Nakamura, Y., Akioka, K., Kitaura, Y., Takeuchi, K. & Yoshikawa, J. (2005). Vascular endothelial growth factor-expressing mesenchymal stem cell transplantation for the treatment of acute myocardial infarction. *Arterioscler Thromb Vasc Biol* 25, 1168-1173.

Moelker, A. D., Baks, T., Wever, K. M., Spitskovsky, D., Wielopolski, P. A., van Beusekom, H. M., van Geuns, R. J., Wnendt, S., Duncker, D. J. & van der Giessen, W. J. (2007). Intracoronary delivery of umbilical cord blood derived unrestricted somatic stem cells is not suitable to improve LV function after myocardial infarction in swine. *J Mol Cell Cardiol* 42, 735-745.

Nakahata, T. & Ogawa, M. (1982). Hemopoietic colony-forming cells in umbilical cord blood with extensive capability to generate mono- and multipotential hemopoietic progenitors. *J Clin Invest* 70, 1324-1328.

Nayar, B., Raju, G. M. & Deka, D. (2002). Hematopoietic stem/progenitor cell harvesting from umbilical cord blood. *Int J Gynaecol Obstet* 79, 31-32.

Newcomb, J. D., Ajmo, C. T., Jr., Sanberg, C. D., Sanberg, P. R., Pennypacker, K. R. & Willing, A. E. (2006). Timing of cord blood treatment after experimental stroke determines therapeutic efficacy. *Cell Transplant* 15, 213-223.

Newcomb, J. D., Sanberg, P. R., Klasko, S. K. & Willing, A. E. (2007). Umbilical cord blood research: current and future perspectives. *Cell Transplant* 16, 151-158.

Newman, M. B., Willing, A. E., Manresa, J. J., Davis-Sanberg, C. & Sanberg, P. R. (2005). Stroke-induced migration of human umbilical cord blood cells: time course and cytokines. *Stem Cells Dev* 14, 576-586.

Newman, M. B., Willing, A. E., Manresa, J. J., Sanberg, C. D. & Sanberg, P. R. (2006). Cytokines produced by cultured human umbilical cord blood (HUCB) cells: implications for brain repair. *Exp Neurol* 199, 201-208.

Ou, Y., Yu, S., Kaneko, Y., Tajiri, N., Bae, E. C., Chheda, S. H., Stahl, C. E., Yang, T., Fang, L., Hu, K., Borlongan, C. V. & Yu, G. (2010). Intravenous infusion of GDNF gene-modified human umbilical cord blood CD34+ cells protects against cerebral ischemic injury in spontaneously hypertensive rats. *Brain Res* 1366, 217-225.

Park, J., Setter, V., Wixler, V. & Schneider, H. (2009). Umbilical cord blood stem cells: induction of differentiation into mesenchymal lineages by cell-cell contacts with various mesenchymal cells. *Tissue Eng Part A* 15, 397-406.

Pranke, P., Failace, R. R., Allebrandt, W. F., Steibel, G., Schmidt, F. & Nardi, N. B. (2001). Hematologic and immunophenotypic characterization of human umbilical cord blood. *Acta Haematol* 105, 71-76.

Qian, H., Yang, Y., Huang, J., Dou, K. & Yang, G. (2006). Cellular cardiomyoplasty by catheter-based infusion of stem cells in clinical settings. *Transpl Immunol* 16, 135-147.

Rainsford, E. & Reen, D. J. (2002). Interleukin 10, produced in abundance by human newborn T cells, may be the regulator of increased tolerance associated with cord blood stem cell transplantation. *Br J Haematol* 116, 702-709.

Riegelsberger, U. M., Deten, A., Posel, C., Zille, M., Kranz, A., Boltze, J., & Wagner, D. C. (2011). Intravenous human umbilical cord blood transplantation for stroke: impact on infarct volume and caspase-3-dependent cell death in spontaneously hypertensive rats. *Exp Neurol* 227, 218-223.

Robinson, S., Niu, T., de Lima, M., Ng, J., Yang, H., McMannis, J., Karandish, S., Sadeghi, T., Fu, P., del Angel, M., O'Connor, S., Champlin, R. & Shpall, E. (2005). Ex vivo expansion of umbilical cord blood. *Cytotherapy* 7, 243-250.

Rocha, V., Wagner, J. E., Jr., Sobocinski, K. A., Klein, J. P., Zhang, M. J., Horowitz, M. M. & Gluckman, E. (2000). Graft-versus-host disease in children who have received a cord-blood or bone marrow transplant from an HLA-identical sibling. Eurocord and International Bone Marrow Transplant Registry Working Committee on Alternative Donor and Stem Cell Sources. *N Engl J Med* 342, 1846-1854.

Roura, S., Farré, J., Hove-Madsen, L., Prat-Vidal, C., Soler-Botija, C., Gálvez-Montón, C., Vilalta, M. & Bayes-Genis, A. (2010). Exposure to cardiomyogenic stimuli fails to transdifferentiate human umbilical cord blood-derived mesenchymal stem cells. *Basic Res Cardiol* 105, 419-430.

Rowe, D. D., Leonardo, C. C., Hall, A. A., Shahaduzzaman, M. D., Collier, L. A., Willing, A. E. & Pennypacker, K. R. (2010). Cord blood administration induces oligodendrocyte survival through alterations in gene expression. *Brain Res* 1366, 172-188.

Sanchez-Ramos, J. R., Song, S., Kamath, S. G., Zigova, T., Willing, A., Cardozo-Pelaez, F., Stedeford, T., Chopp, M. & Sanberg, P. R. (2001). Expression of neural markers in human umbilical cord blood. *Exp Neurol* 171, 109-115.

Senegaglia, A. C., Barboza, L. A., Dallagiovanna, B., Aita, C. A., Hansen, P., Rebelatto, C. L., Aguiar, A. M., Miyague, N. I., Shigunov, P., Barchiki, F., Correa, A., Olandoski, M., Krieger, M. A. & Brofman, P. R. (2010). Are purified or expanded cord blood-derived CD133+ cells better at improving cardiac function? *Exp Biol Med (Maywood)* 235, 119-229.

Shahaduzzaman, M., Golden, J. E., Green, S. M., Rowe, D., Pennypacker, K. R. & Willing, A. E. (2010). Human umbilical cord blood (HUCB) cells protect neurons by altering the neuronal gene expression profile. *Cell Transplantation* 19, 361.

Sirchia, G. & Rebulla, P. (1999). Placental/umbilical cord blood transplantation. *Haematologica* 84, 738-747.

Sondergaard, C. S., Hess, D. A., Maxwell, D. J., Weinheimer, C., Rosová, I., Creer, M. H., Piwnica-Worms, D., Kovacs, A., Pedersen, L. & Nolta, J. A. (2010). Human cord blood progenitors with high aldehyde dehydrogenase activity improve vascular density in a model of acute myocardial infarction. *J Transl Med*. 8, 24.

Song, C. H., Kim, J. J., Kim, J., Kim, S. H. & Kim, Y. G. (2010). Effects of trophogel for adhesion, proliferation, and differentiation of human umbilical cord blood-mesenchymal stem cells (hUCB-MSC). *Cell Transplantation* 19, 363.

Streit, W. J., Walter, S. A. & Pennell, N. A. (1999). Reactive microgliosis. *Prog Neurobiol* 57, 563-581.

Taguchi, A., Soma, T., Tanaka, H., Kanda, T., Nishimura, H., Yoshikawa, H., Tsukamoto, Y., Iso, H., Fujimori, Y., Stern, D. M., Naritomi, H. & Matsuyama, T. (2004). Administration of CD34+ cells after stroke enhances neurogenesis via angiogenesis in a mouse model. *J Clin Invest* 114, 330-338.

Tamaki, S., Eckert, K., He, D., Sutton, R., Doshe, M., Jain, G., Tushinski, R., Reitsma, M., Harris, B., Tsukamoto, A., Gage, F., Weissman, I. & Uchida, N. (2002). Engraftment of sorted/expanded human central nervous system stem cells from fetal brain. *J Neurosci Res* 69, 976-986.

Tan, J., Town, T., Paris, D., Mori, T., Suo, Z., Crawford, F., Mattson, M. P., Flavell, R. A. & Mullan, M. (1999). Microglial activation resulting from CD40-CD40L interaction after beta-amyloid stimulation. *Science* 286, 2352-2355.

Tang, Y. L., Zhao, Q., Qin, X., Shen, L., Cheng, L., Ge, J. & Phillips, M. I. (2005). Paracrine action enhances the effects of autologous mesenchymal stem cell transplantation on vascular regeneration in rat model of myocardial infarction. *Ann Thorac Surg* 80, 229-236; discussion 236-237.

Thomson, B. G., Robertson, K. A., Gowan, D., Heilman, D., Broxmeyer, H. E., Emanuel, D., Kotylo, P., Brahmi, Z. & Smith, F. O. (2000). Analysis of engraftment, graft-versus-host disease, and immune recovery following unrelated donor cord blood transplantation. *Blood* 96, 2703-2711.

Todaro, A. M., Pafumi, C., Pernicone, G., Munda, S., Pilastro, M. R., Russo, A., Farina, M. & Messina, A. (2000). Haematopoietic progenitors from umbilical cord blood. *Blood Purif* 18, 144-147.

Traycoff, C. M., Kosak, S. T., Grigsby, S. & Srour, E. F. (1995). Evaluation of ex vivo expansion potential of cord blood and bone marrow hematopoietic progenitor cells using cell tracking and limiting dilution analysis. *Blood* 85, 2059-2068.

Tse, W. & Laughlin, M. J. (2005). Umbilical cord blood transplantation: a new alternative option. *Hematology Am Soc Hematol Educ Program*, 377-383.

Uchida, N., Buck, D. W., He, D., Reitsma, M. J., Masek, M., Phan, T. V., Tsukamoto, A. S., Gage, F. H. & Weissman, I. L. (2000). Direct isolation of human central nervous system stem cells. *Proc Natl Acad Sci U S A* 97, 14720-14725.

Vanneaux, V., El-Ayoubi, F., Delmau, C., Driancourt, C., Lecourt, S., Grelier, A., Cras, A., Cuccuini, W., Soulier, J., Lataillade, J. J., Lebousse-Kerdiles, M. C., Oury, J. F.,

Sibony, O., Marolleau, J. P., Benbunan, M., Uzan, G. & Larghero, J. (2010). In vitro and in vivo analysis of endothelial progenitor cells from cryopreserved umbilical cord blood: are we ready for clinical application? *Cell Transplant* 19, 1143-1155.

Vaziri, H., Dragowska, W., Allsopp, R. C., Thomas, T. E., Harley, C. B. & Lansdorp, P. M. (1994). Evidence for a mitotic clock in human hematopoietic stem cells: loss of telomeric DNA with age. *Proc Natl Acad Sci U S A* 91, 9857-9860.

Vendrame, M., Cassady, J., Newcomb, J., Butler, T., Pennypacker, K. R., Zigova, T., Sanberg, C. D., Sanberg, P. R. & Willing, A. E. (2004). Infusion of human umbilical cord blood cells in a rat model of stroke dose-dependently rescues behavioral deficits and reduces infarct volume. *Stroke* 35, 2390-2395.

Vendrame, M., Gemma, C., de Mesquita, D., Collier, L., Bickford, P. C., Sanberg, C. D., Sanberg, P. R., Pennypacker, K. R. & Willing, A. E. (2005). Anti-inflammatory effects of human cord blood cells in a rat model of stroke. *Stem Cells Dev* 14, 595-604.

Vendrame, M., Gemma, C., Pennypacker, K. R., Bickford, P. C., Davis Sanberg, C., Sanberg, P. R. & Willing, A. E. (2006). Cord blood rescues stroke-induced changes in splenocyte phenotype and function. *Exp Neurol* 199, 191-200.

Wagner, J. E., Broxmeyer, H. E., Byrd, R. L., Zehnbauer, B., Schmeckpeper, B., Shah, N., Griffin, C., Emanuel, P. D., Zuckerman, K. S., Cooper, S., Carow, C., Bias, W. & Santos, G. W. (1992). Transplantation of umbilical cord blood after myeloablative therapy: analysis of engraftment. *Blood* 79, 1874-1881.

Willing, A. E., Lixian, J., Milliken, M., Poulos, S., Zigova, T., Song, S., Hart, C., Sanchez-Ramos, J. & Sanberg, P. R. (2003a). Intravenous versus intrastriatal cord blood administration in a rodent model of stroke. *J Neurosci Res* 73, 296-307.

Willing, A. E., Vendrame, M., Mallery, J., Cassady, C. J., Davis, C. D., Sanchez-Ramos, J. & Sanberg, P. R. (2003b). Mobilized peripheral blood cells administered intravenously produce functional recovery in stroke. *Cell Transplant* 12, 449-454.

Willing, A. E., Eve, D. J. & Sanberg, P. R. (2007). Umbilical cord blood transfusions for prevention of progressive brain injury and induction of neural recovery: an immunological perspective. *Regen Med* 2, 457-464.

Womble, T. A., Green, S. M., Nelson, A. P., Shahaduzzaman, M. D., Golden, J. E., Sanberg, P. R., Pennypacker, K. R. & Willing, A. E. (2009) CD14+ and CD133+ human umbilical cord blood cells are essential for neurological recovery following MCAO. *Cell Transplant* 18, 240.

Wu, K. H., Cui, B., Yu, C. T. & Liu, Y. L. (2006). Stem cells: new cell source for myocardial constructs tissue engineering. *Med Hypotheses* 67, 1326-1329.

Wu, K. H., Zhou, B., Mo, X. M., Cui, B., Yu, C. T., Lu, S. H., Han, Z. C. & Liu, Y. L. (2007a). Therapeutic potential of human umbilical cord-derived stem cells in ischemic diseases. *Transplant Proc* 39, 1620-1622.

Wu, K. H., Zhou, B., Yu, C. T., Cui, B., Lu, S. H., Han, Z. C. & Liu, Y. L. (2007b). Therapeutic potential of human umbilical cord derived stem cells in a rat myocardial infarction model. *Ann Thorac Surg* 83, 1491-1498.

Xiao, J., Nan, Z., Motooka, Y. & Low, W. C. (2005). Transplantation of a novel cell line population of umbilical cord blood stem cells ameliorates neurological deficits associated with ischemic brain injury. *Stem Cells Dev* 14, 722-733.

Xing, Y. L., Shen, L. H., Li, H. W., Zhang, Y. C., Zhao, L., Zhao, S. M. & Xu, Q. (2009). Optimal time for human umbilical cord blood cell transplantation in rats with myocardial infarction. *Chin Med J (Engl)* 122, 2833-2839.

Yang, S. E., Ha, C. W., Jung, M., Jin, H. J., Lee, M., Song, H., Choi, S., Oh, W. & Yang, Y. S. (2004). Mesenchymal stem/progenitor cells developed in cultures from UC blood. *Cytotherapy* 6, 476-486.

Yang, W. Z., Zhang, Y., Wu, F., Min, W. P., Minev, B., Zhang, M., Luo, X. L., Ramos, F., Ichim, T. E., Riordan, N. H. & Hu, X. (2010). Safety evaluation of allogeneic umbilical cord blood mononuclear cell therapy for degenerative conditions. *J Transl Med* 8, 75.

Yao, C. L., Feng, Y. H., Lin, X. Z., Chu, I. M., Hsieh, T. B. & Hwang, S. M. (2006). Characterization of serum-free ex vivo-expanded hematopoietic stem cells derived from human umbilical cord blood CD133(+) cells. *Stem Cells Dev* 15, 70-78.

Yasuhara, T., Hara, K., Maki, M., Xu, L., Yu, G., Ali, M. M., Masuda, T., Yu, S. J., Bae, E. K., Hayashi, T., Matsukawa, N., Kaneko, Y., Kuzmin-Nichols, N., Ellovitch, S., Cruz, E. L., Klasko, S. K., Sanberg, C. D., Sanberg, P.R. & Borlongan, C. V. (2010). Mannitol facilitates neurotrophic factor up-regulation and behavioural recovery in neonatal hypoxic-ischaemic rats with human umbilical cord blood grafts. *J Cell Mol Med* 14, 914-921.

Yerebakan, C., Sandica, E., Prietz, S., Klopsch, C., Ugurlucan, M., Kaminski, A., Abdija, S., Lorenzen, B., Boltze, J., Nitzsche, B., Egger, D., Barten, M., Furlani, D., Ma, N., Vollmar, B., Liebold, A. & Steinhoff, G. (2009). Autologous umbilical cord blood mononuclear cell transplantation preserves right ventricular function in a novel model of chronic right ventricular volume overload. *Cell Transplant* 18, 855-868.

Yin, A. H., Miraglia, S., Zanjani, E. D., Almeida-Porada, G., Ogawa, M., Leary, A. G., Olweus, J., Kearney, J. & Buck, D. W. (1997). AC133, a novel marker for human hematopoietic stem and progenitor cells. *Blood* 90, 5002-5012.

Zeisberger, S. M., Schulz, J. C., Mairhofer, M., Ponsaerts, P., Wouters, G., Doerr, D., Katsen-Globa, A., Ehrbar, M., Hescheler, J., Hoerstrup, S. P., Zisch, A. H., Kolbus, A. & Zimmermann, H. (2010). Biological and physicochemical characetrization of a serum- and xeno-free chemically defined cryopreservation procedure for adult human progenitor cells. *Cell Transplant* Epub ahead of print. DOI:10.3727/096368910X547426

Zigova, T., Song, S., Willing, A. E., Hudson, J. E., Newman, M. B., Saporta, S., Sanchez-Ramos, J. & Sanberg, P. R. (2002). Human umbilical cord blood cells express neural antigens after transplantation into the developing rat brain. *Cell Transplant* 11, 265-274.

Zola, H., Fusco, M., Macardle, P. J., Flego, L. & Roberton, D. (1995). Expression of cytokine receptors by human cord blood lymphocytes: comparison with adult blood lymphocytes. *Pediatr Res* 38, 397-403.

4

Therapeutic Approaches in Regenerative Medicine of Cardiovascular Diseases: From Bench to Bedside

Antonia Aránega et al.*
Biopathology and Regenerative Medicine Institute (IBIMER), Department of Human Anatomy and Embryology, Centro de Investigación Biomédica, Universidad de Granada Spain

1. Introduction

Despite significant progress in medical research, which has led to lower infant mortality rates and increased life expectancy, cardiovascular diseases (CVDs) continue to be the largest contributors of morbidity and mortality in both developed and developing countries. In fact, they are predicted to be the leading cause of death by 2020 and responsible for 13.4% of total world-wide deaths by 2030. These diseases have a multifactorial origin, notably non-modifiable cardiovascular risk factors (CRFs) such as age, sex, race or family history, and modifiable CRFs, including smoking, hypertension, hypercholesterolemia, diabetes mellitus or hypertriglyceridemia (Picariello et al., 2011; Glynn & Rosner, 2005).
Currently, there are no effective treatments available for many degenerative diseases caused by the death or malfunction of specific cells. In this regard, the main physiopathology of CVDs is related to ischaemic heart disease, and approximately two-thirds of patients surviving an acute myocardial infarction (AMI) are left with debilitating congestive heart failure. AMI causes apoptosis and necrosis of cardiomyocytes, a specialized and differentiated cell population responsible for ventricular contraction, the main event of cardiac function. The low self- renewal rate of these cells (~0.06% of the total population of cardiomyocytes under normal conditions) produces, in cases of acute stress, the ventricular remodelling of non-ischaemic myocardium with scar formation, increasing the likelihood of a new crisis, progressive ventricular dilatation and heart failure, ending in death (Torella et al., 2007).
Organ transplantation is an appropriate therapeutic strategy for patients who have suffered end-stage heart failure and are unresponsive to conventional therapy but has drawbacks related to donor incompatibility, which can lead to rejection and eventually to Graft Versus

---

* Milán Bustamante[1,+], Juan Antonio Marchal [1,+], Macarena Perán[1,2], Elena López [1,2], Pablo Álvarez [1], Fernando Rodríguez-Serrano [1] and Esmeralda Carrillo[1]
[1]*Biopathology and Regenerative Medicine Institute (IBIMER), Department of Human Anatomy and Embryology, Centro de Investigación Biomédica, Universidad de Granada, Spain*
[2]*Departamento de Ciencias de la Salud, Universidad de Jáen, Spain*
Corresponding authors: Juan Antonio Marchal and Antonia Aránega
+*These authors contributed equally to this work*

Host Disease (GVHD), among other conditions. However, there are a number of preoperative and postoperative risks, besides the major challenge of finding an optimal donor heart and the elevated healthcare costs involved (currently ranging from 27,839 € and 51,575 € per patients), especially in view of the uncertain long-term outcomes of the procedure (Pérez -Romero et al., 2010).

1.1 Worldwide costs: The DALY factor

Besides the elevated mortality rates associated with CVDs, they contribute significantly to the burden of disability and disease in developing countries. In an update on the global burden of disease in 2004, the WHO provided a comprehensive assessment of the causes of loss of health in different regions of the world, demonstrating that the leading cost of the death in developed or high-income countries is heart disease, followed by stroke, lung cancer, pneumonia and asthma or bronchitis (WHO, 2004). The most widely-used measure of this burden is the life expectancy adjusted for disability (DALY), defined as the sum of life years lost due to premature death and years lost to disability. CVDs are considered to represent 86% of the total DALY-estimated disease burden among under 70-year-olds worldwide, indicating that global CVD-related deaths will rise from the 16.7 million estimated in 2002 to 23.3 million in 2030.

1.2 Regenerative biomedicine: Breaking down old paradigms?

Over the past decade, new therapeutic strategies have been proposed against diseases with an elevated prevalence world-wide for which standard treatments are not very effective, resulting in the emergence of a new medical science known as regenerative medicine. Scientific evidence has demonstrated that novel therapeutic solutions can be obtained for pathological situations with no current treatment and that existing therapies can be enhanced by the utilization of autologous or allogeneic stem cell transplantation. These derive from the discovery of adult stem cell populations in organs well known for their non-existent regeneration capacity, such as the heart and brain, which correlate with their early development in the embryo stage. Thus, some cardiomyocytes with proliferative capacity have been found in the heart after a myocardial infarction, although inadequate to meet requirements after an extensive AMI.

The discovery of multipotent cell populations in adult tissues has opened up new therapeutic possibilities for diseases that do not respond to conventional medical treatments. The ethical and legal issues involved in the use of embryonic stem cells (ESCs), the recently disclosed plasticity of adult-derived stem cells, their ability to be reprogrammed by defined factors into pluripotent stem cells and a greater understanding, thanks to a multidisciplinary approach, of the mechanisms underlying stem cell differentiation have led to the birth of the "Regenerative Medicine" concept. Unlike many existing products and therapies, this approach offers the potential for regenerative rather than merely palliative or symptomatic treatment. This novel specialty includes gene therapy, stem cell therapy, targeted drug therapy and cell reprogramming.

This emerging field seeks the maintenance, improvement or restoration of cell function, tissues and organs by applying methods mainly related to cell therapy and tissue engineering (Greenwood et al., 2006). The ultimate goal is to repair, replace or regenerate cells, tissues or organs that have lost their normal function. The essential elements for the success of regenerative medicine include cells, soluble molecules and three-dimensional scaffolds that serve as anchor and stimulation for new tissue formation. In relation to CVD, the regeneration of vascular and myocardial tissues is considered a primary endpoint to limit the consequences of acute and chronic ischaemic heart disorders. Current clinical approaches include transdifferentiation techniques and the utilization of organic scaffolds (e.g., vessels or even the whole heart) by means of decellularization methods.

2. Stem cells and cardiovascular diseases

Stem cells are unspecialized cells that can renew themselves through cell division or can be differentiated to become tissue- or organ-specific cells with special functions (Wobus & Boheler, 2005). Up to the 8-cell morulla stage, cells in the embryo are totipotent and can produce the whole organism. Their utilisation is limited by ethical issues and current legislation in developed countries, besides inadequate knowledge of their behaviour and action mechanisms and of their impact on the body homeostasis.

Stem cells derived from the inner cell mass (ICM) of the blastocyst are no longer totipotent but they are pluripotent, having the ability to differentiate into all cell types of the body. These cells include human embryonic stem cells (hESCs) and induced pluripotent stem cells (iPSCs). hESCs are capable of proliferating extensively in an undifferentiated state and in co-culture on fibroblast 'feeder' cells, and have the ability, once removed from feeders and/or grown in suspension as aggregates [called EBs (embryoid bodies)], to differentiate towards all three germ layers, and they can, in principle, give rise to all cell types of the body. hESCs contain factors that can induce reprogramming of the somatic cell nucleus (Allegrucci & Young, 2007; Cowan et al., 2005). Therefore, somatic cell fusion with ESCs regenerates pluripotent cells. However, pluripotent cells obtained by fusion contain both chromosomes from the ESCs and from the somatic cell, causing their post-implantation rejection (Takahashi & Yamanaka, 2006).

iPSC technology was developed to avoid these problems, generating a new type of pluripotent stem cell by directly reprogramming somatic cells rather than using embryos. It allows patient-specific stem cells to be obtained by using the same genome as that of the individual whose cells have been reprogrammed. Besides adult and foetal dermal fibroblasts, iPSCs have been created from multiple human tissues, including lung fibroblasts, keratinocytes (Aasen et al., 2008), fibroblast-like synoviocytes (Takahashi et al., 2007), cord blood (Giorgetti et al., 2009; Haase et al.,2009) , peripheral blood (Loh et al., 2009, Ye et al., 2009) mesenchymal stromal cells (Oda et al., 2010), oral mucosa fibroblasts (Miyoshi et al., 2010) and T-cells (Loh et al., 2010; Seki et al., 2010). These cells can be differentiated and can serve as a cell or disease model or may even lead to future stem cell therapies (Dimos et al., 2008, Park et al., 2008). However, there are a number of drawbacks in the use of retroviral vectors to create iPSCs, and considerable efforts have been made to find alternatives.

In addition, certain organs have multipotent stem cells with more limited differentiation ability (Wobus & Boheler, 2005), including foetal and adult stem cells obtained from bone marrow (BM) and postnatal organs and tissues. Stem cells of foetal origin express stem cell markers similar to hESCs, while their differentiation potential lies between that of hESCs and that of adult stem cells (Pappa & Anagnou, 2009). BM contains two types of multipotent stem cells: hematopoietic stem cells (HSCs), which give rise to all blood cell types (Orkin, 2000) and human mesenchymal stromal cells (hMSCs), which can be differentiated into multiple mesenchymal tissue cell types such as bone, cartilage, fat and muscle cells (Liu et al., 2009).

2.1 The quest for an optimal adult stem cell type: Cardiac Stem Cells (CSC)

Although many different cell types have been considered for myocardial repair, the ideal candidate remains elusive. Adult cardiomyocytes are unable to survive, even when transplanted into normal myocardium, and the resident population of cardiac progenitors is insufficient after AMI. Therefore, researchers continue to search for the optimal cell type for myocardial repair, using various types of adult stem cells obtained from different sources, such as HSCs, endothelial progenitor cells (EPCs), cardiac stem cells (CSCs), muscle-derived stem cells (MdSTs), and MSCs. MSCs are frequently selected for their multiple advantages, such as their multiple anatomical localizations, easy isolation and cryopreservation, and paracrine potential.

The characteristics of CSCs make them an excellent candidate for cardiac regeneration. CSCs isolated from myocardial biopsies (Smith et al., 2009) express stem cell factor receptor (c-Kit) (Bearzi et al., 2007) and stem cell antigen-1 (Sca-1) on their cell surface (Iwakura et al., 2011). A direct relationship has been demonstrated between higher patient age and a smaller population of these cells, which represents a drawback for their therapeutic application, given that the main incidence of CVDs is among the middle-aged. However, intensive work is ongoing to activate these cells to proliferate and differentiate *in situ* using different techniques, such as induction with 5-azacytidine (5-aza), a potent inhibitor of DNA methyltransferase (DNMT1), in combination with TGF-β1 and vitamin C (Smits et al., 2009). Recent proposals include the use of hyperpolarization to differentiate CSCs into the cardiac phenotype, changing charges in the membrane potential (specific for each cell type) to enhance intracellular calcium levels and calcineurin signalling in CSCs and, more importantly, to upregulate the expression of cardiac-specific genes and proteins, leading to the formation of spontaneously beating cardiomyocytes (Van Vliet et al., 2010). Nonetheless, the specific capabilities of functional cardiac cells from these progenitors need to be more thoroughly studied (Gonzales & Pedrazzini, 2009).

2.2 Adult stem cell transplantation: Paracrine effects or change in phenotype?

Recently, cell-based therapeutic strategies have been aimed at replenishing the loss of myocytes by apoptosis and necrosis and at inhibiting the adverse effects of cardiac remodelling (Singla, 2009). Some studies have demonstrated improvement in ventricular function as an achievement of cardiac regeneration (Singla, 2009, Kumar et al., 2005). However, if a real clinical translation is to be achieved, researchers must follow strict criteria for the selection of an appropriate source of stem cells, taking full account of patient safety, production methods, and ethical and legal norms. It is necessary to establish an *in vitro* differentiation strategy to transform these cells into functional cardiomyocytes that can integrate in a damaged heart and exert therapeutic action.

The benefits of stem cell transplantation remain controversial and appear to depend on three mechanisms. First, the transplanted stem cells must be able to differentiate, even at a low rate, into heart or vascular cells. A second mechanism involves the immune and paracrine functions of stem cells and their interactions with an environment subject to cellular damage, with the transplanted cells having the capacity to induce the secretion of growth factors and cytokines that stimulate tissue repair after ischaemic injury, reducing the size of infarction. Finally, the pre-treatment of stem cells by transdifferentiation methods has been shown to produce a population of cells in pre-differentiated state, allowing their acquisition of the cardiac phenotype after transplantation. However, key issues have not yet been elucidated, including identification of the most appropriate cell types or the factors released by stem cells and their role in cardiac repair and regeneration. The methods described below were conducted using hMCSs, which appear to be promising candidates for clinical translation, given their multipotency and ready availability, e.g., *via* liposuction.

2.2.1 Direct transplantation and cell tracking

Cell transplantation is emerging as a potential therapy to treat heart failure. Initial efforts have been focused on the cardiomyocytes and skeletal myoblasts. However, the mechanisms and the outcomes of transplanted cells during cardiac regenerative therapy remain unclear. In order to achieve the optimal concentration of stem cells for repairing the myocardial region of interest, cell delivery strategies need to consider different clinical settings and local milieus. Stem cells are believed to respond differently according to local signalling, and the success of transplantation depends to a large degree on damage signalling in the wounded area. In cardiovascular research stem cells have been delivered through coronary arteries or coronary veins or by means of peripheral-vein infusion. Alternatively, direct intramyocardial injections have been performed, utilizing a surgical, transendocardial, or transvenous approach. Delivery strategies also include the mobilization of stem cells from the bone marrow by means of cytokine therapy, with or without peripheral harvesting. In pre-clinical studies, when BM-derived hematopoietic cells were transplanted directly into the hearts of mice with AMI, no transdifferentiated BM-derived cardiomyocytes were found in the damaged myocardium. However, cell fusion has been found to occur at very low levels between BM-derived cells and host cardiomyocytes outside the infarction area (Nygren et al., 2004). A similar approach has been adopted using skeletal myoblast precursors. BM-stem cell (SC) transplantation has been the most widely used methodology in clinical studies, as shown in Table 1, although its association with arrhythmogenesis, among other adverse effects, has generated major controversy about its benefits.

The main problem with these methods is the lack of knowledge on the behaviour and integration of the transplanted stem cells. Many studies have relied simply on the presence of histological differences between hearts receiving cell transplants compared with non-transplanted patients (control group). However, the preferred approach is to utilize molecular analyses that are able to distinguish donor and host cells. For instance, treating donor cells with fluorescent cell-tracking dyes prior to transplantation has been used to monitor their survival *in vivo*. Because there is typically a high degree of donor cell death following transplantation, caution must be exercised to ensure that the observed signal arises from donor cells rather than from host cells, which have acquired dye released from dead or dying donor cells. Intrinsic genetic differences between donor and host cells as well as gene transfer have also been employed to monitor cell fate following transplantation (Lee & Segers, 2008). Hence, innovative non-invasive imaging techniques to effectively track the cells *in vivo* are vital for the adequate investigation of possible clinical applications, furnishing insights into the mechanisms underlying stem cell-based therapy and addressing some of the major concerns about translational cardiovascular stem cell therapy. Currently, there are three main approaches to stem cell tracking: (1) radioactive labelling for positron emission tomography (PET) and single photon emission computed tomography (SPECT) imaging, (2) iron particle labelling for magnetic resonance imaging (MRI), and (3) reporter gene labelling for bioluminescence, fluorescence, MRI, SPECT and PET imaging.

Clinical trial identifier number	Trial name	Number of patients	Cells	Primary end point	Route of cell delivery	Further Comments
NCT00669227	SCAMI	40	BM-SCs	LVEF	IC	Prior treatment with G-CSF
NCT00395811	CABG	50	BM-SCs	LVEF	IC	Stem Cell Therapy combined CABG
NCT00316381	REGENT	200	Subpopulation of CD34+/CXCR4+ BM-SCs	LVEF	IC	
NCT00587990	PROMETHEUS	45	MSCs	Safety	Transepicar-dial during CABG	Stem Cell Therapy combined CABG
NCT00548613	MESENDO	20	BM-SCs	New blood vessel formation	IC/IM	Combined stem cell therapy
NCT00939042	ALSTER	40	BM-SCs	LVEF	IC	
NCT00442806	APOLLO	48	ADRCs	Safety	Intravenous	
NCT00126100	REVIVAL-2	114	G-CSF-stimulated BMSCs/proge-nitor cells	Efficacy	Subcutaneous	Evaluation of the ↑ of circulating SC after stimulation
NCT00268307		41	BM-SCs	Safety	IC	
NCT00289822		75	BM-SCs	LVEF	IC	
NCT00626145		37	BM-SCs	LVEF	IC day 7 post PCI	Evaluation of reperfusion therapy for STEMI
NCT00810238	C-Cure	240	BM- derived cardiopoietic cells	LVEF	Transendocar dial	Efficiency of vascular regeneration in patients with CAD

ADRCs: Adipose-Derived Stem and Regenerative Cells; EPC: Endothelial progenitor cells; BM-SCs: Bone Marrow stem cells; MSCs: Mesenchymal stromal cells; G-CSF: Granulocyte Colony Stimulating Factor; STEMI: ST-Elevation Myocardial Infarction; PCI: Percutaneous Coronary Intervention; CAD: Chronic Coronary Artery Disease; CABG: Coronary Artery Bypass Graft; IC: Intracoronary; IM: Intramyocardial; LVEF: Left Ventricular Ejection Fraction.

Table 1. Closed/Not recruiting autologous cell therapy trials in patients with coronary heart disease or AMI

2.2.2 Paracrine and immunology features

Two important features of MSCs are their immune-privileged and immunosuppressive functions. These qualities are very attractive for cell-based therapy because they may facilitate the use of allogeneic rather than autologous cells, which offers many potential advantages. Thanks to their immunosuppressive capacity, they may also suppress the immune response to cell injections or to the damaged myocardium itself during myocardial infarction. Findings by Aggarwal and Pittengerin on the lack of T-cell response to allogeneic hMSC transplantation indicated that these cells can modify the immune response in a dose-dependent manner and can reduce the inflammatory response when present in appropriate amounts. Phase III clinical trials are under way on the optimal MSC regimen to prevent or treat GVHD during allogeneic HSC transplantation (Aggarwal & Pittenger 2005).

Secreted growth factors and cytokines are known to be important signalling molecules for modulating acute and chronic immune responses (Table 2). With regard to their immunomodulatory properties, MSCs inhibit T-cell proliferation and influence the maturation and expression profile of antigen-presenting cells such as dendritic cells. Research on the paracrine effect of MSCs in rat AMI models, using intramyocardial injection of culture medium with MSC-secreted growth factors and cytokines (conditioned medium), demonstrated their role in stromal network creation, neovascularization enhancement, angiogenesis promotion, and cardiac function improvement. Paracrine effects of MSCs may include the activation of different mechanisms against cell damage. Therapeutic models have shown that MSCs are recruited to infarction areas by the hepatocyte growth factor (HGF), which is released during necrosis and apoptosis and acts as a "homing signal". This migration of MSCs depends on interaction with c-Met, an HGF receptor (Parekkadan & Milwid, 2010).

Growth factors and cytokines	Functions	References
M-CSF, G-CSF, GM-CSF, SCF-1, LIF, SDF-1, Flt-3, IL 1, 6, 7, 8, 11, 14 and 15.	Signalling molecules and modulation of acute and chronic immune response.	Sumanasinghe et al., 2009 Kim et al., 2005.
VEGF, bFGF, PLGF and MCP-1.	Vasculogenesis and angiogenesis	Kinnaird et al., 2004.

M-CSF: Macrophage Colony-Stimulating Factor; G-CSF: Granulocyte Colony-Stimulating Factor; GM-CSF: Granulocyte Macrophage Colony-Stimulating Factor; SCF-1: Stem Cell Factor; LIF: Leukemia Inhibitory Factor; SDF-1: Stromal Cell-Derived Factor-1; Flt-3: FMS-like tyrosine kinase 3; IL: Interleukin; VEGF: Vascular Endothelial Growth Factor; bFGF: basic Fibroblastic Growth Factor; PLGF: Placental Growth Factor; MCP-1: Monocyte Chemoattractant Protein-1

Table 2. Paracrine activity of MSCs

MSC-derived conditioned medium promotes the proliferation of CSCs and inhibits the apoptosis of CSCs induced by hypoxia and serum starvation. An upregulated expression of cardiomyocyte-related genes was also found in CSCs, including b-myosin heavy chain (b-MHC) and atrial natriuretic peptide (ANP), demonstrating the protective effects of this medium on CSCs and the enhancement of their migration and differentiation (Nakanishi et al., 2008). Preliminary results of *in vitro* assays with human adipose-derived mesenchymal stromal cells (hADSCs) by our group indicate that cytokine- and growth factor-rich conditioned media obtained from human cardiomyocytes in culture influence the acquisition of cardiac phenotype by hADSCs (unpublished data).

2.2.3 Cardiac regeneration: A transdifferentiation phenomenon

Regenerative biomedicine efforts have focused on attempts to reverse heart muscle failure by increasing the amount of human functional heart muscle through the transplantation of various adult stem cell types, notably CSCs, skeletal myoblasts, endothelial stem cells, and MSCs. BM stem cells have been reported to offer a wide range of benefits in blood cell diseases, but the action mechanisms underlying the change in phenotype have not been

clarified. *In vivo* transplantation studies have shown that MSCs can replenish the whole BM system in irradiated rodents and generate not only mesodermal cells but also cells with the characteristics of neural progenitors and neurons (Song & Tuan, 2004).
In this type of cross-lineage differentiation, also known as transdifferentiation, one cell type committed to and progressing along a specific developmental lineage is switched into another cell type from a different lineage by genetic reprogramming. This depends on the ability of adult stem cells to maintain their multidifferentiation potential even after exposure to specific inductive factors (Song & Tuan, 2004).
The conversion from adult differentiated cells must involve both the suppression and regulation of different genes in the cells, implying that genes from both cell types are co-expressed at some point. This phenomenon is currently under investigation by various research groups in an attempt to establish its true therapeutic value. Transdifferentiation studies have supported the notion that cell fate is controlled by master switch genes and that one or two factors can be sufficient to direct cells from one lineage to another (Burke & Tosh, 2005).

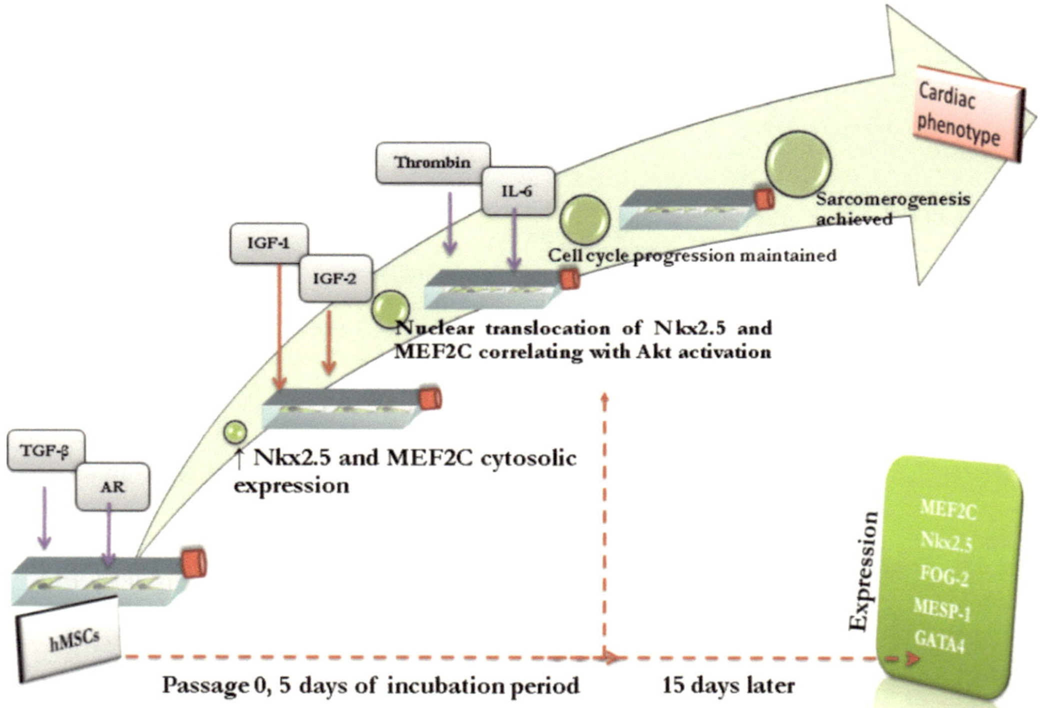

Fig. 1. Guided cardiopoiesis. This diagram depicts the effects of cytokines in MSCs cultures.

This process involves cardiopoiesis, defined as the final engagement of pluripotent or multipotent stem cells in the cardiac differentiation program. Cardiopoietic specification is based on emulation of the natural cardiac development in the embryo, which takes place in the mesoderm under the cardioinductive influence of the neighbouring endoderm (Figure 1). Cardiopoietic guidance emulates natural cardiac differentiation and can be achieved by stimulating stem cells with TGF-β1, BMP-2/4, FGF-2/4, IL-6, IGF-1/2, VEGF-A, EGF and activin-A. The identification of these factors has enhanced our understanding of the mechanisms involved in differentiation and the signals that play a role in heart development (Behfar et al., 2010).

3. From the bench: Current status

Greater knowledge of heart development in the embryo and of the effects of the microenvironment on cardiac maturation has prompted researchers to reproduce these processes in the laboratory by means of different transdifferentiation techniques (Figure 2). Some of them have attempted to recapitulate the activation of cell signalling cascades involved in heart development (e.g., TGF-β and Akt) through the expression of GATA4, Mef2c and FGF 8/10 growth factors. A wide range of cell types and methods have been used to achieve the cardiac phenotype. In this regard, there is major interest in the novel technique developed by Takahashi & Yamanaka (2006), to induce pluripotency by reprogramming fibroblasts with key factors Klf4, Sox2, Oct4 and c-Myc. It is expected that the ability to derive patient-specific hiPSCs by somatic cell reprogramming may be useful for disease modelling and drug discovery as well as basic research.

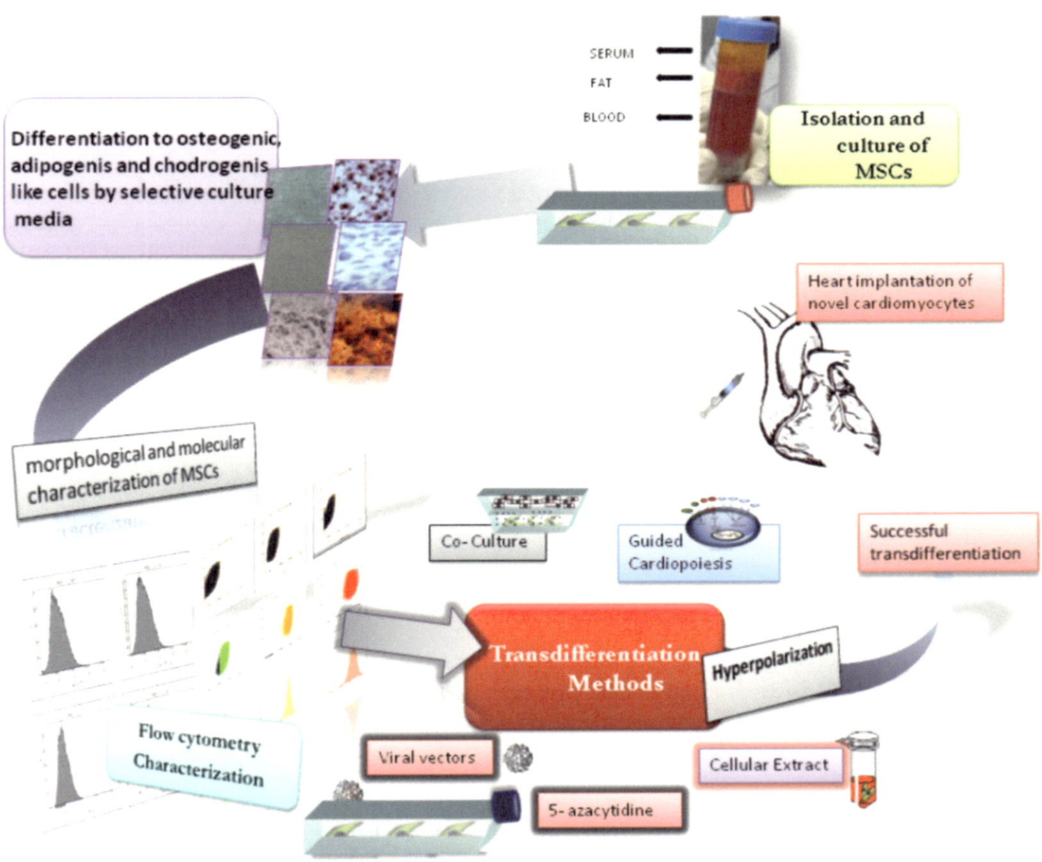

Fig. 2. Methods used to transdifferentiate MSCs to cardiomyocyte like-cells

3.1 Transdifferentiation methods using adult stem cells

3.1.1 Use of allograft cardiac tissue: The co-culture and cell extracts methods

The co-culturing of two different cell types, such as cardiomyocytes isolated from human biospies and hMSCs, which are grown together but separated by a membrane (with 0.4 - 0.3 μm pore size to avoid direct cell-to-cell interaction), has revealed the importance of the microenvironment in stem cell differentiation. This method is used to determine whether soluble chemical factors released from one cell type (which can diffuse though the membrane) are sufficient to induce transdifferentiation of the other cell type. Using this mechanism, Wang et al. (2006) showed *in vitro* the differentiation of BM stem cells into cardiomyocyte-like cells with expression of cardiac-specific genes. Similar results have been obtained by various authors (Antonitsis et al., 2007; Rangappa et al., 2003).

In the cell extract method, the intracellular components of one cell type are introduced into another cell type in order to switch from one cell pattern to another. Purified somatic nuclei or reversibly permeabilized somatic cells are incubated in a nuclear and cytoplasmic extract derived from a different cell type. The extract is believed to provide nuclear regulatory components that mediate alterations in the gene expression profile of the target genome. After exposure to the extract, the cells are resealed in a culture that contains calcium. By this means, hADSCs produced cardiomyocyte proteins after incubation with rat cardiomyocyte extracts. The success of these transdifferentiation strategies was demonstrated by observations of morphological changes, by findings on donor cell-specific genes, and by immunological evidence of alterations in protein expression (Gaustad et al., 2004).

Our group reported that adult cardiomyocytes obtained from human donors retain their capacity to induce cardiomyocyte differentiation of MSCs. Human adipose stem cells (hASCs) isolated from lipoaspirates were transiently permeabilized, exposed to human atrial extracts and then allowed to recover in culture. After 21 days, the cells acquired a cardiomyocyte phenotype, as demonstrated by morphologic changes (appearance of binucleate striated cells and branching fibbers), immunofluorescence detection of cardiac-specific markers (connexin-43, sarcomeric α-actinin, cardiac troponin I and T and desmin) and the presence of cardiomyocyte-related genes (cardiac myosin light chain 1, α-cardiac actin, cardiac troponin T and cardiac β-myosin). The relevance of these findings lies in the potential use of autologous extracts to induce stem cell reprogramming (Perán et al., 2010).

In conclusion, both techniques have an interesting potential in a therapeutic framework, due to their capacity to avoid immunologic responses from patients and produce a good clinical outcome.

3.1.2 Use of 5 –azacytidine and growth factors

The acquisition of a cardiac phenotype has been induced by the use of demethylation agents such as 5-aza, a cytosine analogue that can reduce DNA methyltransferase activity in the cells (Quian et al., 2011). However, *in vivo* studies with 5-aza showed that specific proteins were expressed in myocytes near the injection site in the myocardium, but these cells lacked the complete acquisition of cardiac phenotype and were not capable of symmetric contraction (Valiunas et al., 2004). Similar methods have been reported to produce cells with cardiomyocyte phenotype from human adipose-derived stem cells, and spontaneously beating cells were obtained after co-culture with neonatal rat cardiomyocytes (Choi et al., 2010).

3.1.3 Genetic modifications: Use of viral vectors

The genetic modification of MSCs is an interesting option for improving their therapeutic potential. The transfection of master genes or transcription factors that induce differentiation into the cardiac phenotype was recently developed as a transdifferentiation methodology. In cardiac research, efforts have been made to transfect BM-MSCs with viral constructions, using retroviral vectors that contain the master genes Csx/Nkx2.5 and GATA-4. Single-cell-derived MSCs overexpressing Csx/Nkx2.5 and GATA4 behaved as cardiac transient amplifying cells and retained their plasticity *in vivo*, but 5-aza treatment was required to achieve a complete cardiac phenotype. The frequency of cardiomyogenic differentiation was increased by co-culturing with foetal cardiomyocytes (Yamada et al., 2007).

For human clinical proposes, Ad- vectors are safer than RV- vectors because they do not integrate into the cellular DNA, and therapy with Ad-transduced MSCs is likely to be less immunogenic than the direct administration of Ad-vectors to affected tissues, due to the anti-inflammatory properties of MSCs. Moreover, transgenic expression in proliferating stem cells is only transient, due to the non-integrative properties of Ad-vectors. Nonetheless, short-term therapeutic gene expression may be sufficient or even desirable for the treatment of diseases such as myocardial infarction, in which the transient paracrine effects of MSCs enhance tissue healing and repair responses (Reiser et al., 2005).

Two studies recently reported the direct reprogramming of murine fibroblasts into functional cardiomyocytes. Srivastrava´s group developed a method to recruit and transdifferentiate resident cardiac fibroblasts near the infarction area. This group enforced the expression of developmental cardiac transcription factors Gata4, Mef2c, and Tbx5 and produced cardiomyocytes that matched a normal cardiomyocyte gene expression profile (Ieda et al., 2010). The efficiency of this direct reprogramming was higher in comparison to reprogramming to iPSCs (20% vs. <0.1%) and it was more rapid, but the emergent cardiomyocytes could not be expanded in culture, which is a shortcoming.

In another study, Ding´s group obtained cardiomyocytes by virally transducing mouse embryonic fibroblasts with genes encoding for the transcription factors Oct4, Sox2 and Klf4, followed by modified standard reprogramming medium, in which leukaemia inhibitor factor was removed and foetal bovine serum was added at 1-15% (Efe et al., 2011). This new development may help eliminate some of the obstacles to the use of traditional ESCs or iPSCS, including teratoma formation due to residual undifferentiated cells, and it may lower the cost and delivery time to patients thanks to higher yields and faster production.

3.2 Transdifferentiation methods using iPSCs

Induced pluripotent stem cells (iPSCs) are adult cells that have been genetically reprogrammed to an embryonic stem cell-like state by being forced to express genes and factors important for maintaining the defining properties of embryonic stem cells. The reprogramming of adult cells into ESCs enables the generation of patient-specific stem cells and therefore has enormous potential for the analysis and treatment of degenerative diseases. iPSCs are typically derived by transfection of certain stem cell-associated genes into non-pluripotent cells, such as adult fibroblasts. Transfection is typically achieved through viral vectors, such as retroviruses. Transfected genes include the master transcriptional regulators Oct-3/4 (Pouf51) and Sox2, although some other genes (e.g., Klf4, c-Myc) have been used to enhance the efficiency of induction.

Among various methods used to induce differentiation in pluripotent cell lines, the most widespread is to grow undifferentiated cells as aggregates in suspension, which causes them to form EBs in which cardiac cells spontaneously develop (Son et al., 2011). Cardiomyogenic mesodermal progenitors are normally formed in the embryo during gastrulation as cells of the epiblast pass through the primitive streak (Nury et al., 2009).

Many protocols for directed differentiation to cardiomyocytes are based on these developmental principles. In general, two directed differentiation strategies have been applied. First, the co-culture of hESCs with cardio-inductive cell types, such as endodermal cell line END-2. Cardiac induction by END-2-conditioned medium can be (partly) mimicked by insulin depletion (Freund et al., 2008), inhibition of p38 MAPK (Graichen et al., 2008) and addition of prostaglandin E.

A second method is to induce gastrulation by the use of specific growth factors. Several studies have shown that various combinations of BMP4, WNT3a and Activin A induce gastrulation-like events and meso-/endoderm development in hESCs (Sumi et al., 2008; Vijayagavan et al., 2009). A similar approach uses an efficient cardiac differentiation protocol based on transient stimulation with BMP4, bFGF and activin A followed by VEGF and WNT inhibition through DKK. Both approaches have allowed considerable progress to be made over the past few years with respect to the efficiency of cardiomyocyte differentiation, but its use in clinical therapies remains plagued with numerous problems, including inadequate purity. The same methods are currently used for the generation of iPSCs-derived cardiac cells.

3.2.1 Future applications of iPSCs technology in heart disease

The possibility of creating patient-specific embryonic stem cell-like cells may complement the rather slow development of heart disease models based on hESC. However, their value in elucidating disease mechanisms and as models for drug discovery remains to be demonstrated. The uniformity and reliability of the differentiation of several well-studied hESC lines may make these the most useful near-term research tool; therefore, research on the targeting and genetic modification of hESCs is still vital. In this regard, the use of iPSCs models has opened a new alternative to study illness, especially in countries were legal issues do not allow the use of hESC.

The generation of heart disease models is a highly complicated process, due to the variety of aetiologies and subtypes of CVDs and their complexity. Initial approaches have avoided multi-gene diseases and focussed on heart diseases associated with mutations in single genes, which are most likely to have an effect at single cell level. This is the case of channelopathies, which are related to mutations in a small set of genes that encode cardiac ion channels and sarcomeric proteins. In this way, the phenotypes of these diseases are characterized by disturbed ion channel function (cardiac channelopathies) or impaired contractility (cardiomyopathies). Heritable cardiac channelopathies include long-QT syndrome (LQTS), Brugada syndrome, catecholaminergic polymorphic ventricular tachycardia, cardiac conduction disease and sinus node dysfunction (Priori & Napolitano, 2006.

Recent successes have been reported using human iPSCS as models for human heart disease. Two groups generated human iPSCS lines from heart disease patients that showed a disease phenotype in culture. Carvajal-Vergara et al. (2010) produced cardiomyocytes from two human iPSCs lines derived from two patients with a heterozygous T468M

substitution in *PTPN11* (protein tyrosine phosphate, non-receptor type 11), which leads to Leopard syndrome, offering novel insights into signalling pathways related to this disease.
Another study by Moretti et al. (2010) investigated an ion channel mutation in patients with LQT1. LQTS is a life-threatening congenital condition characterized by cardiac arrhythmias and sudden cardiac death. These patients have a *KCNQ1* gene mutation that results in an elongated QT-interval. The *KCNQ1* gene encodes *I*Ks, the ion channel controlling the slow component of the delayed rectifier K+ current, which is partially responsible for the repolarising phase of the action potential. The authors were able to recapitulate this electrophysiological phenotype in culture by using a whole-cell patch-clamp electrophysiology method after creating human iPSCs lines from two patients in a single family with an R190Q missense mutation in *KCNQ1*. Using a similar method, Itzhaki et al. (2011) developed a patient/disease-specific human iPSCs line from a patient with type-2 LQTS, which is due to the A614V missense mutation in the *KCNH2* gene, coaxing these iPSCs to differentiate into the cardiac lineage. The LQTS human iPSC-derived cardiac-tissue was used as a model to evaluate the potency of existing and novel pharmacological agents that may either aggravate (potassium-channel blockers) or ameliorate (calcium-channel blockers, K_{ATP}-channel openers and late sodium-channel blockers) the disease phenotype.

4. Pre-clinical studies: A step to the bedside

Initial animal experiments found that the transplantation of BM-SCs after AMI and ischaemic cardiomyopathy is associated with a reduction in scar size and improvements in the ejection fraction of the left ventricle (LVEF) and tissue perfusion (Dawn & Bolli, 2005). Other pre-clinical studies demonstrated an improvement in myocardial function after repair of the infarcted area through implantation of EPCs derived from peripheral blood, BM or umbilical cord blood (UCB) (Leor & Marbel, 2006; Perin & López, 2006).
MSCs have also been used in a model of chronic myocardial ischaemia in large animals. Twelve dogs that received intramyocardial injections of 100 million cells showed improved systolic function (by echocardiography) in comparison to controls, both at rest and under stress situations; in addition, the MSCs were able to transdifferentiate into endothelial cells and smooth muscle cells, enhancing vascularization (Perin et al., 2003). Skeletal myoblasts were also studied in the setting of chronic myocardial ischaemia, specifically to treat myocardial scarring, and were found to improve the LVEF and the adverse effects of ventricular remodelling, due to their high capacity to engraft in scarred myocardial segments (Leor et al., 1996).

5. To the bedside: Cardiac cell therapy

The rapid development of stem cell technology has raised hopes for novel and even revolutionary treatments for cardiac and other disorders with tissue damage, awakening the keen interest of the pharmaceutical industry. These expectations are largely based on the 30-year history of successful BM-HSC transplantations in patients with blood diseases and cancer (Thomas et al., 1959; Shizuru et al., 2005). However, few stem cell therapies have been widely accepted by the medical community, and all of these use tissue-specific stem cells. These include BM or UCB stem cell transplantation to treat blood diseases or restore the blood system after cancer treatments, skin stem cell therapies for burns and limbal stem cells for corneal replacement, among others. In each case, stem cells repair the same tissue as that from which they are derived.

Cardiac cell therapy has been attempted in more than 1000 patients worldwide, and the results of the first meta-analysis recently became available. The interest in the use of adult stem cells in cardiac regeneration began after the demonstration by Orlic et al. (2001a) that murine BM-derived cells regenerated heart muscle after direct injection at the site of the previously induced myocardial infarction. These promising results generated significant interest in its clinical application, with the launch of numerous randomized clinical trials on the efficacy of cell therapy in patients with coronary disease, including 37 trials in Europe alone. The interpretation of their results should consider the effectiveness of the cell transplantation method, the delivery time and the cell type used. In this section, the NIH-supported ClinicalTrials website (ClinicalTrials.gov) was searched for closed and ongoing clinical trials in the field. Keywords included "heart", "myocardial infarction", "heart failure" and "cell therapy" (Table 3).

5.1 Bone marrow-derived stem cells

The TOPCARE-AMI study provided the first major clinical report on the transplantation of BM stem cells after myocardial infarction (Leistner et al., 2011). Patients were randomized to receive BM-derived mononuclear cells (29 patients) or EPCs (30 patients) by intracoronary infusion. Both groups showed improved LVEF at 4 months and reduced infarct size at 12 months, but there was no randomized control group with which to compare results. More recently, in a sub-group of the same patients, the authors observed a significant increase in LVEF (by cardiac MRI) and a reduction in infarct size (by delayed enhancement MRI). Interestingly, the migratory capacity of the infused cells proved to be the most important predictor of myocardial remodelling (Britten et al., 2003).

Some trials failed to show a significant increase in cardiac function, whereas others evidenced improvements comparable with those obtained by established treatment regimens (Reffelmann & Kloner, 2009). A range of results has been reported, from a low to a 14% improvement in LVEF (Fernandez-Aviles et al., 2004; Perin & López, 2006; Stamm et al., 2007; Tatsumi et al., 2007).

Many clinical studies have evaluated the therapeutic potential of human BM-derived stem cells (e.g. MSCs or mononuclear cells) to improve cardiac function after myocardial infarction (Gonzales & Pedrazzini, 2009; Mathiasen et al., 2009; Wei et al., 2009). However, no evidence has been reported of cardiac regeneration through differentiation of implanted stem cells into cardiomyocytes and other cardiac cell lineages. Some studies, but not all, have reported beneficial effects on heart function and on symptoms (Janssens et al., 2006; Lunde et al., 2006; Schachinger et al., 2006; Meyert et al., 2006). However, it has been suggested that these benefits are short-term, and a five-year follow-up study found that treatment with BM cells did not achieve sustained improvements in heart function (Meyer et al., 2009). Nonetheless, according to the results of all of these studies, the therapeutic use of human BM stem cells appears to be safe.

A meta-analysis of 18 randomized and non-randomized clinical trials, including 999 patients with AMI and chronic ischaemia, reported encouraging results after the transplantation of BM-MSCs. In comparison to controls, the transplantation improved the LVEF by 5.40% ($P < 0.001$), reduced the post-AMI scar area by 5.49% ($P = 0.003$) and reduced the left ventricular end-systolic volume by 4.80 ml ($P = 0.006$). According to the results of this meta-analysis, BM-MSC therapy appears to be safe and can be used in large-scale randomized trials to assess its impact on cell therapies (Abdel-Latif et al., 2007).

Clinical trial identifier number	Trial name/ Sponsor	Expected enrolment	Acquisition Method/ treatment	Primary end point	Route of cell delivery	Further Comments
Cardiac Stem Cells (CSCs)						
NCT 00474461	SCIPIO	40	Harvested from RAA	LVEF	IC	Evaluation of the regeneration capacity of CSCs in non-viable myocardial segments in patients with ICM
NCT 00981006	ALCADIA	6	Clonally Cell isolation	Safety	IM	Stem Cell Therapy combined with bFGF release and using a gelatin Hydrogel Sheet during CABG
NCT 00893360	CADUCEUS	30	Harvested from biopsies	LVEF	IC	Evaluation of the regeneration capacity of CSCs
Skeletal Myoblasts (SMs)						
NCT 00526253	MARVEL	170	Unknown	LVEF	Catheter delivery system	Evaluation of Myocell™ effects
NCT 00102128	Genzyme	30	Harvested from the thigh muscle	LVEF	Unknown	Evaluation of the regeneration capacity of SM
NCT 00908622	PERCUT ANEO	50	Harvested from biopsies	LVEF	Percutaneous	Cardiomyoplasty benefits with SMs in patients with old AMI.
Endothelial Progenitor Cells (EPCs)						
NCT 00936819	ENACT-AMI	100	Harvested from apheresis, One group transfected with eNOS	LVEF	IC into the IRA	Combined Cell and gene therapy

RAA: Right atrial appendage; IRA: Infarct Related Artery; ICM: Ischaemic Cardiomyopathy; bFGF: Basic Fibroblastic grow factor; CABG: Coronary Artery Bypass Graft; eNOS: human endothelial nitric oxide synthase

Table 3. Ongoing clinical trials using different autologous stem cells.

5.2 Mesenchymal stromal cells

Animal studies are ongoing to determine whether MSCs can be used to treat arthritis, non-healing bone fractures or spinal cord injuries, among other diseases. Although it is also possible that these or similar cells modulate the immune system in response to injury, their use as cardiomyocyte-like cells *via* transdifferentiation methods remains controversial. A major advance in this field was achieved by Behfar et al. (2010), whose results with MSCs using guided cardiopoiesis methods allowed this approach to be evaluated in preclinical studies. They obtained a complete cardiac phenotype by using an endoderm-like protein cocktail with hMSCs harvested from patients with coronary artery disease. The hMSCs were injected into the myocardium of nude infarcted mice and followed up for a year to assess functional and structural endpoints. The recombinant cardiogenic cocktail guidance secured the cardiopoietic phenotype across the patient cohort, achieving a superior functional and structural benefit in comparison to unguided counterparts, with no adverse side effects (Behfar et al., 2010).

However, there is evidence of a low retention of the infused cells in the area of interest, which would reduce the effectiveness of this approach. Current strategies include the use of hMSCs modified with viral vectors carrying genes that increase the affinity of these cells to the sites of cell damage, e.g., through activation of HIF-α factor, which is expressed in states of hypoxia (Xue et al., 2010).

5.3 Skeletal myoblasts

The promising experimental results achieved by Leor et al. (1996) with skeletal myoblasts in a chronic myocardial ischaemia model, described above in section 4, generated significant interest in their possible clinical application. The first clinical trial was conducted at Paris-Descartes University using autologous skeletal myoblasts isolated from muscle biopsies from patients with severe heart failure, expanded *ex vivo* and then transplanted. The clinical status and ejection fraction of the treated patients improved over time, with a strikingly low incidence of hospitalizations for heart failure (0.13/patient-years). Moreover, the risk of arrhythmia could be controlled by medical therapy and/or on-request automatic cardiac defibrillator implantation (Menasche et al., 2001).

Therapy with skeletal myoblast transplantation was tested in the MAGIC trial (Myoblast Autologous Grafting in Ischaemic Cardiomyopathy). An injection of stem cells or culture medium was randomly administered to patients and, although initial fears of serious arrhythmias did not materialize, the results showed no significant benefit from the stem cell treatment.

The ambiguous results of clinical trials reflect differences due to the diversity of cell types and administration routes used. The timing of the assessment of ventricular function post-transplantation may also explain some discrepancies in the results. Several of the benefits achieved by this therapy appear to be transient (Meyer et al, 2009), consistent with the findings of similar studies conducted in animals (van Laake et al., 2008).

5.4 iPSCs technology

The need for deeper knowledge of iPSCs is acknowledged by the scientific community, but only one clinical trial is using this cell type, in an attempt to induce pluripotent stem cells from cell cultures of skin biopsies (Royan Institute, Iran, CI. NCT00953693). This trial is related to hepatic regeneration. The strategy is basically the same for all research purposes:

after isolation of fibroblasts from the patient`s skin biopsy, the cells are transfected after 3-4 weeks with the four factors used by Yamanaka. Transfected cells are isolated by morphological selection in order to generate allogeneic hiPSCs (Seifinejad et al., 2010). In the field of heart disease, basic research includes aspects such as the generation of cardiomyocytes derived from human ES/iPS cells from post-AMI heart failure patients (Itzhaki et al, 2011). These results will be important for drug discovery screening in relation to QT prolongation and cardiac cell transplantation therapy.

6. Funding research through industry: Development of patents in cell therapy

Improved understanding of the development of the human heart has led to proposals for strategies that emulate the cardiogenic programme, which are currently being tested in pre-clinical and clinical studies. There are an increasing number of patents in this area as well as on the establishment of cell lines. A patent operates as a *quid pro quo*: the patent owner obtains the exclusive right to make, use and sell an invention in exchange for its public disclosure. Some scientists believe that patents harm the research environment by increasing secrecy and costs. Others consider that the regenerative medicine sector requires access to funding and resources for the protracted procedures required, including not only the research and clinical trials but also the legal issues that must be addressed with respect to patent protection for regenerative medicine-related intellectual property. Interestingly, regenerative medicine may change the conventional reliance on patents and their finite exclusivity, which is a truly major issue for the pharmacy industry. Patent registers in the USA show a growing number of patents for the treatment of CVDs using stem cells, representing a validation of the therapeutic potential of this approach. In Europe, there is a slight trend towards the creation of patents related to methods for isolating cell populations with cardiopoietic potential and for preparing enrichment culture media, which are in pre-clinical stages of application.

A search of the sources "Patent Lens" of the organization "Initiative for Open Innovation (IOI)", "Patentscope" of the World Intellectual Property Organization (WIPO), and the "Invenes", "Esp@cenet", "Free patents online" and "Latipat" databases of the *Spanish Patents Office*, shows that patent applications to commercialise cell therapies for CVDs centre on three main topics: i) methods to deliver stem cells to the damaged myocardium, ii) methods to isolate and differentiate autologous cells harvested from patients, and, finally, iii) viral vectors for gene delivery.

6.1 Achieving the cardiac cell phenotype: Producing stem cell- derived cardiomyocytes

Given the difficulty of isolating and culturing heart cells, many studies have focused on developing methods to induce the differentiation of stem cells into the myocardial phenotype. For this purpose, MSCs are the most widely used cells in the patents and clinical trials reviewed, followed by HSCs, but knowledge of the existence of cardiac progenitor populations has made these the object of the most recent studies. One example is the development by Chien et al. (2011) (Patent application US 2011/0003327) of methods to identify and isolate atrial progenitors and, in some cases, atrial progenitors that are positive for Islet 1 (Isl1) and sarcolipin, which are growth factors responsible for the negative regulation of second heart field progenitor cell differentiation (Buckingham et al., 2005).

Another invention of this team relates to methods for isolating cardiovascular stem cells, placing cell populations in contact with agents reactive to Isl 1, Nkx2.5 and Flk-l and separating reactive from non-reactive cells (Chien & Chang 2008; Patent application WO 2008/054819 A2).

The use of MSCs from different sources is reflected in the patent of Büscher et al. (2010) (US patent application 20100304477), who characterized a novel population of human cardiac adipose tissue-derived adult stem cells from the epicardial area. These cells are isolated by their surface marker profile and their gene and protein expression of different cardiac muscle cell components, including the cardiac factors GATA-4 and Nkx2.5, the sarcomere components beta-myosin heavy chain and α-actinin, and the regulators of electrochemical connection and intracellular calcium distribution, connexin-43 (Cx43) and SERCA-2, respectively.

6.2 How can stem cells be guided to the damaged myocardium?

Numerous techniques have been developed for this purpose, including the use of growth factors and molecules that deliver and mobilize stem cells to the injured site. This paracrine effect has been described in MSCs and also includes the activation of different mechanisms against cell damage. In this respect, granulocyte colony-stimulating factor (G-CSF) has proven useful alone or in combination with the stem cell factor (SCF) in the mobilization of BM stem cells. The pre-AMI mobilization of primitive BMCs by the administration of G-CSF and SCF (US Patent 7220407) results in a significant degree of tissue regeneration at the ischaemic site (Orlic et al., 2001b). It is also contemplated that human G-CSF therapy may be used in a similar manner along with other agents commonly used for the treatment of AMI in conjunction with reperfusion.

In the field of gene therapy, several methods have been developed for treating patients with CVDs, including heart disease and peripheral vascular disease. The preferred methods involve the *in vivo* delivery to the myocardium or peripheral ischaemic tissue of genes encoding angiogenic proteins or peptides by the introduction of a vector containing the gene into a blood vessel supplying the heart or into peripheral ischaemic tissue. The patent "Techniques and compositions for treating cardiovascular disease by in vivo gene delivery" (Hammond et al., 2003; US Patent Application 20030148968), describes a method that employs a replication-deficient adenovirus as vector. By injecting the viral vector stock with angiogenic protein- or peptide-encoding genes deeply into the lumen of one or both coronary arteries or grafts, it is possible to locally transfect an adequate number of cells, especially cardiac myocytes, into the affected myocardium. This methodology maximizes the therapeutic efficacy of gene transfer and minimizes undesirable angiogenesis at extracardiac sites and the possibility of an inflammatory response to viral proteins.

7. Clinical translation

The translation of research findings to the clinical setting can be a complex process. This discipline, based on interventional epidemiology, aims to increase the throughput of findings from the basic or early phase of clinical research in ageing and to use them to drive innovation in healthcare settings. The success of clinical translation depends on the capacity of scientists, clinicians, regulators and other experts to share their expertise with one other and on their ability to respond adequately to negative or positive clinical outcomes.

In the field of CVD, considerable efforts have been made to establish feedback between research groups and clinical practice. Universities and hospitals have developed clinical translation units in which stem cell-related issues are focused on the utility of plasma biomarkers of acute brain injury to improve the early diagnosis of acute stroke and on the prevalence and specific causes of ventricular repolarisation abnormalities in older patients undergoing psychoactive drug treatment, among others.

7.1 Cell therapy

As reported above, human cell-based medicine is proving increasingly attractive to the pharmaceutical and medical equipment industries, opening up the way for the self-funding needed for scientific development and the implementation of clinical trials. This is of particular importance at a time of intense pressure to reduce public health costs.

Progress has been made in developing protocols for obtaining and isolating stem cells, in establishing legislation that specifies the parameters of good manufacturing practice for large-scale production, and in accumulating pre-clinical data that validating the effects of this therapy. As a result, several types of stem/progenitor cells have been proposed for use in regenerative medicine, leading to the consideration of stem cells as pharmaceuticals.

Large pharmaceutical companies apply stem cell-based technologies and other proprietary methods in the area of regenerative medicine to bring patient-specific therapies from the laboratory to the bedside. This is the case of products being developed by Advanced Cell Technology, Bioheart Inc. and Cardio3 Bioscience companies (Figure 3).

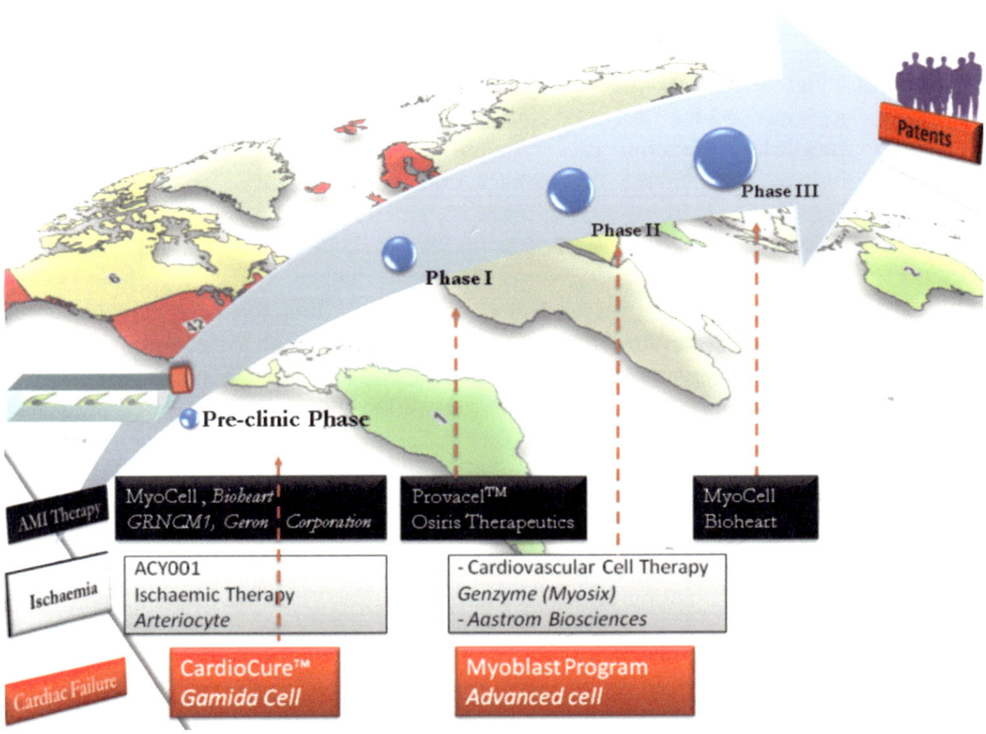

Fig. 3. Examples of products in development for cell therapy in diseases related to the cardiovascular system.

We highlight the initiatives taken by some U.S. drug companies, who have begun to market cell therapies after successful clinical trials. One example is Athersys Inc., whose MultiStem cell therapy has a broad potential to treat diseases such as AMI, stroke, and inflammatory bowel disease. MultiStem promotes tissue repair through the induction of therapeutic factors produced in response to signs of inflammation and tissue damage. Likewise, the Bioheart company is now marketing MyoCell as a result of the MARVEL trial, a randomized, double-blind, placebo-controlled, multicenter, phase II/III trial involving 330 patients in North America and Europe. MyoCell uses muscle tissue samples from the patient, isolating and expanding stem cells and then transplanting them by endoventricular injection into the patient's scar tissue (MyoStar ™).

7.2 Scaffolding: The development and translation of the tissue-engineered vascular graft

The post-mitotic and myogenic nature of myocardial tissue hampers the *in vitro* isolation and maintenance of cardiomyocytes. *In vitro* dedifferentiation is evidenced by the loss of sarcomeric distribution, observed in the distribution pattern of protein alpha actinin, troponin and connexin 43. These proteins are essential for muscle contraction. The same difficulty arises with the generation of cardiomyocytes *in vitro,* demonstrating that an effective approach for cardiac regeneration must include both specialized cells in a three-dimensional support or scaffold that provides the physical properties necessary for contraction. However, because of the heterogeneous population residing in the myocardium, attempts have focused on the generation of vascular replacements for larger vessels such as arteries. The utilisation by Dr. Taylor's group of a decellularisation method in mouse heart yielded encouraging findings in relation to the generation of bio-artificial hearts. They decellularised the hearts by coronary perfusion with detergents, preserving the underlying extracellular matrix and producing an acellular, perfusable vascular architecture with competent acellular valves and intact chamber geometry. They mimicked the composition of cardiac cells by reseeding these constructs with cardiac or endothelial cells. The group maintained eight constructs for up to 28 days by coronary perfusion in a bioreactor that simulated cardiac physiology, observing macroscopic contractions by day 4. By day 8, under physiological load and electrical stimulation, the constructs generate a pump function equivalent to around 2% of adult or 25% of 16-week foetal heart function in a modified working heart preparation (Ott et al., 2008).

Research by Dr. Christopher Breuer and Dr. Toshiharu Shinoka on tissue engineered vascular grafts (TEVG) is in the process of clinical translation for congenital cardiovascular diseases. In the United States, around 1 in 100 infants is expected to have heart defects, i.e., around 36,000 infants per year. Ten percent of these cases—over 3,600 infants—result in death. TVEGs are biodegradable synthetic scaffolds made of the same material as absorbable sutures and seeded with the individual's own cells. The scaffold degrades by hydrolysis, eventually leaving only the living vessel in the patient. Bone marrow aspiration is used to harvest the cells, inserting the needle through the cortex of the spinal bone and marrow. The marrow is drawn up and separated by density centrifugation, yielding BMCs that are directly seeded onto the scaffold by pipetting. The seeded scaffold is then incubated in the patient's plasma for two hours to attach cells to the scaffold. The graft is then ready for implantation. The whole process takes only a few hours, allowing a viable TEVG to be obtained on the same day as the cell harvesting (Nuti, 2010).

8. Conclusion

There have been numerous clinical trials of cell therapies for CVDs, and several problems have been identified. Questions that remain to be addressed in translational research from animals to humans concern the cell type, preparation time, delivery, cell number and optimization. The measurement of results must also be resolved, including the efficacy assessment method and length of follow-up. Nevertheless, the potential therapeutic contribution of adult stem cells is enormous. hESCs have been widely studied in relation to heart disease, but their use in cell therapy has been limited, in part due to ethically-based restrictions on research with this cell type in some countries and the lack of adequate funding.

With regard to iPSCs, the most important and immediate advances have been in relation to drug discovery, in the context of an urgent need for new approaches to meet the increasing incidence of myocardial infarction and heart failure in ageing populations. The number of novel drugs entering the market for heart disease is decreasing, whereas investment by the pharmaceutical industry in this area continues to increase, underlining the need for patents to safeguard the intellectual property of research groups and provide them with sufficient financial support to run clinical trials.

Criteria established by international scientific organizations such as the European Society of Cardiology and the American Society of Cardiology must be followed by researchers who address these issues. The correct design of future clinical trials is of particular importance before clinical stem cell therapy can be widely applied. In this regard, the International Society of Stem Cell Research (ISSCR, 2011) is taking the exceptional step of developing guidelines for patients, researchers and physicians, establishing safety procedures and providing patients worldwide with information to meet their concerns and describe their legal rights.

9. Acknowledgment

This work was supported in part by grants from the Instituto de Salud Carlos III (Fondo de Investigación Sanitaria, grant number PI10/02295), the Consejería de Salud (Junta de Andalucía, grant number PI-0384/2008), and the Consejería de Economía, Innovación y Ciencia (Junta de Andalucía, excellence project number CTS-6568).

10. References

Aasen, T.; Raya, A.; Barrero, M. J.; Garreta, E.; Consiglio, A.; Gonzalez, F.; Vassena, R.; Bilic, J.; Pekarik, V. & Tiscornia, G. (2008). *Efficient and rapid generation of induced pluripotent stem cells from human keratinocytes.* Nat. Biotechnol. 26, 1276-1284.

Abdel-Latif, A.; Bolli, R.; Tleyjeh, I.M.; Montori, V.M.; Perin, E.C.; Hornung, C.A.; Zuba-Surma, E.K.; Al-Mallah, M. & Dawn, B.(2007). *Adult bone marrow-derived cells for cardiac repair: a systematic review and meta-analysis.* Arch Intern Med. 167:989-97.

Aggarwal, S. & Pittenger, MF. (2005). *Human mesenchymal stem cells modulate allogeneic immune cell responses.* Blood 105:1815–1822

Antonitsis, P.; Ioannidou-Papagiannaki, E.; Kaidoglou, A. & Papakonstantinou, C.H. (2007). *In vitro cardiomyogenic differentiation of adult human bone marrow mesenchymal stem cells. The role of 5-azacytidine.* Interact CardioVasc Thorac Surg. 6:593-597.

Bearzi, C.; Rota, M.; Hosoda, T.; Tillmanns, J.; Nascimbene, A.; De Angelis, A.; Yasuzawa-Amano, S.; Trofimova, I.; Siggins, RW.; Lecapitaine, N.; Cascapera, S.; Beltrami, A.P.; D'Alessandro, D.A.; Zias, E.; Quaini, F.; Urbanek, K.; Michler, R.E.; Bolli, R.; Kajstura, J.; Leri, A. & Anversa, P. (2007) *Human cardiac stem cells. Proc. Natl. Acad. Sci. USA*, 104, 14068-14073.

Behfar, A.; Yamada, S.; Crespo-Diaz, R.; Nesbitt, J.J.; Rowe, L.A.; Perez-Terzic, C.; Gaussin, V.; Homsy, C.; Bartunek, J. & Terzic A. (2010). *Guided cardiopoiesis enhances therapeutic benefit of bone marrow human mesenchymal stem cells in chronic myocardial infarction.* J. Am. Coll. Cardiol. 56:721-34.

Britten, M.B. Abolmaali, N.D.; Assmus, B.; Lehmann, R.; Honold, J.; Schmitt, J.; Vogl, T.J.; Martin, H.; Schächinger, V.; Dimmeler, S. & Zeiher, A.M. (2003). *Infarct Remodeling After Intracoronary Progenitor Cell Treatment in Patients with Acute Myocardial Infarction (TOPCARE-AMI. Mechanistic Insights From Serial Contrast-Enhanced Magnetic Resonance Imaging.* Circulation. 108:2212-2218.

Buckingham, M.; Meilhac, S. & Zaffran S. (2005).*Building the mammalian heart from two sources of myocardial cells.* Nat. Rev. Genet. 6:826–835.

Büscher, D.; Bayes, A.; Roura, S.; Farré J. & Prat, C. *Population of adult stem cells derived from cardiac adipose tissue and use thereof in cardiac regeneration.* 20100304477, (2010).

Burke, D.Z. & Tosh, D. (2005). *Therapeutic Potential of transdifferentiated cells.* Science 108:309–321.

Carvajal-Vergara, X.; Sevilla, A.; D'Souza, S.L.; Ang ,Y.S.; Schaniel, C.; Lee, D.F.; Yang, L.; Kaplan, A.D.; Adler, E.D.; Rozov, R.; Ge, Y.; Cohen, N.; Edelmann, L.J.; Chang, B.; Waghray, A.; Su, J.; Pardo, S.; Lichtenbelt, K.D.; Tartaglia, M.; Gelb, B.D. & Lemischka, I.R. (2010). *Patient-specific induced pluripotent stem-cell-derived models of LEOPARD syndrome.* Nature 465:808–812

Chien, K.; Atsushi, N. & Haruko N. *Methods for production of atrial progenitors and their differentiation into smooth muscle cells and cardiomyocytes.* US Patent Application 0003327 (2011).

Chien, K.R. & Chang, Y.S. *Regeneration and survival of cardiac progenitors and cardiomyocytes with a stretch activated transcription factor.*WO 2008/054819 A2 (2008).

Choi, Y. S.; Dusting, G. J.; Stubbs, S.; Arunothayaraj, S.; Han, X. L.; Collas, P.; Morrison, W.A. & Dilley, R.J. (2010). *Differentiation of human adiposederived stem cells into beating cardiomyocytes.* J Cell Mol Med. 14:878-89.

Cowan, C.A.; Atienza, J.; Melton, D.A. & Eggan, K. (2005). *Nuclear reprogramming of somatic cells after fusion with human embryonic stem cells.* Science 309:1369-1373

Dawn, B. & Bolli, R. (2005). *Adult bone marrow-derived cells: regenerative potential, plasticity, and tissue commitment.* Basic Res Cardiol 100: 494–503.

Dimos, J.T.; Rodolfa, K.T.; Niakan, K.K.; Weisenthal, L.M.; Mitsumoto, H.; Chung, W.; Croft, G.F.; Saphier, G.; Leibel, R.; Goland, R.; Wichterle, H.; Henderson, C.E. & Eggan, K.(2008). *Induced pluripotent stem cells generated from patients with ALS can be differentiated into motor neurons.* Science 321:1218–1221.

Efe, J.A.; Hilcove, S.; Kim, J.; Zhou, H.; Ouyang, K.; Wang, G.; Chen, J. & Ding, S. (2011). *Conversion of mouse fibroblasts into cardiomyocytes using a direct reprogramming strategy.* Nat Cell Biol. 13:215-22.

Fernandez-Aviles, F.; San Roman, J.A.; Garcia-Frade, J.; Fernández, M.E.; Peñarrubia, M.J.; de la Fuente, L.; Gómez-Bueno, M.; Cantalapiedra, A., Fernández, J.; Gutierrez, O.; Sánchez, P.L.; Hernández, C.; Sanz, R.; García-Sancho, J. & Sánchez A.(2004).

Experimental and clinical regenerative capacity of human bone marrow cells after myocardial infarction. Circ Res 95: 742–748.

Freund, C.; Ward-van Oostwaard, D.; Monshouwer-Kloots, J.; van den Brink, S.; van Rooijen, M.; Xu, X.; Zweigerdt, R.; Mummery, C. & Passier, R. (2008). *Insulin Redirects Differentiation from Cardiogenic Mesoderm and Endoderm to Neuroectoderm in Differentiating Human Embryonic Stem Cells.* Stem Cells, 26:724–733.

Gaustad, K.G.; Boquest, A.C.; Anderson, B.E.; Gerdes, A.M. & Collas, P. (2004). *Differentiation of human adipose tissue stem cells using extracts of rat cardiomyocytes.* Biochem Biophys Res Commun. 314: 420-7.

Giorgetti, A.; Montserrat, N.; Aasen, T.; Gonzalez, F.; Rodriguez-Piza, I.; Vassena, R.; Raya, A.; Boue, S.; Barrero, M. J. & Corbella, B. A. (2009). *Generation of induced pluripotent stem cells from human cord blood using OCT4 and SOX2.* Cell Stem Cell 5:353-357.

Glynn, R.J. & Rosner, B. (2005). *Comparison of Risk Factors for the Competing Risks of Coronary Heart Disease, Stroke, and Venous Thromboembolism.* Am. J. Epidemiol. 162: 975-982.

Gonzales, C. & Pedrazzini, T. (2009). *Progenitor cell therapy for heart disease.* Experimental Cell Research, 315: 3077–3085.

Graichen, R.; Xu, X.; Braam, S. R.; Balakrishnan, T.; Norfiza, S.; Sieh, S.; Soo, S. Y.; Tham, S. C.; Mummery, C.; Colman, A.; Zweigerdt, R. & Davidson, B. P. (2008). *Enhanced cardiomyogenesis of human embryonic stem cells by a small molecular inhibitor of p38 MAPK.* Differentiation, 76:357–370.

Greenwood, H.L.; Thorsteinsdottir, H.; Perry, G.; Renihan, J.; Singer, P.A. & Daar, A.S. *Regenerative medicine: new opportunities for developing countries.* (2006). International Journal of Biotechnology, 8: 60-77.

Haase, A.; Olmer, R.; Schwanke, K.; Wunderlich, S.; Merkert, S.; Hess, C.; Zweigerdt, R.; Gruh, I.; Meyer, J. & Wagner, S. (2009). *Generation of induced pluripotent stem cells from human cord blood.* Cell Stem Cell 5, 434-441.

Hammond, K.H.; Dillmann, W. & Giordano, F.J. *Techniques and compositions for treating cardiovascular disease by in vivo gene delivery* (2003).

Ieda, M.; Fu, J.D.; Delgado-Olguin, P.; Vedantham, V.; Hayashi, Y.; Bruneau, B.G. & Srivastava, D. (2010). *Direct reprogramming of fibroblasts into functional cardiomyocytes by defined factors.* Cell. 142:375-86.

International Society of Stem Cells Research. (n.d). Guidelines of clinical translation, In: *A closer look to stem cells treatments* , 10.06.2011, Available from http://www.closerlookatstemcells.org//AM

Itzhaki, I.; L. Maizels, Huber, I.; Zwi-Dantsis, L.; Caspi, O.; Winterstern, A.; Feldman, O.; Gepstein, A.; Arbel, G.; Hammerman, H.; Boulos, M. & Gepstein, L.(2011). *Modelling the long QT syndrome with induced pluripotent stem cells.* Nature 471(7337): 225-229.

Iwakura, T.; Mohri, T.; Hamatani, T.; Obana, M.; Yamashita, T.; Maeda, M.; Katakami, N.; Kaneto, H.; Oka, T.; Komuro, I.; Azuma, J.; Nakayama, H.& Fujio, Y. (2011) *.STAT3/Pim-1 signaling pathway plays a crucial role in endothelial differentiation of cardiac resident Sca-1+ cells both in vitro and in vivo.* J Mol Cell Cardiol. 51:207-214.

Janssens, S.; Dubois, C.; Bogaert, J.; Theunissen, K.; Deroose, C.; Desmet, W.; Kalantzi, M.; Herbots, L.; Sinnaeve, P.; Dens, J.; Maertens, J.; Rademakers, F.; Dymarkowski, S.; Gheysens, O.; Van Cleemput, J.;Bormans, G.; Nuyts, J.; Belmans, A.; Mortelmans, L.; Boogaerts, M. & Van de Werf, F. (2006).*Autologous bone marrowderived stem-cell transfer in patients with ST-segment elevation myocardial infarction: double-blind, randomised controlled trial.* Lancet. 367:113–21.

Kinnaird, T.; Stabile, E.; Burnett, M.S.; Lee, C.W.; Barr, S.; Fuchs, S. & Epstein, S.E. (2004). *Marrow derived stromal cells express genes encoding a broad spectrumof arteriogenic cytokines and promote in vitro and in vivo arteriogenesis through paracrine mechanisms.* Circ Res 94:678–685.

Kim, D.H.; Yoo, K.H.; Choi, K.S.; Choi, J.; Choi, S.Y.; Yang, S.E.; Yang, Y.S.; Im, H.J.; Kim, K.H.; Jung, H.L., Sung, K.W. & Koo, H.H. (2005). *Gene expression profile of cytokine and growth factor during differentiation of bone marrow-derived mesenchymal stem cell.* Cytokine. 31:119-26.

Kumar, D.; Kamp, T.J. & LeWinter, M.M. (2005). *Embryonic stem cells: differentiation into cardiomyocytes and potential for heart repair and regeneration.* Coron Artery Dis 16: 111-116.

Leistner, D.M.; Fischer-Rasokat, U.; Honold, J.; Seeger, F.H.; Schächinger, V.; Lehmann, R.; Martin, H.; Burck, I.; Urbich, C., Dimmeler, S.; Zeiher, A.M. & Assmus B. (2011). Transplantation of progenitor cells and regeneration enhancement in acute myocardial infarction (TOPCARE-AMI): final 5-year results suggest long-term safety and efficacy. Clin Res Cardiol. 2011 Jun 3. [Epub ahead of print]

Leor, J. & Marber, M. (2006). *Endothelial progenitors: a new Tower of Babel?.* J Am Coll Cardiol. 17;48(8):1579-87.

Leor, J.; Patterson, M.; Quinones, M.J.; Kedes, L.H. & Kloner R.A. (1996). *Transplantation of fetal myocardial tissue into the infarcted myocardium of rat. A potential method for repair of infarcted myocardium?.* Circulation. 1;94(9 Suppl):II332-6.

Liu, Z.J.; Zhuge, Y. & Velazquez, O.C. (2009). *Trafficking and differentiation of mesenchymal stem cells.* J Cell Biochem. 106:984-991.

Loh, Y. H.; Agarwal, S.; Park, I. H.; Urbach, A.; Huo, H.; Heffner, G. C.; Kim, K.; Miller, J. D.; Ng, K. & Daley, G. Q. (2009). *Generation of induced pluripotent stem cells from human blood.* Blood 113, 5476-5479.

Loh, Y.-H.; Hartung, O.; Li, H.; Guo, C.; Sahalie, J. M.; Manos, P. D.; Urbach, A.; Heffner, G. C.; Grskovic, M. & Vigneault, F. (2010). *Reprogramming of T-cells from human peripheral blood.* Cell Stem Cell 7, 15-19.

Lunde, K.; Solheim, S.; Aakhus, S.; Arnesen, H.; Abdelnoor, M.; Egeland, T.; Ilebekk, A.; Endresen, K.; Mangschau, A., Fjeld, J.G.; Smith, H.J.; Taraldsrud, E.; Grøgaard, H.K.; Bjørnerheim, R.; Brekke, M.; Müller, C.; Hopp, H.; Ragnarsson, A.; Brinchmann, J.E. & Forfang, K. (2006). *Intracoronary injection of mononuclear bone marrow cells in acute myocardial infarction.* N Engl J Med. 355:1199–209.

Mathiasen, A.B.; Haack-Sørensen, M. & Kastrup, J. (2009). *Mesenchymal stromal cells for cardiovascular repair: current status and future challenges.* Future Cardiol. 5:605–17.

Menasche, P.; Alfieri, O.; Janssens, S.; McKenna, W.; Reichenspurner, H.; Trinquart, L.; Vilquin, J.T.; Marolleau, J.P.; Seymour, B., Larghero, J., Lake, S.; Chatellier, G.; Solomon, S.; Desnos, M. & Hagège, A.A. (2008).*The myoblast autologous grafting in ischemic cardiomyopathy (MAGIC) trial: first randomized placebo-controlled study of myoblast transplantation.* Circulation. 117(9):1189–200.

Meyer, G.P.; Wollert, K.C.; Lotz, J.; Steffens, J.; Lippolt, P., Fichtner, S.; Hecker, H.; Schaefer, A.; Arseniev, L.; Hertenstein, B.; Ganser, A. & Drexler, H. (2006). *Intracoronary bone marrow cell transfer after myocardial infarction: eighteen months' follow-up data from the randomized, controlled BOOST (BOne marrOw transfer to enhance ST-elevation infarct regeneration) trial.* Circulation. 113:1287-94.

Meyer, G.P.; Wollert, K.C.; Lotz, J.; Pirr, J.; Rager, U.; Lippolt, P.; Hahn, A.; Fichtner, S.; Schaefer, A.; Arseniev, L.; Ganser, A. & Drexler, H. (2009). *Intracoronary bone marrow cell transfer after myocardial infarction: 5-year follow-up from the randomized-controlled BOOST trial.* Eur Heart J. 30:2978-84.

Miyoshi, K.; Tsuji, D.; Kudoh, K.; Satomura, K.; Muto, T.; Itoh, K. & Noma, T. (2010). *Generation of human induced pluripotent stem cells from oral mucosa.* J. Biosci. Bioeng. 110, 345-50.

Moretti, A.; Bellin, M.; Welling, A.; Jung, C. B.; Lam, J. T;, Bott-Flugel, L.; Dorn, T.; Goedel, A.; Hohnke, C.; Hofmann, F; Seyfarth, M.; Daniel Sinnecker, D.; Schömig, A. & Laugwitz, K.L. (2010). *Patient-specific induced pluripotent stem-cell models for long-QT syndrome.* N. Engl. J. Med. 363, 1397–1409.

Nakanishi, C.; Yamagishi, M.; Yamahara, K.; Hagino, I.; Mori, H.; Sawa, Y.; Yagihara, T.; Kitamura, S. & Nagaya, N. (2008).*Activation of cardiac progenitor cells through paracrine effects of mesenchymal stem cells.* Biochem. Biophys. Res. Commun. 374:11-16.

Nury, D.; Neri, T. & Pucéat, M. (2009). *Human embryonic stem cells and cardiac cell fate. Journal of Cellular Physiology.* 218:455–459.

Nuti, S. (October 2010). Tissue Engineered Vascular Grafts: The Bright Future of Heart Health. In: *Yale scientific Magazine,* 25.06.2011, Available from http://www.yalescientific.org/2010/10/tissue-engineered-vascular-grafts-the-bright-future-of-heart-health/

Nygren, J.M.; Jovinge, S.; Breitbach, M.; Säwén, P.; Röll, W.; Hescheler, J.; Taneera, J.; Fleischmann, B.K. & Jacobsen, S.E.(2004). *Bone marrow-derived hematopoietic cells generate cardiomyocytes at a low frequency through cell fusion, but not transdifferentiation.* Nat Med10: 494–501.

Oda, Y.; Yoshimura, Y.; Ohnishi, H.; Tadokoro, M.; Katsube, Y.; Sasao, M.; Kubo, Y.; Hattori, K.; Saito, S. & Horimoto, (2010) *Induction of pluripotent stem cells from human third molar mesenchymal stromal cells.* J. Biol. Chem. 285, 29270-29278.

Orkin, S.H. (2000). *Diversification of haematopoietic stem cells to specific lineages.* Nat. Rev. Genet. 1, 57–64.

Orlic, D.; Kajstura, J.; Chimenti, S.; Jakoniuk, I.; Anderson, S.M.; Li, B.; Pickel, J.; McKay, R.; Nadal-Ginard, B.; Bodine D.M.; Leri1, A. & Anversa, P. (2001a). *Bone marrow cells regenerate infarcted myocardium.* Nature 410: 701–705.

Orlic, D.; Kajstura, J; Chimenti, S.; Limana, F.; Jakoniuk, I.; Quaini, F.; Nadal-Ginard, B.; Bodine, D.M.; Leri, A. & Anversa, P. (2001b). *Mobilized bone marrow cells repair the infarcted heart, improving function and survival.* Proc Natl Acad Sci U S A. 28;98:10344-9.

Ott, H.C.; Matthiesen, T.D.; Goh, S-K.; Black,L.D.; Kren, S.M.; Netoff, T-I & Taylor, D. (2008).*Perfusion decellularized matrix: using nature's platform to engineer a bioartificial heart.* Nature Medicine 14, 213 - 221.

Pappa, K. & Anagnou, N. (2009).*Novel sources of fetal stem cells: where do they fit on the developmental continuum?* Regen. Med. 4:423–433.

Parekkadan, B. & Milwid, J.M. (2010). *Mesenchymal Stem Cells as Therapeutic.* Ann Rev Biomed Eng 12: 87-117.

Park, IH.; Zhao, R.; West, J.A.; Yabuuchi, A.; Huo, H.; Ince, T.A.; Lerou, P.H.; Lensch, M.W.& Daley, G.Q.(2008). *Reprogramming of human somatic cells to pluripotency with defined factors.* Nature 10;451(7175):141-6.

Perán, M.; Marchal, J.A.; López, E.; Jiménez-Navarro, M.; Boulaiz, H.; Rodríguez- Serrano, F.; Carrillo, E.; Sanchez-Espin, G.; De Teresa, E.; Tosh, D. & Aránega, A. (2010). *Human cardiac tissue induces transdifferentiation of adult stem cells towards cardiomyocytes.* Cytotherapy 12 332–337.

Pérez Romero, C.; Martín, J.J.; López del Amo González, M.P.; Miranda Serrano, B.; Burgos Rodríguez, R. & Alonso Gil M. *Costes basados en actividades de los programas de trasplantes de riñón, hígado y corazón en siete hospitales españoles.*(2010). Escuela Andaluza de Salud Pública. <http://www.fundacionsigno.com> .Accessed: 19 de Abril de 2010].

Perin, E.C. & López, J. (2006). *Methods of stem cell delivery in cardiac diseases.* Nat Clin Pract Cardiovasc Med. 3 Suppl 1:S110-3.

Perin, E.C.; Dohmann, H.F.R.; Borojevic, R.; Silva, S.A.; Sousa, A.L.S.; Mesquita, C.T.; Rossi, M.I.D.; Carvalho, A.C.; Dutra, H.S.; Dohmann, H.J.F.; Silva, G.V.; Belem, L.; Vivacqua, R.; Rangel, F.O.D.; Esporcatte, R.; Geng, Y.J.; Vaughn, W.K.; Assad, J.A.R.; Mesquita, E.T. & Willerson, J.T. (2003). *Transendocardial, autologous bone marrow cell transplantation for severe, chronic ischemic heart failure.* Circulation. 107:2294-2302.

Picariello, C.; Lazzeri, C.; Attanà, P.; Chiostri, M.; Gensini, G..F. & Valente, S. (2011). *The impact of hypertension on patients with acute coronary syndromes.* Int J Hypertens. 2011:563657. Epub 2011 Jun 22.

Priori, S.G. & Napolitano, C. (2006) *Role of genetic analyses in cardiology: part I: mendelian diseases: cardiac channelopathies.* Circulation 113, 1130–1135.

Qian, Q.; Qian, H.; Zhang, X.; Zhu, W.; Yan, Y.; Ye, S.; Peng, X.; Li, W.; Xu, Z.; Sun, L. & Xu, W. (2011). *5-Azacytidine Induces Cardiac Differentiation of Human Umbilical Cord-Derived Mesenchymal Stem Cells by Activating Extracellular Regulated Kinase.* Stem Cells Dev. [Epub ahead of print]

Rangappa, S.; Fen,C.; Lee, E.H.; Bongso, A. & Wei, E.S. (2003). *Transformation of adult mesenchymal stem cells isolated from the fatty tissue into cardiomyocytes.* Ann Thorac Surg. 75: 775-779.

Reffelmann, T. & Kloner, R.A. (2009). *Intracoronary blood- or bone marrow-derived cell transplantation in patients with ischemic heart disease.* Regen Med. 4(5):709-19.

Reiser, J.; Zhang. X; Hemenway, C.H.; Mondal, D.; Pradhan, L. & La Russa, V.F. (2005). *Potential of mesenchymal stem cells in gene therapy approaches for inherited and acquired diseases.* Expert Opinion on Biological Therapy 5, 12: 571-1584.

Schachinger, V.; Erbs, S.; Elsässer, A.; Haberbosch,W.; Hambrecht, R.; Hölschermann, H.; Yu, J.; Corti, R.; Mathey, D.G.; Hamm, C.W.; Süselbeck, T.; Assmus, B.; Tonn, T.; Dimmeler, S. & Zeiher, A.M. (2006).*Intracoronary bone marrow-derived progenitor cells in acute myocardial infarction.* N Engl J Med. 355:1210–21.

Seifinejad, A.; Taei, A.; Totonchi, M.; Vazirinasab, H.; Hassani, S.N.; Aghdami, N.; Shahbazi, E.; Yazdi, R.S.; Salekdeh G.H. & Baharvand, H. (2010).*Generation of human induced pluripotent stem cells from a Bombay individual: moving towards "universal-donor" red blood cells.* Biochem Biophys Res Commun. 1;391(1):329-34.

Seki, T.; Yuasa, S.; Oda, M.; Egashira, T.; Yae, K.; Kusumoto, D.; Nakata, H.; Tohyama, S.; Hashimoto, H. & Kodaira, M. (2010). *Generation of induced pluripotent stem cells from human terminally differentiated circulating T-cells.* Cell Stem Cell 7, 11-14.

Segers, V.F. & Lee, R.T. (2008). *Stem-cell therapy for cardiac disease.* Nature 451, 937-942.

Shizuru, J.A.; Negrin, R.S. & Weissman, I.L. (2005). *Hematopoietic stem and progenitor cells: clinical and preclinical regeneration of the hematolymphoid system.* Annu Rev Med.56:509-538.

Singla, D.K. (2010). *Stem cells in the infarcted heart.* J Cardiovasc Transl Res. 3(1):73-8.

Smits A.M.; Van Vliet P.; Metz C.H.; Korfage T.; Sluijter J.P.G.; Doevendans P.A.; & Goumans M.J. (2009). *Human Cardiomyocyte progenitor cells differentiate into functional mature cardiomyocytes: an in vitro model for studying human cardiac physiology and pathophysiology.* Nature Protocols 4, 232 – 243.

Son, M.Y.; Kim, H.J.; Kim, M.J. & Cho, Y.S. (2011). *Physical passaging of embryoid bodies generated from human pluripotent stem cells.* PLoS One. 2011 6:e19134.

Song, L. & Tuan, R.S. (2004). *Transdifferentiation potential of human mesenchymal stem cells derived from bone marrow.* The FASEB Journal express article 10.1096/fj.03-1100fje.

Stamm, C.; Kleine, H.D.; Choi, Y.H.; Dunkelmann, S.; Lauffs, J.A.; Lorenzen, B.; David, A.; Liebold, A.; Nienaber, C.; Zurakowski, D.; Freund, M. & Steinhoff, G. (2007). *Intramyocardial delivery of CD133+ bone marrow cells and coronary artery bypass grafting for chronic ischemic heart disease: safety and efficacy studies.* J Thorac Cardiovasc Surg. 133:717-725.

Sumanasinghe, R.D.; Pfeiler, T.W.; Monteiro-Riviere, N.A. & Loboa, E.G. (2009). *Expression of proinflammatory cytokines by human mesenchymal stem cells in response to cyclic tensile strain.* J Cell Physiol. 219:77-83.

Sumi, T.; Tsuneyoshi, N.; Nakatsuji, N. & Suemori, H. (2008). *Defining early lineage specification of human embryonic stem cells by the orchestrated balance of canonical Wnt/β-catenin, Activin/Nodal and BMP signalling.* Development 135:2969-2979.

Takahashi, K. & Yamanaka, S. (2006). *Induction of pluripotent stem cells from mouse embryonic and adult fibroblast cultures by defined factors.* Cell. 126, 663-676.

Takahashi, K.; Tanabe, K..; Ohnuki, M.; Narita, M., Ichisaka, T.; Tomoda, K. & Yamanaka, S. (2007). *Induction of pluripotent stem cells from adult human fibroblasts by defined factors.* Cell 131, 861-872.

Tatsumi, T.; Ashihara, E.; Yasui, T.; Matsunaga, S.; Kido, A.; Sasada, Y.; Nishikawa, S.; Hadase, M.; Koide, M ; Nakamura, R.; Irie, H.; Ito, K.; Matsui, A.; Matsui, H.; Katamura, M.; Kusuoka, S.; Matoba, S.; Okayama, S.; Horii, M.; Uemura, S.; Shimazaki, C.; Tsuji, H.; Saito, Y. & Matsubara, H.(2007). *Intracoronary transplantation of non-expanded peripheral blood-derived mononuclear cells promotes improvement of cardiac function in patients with acute myocardial function.* Circ J 71:1199–207.

Thomas, E.D.; Lochte, H.L.; Jr., Cannon, J.H.; Sahler, O.D. & Ferrebee, J.W. (1959). *Supralethal whole body irradiation and isologous marrow transplantation in man.* J Clin Invest *38*, 1709-1716.

Torella, D.; Ellison, G.M.; Karakikes, I. & Nadal-Ginard B. (2007). *Growth-factor-mediated cardiac stem cell activation in myocardial regeneration.* Nat. Clin. Pract. Cardiovasc. Med. 4: S46-51.

Valiunas, V.; Doronin, S.; Valiuniene, L.; Potapova, I.; Zuckerman, J.; Walcott, B.; Robinson, R.B.; Rosen, M.R.; Brink, P.R. & Cohen, I.S. (2004). *Human mesenchymal stem cells make cardiac connexins and form functional gap junctions.* J Physiol 555:617–626.

Van Laake, L.W.; Passier, R.; Doevendans, P.A. & Mummery, C.L.(2008). *Human embryonic stem cell-derived cardiomyocytes and cardiac repair in rodents.* Circ Res. 9; 102:1008-10.

Van Vliet, P.; De Boer, T.P.; Van der Heyden, M.A.G.; El Tamer, M.K.; Sluijter, J.P.G.; Doevendans, P.A. & Goumans, M.J. (2010). *Hyperpolarization Induces Differentiation in Human Cardiomyocyte Progenitor Cells.* Stem Cell Rev. and Rep. 6:178–185.

Vijayaragavan, K.; Szabo, E.; Bossé, M.; Ramos-Mejia, V.; Moon, R.T. & Bhatia, M. (2009). *Noncanonical Wnt Signaling Orchestrates Early Developmental Events toward Hematopoietic Cell Fate from Human Embryonic Stem Cells.* Cell Stem Cell, 4 (3), pp. 248-262.

Wang, T.; Xu, Z.; Jiang, W. & Ma, A. (2006). *Cell-to-cell contact induces mesenchymal stem cell to differentiate into cardiomyocyte and smooth muscle cell.* International Journal of Cardiology 109 74 - 81.

Wei, H.M.; Wong, P.; Hsu, L.F. & Shim, W. (2009). *Human bone marrow-derived adult stem cells for post-myocardial infarction cardiac repair: current status and future directions.* Singapore Med J 50:935-42.

Wobus, A. M. & Boheler, K. R.. (2005). *Embryonic stem cells: prospects for developmental biology and cell therapy.* Physiol Rev. 85, 635-678.

World Health Organization (WHO). (2004). *The global burden of disease: 2004 update.* ISBN 978 92 4 156371,WHO, Retrieved from http://www.who.int/healthinfo/global_burden_disease/GBD_report_2004update_full.pdf

Xue, J.J.; Wang. Y.S.; Ma. H.; Hu, Y. & Cheng, K.L. (2010). *Effects of a recombinant adenovirus expressing human hypoxia-inducible factor 1a double-mutant on the in vitro differentiation of bone marrow mesenchymal stem cells to cardiomyocytes.* Zhonghua Xin Xue Guan Bing Za Zhi. 38:638-43.

Yamada, Y.; Sakurada, K.; Takeda, Y.; Gojo, Z. & Umezawa, A. (2007). *Single-cell-derived mesenchymal stem cells overexpressing Csx/Nkx2.5 and GATA4 undergo the stochastic cardiomyogenic fate and behave like transient amplifying cells.* Experimental cell Research 698-706.

Ye, Z.; Zhan, H.; Mali, P.; Dowey, S.; Williams, D. M.; Jang, Y. Y.; Dang, C. V.; Spivak, J. L.; Moliterno, A. R. & Cheng, L. (2009). *Human-induced pluripotent stem cells from blood cells of healthy donors and patients with acquired blood disorders.* Blood 114, 5473-5480.

5

Cytoprotection and Preconditioning for Stem Cell Therapy

S. Y. Lim[1], R. J. Dilley[1] and G. J. Dusting[1,2]
[1]*O'Brien Institute, University of Melbourne*
[2]*Centre for Eye Research Australia, University of Melbourne*
Australia

1. Introduction

Coronary heart disease is the leading cause of morbidity and mortality worldwide. To date, management of myocardial infarction (MI) has been limited to timely revascularization and drug therapy aimed to restore coronary blood flow and to reduce myocardial workload. When disease progresses to life-threatening end-stage heart failure, heart transplantation is the only effective therapeutic option available. However, its usage is very much restricted by the severe shortage of heart donors and the complications associated with enduring immune suppressive drug treatments (Miniati & Robbins, 2002). Therefore, innovative treatment strategies are clearly needed to improve patient outcomes. Recent advances in stem cell medicine have shed new light on potential MI therapies by exploiting the pluripotency of stem cells for cardiac repair and regeneration. Although immense progress has been made on the choice of cells and optimizing transplantation conditions, these remain critical issues when translating into the clinical setting of MI. In particular, poor survival of transplanted cells in the hostile microenvironment of the ischaemic myocardium and hence lack of significant engraftment in the heart has been a major impediment for achieving an effective stem cell therapy for MI (Pagani et al., 2003). Various cytoprotection strategies have been developed over the past decade to circumvent this limitation and the non-genetic approach of preconditioning has emerged as one of the most promising clinically adaptable strategies to promote stem cell survival and function under various ischaemic conditions. Although genetic enhancement of stem cells has been very successful in pre-clinical studies, the technical complexity and safety concerns (oncogenicity and mutagenesis) associated with this alternative approach have precluded its application in clinical translation (Bonaros et al., 2008; Penn & Mangi, 2008). This review will focus on current pre-clinical development of non-genetic preconditioning approaches to improve the therapeutic potential of stem and progenitor cells for repair of the heart after MI.

2. History of ischaemic and pharmacological preconditioning

The protective phenomenon of preconditioning was first described by Murry *et al.* in 1986 whereby exposure to brief cycle(s) of sub-lethal ischaemia with intermittent reperfusion, which in itself does not induce injury, render the heart more resistant to subsequent lethal ischaemic insults; this phenomenon was termed ischaemic preconditioning (IPC) (Murry et al., 1986). Subsequent studies by various laboratories have quickly established IPC as the most powerful and effective means of endogenous protection against ischaemic injury. Although this protective intervention can be easily reproduced in various pre-clinical studies, successful translation into clinical practice has been limited by the safety consideration of needing to manipulate the already injured heart. To circumvent this limitation, effort has been concentrated on clarifying the underlying molecular mechanisms governing the cardioprotective effect of IPC which have lead to the discovery of various pharmacological agents that can directly activate the protective signalling pathways to achieve myocardial protection without ischaemia, an intervention called pharmacological preconditioning. Despite extensive research, the mechanism(s) underlying the protective effect of preconditioning remain to be fully elucidated. It is believed to involve multiple intricate endogenous signalling pathways (Yellon & Downey, 2003; Huffmyer & Raphael, 2009) including agonists of G-protein coupled receptors (adenosine, bradykinin, opioids, etc), growth factors (IGF, TGFβ, VEGF, etc), phosphodiesterase inhibitors, mitochondrial K_{ATP} channel openers, cytokines (TNFα, IL-1β, IL-6, etc), nitric oxide (NO), and others. Some of these have been promoted into the clinical arena, for example adenosine in AMISTAD I (Mahaffey et al., 1999) and II (Ross et al., 2005) trials. In general, non-genetic preconditioning strategies employed by current pre-clinical studies to improve survival and function of stem and progenitor cells can be categorized into ischaemic/hypoxic and pharmacological preconditioning.

3. Ex vivo ischaemic and hypoxic preconditioning of stem cells

The ischaemic conditions used to simulate IPC *in vitro* are quite diverse. The majority of studies have experimented with hypoxia or anoxia alone, termed hypoxic preconditioning (HPC), while others include nutrient deprivation. In some studies, hydrogen peroxide (H_2O_2) was used to simulate the ischaemic condition of oxidative stress (Li et al., 2009; Sharma et al., 2008). Furthermore, different HPC protocols, from the classical multiple cycles of brief hypoxia with intermittent reoxygenation to a single long-term exposure to hypoxia, have been employed to demonstrate the cytoprotective effect of HPC on stem and progenitor cells *in vitro*. It is also interesting to note that subjecting stem and progenitor cells to heat shock, as a form of sublethal cell stress, also capable of promoting their survival and *in vivo* engraftment (Laflamme et al., 2005; Maurel et al., 2005; Suzuki et al., 2000). Despite these differences and the lack of an optimal protocol definition, the beneficial effects of HPC on stem and progenitor cell function were unequivocally demonstrated in all these studies and involved multiple signalling molecules (Table 1) (Figure 1).

3.1 Survival

The poor survival and retention of transplanted stem and progenitor cells has driven the investigation towards effective cytoprotective strategies which aim to enhance their survival in the ischaemic environment. The extent of retention of the delivered cells was documented to be rather low with recent studies suggesting that more than 90% are lost partly because of necrosis and apoptosis in the ischaemic myocardium following their delivery by intramyocardial, retrograde transvenous, intracoronary or systemic routes (Terrovitis et al., 2010; Aicher et al., 2003; Goussetis et al., 2006). One means of overcoming this limitation would be to increase the survival of transplanted cells, thus avoiding the impractical and costly alternative of delivering large excesses of stem cells into the injured myocardium.

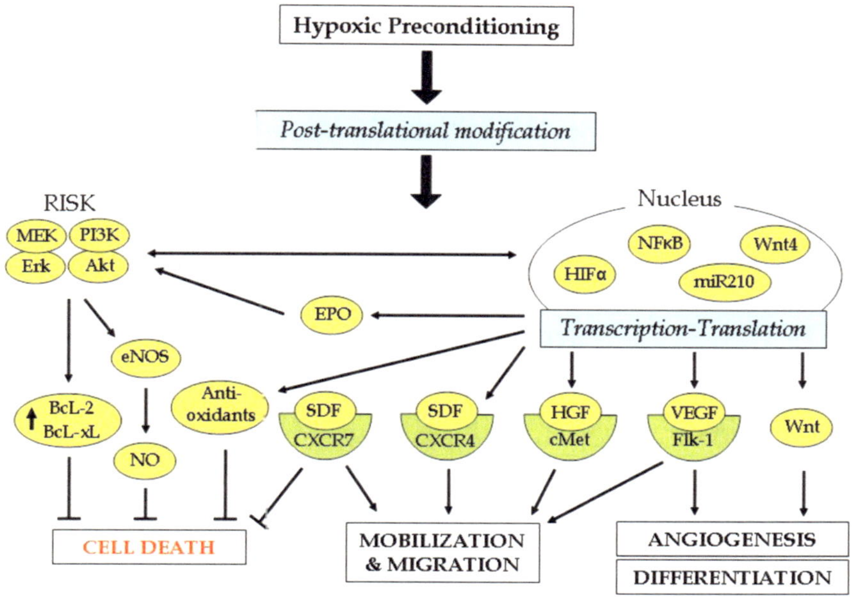

Fig. 1. Mechanisms underlying the cytoprotective effect of hypoxic preconditioning in stem and progenitor cells.

HPC by sublethal hypoxia has been shown to enhance the survival of stem and progenitor cells isolated from various species including humans *in vitro* (Table 1). Similarly, the cytoprotective effect of HPC was demonstrated in a number of *in vivo* studies using clinically relevant ischaemic models (Table 2). In these studies, *ex vivo* HPC rendered stem and progenitor cells more resilient to cell death when transplanted into the infarcted myocardium (He et al., 2009; Hu et al., 2008; Tang et al., 2009; Uemura et al., 2006), ischaemic limb (Akita et al., 2003; Kubo et al., 2008; Li et al., 2002; Rosova et al., 2008) or ischaemic brain (Theus et al., 2008), and this finding was significantly correlated with improved functional recovery of the ischaemic tissues. For instance, transplantation of hypoxic preconditioned mesenchymal stem cells (MSCs) into the ischaemic myocardium showed enhanced therapeutic benefits in terms of infarct size reduction, increased angiogenesis, improved ventricular function and less adverse cardiac remodelling (Hu et al., 2008). These beneficial effects of hypoxic preconditioned stem cells are attributed to enhanced pleiotropic paracrine activities instead of transdiffentiation and cell fusion, which occurred at insignificantly low frequency. Convincing evidence in support of the paracrine paradigm were provided by *in vitro* studies with conditioned media, where media from hypoxia-conditioned MSCs was shown to be cytoprotective in cultured human aortic endothelial cells (Hung et al., 2007a) and primary rat cardiomyocytes (Gnecchi et al., 2006) subjected to hypoxic injury.

Mechanistic evaluations of the cytoprotective effect of HPC in stem and progenitor cells have implicated the up-regulation of a diverse array of soluble survival proteins such as growth factor VEGF (Akita et al., 2003; Potier et al., 2007), anti-apoptotic proteins Bcl-2 and Bcl-xL (Francis & Wei, 2010; Hu et al., 2008; Theus et al., 2008; Wang et al., 2008a), antioxidants heme oxygenase-1, hexokinase-2, catalase and superoxide dismutase (Kubo et

Stem cells	PC stimulus	End points	Mechanisms	References
Hypoxia				
MSCs (mouse)	<0.1% O_2, 4 h	↑ survival of co-cultured adult CMs	Akt; VEGF; SDF-1; eNOS	Uemura et al., 2006
MSCs (mouse)	0.5% O_2, 24 h	↑ survival	HIF-1α; Ang-1; VEGF/Flk1; Bcl-2; Bcl-xL; p105; EPO; NFκB	Hu et al., 2008
MSCs (mouse)	0.5% O_2, 24 h	↑ survival	HIF-1α; EPO; Bcl-2; Bcl-xL	Theus et al., 2008
MSCs (mouse)	1% O_2, 36 h	Conditioned media ↑ cell migration	Wnt4	Leroux et al., 2010
MSCs (mouse)	3% O_2, 3-24 h	↑ survival; ↑ cell migration; ↑ cell adhesion	Akt; HIF-1α; CXCR4; CXCR7	Liu et al., 2010
MSCs (rat)	<0.1% O_2 & serum-free, 3 h	↑ survival of co-cultured neonatal CMs	-	He et al., 2009
MSCs (rat)	0.5% O_2, 12 h	Conditioned media ↑ adult CMs survival	VEGF; HGF; bFGF; Thymosin β4	Gnecchi et al., 2006
MSCs (rat)	0.5% O_2, 24 h	↑ survival	Akt; VEGF; HIF-1α	Chacko et al., 2010
MSCs (rat)	1% O_2, 24 h	↑ survival	Catalase, Mn-SOD, p38MAPK, Bcl-2	Peterson et al., 2011
MSCs (rat)	2% O_2, 4-48 h	↑ angiogenic factors secretion	VEGF/Flk1; VE-cadherin	Li et al., 2002
MSCs (rat)	8% O_2, 10-30 min	↑ survival	Akt; ERK1/2; Bcl-2; Bax; VEGF	Wang et al., 2008a
MSCs (rat)	<0.1% O_2, 10/30 min x 1-3	↑ survival	Akt; ERK1/2; Bcl-xL; HIF-1α; miR-210	Kim et al., 2009
MSCs (human)	1% O_2, 22 h	↑ cell migration	HIF-1α; CX3CR1; CXCR4	Hung et al., 2007b
MSCs (human)	1% O_2, 2 d	↑ angiogenic factors secretion ↓ osteogenic diff.	VEGF	Potier et al., 2007
MSCs (human)	1% O_2 & serum-free, 2 d	Conditioned media ↑ endothelial cell survival & angiogenesis	Akt; IL-6; eNOS	Hung et al., 2007a
MSCs (human)	2% O_2, 3d	Restore hypoxia-induced ↓ of osteogenic diff.	-	Volkmer et al., 2010
MSCs (human)	1-3% O_2, 16 h	↑ cell migration	Akt; HGF/cMet	Rosova et al., 2008
PB-MNCs (mouse)	2% O_2, 24 h	↑ survival	HO-1, autocrine motility factor, hexokinase-2	Kubo et al., 2008

PB-MNCs (mouse)	2% O_2, 24 h	↑ cell migration	VEGF; NOS	Ii et al., 2005
PB-MNCs (human)	pO_2 of 35mmHg, 7 d	↑ endothelial diff.; ↑ cell migration	VEGF; KDR tyrosine kinase	Akita et al., 2003
ESNPCs (mouse)	1% O_2, 4-12 h + 24 h Reoxy.	↑ survival; ↑ neuronal diff.	HIF-1α; EPO; Bcl-2	Theus et al., 2008
ESNPCs (human)	0.1% O_2, 12 h + 0-5 d Reoxy.	↑ survival; ↑ neuronal diff.	HIF-1α; HIF-2α; EPO; VEGF; Bcl-2; Bax; Akt	Francis & Wei, 2010
ASCs (human)	1% O_2, 24 h + 24 h Reoxy.	↑ survival of co-cultured NSCs	-	Oh et al., 2010
CLK (mouse)	0.1% O_2, 4-24 h	↑ cell migration	HIF-1α; SDF-1/CXCR4	Tang et al., 2009
Bone marrow CD133+ (human)	1.5% O_2, 24 h + 2 d Reoxy.	↑ endothelial diff.; ↑ angiogenic-related genes	-	Ong et al., 2010
NSCs (mouse)	0.5% O_2, 3 h	↑ functional engraftment	Connexin-43	Jaderstad et al., 2010
Hydrogen peroxide				
MSCs (rat)	20 μM H_2O_2, 24 h	↑ survival ↑ cell migration	SDF-1/CXCR4; ERK1/2	Li et al., 2009
NPCs (mouse)	0.05-0.5 μM H_2O_2, 24 h	↑ survival	-	Sharma et al., 2008
Heat shock				
Skeletal Mb (rat)	42^{0}C, 1 h	↑ survival	HSP72	Suzuki et al., 2000
Skeletal Mb (rat)	42^{0}C, 70 min	↑ survival	HSP70	Maurel et al., 2005
ESCM (human)	43^{0}C, 30 min	↑ survival	HSP60, HSP70, HSP90	Laflamme et al., 2005

Table 1. Effect of ischaemic or hypoxic preconditioning on stem and progenitor cells *in vitro*. CLK (cardiosphere-derived $Lin^-c\text{-}kit^+$ progenitor cells), CMs (cardiomyocytes), diff. (differentiation), ESCM (embryonic stem cell-derived cardiomyocytes), ESNPCs (embryonic stem cell-derived neural progenitor cells), HO-1 (heme oxygenase-1), HSP (Heat shock protein), KDR (kinase insert domain receptor), Mb (myoblasts), NPCs (neural progenitor cells), NSCs (neural stem cells), PB-MNCs (peripheral blood mononuclear cells), Reoxy (reoxygenation), Wnt4 (wingless-related MMTV integration site 4).

al., 2008; Peterson et al., 2011) , erythropoietin (EPO) (Hu et al., 2008; Theus et al., 2008) and NO (Uemura et al., 2006; Ii et al., 2005) as the contributing factors. Other potential cytokines and growth factors that have been suggested are basic fibroblast growth factor (bFGF) (Gnecchi et al., 2006), hepatocyte growth factor (HGF) (Gnecchi et al., 2006), IL-1β (Kubo et al. 2008), IL-6 (Hung et al., 2007a) and thymosin β4 (Gnecchi et al., 2006), though more supporting evidence for these factors in mediating the pro-survival effect of HPC in the context of stem cell preconditioning are warranted. In addition, HPC has also been shown to activate several transcription factors and signal transduction cascades that are known to be

In vivo models	Stem cells	PC stimulus	End points	References
MI (mouse)	MSCs (mouse)	Anoxia, 4 h + 2 h reoxygenation	↓ infarct size & apoptosis; ↑ LV function	Uemura et al., 2006
MI (mouse)	CLK (mouse)	0.1% O_2, 6 hours	↓ infarct size; ↑ angiogenesis; ↓ LV remodelling; ↑ LV function	Tang et al., 2009
MI (rat)	MSCs (mouse)	0.5% O_2, 24 h + 2 h reoxygenation	↑ cell survival; ↓ infarct size; ↑ angiogenesis; ↑ LV function	Hu et al., 2008
MI (rat)	MSCs (rat)	Anoxia & serum-free, 3 h	↓ infarct size & apoptosis; ↑ LV function	He et al., 2009
MI (rat)	MSCs (rat)	10/30 min anoxia-reoxygenation	↑ cell survival	Kim et al., 2009
Limb ischaemia (mouse)	MSCs (mouse)	1% O_2, 36 h	↑ cell survival; ↑ skeletal muscle regeneration; ↑ limb perfusion; ↑ neovascularization	Leroux et al., 2010
Limb ischaemia (mouse)	MSCs (human)	1-3% O_2, 16 h	↑ limb perfusion	Rosova et al., 2008
Limb ischaemia (mouse)	PB-MNCs (mouse)	2% O_2, 24 h	↑ cell survival; ↑ limb perfusion; ↑ neovascularization	Kubo et al., 2008
Limb ischaemia (rat)	MSCs (rat)	2% O_2, 24 h	↑ limb perfusion; ↑ neovascularization	Li et al., 2002
Limb ischaemia (rat)	PB-MNCs (human)	pO_2 of 35mmHg, 7 d	↑ limb perfusion; ↑ neovascularization	Akita et al., 2003
Brain ischaemia (rat)	ESNPCs (mouse)	1% O_2, 8 h	↑ cell survival; ↑ recovery of sensorimotor function	Theus et al., 2008
Diabetic cardiomyopathy (rat)	MSCs (rat)	Anoxia	↓ apoptosis; ↑ angiogenesis; ↓ LV remodelling; ↑ LV function	Li et al., 2008
Spinal cord injury (rat)	ASCs (human)	1% O_2, 24 h + 24 h reoxygenation	↑ survival of co-transplanted NSCs	Oh et al., 2010

Table 2. Therapeutic potential of hypoxic-preconditioned stem and progenitor cells.

protective and functionally beneficial including the survival kinase Akt (Hung et al., 2007a; Kim et al., 2009), ERK1/2 (Wang et al., 2008a), p38MAPK and survivin (Peterson et al., 2011), SDF-1/CXCR4 and CXCR7 chemokine signalling pathway (Liu et al., 2010), microRNA(miR)-210 (Kim et al., 2009), transcription factors HIF-1α (Kim et al., 2009; Francis & Wei, 2010) and NFκB (Hu et al., 2008). It is also important to note that these mechanistic pathways and paracrine factors interact with each other and are not mutually exclusive. For example, stabilisation of HIF-1α by HPC, possibly through activation of the PI3K/Akt pathway (Francis & Wei, 2010; Liu et al., 2010), allows its translocation into the nucleus to up-regulate the transcription and translation of various anti-apoptotic proteins such as CXCR7 (Liu et al., 2010), Bcl-2 (Francis & Wei, 2010) and miR-210 (Kim et al., 2009) in stem and progenitor cells. Furthermore, up-regulation of miR-210 has been demonstrated to down-regulate the expression of CAP8AP2, a pro-apoptotic protein that activates death-effector caspase-8 and promotes Fas-induced apoptosis (Kim et al., 2009).

3.2 Differentiation and engraftment

In addition to cell survival and retention, lack of significant functional cell engraftment of transplanted cells in the injured tissues has posed another significant challenge for cell-based therapy. In most studies, transplanted stem and progenitor cells do not appear to be trans-differentiated and incorporated into host tissues. Instead, the functional improvement of the ischaemic conditions is likely attributed to the paracrine activities of transplanted cells. Therefore, interventions that can promote stem cell differentiation and functional engraftment in the target tissues post-transplantation should deserve much attention. HPC has been shown not only to enhance stem cell survival but also to promote their differentiation and engraftment. Hypoxia is a potent differentiation inducer of stem cells and studies have demonstrated an acceleration of MSC differentiation when cultured under hypoxic conditions (5-8% O_2) compared with that in normoxic culture, possibly through stabilisation of the oxygen sensitive transcription factor HIF-1α (Lennon et al., 2001; Ren et al., 2006). In contrast, a number of studies have indicated that hypoxia strongly inhibits the differentiation capacity of human bone marrow-derived MSCs (Hung et al., 2007b; Potier et al., 2007; Salim et al., 2004; Volkmer et al., 2010) and adipose-derived mesenchymal stem cells (ASCs) (Malladi et al., 2006; Wang et al., 2005), without affecting the cell viability. Similar conflicting results on stem cell differentiation potential were demonstrated by studies on short-term exposure to hypoxia. Studies on mouse (Theus et al., 2008) and human (Francis & Wei, 2010) embryonic stem cells (ESCs) have indicated a favourable effect of HPC in promoting their neuronal differentiation. A recent study by Volkmer *et al.* has also reported that HPC can restore the osteogenic differentiation capacity of human MSCs which was otherwise compromised under hypoxic conditions (Volkmer et al., 2010). Conversely, Potier et al. showed that short-term exposure of human MSCs to hypoxia (<1% O_2) has a negative impact on their osteogenic differentiation under normal *in vitro* culture condition (21% O_2) (Potier et al., 2007). Although the reason for this discrepancy remains unclear, the differences in cell type and species, oxygen tension, duration of exposure to hypoxic conditions and culture conditions could be the answers to these contradictory results.

In terms of functional engraftment, a recent *in vitro* study by Jaderstad and associates has reported an increased in gap-junctional intercellular communication between hypoxic preconditioned neural stem cells and host cells *in vitro*, a consequence of increased expression of connexin 43 (Jaderstad et al., 2010). In support of this finding was a previous study reporting that HPC of human MSCs enhanced their xenografting efficiency into chick embryos, a model employed to examine the *in vivo* engraftment and differentiation potential of stem cells (Hung et al., 2007b). In addition, Xie *et al.* has reported that conditioned medium from rat neonatal cardiomyocytes subjected to 2 hours of hypoxia followed by overnight reoxygenation can induce MSC differentiation into cardiomyocyte lineage as indicated by an increase in cardiac myosin heavy chain and troponin T expression (Xie et al., 2006). The latter study has also suggested that HPC may induce secretion of various soluble differentiation factors, whose identity remains to be determined and depend on the cell types. However, what remains unknown is whether these effects of HPC in promoting stem cell differentiation and *in vivo* engraftment can be translated when preconditioned stem cells are transplanted into adult tissues. To this end, early studies on peripheral blood mononuclear cells (Akita et al., 2003; Kubo et al., 2008) and MSCs (Leroux et al., 2010; Li et al., 2002) have indicated that HPC enhances not only their differentiation into endothelial progenitor cells (EPCs) *in vitro* but also promotes neovascularisation when transplanted into the ischaemic hindlimb. This effect was associated with improvement of blood perfusion and acceleration of tissue repair. Nevertheless, the lack of detailed histological analysis of angiogenesis in the host tissues, i.e. quantifying the blood vessels derived from implanted cells, has cast doubt on the enhanced functional integration of these transplanted cells as the contributing mechanism (Akita et al., 2003). Instead, the improvement in overall neovascularisation in these studies can be interpreted as a result of increased angiogenic cytokines released by preconditioned cells, such as VEGF (Akita et al., 2003; Leroux et al., 2010; Li et al., 2002), thus promoting intrinsic angiogenesis in the host. Supporting this notion is a study that showed MSCs subjected to *ex vivo* HPC expressed a higher level of VEGF mRNA and induced greater local VEGF production in the ischaemic hindlimb after implantation, possibly through activation of the canonical Wnt (wingless-related MMTV integration site) pathway (Li et al., 2002; Leroux et al., 2010).

3.3 Cell migration

In cell-based therapy, effective treatment also relies on the ability of transplanted stem and progenitor cells to migrate to the site of injured tissues to exert reparative and regenerative effects. Short-term exposure to hypoxia has been shown to enhance the migratory capacity of stem and progenitor cells *ex vivo* by modulating the expression of various chemokines and cytokines receptors. Hung *et al.* reported an upregulation of CXC3R1 and CXCR4 expression on both mRNA and protein levels when MSCs were cultured under hypoxic condition compared with normoxia, resulted in an increased cell migration in response to the fractalkine/CX3CL1 and SDF-1α/CXCL12, respectively (Hung et al., 2007b). A recent study also showed that HPC enhances MSC adhesion, an important step during cell trafficking *in vivo*, through upregulation of CXCR4 and CXCR7 (Liu et al., 2010). The induction of these chemokine receptors has been shown to be driven mainly by transcription factor HIF-1α (Liu et al., 2010; Hung et al., 2007b; Tang et al., 2009). Using low dose of H_2O_2 as a preconditioning stimulus, Li *et al.* also showed that the enhanced chemotaxis of preconditioned MSCs was attributed to the up-regulation of CXCR4 in an ERK-dependent manner (Li et al., 2009). Translating these *in vitro* findings into an *in vivo* setting, Tang *et al.* has shown that short-term exposure of murine cardiac progenitor cells to hypoxia not only enhanced their migratory activity *in vitro* but also *in vivo* recruitment to the ischaemic myocardium when administered intravenously, through a CXCR4-dependent manner (Tang et al., 2009). In addition to the chemokine mechanisms, other studies have suggested that HPC enhanced migratory function of stem and progenitor cells through regulation of cytokine signalling (Akita et al., 2003; Rosova et al., 2008; Ii et al., 2005). For instance, hypoxia enhanced the migratory function of human EPCs in response to VEGF, possibly through up-regulation of KDR/VEGFR2 expression (Akita et al., 2003). In another study,

Rosova and colleagues showed that HPC increased the expression of the tyrosine kinase receptor, c-Met, in preconditioned MSCs rendered the cells more responsive to HGF (Rosova et al., 2008). Interestingly, HPC can also induce the secretion of chemo-attractants from preconditioned MSCs to promote endothelial cell migration through Wnt4-dependent signalling pathway (Leroux et al., 2010).

3.4 Cell proliferation

Stem and progenitor cells self-renew and this is one of the properties that make them an attractive autologous cell source for cell-based therapy and tissue engineering, where success is highly dependent on abundant cell supply. *Ex vivo* cell expansion is traditionally performed under ambient oxygen concentration of 20% O_2, which is considered to be hyperoxia compared to their physiological niches (2-7% O_2). Thus, it is imperative to simulate various aspects of the stem and progenitor cells' endogenous microenvironment, including hypoxia, in order to maintain their native characteristics and to comprehend how they respond to a hypoxic environment in injured ischaemic tissues. Studies investigating the effect of hypoxia on stem cell proliferation potential have yielded contradictory results, possibly due to the differences in hypoxic conditions, cell type, serum concentration and culture duration (Das et al., 2010). Compared to the routine normoxic culture of 20% O_2, long-term culture of human MSCs in 1% O_2 has been shown to reduce their proliferative potential (Hung et al., 2007b). Conversely, bone marrow-derived MSCs (D'Ippolito et al., 2006; Grayson et al., 2007; Lennon et al., 2001; Ren et al., 2006), but not ASCs (Wang et al., 2005), cultured under hypoxic conditions with slightly higher oxygen tension (≥2% O_2), showed increased cell proliferation. Importantly, short-term exposure to hypoxia did not negatively affect the proliferative potential of stem cells (Francis & Wei, 2010; Leroux et al., 2010; Rosova et al., 2008), an observation that will alleviate the safety concerns of HPC when clinical applications are being considered.

4. Ex vivo pharmacological preconditioning of stem cells

While ischaemic or hypoxic preconditioning has been shown to regulate multiple stress-responsive mechanisms that promote stem and progenitor cell survival under various ischaemic conditions, preconditioning with specific pharmacological agents seems to target a more linear signalling pathway. This has been explored in various studies on stem and progenitor cells (Table 3).

4.1 Diazoxide

The mitochondrial ATP-sensitive potassium (mitoK_{ATP}) channel is an important mediator of cardioprotection (Yellon & Downey, 2003; O'Rourke, 2004) where opening of the channels has been shown to induce protection by preventing calcium overload, inhibiting mitochondrial permeability transition pore (mPTP) opening, preserving ATP production, uncoupling of mitochondrial oxidative phosphorylation, succinate dehydrogenase inhibition, reducing detrimental reactive oxygen species (ROS) production at reperfusion and PKC activation. Using the mitoK_{ATP} opener diazoxide, Baines and colleagues were the first to demonstrate that opening of mitoK_{ATP} channels prior to ischaemia could mimic the infarct-limiting effect of IPC in the setting of myocardial ischaemia-reperfusion injury (Baines et al., 1999). Recently, diazoxide has also featured in a number of studies by Ashraf's group to precondition stem and progenitor cells. They have shown that *ex vivo* preconditioning with diazoxide can promote skeletal myoblasts (Haider et al., 2010; Niagara

PC stimulus	Cells	In vitro	In vivo	Mechanisms	References
Diazoxide (200 µM), 30 min	Skeletal Mb (rat)	↑ survival	(MI) ↑ survival; ↑ angiomyogenesis; ↑ LV function	Akt; bFGF; HGF	Niagara et al., 2007
Diazoxide (200 µM), 30 min	Skeletal Mb (rat)	↑ survival	(MI) ↑ survival & proliferation; ↑ angiomyogenesis; ↑ LV function	Akt; ERK1/2; STAT3; IL-11; miR-21	Haider et al., 2010
Diazoxide (200 µM), 30 min	MSCs (rat)	↑ survival	(MI) ↑ survival; ↓ infarct size; ↑ angiomyogenesis; ↑ LV function	Akt; GSK3β; NFκB; HGF; IGF; FGF-2; Ang-2	Afzal et al., 2010
Diazoxide (200 µM), 1-3 h	MSCs (rat)	↑ survival	-	NFκB; Fas; miR-146a	Suzuki et al., 2010
Diazoxide (200 µM), 30 min	MSCs (rat)	↑ survival	(MI) ↑ survival; ↓ infarct size; ↑ LV function	Akt; bFGF; HGF	Cui et al., 2010
SDF-1 (50 ng/mL), 1 h	MSCs (rat)	↑ survival; ↑ proliferation	(MI) ↑ survival; ↓ infarct size; ↑ angiomyogenesis; ↑ LV function	Akt; CXCR4; VEGF	Pasha et al., 2008
SDF-1 (10-100 ng/mL), 24 h	MSCs (rat)	↑ survival	-	CXCR4	Chen et al., 2009
IGF-1 (100 nM), 30 min	BM Sca-1+ (mouse)	↑ survival; ↑ myogenic diff.	(MI) ↑ survival; ↓ infarct size; ↑ myogenic diff.; ↑ angiogenesis; ↑ LV function	Akt; Cx43	Lu et al., 2009
IGF-1 (100 nM), 30 min	BM Sca-1+ (mouse)	↑ survival	-	ERK1/2; Cx43	Lu et al., 2010
TGFα (0.25 µg/mL), 24 h	MSCs (mouse)	-	(MI) ↓ apoptosis; ↑ LV function	VEGF; p38MAPK	Herrmann et al., 2010b
TGFα (0.25-1 µg/mL), 24 h	MSCs (human)	↑ VEGF production	-	PI3K; MEK	Wang et al., 2008b
TGFα (0.01-1 µg/mL), 24 h	MSCs (human)	↑ HGF production	-	PI3K; MEK; TNFR; p38MAPK	Wang et al., 2009b
PDGF (0.4 nM), 5 d	MSCs (rat)	↑ adipogenic diff.	-	-	Tamama et al., 2006
PDGF (0.4 nM), 5 d	MSCs (human)	↓ adipogenic-; ↑ osteogenic- diff.	-	-	Tamama et al., 2006
BMP-2 + bFGF + IGF-1 (10, 50 & 2 ng/mL), 1-7 d	MSCs (rat)	↑ survival; ↑ myogenic diff.	(MI) ↑ survival; ↓ infarct size; ↑ LV function	Cx43; Akt; CREB	Hahn et al., 2008
IGF-1 + bFGF (50 & 50 ng/mL), 1 h	MSCs (mouse)	↑ survival & proliferation; ↑ angiogenic potential	-	Akt; SOD; Ang-1; Bax; Bak; 16^{INK4a}; $p66^{shc}$; p53	Khan et al., 2011

PC stimulus	Cells	In vitro	In vivo	Mechanisms	References
Sevoflurane (2%), 30 min x 3	EPCs (human)	↑ CFC	-	VEGF	Lucchinetti et al., 2009
Isoflurane (0.5 mM), 10 min	ESCM (human)	↑ survival	-	mitoK_{ATP}; ΔΨm; ROS	Sepac et al., 2010
CsA (0.5-5 μM), 30 min	MSCs (rat)	↑ survival	-	Bcl-2; BAD; ΔΨm	Wang et al., 2008a
LPS (1 μg/mL), 12 h	MSCs (mouse)	↑ survival	-	TRL-4; Akt; NFκB	Wang et al., 2009c
LPS (1 μg/mL), 2 d	MSCs (mouse)	-	(MI) ↑ survival; ↓ cardiac fibrosis; ↑ angiogenesis; ↑ LV function	TRL-4; Akt; VEGF; NFκB	Yao et al., 2009
Melatonin (5 μM), 24 h	MSCs (rat)	↑ survival ; Conditioned media ↑EPCs proliferation & angiogenesis	(Renal IR) ↑ survival; ↑ renal function; ↑ renal cell proliferation; ↑ angiogenesis	Catalase; SOD, bFGF; HGF	Mias et al., 2008
Trimetazidine (10 μM), 6 h	MSCs (rat)	↑ survival	(MI) ↓ infarct size; ↑ LV function	Akt; HIF-1α; survivin; Bcl-2	Wisel et al., 2009
Fucoidan (10 μg/mL), 36 h	EPCs (human)	↑ cell migration; ↑ angiogenic potential	-	-	Zemani et al., 2005
LPA (10 μM), 1 h	MSCs (rat)	-	(MI) ↑ survival; ↑ angiogenesis	VEGF	Liu et al., 2009
Lithium Chloride (5-20 mM), 24 h	Skeletal Mb (rat)	↑ survival & proliferation; ↑ gap junction formation	-	Cx43; VEGF; β-catenin; GSK-3β	Du et al., 2009
rhHsp90α (0.1-10 μM), 24 h	MSCs (rat)	↑ survival	-	Akt; ERK1/2; NO; Bcl-2; Bcl-xL; Bax	Gao et al., 2010
β-met (2 mM), 1 h	MSCs (rat)	↑ survival	-	HSP72	Cizkova et al., 2006
rhEPO (10 U/mL), 24 h	ESNPCs (mouse)	↑ survival	-	Bcl-2	Theus et al., 2008
Carbamylated EPO (100 ng/mL), 30 min	ESCM (human)	-	(MI) ↑ survival	Akt	Robey et al., 2008
Simvastatin (25 μM), 24 h	EPCs (human)	↑ survival	-	-	Henrich et al., 2007

Ang (angiopoietin), β-met (β-mercaptoethanol), CFC (colony forming capacity), CREB (cAMP response element binding protein), Mb (myoblasts), rh (recombinant human), ΔΨm (mitochondrial membrane potential).

Table 3. Pharmacological preconditioning of stem and progenitor cells.

et al., 2007) and MSCs (Afzal et al., 2010; Suzuki et al., 2010) survival both *in vitro* and *in vivo* post-transplantation. Furthermore, transplantation of these preconditioned cells into the infarcted myocardium was associated with smaller infarct size, improved LV function, myogenic differentiation and angiogenesis (Afzal et al., 2010; Haider et al., 2010; Niagara et al., 2007). The potential mechanisms responsible for the cytoprotective effect of diazoxide in skeletal myoblasts include enhanced release of paracrine growth factors such as bFGF and HGF, and activation of survival kinase PI3K/Akt (Niagara et al., 2007). A later study by the same group expanded the mechanistic finding to include IL-11, the ERK1/2-STAT3 signalling pathway and up-regulation of miR-21 (Haider et al., 2010). In MSCs, Afzal *et al.* reported NFκB activation as another important underlying mechanism of diazoxide-induced protection especially during the late phase preconditioning (Afzal et al., 2010). The activation of NFκB in diazoxide preconditioned MSCs was subsequently implicated to regulate the expression of miR-146a, which in turn acts as a negative regulator of the Fas gene, a death receptor of apoptosis (Suzuki et al., 2010).

4.2 Stromal cell-derived factor-1 (CXCR12)

SDF-1 or CXCL12 is a cytokine belonging to the CXC chemokine subfamily. Specific binding of SDF-1 to CXCR4 induces dimerization of the receptor and activates multiple signalling pathways to regulate trafficking and differentiation of stem and progenitor cells (Chen et al., 2011; Kucia et al., 2004). In addition, activation of the SDF-1/CXCR4 axis has been shown to promote cell survival and proliferation (Broxmeyer et al., 2003; Kucia et al., 2004; Hu et al., 2007). The therapeutic potential of SDF-1 was later illustrated in an experimental model of myocardial infarction (Takahashi, 2010). For example, SDF-1 administration conferred cardioprotection through the PI3K/Akt signalling pathway in the setting of acute ischaemia-reperfusion injury (Hu et al., 2007) and chronic ischaemic heart failure (Saxena et al., 2008). In the setting of hindlimb ischaemia, intramuscular injection of SDF-1 increased angiogenic factor VEGF expression, and enhanced the retention and neovascularisation efficacy of transplanted EPCs (Yamaguchi et al., 2003). Other delivery methods for SDF-1 include direct gene delivery with adenoviral vector (Abbott et al., 2004) and plasmid DNA (Hiasa et al., 2004) encoding for SDF-1, or transplantation of genetically modified cardiac fibroblasts (Askari et al., 2003), MSCs (Zhang et al., 2007) or skeletal myoblasts (Elmadbouh et al., 2007) that over-expressed SDF-1 into the ischaemic tissues. Therefore, it is not surprising that SDF-1 has been exploited as a potential preconditioning agent to enhance stem and progenitor cells survival and function without the long-term concern of genetic manipulation. In this regard, pre-treatment with SDF-1 has been shown to enhance MSC survival both *in vitro* (Chen et al., 2009; Pasha et al., 2008) and following intramyocardial transplantation in the infarcted myocardium through activation of PI3K/Akt signalling pathway and is dependent on CXCR4 (Pasha et al., 2008). In this study, the authors also demonstrated that SDF-1 preconditioning enhances paracrine activities of transplanted MSCs in the infarcted myocardium and contributes to a smaller infarct size, improved cardiac function and promoted revascularization. Interestingly, the cytoprotective effect of SDF-1 appears not to be limited to pre-treatment. Treatment of EPCs with SDF-1 at the end of lethal serum deprivation also significantly reduced apoptotic cell death (Yamaguchi et al., 2003). However, *in vivo* application of SDF-1 requires vigilant safety evaluation before translating into the clinical setting because of the potentially detrimental side effects of its cleavage products. For instance, SDF-1(5-68), a toxic product of SDF-1 cleavage by exopeptidases and metalloproteinase-2, has been implicated in neuroinflammation and neuronal death (Zhang et al., 2003). In which case, SSDF-1(S4V), a modified SDF-1 that is resistant to endogenous proteases cleavage, may have a better clinical safety profile (Segers et al., 2007).

4.3 Growth factors

Various growth factors have been employed to precondition stem and progenitor cells in order to enhance their paracrine activity, cell survival and differentiation (Abarbanell et al., 2009). For example, Lu and colleagues showed that preconditioning with IGF-1 improved bone marrow Sca-1$^+$ stem cell survival against simulated ischaemia *in vitro* and after *in vivo* transplantation. This pro-survival effect of IGF-1 was shown to be dependent on the activation of PI3K/Akt (Lu et al., 2009) and ERK1/2 (Lu et al., 2010) signalling pathways, and a down-stream mediator connexin-43. A subsequent study indicated that pre-treatment with IGF-1 also enhances Sca-1$^+$ cell cardiomyogenesis potential and transplantation of these preconditioned cells into infarcted heart mitigated myocardial infarction and ventricular dysfunction (Lu et al., 2009). Moreover, MSCs exposed to IGF-1 had an enhanced migratory response to SDF-1, a response dependent on the PI3K/Akt signalling pathway (Li et al., 2007).
It has been well documented that epidermal growth factor (EGF), the prototypical growth factor with intrinsic protein tyrosine kinase activity, plays an important role in the regulation of cell growth, proliferation and differentiation by binding to its receptor (Wells, 1999). In MSCs, EGF has been shown to promote cell proliferation and motility which are beneficial for *ex vivo* cell expansion prior to *in vivo* transplantation, but fail to rescue MSCs from low serum-induced apoptosis despite an elevation of Akt. Similar effects were observed with platelet-derived growth factor (PDGF) (Tamama et al., 2006). On the other hand, pre-treatment with TGFα, another member of the EGF superfamily and a potent activator of EGF receptor, has been shown to enhance the therapeutic potential of MSCs in an experimental model of acute myocardial ischaemia-reperfusion injury in terms of myocardial function recovery and inflammation (Herrmann et al., 2010b). These beneficial effects of TGFα have been attributed to the reduction in pro-inflammatory cytokine production and increased VEGF production. The latter effect was shown to be mediated by MEK and PI3K signalling pathways (Herrmann et al., 2010b). It is interesting to note that the effect of TGFα on VEGF production in MSCs has been demonstrated as bimodal, i.e. production of VEGF was suppressed by low concentrations of TGFα whereas high concentrations increased the secretion (Wang et al., 2008b). In addition to VEGF, TGFα treatment also increases the production of HGF from MSCs (Wang et al., 2009b). HGF is an important signalling factor in stem cell-mediated repair where it promotes stem cell adhesion, migration and survival (Vandervelde et al., 2005). Although the potential of HGF as a preconditioning agent to promote stem and progenitor cell survival is yet to be confirmed, existing studies have showed that short-term exposure of MSCs to HGF can activate Akt and ERK1/2 pro-survival kinases, and induce expression of anti-apoptotic protein, Bcl-2 and Bcl-xL (Forte et al., 2006). Additionally, HGF treatment also increased MSC motility function through up-regulation of c-Met receptors and induced their differentiation along a myogenic lineage, effects that can enhance their therapeutic potential in cardiac repair (Forte et al., 2006; Neuss et al., 2004; Rosova et al., 2008). Regarding cell proliferation, HGF treatment was reported to have a negative impact on MSC proliferation by blocking cells in the G_0-G_1 phase through p38 MAPK pathway and concomitant up-regulation of cell cycle progression inhibitors, p21^{waf1} and p21^{kip} (Neuss et al., 2004; Forte et al., 2006).

Although different growth factors may share some common downstream signalling pathways in modulating stem and progenitor cell functions, they also independently activate distinct signalling cascades (Abarbanell et al., 2009). This rationale that a combination of multiple growth factors may have additive and synergistic effects in promoting stem and progenitor cell survival and functional improvement, offers a strategic advantage over the approach involving the use of a single factor. In this respect, a recent study by Hahn *et al* has demonstrated the pro-survival effect of multiple growth factors in MSCs using a combination of IGF-1, bFGF and BMP-2 (Hahn et al., 2008). Treatment with this growth factor cocktail also enhanced MSC cardiac differentiation efficiency and incurred cytoprotection of co-cultured adult cardiomyocytes. This effect was dependent on gap-junction communication leading to phosphorylation of Akt and c-AMP response element binding protein (CREB) in cardiomyocytes. Moreover, the therapeutic efficacy of MSCs in infarcted myocardium was also significantly potentiated with this multiple growth factors preconditioning method (Hahn et al., 2008). Similarly, Khan *et al* preconditioned MSCs isolated from diabetic mice with a combination of IGF-1 and bFGF and showed enhanced cell survival, proliferation, motility and angiogenic potential compared to untreated cells or cells treated with single growth factor (Khan et al., 2011). Furthermore, a potent synergistic effect between TGFα and TNFα has been demonstrated to enhance the MSC-derived VEGF and HGF production that may be essential for cell survival, migration and angiogenesis (Herrmann et al., 2010a; Wang et al., 2008b; Wang et al., 2009b).

4.4 Anaesthetics

Anaesthetics, routinely administered in patients during surgery, are widely recognised to have preconditioning properties protecting the heart and many other organs against ischaemia and reperfusion injury in laboratory settings as well as in humans with coronary heart diseases (Huffmyer & Raphael, 2009; Yellon & Downey, 2003). The efficacy of anaesthetic-induced preconditioning has been shown to be similar to IPC in terms of infarct size reduction, increased collateral blood flow and attenuated inflammatory responses during ischaemia. Although the mechanisms underlying the protective effect of anaesthetic-induced preconditioning have been shown to resemble those responsible for IPC (Zaugg et al., 2003), Mullenheim *et al* showed that administration of sevoflurane combined with IPC can synergistically protect rabbit hearts against myocardial infarction indicating the existence of parallel protective mechanisms (Mullenheim et al., 2003). In ESC-derived cardiomyocytes, brief exposure to the commonly used volatile anaesthetic, **isoflurane**, significantly attenuated cell death against oxidative stress through opening of mitoK_{ATP} channels, production of signalling ROS and inhibition of mPTP opening (Sepac et al., 2010). Other benefits include enhanced cell growth capacity and increased expression of pro-angiogenic VEGF as demonstrated in human EPCs preconditioned with **sevoflurane** (Lucchinetti et al., 2009).

4.5 Cyclosporin-A (CsA)

The mPTP is a non-specific channel of the inner mitochondrial membrane, whose opening at the onset of reperfusion is a critical mediator of lethal myocardial ischaemia-reperfusion injury. IPC exerts its cardioprotective effect by inhibiting the opening of the mPTP and pharmacological inhibition of mPTP with CsA has been shown to confer cytoprotection both *in vitro* and *in vivo* (Hausenloy & Yellon, 2003; Lim et al., 2007). In addition to inhibiting mPTP opening, CsA can also inhibit calcineurin-mediated dephosphorylation of the apoptogenic protein BAD and uncouple the mitochondrial respiratory chain, which might, in itself, result in protection (Wilkins et al., 2004). Interestingly, mPTP can also open reversibly under basal conditions without causing cell death. This form of reversible non-pathological mPTP opening has been noted to contribute to the cardioprotection elicited by IPC through the mitochondrial generation of signalling ROS and the subsequent activation of the pro-survival kinases Akt and ERK1/2 (Hausenloy et al., 2010). In stem cells, pre-treatment with CsA has been shown to protect MSCs from hypoxia-reoxygenation induced apoptosis through stabilizing mitochondrial membrane potential and promoting Bcl-2 and phosphorylated BAD protein expression (Wang et al., 2008a). In line with this study, we recently showed that pre-treatment with 0.2 μM of CsA for 30 minutes effectively increased the resistance of human ASCs to subsequent simulated ischaemia-induced cell death (Figure 2, unpublished data).

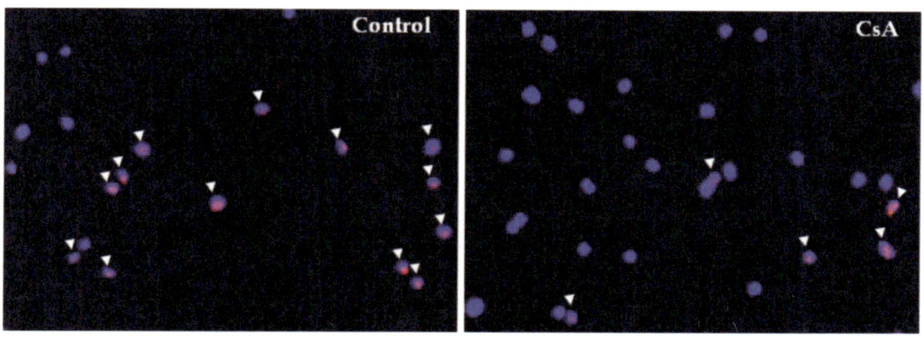

Fig. 2. Preconditioning with CsA confers cytoprotection in stem cells. Pre-treatment with CsA (0.2 μM, 30 min) reduced cell death (determined by propidium iodide staining, arrow heads) in hASCs subjected to 15 hours of simulated ischaemia (<1% O_2 and ischaemic buffer as described previously (Lim et al., 2008)) from 58.5±2.6% to 42.5±1.8% (n=4-5, $P < 0.05$).

4.6 Lipopolysaccharide

LPS is an antigenic component of the outer membrane of gram-negative bacteria and an agonist of Toll-like receptor-4 (TLR4) capable of eliciting immune responses in animals. The cardioprotective effect of LPS preconditioning was previously demonstrated in an *in vivo* murine model of myocardial ischaemia-reperfusion injury (Ha et al., 2008). Evidence supporting the pro-survival effect of LPS preconditioning in MSCs was recently provided by Wang and colleagues. The authors reported that activation of Toll-like receptor-4 (TLR4) with low dose of LPS can prevent MSCs from apoptotic cell death induced by oxidative stress and serum deprivation through PI3K/Akt- and NFκB-dependent mechanisms (Wang et al., 2009c). Furthermore, transplantation of these LPS-preconditioned MSCs into infarcted rat hearts significantly improved cardiac function, reduced apoptosis and fibrosis, and enhanced angiogenesis (Yao et al., 2009).

4.7 Other agents

Other drugs and hormones have also been employed as preconditioning agents to promote stem and progenitor cell survival and function. For example, pre-treatment with the pineal hormone **melatonin** can improve the therapeutic effectiveness of MSCs in the setting of acute renal ischemia-reperfusion injury by potentiating their survival and paracrine activity

(Mias et al., 2008). **Trimetazidine** is a cytoprotective anti-ischaemic agent that acts to reduce ischaemia-induced metabolic damage by shifting the energy substrate preference from fatty acid oxidation to glucose oxidation. It has been employed recently to precondition MSCs and showed increased cell survival and enhanced therapeutic potential in reducing myocardial ischaemic injury (Wisel et al., 2009). **Fucoidan**, a high molecular weight sulphated polysaccharide, also favourably enhanced the migratory potential of pre-treated EPCs *in vitro*. Although alone it did not promote angiogenesis, treatment with fucoidan potentiated the angiogenenic effect of bFGF in EPCs (Zemani et al., 2005). Du *et al.* preconditioned skeletal myoblasts with **lithium chloride** and showed enhanced cell survival and increased gap-junctional coupling with co-cultured neonatal cardiomyocytes, a result of increased Cx43 expression (Du et al., 2009). Similarly, pre-treatment with the endogenous phospholipid signalling molecule, **lysophosphatidic acid (LPA)**, can improve MSC survival *in vivo* in ischaemic myocardium and enhance their angiogenic effects (Liu et al., 2009). Interestingly, an *in vitro* study by the same group indicated that LPA failed to precondition MSCs (Chen et al., 2008). In this study, the anti-apoptotic effect of LPA was only evident when the drug was present throughout the hypoxia and serum deprivation insult but not when LPA was removed after the pre-treatment period. Nevertheless, the cytoprotective effect of LPA was demonstrated to be dependent on the activation of LPA receptor-1, and pertussis toxin-sensitive PI3K/Akt and ERK pathways (Chen et al., 2008).

Heat shock proteins (HSP) are known protective mediators of preconditioning induced by ischaemia, hypoxia, heat stress and oxidative stress (Das & Maulik, 2006) (Table 1). In MSCs, preconditioning with recombinant human HSP90α produced an anti-apoptotic effect via activation of PI3K/Akt and ERK signalling pathways (Gao et al., 2010). Prior to this study, genetic modification to over-express HSP20 had been shown to enhance the survival of MSCs against oxidative stress and improve their therapeutic potential in ischaemic rat heart (Wang et al., 2009a). Therefore, it is not surprising that drugs capable of increasing HSP expression are potential preconditioning agents. Indeed, short-term exposure to **β-mercaptoethanol** has been shown to protect MSCs from oxidative injury, a cytoprotective effect associated with an elevation of HSP72 expression (Cizkova et al., 2006).

The haematopoietic cytokine **erythropoietin** (EPO) also exerts cardioprotection in both animal and clinical studies with many intracellular signalling pathways implicated, including PI3K/Akt, ERK1/2, p38 MAPK, PKC, eNOS and guanylyl cyclase and the opening of mitochondrial K_{ATP} channels (Riksen et al., 2008). In MSCs (Hu et al., 2008; Theus et al., 2008) and embryonic stem cell-derived neural progenitor cells (NPCs) (Francis & Wei, 2010; Theus et al., 2008), EPO is already involved in the cytoprotective effect of HPC. As a preconditioning agent, recombinant human EPO mimicked the cytoprotective effect of HPC in protecting mouse NPCs against serum deprivation-induced apoptotic cell death (Theus et al., 2008). Similarly, ESC-derived cardiomyocytes survived better in the infarcted mouse hearts when they were preconditioned with carbamylated EPO prior to implantation (Robey et al., 2008). The hydroxyl-methylglutaryl coenzyme A reductase inhibitors, also known as statins, are effective for lowering serum cholesterol and have been widely prescribed for patients with coronary heart diseases as the primary and secondary preventive treatment of cardiovascular events. In addition to lipid-lowering effects, statins have been shown to exert multiple pleiotropic effects including protection from MI, improved endothelial function and reduced platelet adhesion and atherosclerotic plaque rupture (Ludman et al., 2009). In MSCs, treatment with **lovastatin** has been shown to improve cell survival when challenged with hypoxia and serum deprivation, a protection mediated by PI3K/Akt and ERK1/2 pathways (Xu et al., 2008). Similarly, human EPCs pre-treated with **simvastatin** also exhibit higher resistance against TNFα induced apoptotic cell death (Henrich et al., 2007).

5. Ischaemic or hypoxic preconditioning in vivo mobilises endogenous stem and progenitor cells

Endogenous stem and progenitor cells can be mobilized from their niches in various organs and tissues, including bone marrow, skeletal muscle, heart, brain, skin, liver, adipose, blood vessels and others, and then home to target tissues. This property of stem and progenitor cells has fuelled investigations of potential clinically adaptable strategies to actively recruit endogenous stem and progenitor cells to serve as integrated participants in regenerating the injured tissues through stem cell trans-diffentiation and/or as supportive players via pleiotropic paracrine effects (Chen et al., 2011; Krankel et al., 2011). By subjecting rats to 3 weeks of chronic hypoxia, Rochefort *et al* have unveiled the potential of HPC in mobilising endogenous MSCs. In this study, circulating MSCs were higher in rats subjected to chronic hypoxia compared to the control normoxic cohort (Rochefort et al., 2006) (Table 4). Interestingly, this hypoxic condition did not affect the circulating level of haematopoietic stem cells (HSCs) indicating a possible cell-type specific effect of HPC (Rochefort et al., 2006). In line with this study, rats preconditioned with 6 hours of hypoxia daily for 6 weeks also have a higher level of $CD34^+CXCR4^+$ cells in their blood circulation and in the infarcted hearts with concomitant reduction in acute myocardial ischaemia-reperfusion injury (Lin et al., 2008). Using a relatively more invasive preconditioning protocol of transient coronary artery occlusion and reperfusion, Ii and colleagues have showed that IPC modulates endogenous EPC kinetic and increases their recruitment to the infarcted myocardium (Ii et al., 2005). This observation was associated with infarct size limitation, increased angiogenesis and cardiac function improvement, beneficial effects that were shown to be strongly dependent on the iNOS and eNOS activities of the EPCs (Ii et al., 2005). The mobilizing and homing effect of IPC was subsequently illustrated on other cell types such as MSCs and HSCs using a clinically relevant porcine myocardial ischaemia-reperfusion injury experimental model (Gyongyosi et al., 2010). Excitingly, Kamota *et al* showed that preconditioning applied on the abdominal aorta can also increase the accumulation of bone marrow-derived sca-1^+ and c-kit^- stem cells in infarcted hearts through a SDF-1/CXCR4-dependent mechanism, thus protecting the hearts against injury (Kamota et al., 2009). This finding is clinically important as it supported the translation of the non-invasive strategy of remote IPC into clinical practise. Remote IPC is a clinically amenable strategy which can be induced by simple transient limb ischaemia. This cardioprotective strategy has been extensively trialled in patients undergoing cardiac surgeries such as coronary angioplasty and coronary artery bypass surgery, and has thus far showed tangible beneficial effects in reducing myocardial injury with no known adverse risks (Hausenloy & Yellon, 2008).

Although the precise molecular and cellular mechanisms governing the homing effect of ischaemic/hypoxic preconditioning on endogenous stem and progenitor cells remain to be fully addressed, evidence of a role for the chemokine axis SDF-1/CXCR4 (Kamota et al., 2009; Lin et al., 2008) and EPO (Lin et al., 2008) has been found in recent studies. Other possible stem cell homing factors include VEGF, colony-stimulating factor, monocyte

PC stimulus	Models	End points	Mechanisms	References
Hypoxia; 24 h of 50kPa O_2 daily for 3 weeks	Rat, Sham	↑ circulating MSCs	-	Rochefort et al., 2006
Hypoxia; 6 h of 10% O_2 daily for 6 weeks	Rat, MI	↑ circulating & heart $CD43^+CXCR4^+$ cells; ↓ infarct size; ↑ LV function; ↓ plasma CK-MB	SDF-1/CXCR4; EPO; VEGF	Lin et al., 2008
Ischaemia; 4 x 4 min/4 min coronary artery O/R	Mouse, MI	↑ circulating & heart EPCs; ↓ infarct size; ↑ LV function; ↑ angiogenesis	eNOS; iNOS; VEGF	Ii et al., 2005
Ischaemia; 2 x 5 min/5 min coronary artery O/R	Pig, MI	↑ circulating HSCs; ↑ heart MSCs & HSCs; ↓ infarct size & apoptosis; ↑ LV function	SDF-1α; VEGF; TNFα; IL-8	Gyongyosi et al., 2010
Ischaemia; 4 x 5 min/5 min abdominal artery O/R	Mouse, MI	↑ circulating $CD43^+flk\text{-}1^+$ cells; ↑ heart $Sca\text{-}1^+$ & $c\text{-}kit^+$ BMSCs; ↓ infarct size & apoptosis; ↑ LV function	SDF-1/CXCR4; VEGF	Kamota et al., 2009
Ischaemia; 25 min/7 d renal artery O/R	Mouse, RI	↑ renal EPCs	-	Patschan et al., 2006

Table 4. Effect of in vivo ischaemic/hypoxic preconditioning on endogenous stem and progenitor cells. BMSCs (bone marrow stem cells), O/R (occlusion/reperfusion), RI (renal ischaemia).

chemotactic protein-3, HGF, IGF-1, IL-8/growth regulated oncogene-1, stem cell factor, TGF-β3, Wnt antagonist and other chemokines (Krankel et al., 2011; Binger et al., 2009; Chen et al., 2011). However, whether these navigational factors govern the homing effect of *in vivo* IPC warrants further investigation. In perspective, a living body appears to host a great reservoir of various stem and progenitor cells ready to be recruited for regeneration and repair, and can be catalysed by external stimuli such as IPC. This is of great clinical importance because *in vivo* IPC can be readily and non-invasively achieved in patients. However, given the fact that majority of ischaemic disease sufferers are elderly patients, and endogenous stem and progenitor populations and functions are known to decline with age and are negatively affected by other co-morbidity such as diabetes and hypertension (Krankel et al., 2011), it is important to determine whether there are sufficient populations of functionally competent resident stem and progenitor cells to be mobilised and recruited to the target tissues to exert significant repair in older patient cohorts. Otherwise, a combination of exogenously administered cells and potent homing factors might be utilised to supplement the endogenous reservoirs for effective cell-based therapy. Therefore, detailed characterisation of various endogenous stem cell niches and deciphering the mechanisms governing endogenous stem cell repopulation, mobilisation and homing to target tissues are pivotal to the future development of clinically sound pharmacological interventions to harness fully the host's innate regenerative capacity.

6. Conclusion

Based on the existing literature, it is undeniable that non-genetic approaches of preconditioning techniques offer much promise as cytoprotective strategies for stem cell therapy, without the long-term concern of genetic manipulation. These pro-survival strategies are therefore well suited to clinical translation. Collectively, preconditioning of stem and progenitor cells elicits multiple beneficial effects including (1) promotion of cell survival in the hostile ischaemic environment, (2) enhancement of paracrine activity to create a supportive environment that is rich in trophic and angiogenic factors, (3) increase of cell motility and trafficking, (4) increase of cell proliferative potential, (5) promotion of cell differentiation allowing functional integration, and (6) enhancement of therapeutic efficacy in ischaemic tissues *in vivo*.

In cell based therapy for ischaemic diseases, the ability of preconditioning *ex vivo* to enhance stem and progenitor cell survival and function means more implanted cells will be available for tissue repair and thus fewer donor cells may be needed to achieve the same functional outcome. In parallel, *in vivo* preconditioning is capable of harnessing the host's inherent regeneration mechanisms through activation of various paracrine signalling cascades, and mobilizing and recruiting resident stem and progenitor cells for effective therapeutics (Figure 3). Therefore, it will be of great therapeutic interest to determine whether there is an additive or synergistic effect of *ex vivo* preconditioning of implanted cells and *in vivo* preconditioning of host tissues, which may provide an optimal regenerative environment for tissue repair and regeneration, and contribute to successful stem cell therapy and tissue engineering.

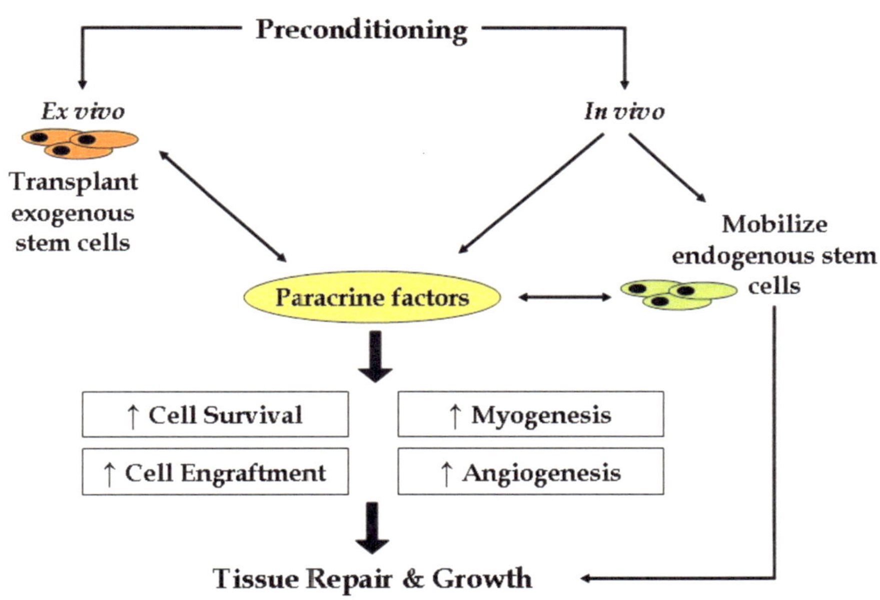

Fig. 3. Potential additive effect of *ex vivo* preconditioning (transplanted exogenous stem cells) and *in vivo* preconditioning (patients) to enhance tissue repair and tissue engineering with stem cells.

7. Acknowledgement

This work was supported by NHMRC Principal Research Fellowship (to GJD), Project grant #509271, the Wicking Trust, and the State Government's Department of Innovation, Industry and Regional Development's Operational Infrastructure Support Program.

8. References

Abarbanell, A.M.; Coffey, A.C.; Fehrenbacher, J.W.; Beckman, D.J.; Herrmann, J.L.; Weil, B. & Meldrum, D.R. (2009). Proinflammatory cytokine effects on mesenchymal stem cell therapy for the ischemic heart. *Ann Thorac Surg,* Vol.88, No.3, pp. 1036-1043, ISSN 1552-6259

Abbott, J.D.; Huang, Y.; Liu, D.; Hickey, R.; Krause, D.S. & Giordano, F.J. (2004). Stromal cell-derived factor-1alpha plays a critical role in stem cell recruitment to the heart after myocardial infarction but is not sufficient to induce homing in the absence of injury. *Circulation,* Vol.110, No.21, pp. 3300-3305, ISSN 1524-4539

Afzal, M.R.; Haider, H.; Idris, N.M.; Jiang, S.; Ahmed, R.P. & Ashraf, M. (2010). Preconditioning promotes survival and angiomyogenic potential of mesenchymal stem cells in the infarcted heart via NF-kappaB signaling. *Antioxid Redox Signal,* Vol.12, No.6, pp. 693-702, ISSN 1557-7716

Aicher, A.; Brenner, W.; Zuhayra, M.; Badorff, C.; Massoudi, S.; Assmus, B.; Eckey, T.; Henze, E.; Zeiher, A.M. & Dimmeler, S. (2003). Assessment of the tissue distribution of transplanted human endothelial progenitor cells by radioactive labeling. *Circulation,* Vol.107, No.16, pp. 2134-2139, ISSN 1524-4539

Akita, T.; Murohara, T.; Ikeda, H.; Sasaki, K.; Shimada, T.; Egami, K. & Imaizumi, T. (2003). Hypoxic preconditioning augments efficacy of human endothelial progenitor cells for therapeutic neovascularization. *Lab Invest,* Vol.83, No.1, pp. 65-73, ISSN 0023-6837

Askari, A.T.; Unzek, S.; Popovic, Z.B.; Goldman, C.K.; Forudi, F.; Kiedrowski, M.; Rovner, A.; Ellis, S.G.; Thomas, J.D.; DiCorleto, P.E.; Topol, E.J. & Penn, M.S. (2003). Effect of stromal-cell-derived factor 1 on stem-cell homing and tissue regeneration in ischaemic cardiomyopathy. *Lancet,* Vol.362, No.9385, pp. 697-703, ISSN 1474-547X

Baines, C.P.; Liu, G.S.; Birincioglu, M.; Critz, S.D.; Cohen, M.V. & Downey, J.M. (1999). Ischemic preconditioning depends on interaction between mitochondrial KATP channels and actin cytoskeleton. *Am J Physiol,* Vol.276, No.4 Pt 2, pp. H1361-1368, ISSN 0002-9513

Binger, T.; Stich, S.; Andreas, K.; Kaps, C.; Sezer, O.; Notter, M.; Sittinger, M. & Ringe, J. (2009). Migration potential and gene expression profile of human mesenchymal stem cells induced by CCL25. *Exp Cell Res,* Vol.315, No.8, pp. 1468-1479, ISSN 1090-2422

Bonaros, N.; Rauf, R.; Schachner, T.; Laufer, G. & Kocher, A. (2008). Enhanced cell therapy for ischemic heart disease. *Transplantation,* Vol.86, No.9, pp. 1151-1160, ISSN 1534-6080

Broxmeyer, H.E.; Kohli, L.; Kim, C.H.; Lee, Y.; Mantel, C.; Cooper, S.; Hangoc, G.; Shaheen, M.; Li, X. & Clapp, D.W. (2003). Stromal cell-derived factor-1/CXCL12 directly enhances survival/antiapoptosis of myeloid progenitor cells through CXCR4 and

G(alpha)i proteins and enhances engraftment of competitive, repopulating stem cells. *J Leukoc Biol*, Vol.73, No.5, pp. 630-638, ISSN 0741-5400

Chen, F.M.; Wu, L.A.; Zhang, M.; Zhang, R. & Sun, H.H. (2011). Homing of endogenous stem/progenitor cells for in situ tissue regeneration: Promises, strategies, and translational perspectives. *Biomaterials*, Vol.32, No.12, pp. 3189-3209, ISSN 1878-5905

Chen, J.; Baydoun, A.R.; Xu, R.; Deng, L.; Liu, X.; Zhu, W.; Shi, L.; Cong, X.; Hu, S. & Chen, X. (2008). Lysophosphatidic acid protects mesenchymal stem cells against hypoxia and serum deprivation-induced apoptosis. *Stem Cells*, Vol.26, No.1, pp. 135-145, ISSN 1549-4918

Chen, J.; Du, X. & Zhang, K. (2009). Effects of stromal-derived factor 1 preconditioning on apoptosis of rat bone mesenchymal stem cells. *J Huazhong Univ Sci Technolog Med Sci*, Vol.29, No.4, pp. 423-426, ISSN 1672-0733

Chacko, S.M.; Ahmed, S.; Selvendiran, K.; Kuppusamy, M.L.; Khan, M. & Kuppusamy, P. (2010). Hypoxic preconditioning induces the expression of pro-survival and pro-angiogenic markers in mesenchymal stem cells. *Am J Physiol Cell Physiol*, ISSN 1522-1563

Cizkova, D.; Rosocha, J.; Vanicky, I.; Radonak, J.; Galik, J. & Cizek, M. (2006). Induction of mesenchymal stem cells leads to HSP72 synthesis and higher resistance to oxidative stress. *Neurochem Res*, Vol.31, No.8, pp. 1011-1020, ISSN 0364-3190

Cui, X.; Wang, H.; Guo, H.; Wang, C.; Ao, H.; Liu, X. & Tan, Y.Z. (2010). Transplantation of mesenchymal stem cells preconditioned with diazoxide, a mitochondrial ATP-sensitive potassium channel opener, promotes repair of myocardial infarction in rats. *Tohoku J Exp Med*, Vol.220, No.2, pp. 139-147, ISSN 1349-3329

D'Ippolito, G.; Diabira, S.; Howard, G.A.; Roos, B.A. & Schiller, P.C. (2006). Low oxygen tension inhibits osteogenic differentiation and enhances stemness of human MIAMI cells. *Bone*, Vol.39, No.3, pp. 513-522, ISSN 8756-3282

Das, D.K. & Maulik, N. (2006). Cardiac genomic response following preconditioning stimulus. *Cardiovasc Res*, Vol.70, No.2, pp. 254-263, ISSN 0008-6363

Das, R.; Jahr, H.; van Osch, G.J. & Farrell, E. (2010). The role of hypoxia in bone marrow-derived mesenchymal stem cells: considerations for regenerative medicine approaches. *Tissue Eng Part B Rev*, Vol.16, No.2, pp. 159-168, ISSN 1937-3376

Du, W.J.; Li, J.K.; Wang, Q.Y.; Hou, J.B. & Yu, B. (2009). Lithium chloride preconditioning optimizes skeletal myoblast functions for cellular cardiomyoplasty in vitro via glycogen synthase kinase-3beta/beta-catenin signaling. *Cells Tissues Organs*, Vol.190, No.1, pp. 11-19, ISSN 1422-6421

Elmadbouh, I.; Haider, H.; Jiang, S.; Idris, N.M.; Lu, G. & Ashraf, M. (2007). Ex vivo delivered stromal cell-derived factor-1alpha promotes stem cell homing and induces angiomyogenesis in the infarcted myocardium. *J Mol Cell Cardiol*, Vol.42, No.4, pp. 792-803, ISSN 0022-2828

Forte, G.; Minieri, M.; Cossa, P.; Antenucci, D.; Sala, M.; Gnocchi, V.; Fiaccavento, R.; Carotenuto, F.; De Vito, P.; Baldini, P.M.; Prat, M. & Di Nardo, P. (2006). Hepatocyte growth factor effects on mesenchymal stem cells: proliferation, migration, and differentiation. *Stem Cells*, Vol.24, No.1, pp. 23-33, ISSN 1066-5099

Francis, K.R. & Wei, L. (2010). Human embryonic stem cell neural differentiation and enhanced cell survival promoted by hypoxic preconditioning. *Cell Death Dis,* Vol.1, No.2, pp. e22, ISSN 2041-4889

Gao, F.; Hu, X.Y.; Xie, X.J.; Xu, Q.Y.; Wang, Y.P.; Liu, X.B.; Xiang, M.X.; Sun, Y. & Wang, J.A. (2010). Heat shock protein 90 protects rat mesenchymal stem cells against hypoxia and serum deprivation-induced apoptosis via the PI3K/Akt and ERK1/2 pathways. *J Zhejiang Univ Sci B,* Vol.11, No.8, pp. 608-617, ISSN 1862-1783

Gnecchi, M.; He, H.; Noiseux, N.; Liang, O.D.; Zhang, L.; Morello, F.; Mu, H.; Melo, L.G.; Pratt, R.E.; Ingwall, J.S. & Dzau, V.J. (2006). Evidence supporting paracrine hypothesis for Akt-modified mesenchymal stem cell-mediated cardiac protection and functional improvement. *FASEB J,* Vol.20, No.6, pp. 661-669, ISSN 1530-6860

Goussetis, E.; Manginas, A.; Koutelou, M.; Peristeri, I.; Theodosaki, M.; Kollaros, N.; Leontiadis, E.; Theodorakos, A.; Paterakis, G.; Karatasakis, G.; Cokkinos, D.V. & Graphakos, S. (2006). Intracoronary infusion of CD133+ and CD133-CD34+ selected autologous bone marrow progenitor cells in patients with chronic ischemic cardiomyopathy: cell isolation, adherence to the infarcted area, and body distribution. *Stem Cells,* Vol.24, No.10, pp. 2279-2283, ISSN 1066-5099

Grayson, W.L.; Zhao, F.; Bunnell, B. & Ma, T. (2007). Hypoxia enhances proliferation and tissue formation of human mesenchymal stem cells. *Biochem Biophys Res Commun,* Vol.358, No.3, pp. 948-953, ISSN 0006-291X

Gyongyosi, M.; Posa, A.; Pavo, N.; Hemetsberger, R.; Kvakan, H.; Steiner-Boker, S.; Petrasi, Z.; Manczur, F.; Pavo, I.J.; Edes, I.F.; Wojta, J.; Glogar, D. & Huber, K. (2010). Differential effect of ischaemic preconditioning on mobilisation and recruitment of haematopoietic and mesenchymal stem cells in porcine myocardial ischaemia-reperfusion. *Thromb Haemost,* Vol.104, No.2, pp. 376-384, ISSN 0340-6245

Ha, T.; Hua, F.; Liu, X.; Ma, J.; McMullen, J.R.; Shioi, T.; Izumo, S.; Kelley, J.; Gao, X.; Browder, W.; Williams, D.L.; Kao, R.L. & Li, C. (2008). Lipopolysaccharide-induced myocardial protection against ischaemia/reperfusion injury is mediated through a PI3K/Akt-dependent mechanism. *Cardiovasc Res,* Vol.78, No.3, pp. 546-553, ISSN 0008-6363

Hahn, J.Y.; Cho, H.J.; Kang, H.J.; Kim, T.S.; Kim, M.H.; Chung, J.H.; Bae, J.W.; Oh, B.H.; Park, Y.B. & Kim, H.S. (2008). Pre-treatment of mesenchymal stem cells with a combination of growth factors enhances gap junction formation, cytoprotective effect on cardiomyocytes, and therapeutic efficacy for myocardial infarction. *J Am Coll Cardiol,* Vol.51, No.9, pp. 933-943, ISSN 1558-3597

Haider, K.H.; Idris, N.M.; Kim, H.W.; Ahmed, R.P.; Shujia, J. & Ashraf, M. (2010). MicroRNA-21 is a key determinant in IL-11/Stat3 anti-apoptotic signalling pathway in preconditioning of skeletal myoblasts. *Cardiovasc Res,* Vol.88, No.1, pp. 168-178, ISSN 1755-3245

Hausenloy, D.J. & Yellon, D.M. (2003). The mitochondrial permeability transition pore: its fundamental role in mediating cell death during ischaemia and reperfusion. *J.Mol.Cell Cardiol.,* Vol.35, No.4, pp. 339-341, ISSN 0022-2828

Hausenloy, D.J. & Yellon, D.M. (2008). Remote ischaemic preconditioning: underlying mechanisms and clinical application. *Cardiovasc.Res.,* Vol.79, No.3, pp. 377-386, ISSN 0008-6363

Hausenloy, D.J.; Lim, S.Y.; Ong, S.G.; Davidson, S.M. & Yellon, D.M. (2010). Mitochondrial cyclophilin-D as a critical mediator of ischaemic preconditioning. *Cardiovasc Res,* Vol.88, No.1, pp. 67-74, ISSN 1755-3245

He, A.; Jiang, Y.; Gui, C.; Sun, Y.; Li, J. & Wang, J.A. (2009). The antiapoptotic effect of mesenchymal stem cell transplantation on ischemic myocardium is enhanced by anoxic preconditioning. *Can J Cardiol,* Vol.25, No.6, pp. 353-358, ISSN 1916-7075

Henrich, D.; Seebach, C.; Wilhelm, K. & Marzi, I. (2007). High dosage of simvastatin reduces TNF-alpha-induced apoptosis of endothelial progenitor cells but fails to prevent apoptosis induced by IL-1beta in vitro. *J Surg Res,* Vol.142, No.1, pp. 13-19, ISSN 0022-4804

Herrmann, J.L.; Abarbanell, A.M.; Weil, B.R.; Manukyan, M.C.; Poynter, J.A.; Brewster, B.J.; Wang, Y. & Meldrum, D.R. (2010a). Optimizing Stem Cell Function for the Treatment of Ischemic Heart Disease. *J Surg Res,* ISSN 1095-8673

Herrmann, J.L.; Wang, Y.; Abarbanell, A.M.; Weil, B.R.; Tan, J. & Meldrum, D.R. (2010b). Preconditioning mesenchymal stem cells with transforming growth factor-alpha improves mesenchymal stem cell-mediated cardioprotection. *Shock,* Vol.33, No.1, pp. 24-30, ISSN 1540-0514

Hiasa, K.; Ishibashi, M.; Ohtani, K.; Inoue, S.; Zhao, Q.; Kitamoto, S.; Sata, M.; Ichiki, T.; Takeshita, A. & Egashira, K. (2004). Gene transfer of stromal cell-derived factor-1alpha enhances ischemic vasculogenesis and angiogenesis via vascular endothelial growth factor/endothelial nitric oxide synthase-related pathway: next-generation chemokine therapy for therapeutic neovascularization. *Circulation,* Vol.109, No.20, pp. 2454-2461, ISSN 1524-4539

Hu, X.; Dai, S.; Wu, W.J.; Tan, W.; Zhu, X.; Mu, J.; Guo, Y.; Bolli, R. & Rokosh, G. (2007). Stromal cell derived factor-1 alpha confers protection against myocardial ischemia/reperfusion injury: role of the cardiac stromal cell derived factor-1 alpha CXCR4 axis. *Circulation,* Vol.116, No.6, pp. 654-663, ISSN 1524-4539

Hu, X.; Yu, S.P.; Fraser, J.L.; Lu, Z.; Ogle, M.E.; Wang, J.A. & Wei, L. (2008). Transplantation of hypoxia-preconditioned mesenchymal stem cells improves infarcted heart function via enhanced survival of implanted cells and angiogenesis. *J Thorac Cardiovasc Surg,* Vol.135, No.4, pp. 799-808, ISSN 1097-685X

Huffmyer, J. & Raphael, J. (2009). Physiology and pharmacology of myocardial preconditioning and postconditioning. *Semin Cardiothorac Vasc Anesth,* Vol.13, No.1, pp. 5-18, ISSN 1089-2532

Hung, S.C.; Pochampally, R.R.; Chen, S.C.; Hsu, S.C. & Prockop, D.J. (2007a). Angiogenic effects of human multipotent stromal cell conditioned medium activate the PI3K-Akt pathway in hypoxic endothelial cells to inhibit apoptosis, increase survival, and stimulate angiogenesis. *Stem Cells,* Vol.25, No.9, pp. 2363-2370, ISSN 1549-4918

Hung, S.C.; Pochampally, R.R.; Hsu, S.C.; Sanchez, C.; Chen, S.C.; Spees, J. & Prockop, D.J. (2007b). Short-term exposure of multipotent stromal cells to low oxygen increases their expression of CX3CR1 and CXCR4 and their engraftment in vivo. *PLoS One,* Vol.2, No.5, pp. e416, ISSN 1932-6203

Ii, M.; Nishimura, H.; Iwakura, A.; Wecker, A.; Eaton, E.; Asahara, T. & Losordo, D.W. (2005). Endothelial progenitor cells are rapidly recruited to myocardium and mediate protective effect of ischemic preconditioning via "imported" nitric oxide synthase activity. *Circulation,* Vol.111, No.9, pp. 1114-1120, ISSN 1524-4539

Jaderstad, J.; Brismar, H. & Herlenius, E. (2010). Hypoxic preconditioning increases gap-junctional graft and host communication. *Neuroreport*, Vol.21, No.17, pp. 1126-1132, ISSN 1473-558X

Kamota, T.; Li, T.S.; Morikage, N.; Murakami, M.; Ohshima, M.; Kubo, M.; Kobayashi, T.; Mikamo, A.; Ikeda, Y.; Matsuzaki, M. & Hamano, K. (2009). Ischemic pre-conditioning enhances the mobilization and recruitment of bone marrow stem cells to protect against ischemia/reperfusion injury in the late phase. *J Am Coll Cardiol*, Vol.53, No.19, pp. 1814-1822, ISSN 1558-3597

Khan, M.; Akhtar, S.; Mohsin, S.; S, N.K. & Riazuddin, S. (2011). Growth factor preconditioning increases the function of diabetes-impaired mesenchymal stem cells. *Stem Cells Dev*, Vol.20, No.1, pp. 67-75, ISSN 1557-8534

Kim, H.W.; Haider, H.K.; Jiang, S. & Ashraf, M. (2009). Ischemic preconditioning augments survival of stem cells via miR-210 expression by targeting caspase-8-associated protein 2. *J Biol Chem*, Vol.284, No.48, pp. 33161-33168, ISSN 1083-351X

Krankel, N.; Spinetti, G.; Amadesi, S. & Madeddu, P. (2011). Targeting stem cell niches and trafficking for cardiovascular therapy. *Pharmacol Ther*, Vol.129, No.1, pp. 62-81, ISSN 1879-016X

Kubo, M.; Li, T.S.; Suzuki, R.; Shirasawa, B.; Morikage, N.; Ohshima, M.; Qin, S.L. & Hamano, K. (2008). Hypoxic preconditioning increases survival and angiogenic potency of peripheral blood mononuclear cells via oxidative stress resistance. *Am J Physiol Heart Circ Physiol*, Vol.294, No.2, pp. H590-595, ISSN 0363-6135

Kucia, M.; Jankowski, K.; Reca, R.; Wysoczynski, M.; Bandura, L.; Allendorf, D.J.; Zhang, J.; Ratajczak, J. & Ratajczak, M.Z. (2004). CXCR4-SDF-1 signalling, locomotion, chemotaxis and adhesion. *J Mol Histol*, Vol.35, No.3, pp. 233-245, ISSN 1567-2379

Laflamme, M.A.; Gold, J.; Xu, C.; Hassanipour, M.; Rosler, E.; Police, S.; Muskheli, V. & Murry, C.E. (2005). Formation of human myocardium in the rat heart from human embryonic stem cells. *Am J Pathol*, Vol.167, No.3, pp. 663-671, ISSN 0002-9440

Lennon, D.P.; Edmison, J.M. & Caplan, A.I. (2001). Cultivation of rat marrow-derived mesenchymal stem cells in reduced oxygen tension: effects on in vitro and in vivo osteochondrogenesis. *J Cell Physiol*, Vol.187, No.3, pp. 345-355, ISSN 0021-9541

Leroux, L.; Descamps, B.; Tojais, N.F.; Seguy, B.; Oses, P.; Moreau, C.; Daret, D.; Ivanovic, Z.; Boiron, J.M.; Lamaziere, J.M.; Dufourcq, P.; Couffinhal, T. & Duplaa, C. (2010). Hypoxia preconditioned mesenchymal stem cells improve vascular and skeletal muscle fiber regeneration after ischemia through a Wnt4-dependent pathway. *Mol Ther*, Vol.18, No.8, pp. 1545-1552, ISSN 1525-0024

Li, J.H.; Zhang, N. & Wang, J.A. (2008). Improved anti-apoptotic and anti-remodeling potency of bone marrow mesenchymal stem cells by anoxic pre-conditioning in diabetic cardiomyopathy. *J Endocrinol Invest*, Vol.31, No.2, pp. 103-110, ISSN 1720-8386

Li, S.; Deng, Y.; Feng, J. & Ye, W. (2009). Oxidative preconditioning promotes bone marrow mesenchymal stem cells migration and prevents apoptosis. *Cell Biol Int*, Vol.33, No.3, pp. 411-418, ISSN 1095-8355

Li, T.S.; Hamano, K.; Suzuki, K.; Ito, H.; Zempo, N. & Matsuzaki, M. (2002). Improved angiogenic potency by implantation of ex vivo hypoxia prestimulated bone marrow cells in rats. *Am J Physiol Heart Circ Physiol*, Vol.283, No.2, pp. H468-473, ISSN 0363-6135

Li, Y.; Yu, X.; Lin, S.; Li, X.; Zhang, S. & Song, Y.H. (2007). Insulin-like growth factor 1 enhances the migratory capacity of mesenchymal stem cells. *Biochem Biophys Res Commun,* Vol.356, No.3, pp. 780-784, ISSN 0006-291X

Lim, S.Y.; Davidson, S.M.; Hausenloy, D.J. & Yellon, D.M. (2007). Preconditioning and postconditioning: The essential role of the mitochondrial permeability transition pore. *Cardiovasc.Res.,* Vol.75, No.3, pp. 530-535, ISSN 0008-6363

Lim, S.Y.; Davidson, S.M.; Paramanathan, A.J.; Smith, C.C.; Yellon, D.M. & Hausenloy, D.J. (2008). The novel adipocytokine visfatin exerts direct cardioprotective effects. *J.Cell Mol.Med.,* Vol.12, No.4, pp. 1395-1403, ISSN 1582-4934

Lin, J.S.; Chen, Y.S.; Chiang, H.S. & Ma, M.C. (2008). Hypoxic preconditioning protects rat hearts against ischaemia-reperfusion injury: role of erythropoietin on progenitor cell mobilization. *J Physiol,* Vol.586, No.Pt 23, pp. 5757-5769, ISSN 1469-7793

Liu, H.; Xue, W.; Ge, G.; Luo, X.; Li, Y.; Xiang, H.; Ding, X.; Tian, P. & Tian, X. (2010). Hypoxic preconditioning advances CXCR4 and CXCR7 expression by activating HIF-1alpha in MSCs. *Biochem Biophys Res Commun,* Vol.401, No.4, pp. 509-515, ISSN 1090-2104

Liu, X.; Hou, J.; Shi, L.; Chen, J.; Sang, J.; Hu, S.; Cong, X. & Chen, X. (2009). Lysophosphatidic acid protects mesenchymal stem cells against ischemia-induced apoptosis in vivo. *Stem Cells Dev,* Vol.18, No.7, pp. 947-954, ISSN 1557-8534

Lu, G.; Haider, H.K.; Jiang, S. & Ashraf, M. (2009). Sca-1+ stem cell survival and engraftment in the infarcted heart: dual role for preconditioning-induced connexin-43. *Circulation,* Vol.119, No.19, pp. 2587-2596, ISSN 1524-4539

Lu, G.; Haider, H.; Porollo, A. & Ashraf, M. (2010). Mitochondria-specific transgenic overexpression of connexin-43 simulates preconditioning-induced cytoprotection of stem cells. *Cardiovasc Res,* Vol.88, No.2, pp. 277-286, ISSN 1755-3245

Lucchinetti, E.; Zeisberger, S.M.; Baruscotti, I.; Wacker, J.; Feng, J.; Zaugg, K.; Dubey, R.; Zisch, A.H. & Zaugg, M. (2009). Stem cell-like human endothelial progenitors show enhanced colony-forming capacity after brief sevoflurane exposure: preconditioning of angiogenic cells by volatile anesthetics. *Anesth Analg,* Vol.109, No.4, pp. 1117-1126, ISSN 1526-7598

Ludman, A.; Venugopal, V.; Yellon, D.M. & Hausenloy, D.J. (2009). Statins and cardioprotection--more than just lipid lowering? *Pharmacol Ther,* Vol.122, No.1, pp. 30-43, ISSN 1879-016X

Mahaffey, K.W.; Puma, J.A.; Barbagelata, N.A.; DiCarli, M.F.; Leesar, M.A.; Browne, K.F.; Eisenberg, P.R.; Bolli, R.; Casas, A.C.; Molina-Viamonte, V.; Orlandi, C.; Blevins, R.; Gibbons, R.J.; Califf, R.M. & Granger, C.B. (1999). Adenosine as an adjunct to thrombolytic therapy for acute myocardial infarction: results of a multicenter, randomized, placebo-controlled trial: the Acute Myocardial Infarction STudy of ADenosine (AMISTAD) trial. *J Am Coll Cardiol,* Vol.34, No.6, pp. 1711-1720, ISSN 0735-1097

Malladi, P.; Xu, Y.; Chiou, M.; Giaccia, A.J. & Longaker, M.T. (2006). Effect of reduced oxygen tension on chondrogenesis and osteogenesis in adipose-derived mesenchymal cells. *Am J Physiol Cell Physiol,* Vol.290, No.4, pp. C1139-1146, ISSN 0363-6143

Maurel, A.; Azarnoush, K.; Sabbah, L.; Vignier, N.; Le Lorc'h, M.; Mandet, C.; Bissery, A.; Garcin, I.; Carrion, C.; Fiszman, M.; Bruneval, P.; Hagege, A.; Carpentier, A.;

Vilquin, J.T. & Menasche, P. (2005). Can cold or heat shock improve skeletal myoblast engraftment in infarcted myocardium? *Transplantation,* Vol.80, No.5, pp. 660-665, ISSN 0041-1337

Mias, C.; Trouche, E.; Seguelas, M.H.; Calcagno, F.; Dignat-George, F.; Sabatier, F.; Piercecchi-Marti, M.D.; Daniel, L.; Bianchi, P.; Calise, D.; Bourin, P.; Parini, A. & Cussac, D. (2008). Ex vivo pretreatment with melatonin improves survival, proangiogenic/mitogenic activity, and efficiency of mesenchymal stem cells injected into ischemic kidney. *Stem Cells,* Vol.26, No.7, pp. 1749-1757, ISSN 1549-4918

Miniati, D.N. & Robbins, R.C. (2002). Heart transplantation: a thirty-year perspective. *Annu Rev Med,* Vol.53, pp. 189-205, ISSN 0066-4219

Mullenheim, J.; Ebel, D.; Bauer, M.; Otto, F.; Heinen, A.; Frassdorf, J.; Preckel, B. & Schlack, W. (2003). Sevoflurane confers additional cardioprotection after ischemic late preconditioning in rabbits. *Anesthesiology,* Vol.99, No.3, pp. 624-631, ISSN 0003-3022

Murry, C.E.; Jennings, R.B. & Reimer, K.A. (1986). Preconditioning with ischemia: a delay of lethal cell injury in ischemic myocardium. *Circulation,* Vol.74, No.5, pp. 1124-1136, ISSN 1524-4539

Neuss, S.; Becher, E.; Woltje, M.; Tietze, L. & Jahnen-Dechent, W. (2004). Functional expression of HGF and HGF receptor/c-met in adult human mesenchymal stem cells suggests a role in cell mobilization, tissue repair, and wound healing. *Stem Cells,* Vol.22, No.3, pp. 405-414, ISSN 1066-5099

Niagara, M.I.; Haider, H.; Jiang, S. & Ashraf, M. (2007). Pharmacologically preconditioned skeletal myoblasts are resistant to oxidative stress and promote angiomyogenesis via release of paracrine factors in the infarcted heart. *Circ Res,* Vol.100, No.4, pp. 545-555, ISSN 1524-4571

O'Rourke, B. (2004). Evidence for mitochondrial K+ channels and their role in cardioprotection. *Circ Res,* Vol.94, No.4, pp. 420-432, ISSN 1524-4571

Oh, J.S.; Ha, Y.; An, S.S.; Khan, M.; Pennant, W.A.; Kim, H.J.; Yoon do, H.; Lee, M. & Kim, K.N. (2010). Hypoxia-preconditioned adipose tissue-derived mesenchymal stem cell increase the survival and gene expression of engineered neural stem cells in a spinal cord injury model. *Neurosci Lett,* Vol.472, No.3, pp. 215-219, ISSN 1872-7972

Ong, L.L.; Li, W.; Oldigs, J.K.; Kaminski, A.; Gerstmayer, B.; Piechaczek, C.; Wagner, W.; Li, R.K.; Ma, N. & Steinhoff, G. (2010). Hypoxic/Normoxic Preconditioning Increases Endothelial Differentiation Potential of Human Bone Marrow CD133+ Cells. *Tissue Eng Part C Methods,* Vol.16, No.5, pp. 1069-1081, ISSN 1937-3392

Pagani, F.D.; DerSimonian, H.; Zawadzka, A.; Wetzel, K.; Edge, A.S.; Jacoby, D.B.; Dinsmore, J.H.; Wright, S.; Aretz, T.H.; Eisen, H.J. & Aaronson, K.D. (2003). Autologous skeletal myoblasts transplanted to ischemia-damaged myocardium in humans. Histological analysis of cell survival and differentiation. *J Am Coll Cardiol,* Vol.41, No.5, pp. 879-888, ISSN 0735-1097

Pasha, Z.; Wang, Y.; Sheikh, R.; Zhang, D.; Zhao, T. & Ashraf, M. (2008). Preconditioning enhances cell survival and differentiation of stem cells during transplantation in infarcted myocardium. *Cardiovasc Res,* Vol.77, No.1, pp. 134-142, ISSN 0008-6363

Patschan, D.; Krupincza, K.; Patschan, S.; Zhang, Z.; Hamby, C. & Goligorsky, M.S. (2006). Dynamics of mobilization and homing of endothelial progenitor cells after acute renal ischemia: modulation by ischemic preconditioning. *Am J Physiol Renal Physiol,* Vol.291, No.1, pp. F176-185, ISSN 1931-857X

Penn, M.S. & Mangi, A.A. (2008). Genetic enhancement of stem cell engraftment, survival, and efficacy. *Circ Res,* Vol.102, No.12, pp. 1471-1482, ISSN 1524-4571

Peterson, K.M.; Aly, A.; Lerman, A.; Lerman, L.O. & Rodriguez-Porcel, M. (2011). Improved survival of mesenchymal stromal cell after hypoxia preconditioning: role of oxidative stress. *Life Sci,* Vol.88, No.1-2, pp. 65-73, ISSN 1879-0631

Potier, E.; Ferreira, E.; Andriamanalijaona, R.; Pujol, J.P.; Oudina, K.; Logeart-Avramoglou, D. & Petite, H. (2007). Hypoxia affects mesenchymal stromal cell osteogenic differentiation and angiogenic factor expression. *Bone,* Vol.40, No.4, pp. 1078-1087, ISSN 8756-3282

Ren, H.; Cao, Y.; Zhao, Q.; Li, J.; Zhou, C.; Liao, L.; Jia, M.; Cai, H.; Han, Z.C.; Yang, R.; Chen, G. & Zhao, R.C. (2006). Proliferation and differentiation of bone marrow stromal cells under hypoxic conditions. *Biochem Biophys Res Commun,* Vol.347, No.1, pp. 12-21, ISSN 0006-291X

Riksen, N.P.; Hausenloy, D.J. & Yellon, D.M. (2008). Erythropoietin: ready for prime-time cardioprotection. *Trends Pharmacol Sci,* Vol.29, No.5, pp. 258-267, ISSN 0165-6147

Robey, T.E.; Saiget, M.K.; Reinecke, H. & Murry, C.E. (2008). Systems approaches to preventing transplanted cell death in cardiac repair. *J Mol Cell Cardiol,* Vol.45, No.4, pp. 567-581, ISSN 1095-8584

Rochefort, G.Y.; Delorme, B.; Lopez, A.; Herault, O.; Bonnet, P.; Charbord, P.; Eder, V. & Domenech, J. (2006). Multipotential mesenchymal stem cells are mobilized into peripheral blood by hypoxia. *Stem Cells,* Vol.24, No.10, pp. 2202-2208, ISSN 1066-5099

Rosova, I.; Dao, M.; Capoccia, B.; Link, D. & Nolta, J.A. (2008). Hypoxic preconditioning results in increased motility and improved therapeutic potential of human mesenchymal stem cells. *Stem Cells,* Vol.26, No.8, pp. 2173-2182, ISSN 1549-4918

Ross, A.M.; Gibbons, R.J.; Stone, G.W.; Kloner, R.A. & Alexander, R.W. (2005). A randomized, double-blinded, placebo-controlled multicenter trial of adenosine as an adjunct to reperfusion in the treatment of acute myocardial infarction (AMISTAD-II). *J Am Coll Cardiol,* Vol.45, No.11, pp. 1775-1780, ISSN 0735-1097

Salim, A.; Nacamuli, R.P.; Morgan, E.F.; Giaccia, A.J. & Longaker, M.T. (2004). Transient changes in oxygen tension inhibit osteogenic differentiation and Runx2 expression in osteoblasts. *J Biol Chem,* Vol.279, No.38, pp. 40007-40016, ISSN 0021-9258

Saxena, A.; Fish, J.E.; White, M.D.; Yu, S.; Smyth, J.W.; Shaw, R.M.; DiMaio, J.M. & Srivastava, D. (2008). Stromal cell-derived factor-1alpha is cardioprotective after myocardial infarction. *Circulation,* Vol.117, No.17, pp. 2224-2231, ISSN 1524-4539

Segers, V.F.; Tokunou, T.; Higgins, L.J.; MacGillivray, C.; Gannon, J. & Lee, R.T. (2007). Local delivery of protease-resistant stromal cell derived factor-1 for stem cell recruitment after myocardial infarction. *Circulation,* Vol.116, No.15, pp. 1683-1692, ISSN 1524-4539

Sepac, A.; Sedlic, F.; Si-Tayeb, K.; Lough, J.; Duncan, S.A.; Bienengraeber, M.; Park, F.; Kim, J. & Bosnjak, Z.J. (2010). Isoflurane preconditioning elicits competent endogenous mechanisms of protection from oxidative stress in cardiomyocytes derived from human embryonic stem cells. *Anesthesiology,* Vol.113, No.4, pp. 906-916, ISSN 1528-1175

Sharma, R.K.; Zhou, Q. & Netland, P.A. (2008). Effect of oxidative preconditioning on neural progenitor cells. *Brain Res,* Vol.1243, pp. 19-26, ISSN 1872-6240

Suzuki, K.; Smolenski, R.T.; Jayakumar, J.; Murtuza, B.; Brand, N.J. & Yacoub, M.H. (2000). Heat shock treatment enhances graft cell survival in skeletal myoblast transplantation to the heart. *Circulation,* Vol.102, No.19 Suppl 3, pp. III216-221, ISSN 1524-4539

Suzuki, Y.; Kim, H.W.; Ashraf, M. & Haider, H.K. (2010). Diazoxide potentiates mesenchymal stem cell survival via NF-{kappa}B-dependent miR-146a expression by targeting Fas. *Am J Physiol Heart Circ Physiol,* ISSN 1522-1539

Takahashi, M. (2010). Role of the SDF-1/CXCR4 system in myocardial infarction. *Circ J,* Vol.74, No.3, pp. 418-423, ISSN 1347-4820

Tamama, K.; Fan, V.H.; Griffith, L.G.; Blair, H.C. & Wells, A. (2006). Epidermal growth factor as a candidate for ex vivo expansion of bone marrow-derived mesenchymal stem cells. *Stem Cells,* Vol.24, No.3, pp. 686-695, ISSN 1066-5099

Tang, Y.L.; Zhu, W.; Cheng, M.; Chen, L.; Zhang, J.; Sun, T.; Kishore, R.; Phillips, M.I.; Losordo, D.W. & Qin, G. (2009). Hypoxic preconditioning enhances the benefit of cardiac progenitor cell therapy for treatment of myocardial infarction by inducing CXCR4 expression. *Circ Res,* Vol.104, No.10, pp. 1209-1216, ISSN 1524-4571

Terrovitis, J.V.; Smith, R.R. & Marban, E. (2010). Assessment and optimization of cell engraftment after transplantation into the heart. *Circ Res,* Vol.106, No.3, pp. 479-494, ISSN 1524-4571

Theus, M.H.; Wei, L.; Cui, L.; Francis, K.; Hu, X.; Keogh, C. & Yu, S.P. (2008). In vitro hypoxic preconditioning of embryonic stem cells as a strategy of promoting cell survival and functional benefits after transplantation into the ischemic rat brain. *Exp Neurol,* Vol.210, No.2, pp. 656-670, ISSN 0014-4886

Uemura, R.; Xu, M.; Ahmad, N. & Ashraf, M. (2006). Bone marrow stem cells prevent left ventricular remodeling of ischemic heart through paracrine signaling. *Circ Res,* Vol.98, No.11, pp. 1414-1421, ISSN 1524-4571

Vandervelde, S.; van Luyn, M.J.; Tio, R.A. & Harmsen, M.C. (2005). Signaling factors in stem cell-mediated repair of infarcted myocardium. *J Mol Cell Cardiol,* Vol.39, No.2, pp. 363-376, ISSN 0022-2828

Volkmer, E.; Kallukalam, B.C.; Maertz, J.; Otto, S.; Drosse, I.; Polzer, H.; Bocker, W.; Stengele, M.; Docheva, D.; Mutschler, W. & Schieker, M. (2010). Hypoxic preconditioning of human mesenchymal stem cells overcomes hypoxia-induced inhibition of osteogenic differentiation. *Tissue Eng Part A,* Vol.16, No.1, pp. 153-164, ISSN 1937-335X

Wang, D.W.; Fermor, B.; Gimble, J.M.; Awad, H.A. & Guilak, F. (2005). Influence of oxygen on the proliferation and metabolism of adipose derived adult stem cells. *J Cell Physiol,* Vol.204, No.1, pp. 184-191, ISSN 0021-9541

Wang, J.A.; Chen, T.L.; Jiang, J.; Shi, H.; Gui, C.; Luo, R.H.; Xie, X.J.; Xiang, M.X. & Zhang, X. (2008a). Hypoxic preconditioning attenuates hypoxia/reoxygenation-induced apoptosis in mesenchymal stem cells. *Acta Pharmacol Sin,* Vol.29, No.1, pp. 74-82, ISSN 1671-4083

Wang, X.; Zhao, T.; Huang, W.; Wang, T.; Qian, J.; Xu, M.; Kranias, E.G.; Wang, Y. & Fan, G.C. (2009a). Hsp20-engineered mesenchymal stem cells are resistant to oxidative stress via enhanced activation of Akt and increased secretion of growth factors. *Stem Cells,* Vol.27, No.12, pp. 3021-3031, ISSN 1549-4918

Wang, Y.; Crisostomo, P.R.; Wang, M.; Markel, T.A.; Novotny, N.M. & Meldrum, D.R. (2008b). TGF-alpha increases human mesenchymal stem cell-secreted VEGF by MEK- and PI3-K- but not JNK- or ERK-dependent mechanisms. *Am J Physiol Regul Integr Comp Physiol,* Vol.295, No.4, pp. R1115-1123, ISSN 0363-6119

Wang, Y.; Weil, B.R.; Herrmann, J.L.; Abarbanell, A.M.; Tan, J.; Markel, T.A.; Kelly, M.L. & Meldrum, D.R. (2009b). MEK, p38, and PI-3K mediate cross talk between EGFR and TNFR in enhancing hepatocyte growth factor production from human mesenchymal stem cells. *Am J Physiol Cell Physiol,* Vol.297, No.5, pp. C1284-1293, ISSN 1522-1563

Wang, Z.J.; Zhang, F.M.; Wang, L.S.; Yao, Y.W.; Zhao, Q. & Gao, X. (2009c). Lipopolysaccharides can protect mesenchymal stem cells (MSCs) from oxidative stress-induced apoptosis and enhance proliferation of MSCs via Toll-like receptor(TLR)-4 and PI3K/Akt. *Cell Biol Int,* Vol.33, No.6, pp. 665-674, ISSN 1095-8355

Wells, A. (1999). EGF receptor. *Int J Biochem Cell Biol,* Vol.31, No.6, pp. 637-643, ISSN 1357-2725

Wilkins, B.J.; Dai, Y.S.; Bueno, O.F.; Parsons, S.A.; Xu, J.; Plank, D.M.; Jones, F.; Kimball, T.R. & Molkentin, J.D. (2004). Calcineurin/NFAT coupling participates in pathological, but not physiological, cardiac hypertrophy. *Circ Res,* Vol.94, No.1, pp. 110-118, ISSN 1524-4571

Wisel, S.; Khan, M.; Kuppusamy, M.L.; Mohan, I.K.; Chacko, S.M.; Rivera, B.K.; Sun, B.C.; Hideg, K. & Kuppusamy, P. (2009). Pharmacological preconditioning of mesenchymal stem cells with trimetazidine (1-[2,3,4-trimethoxybenzyl]piperazine) protects hypoxic cells against oxidative stress and enhances recovery of myocardial function in infarcted heart through Bcl-2 expression. *J Pharmacol Exp Ther,* Vol.329, No.2, pp. 543-550, ISSN 1521-0103

Xie, X.J.; Wang, J.A.; Cao, J. & Zhang, X. (2006). Differentiation of bone marrow mesenchymal stem cells induced by myocardial medium under hypoxic conditions. *Acta Pharmacol Sin,* Vol.27, No.9, pp. 1153-1158, ISSN 1671-4083

Xu, R.; Chen, J.; Cong, X.; Hu, S. & Chen, X. (2008). Lovastatin protects mesenchymal stem cells against hypoxia- and serum deprivation-induced apoptosis by activation of PI3K/Akt and ERK1/2. *J Cell Biochem,* Vol.103, No.1, pp. 256-269, ISSN 0730-2312

Yamaguchi, J.; Kusano, K.F.; Masuo, O.; Kawamoto, A.; Silver, M.; Murasawa, S.; Bosch-Marce, M.; Masuda, H.; Losordo, D.W.; Isner, J.M. & Asahara, T. (2003). Stromal cell-derived factor-1 effects on ex vivo expanded endothelial progenitor cell recruitment for ischemic neovascularization. *Circulation,* Vol.107, No.9, pp. 1322-1328, ISSN 1524-4539

Yao, Y.; Zhang, F.; Wang, L.; Zhang, G.; Wang, Z.; Chen, J. & Gao, X. (2009). Lipopolysaccharide preconditioning enhances the efficacy of mesenchymal stem cells transplantation in a rat model of acute myocardial infarction. *J Biomed Sci,* Vol.16, pp. 74, ISSN 1423-0127

Yellon, D.M. & Downey, J.M. (2003). Preconditioning the myocardium: from cellular physiology to clinical cardiology. *Physiol Rev.,* Vol.83, No.4, pp. 1113-1151, ISSN 0031-9333

Zaugg, M.; Lucchinetti, E.; Uecker, M.; Pasch, T. & Schaub, M.C. (2003). Anaesthetics and cardiac preconditioning. Part I. Signalling and cytoprotective mechanisms. *Br J Anaesth,* Vol.91, No.4, pp. 551-565, ISSN 0007-0912

Zemani, F.; Benisvy, D.; Galy-Fauroux, I.; Lokajczyk, A.; Colliec-Jouault, S.; Uzan, G.; Fischer, A.M. & Boisson-Vidal, C. (2005). Low-molecular-weight fucoidan enhances the proangiogenic phenotype of endothelial progenitor cells. *Biochem Pharmacol,* Vol.70, No.8, pp. 1167-1175, ISSN 0006-2952

Zhang, K.; McQuibban, G.A.; Silva, C.; Butler, G.S.; Johnston, J.B.; Holden, J.; Clark-Lewis, I.; Overall, C.M. & Power, C. (2003). HIV-induced metalloproteinase processing of the chemokine stromal cell derived factor-1 causes neurodegeneration. *Nat Neurosci,* Vol.6, No.10, pp. 1064-1071, ISSN 1097-6256

Zhang, M.; Mal, N.; Kiedrowski, M.; Chacko, M.; Askari, A.T.; Popovic, Z.B.; Koc, O.N. & Penn, M.S. (2007). SDF-1 expression by mesenchymal stem cells results in trophic support of cardiac myocytes after myocardial infarction. *FASEB J,* Vol.21, No.12, pp. 3197-3207, ISSN 1530-6860

6

Degeneration and Regeneration in the Vertebrate Retina

Gabriele Colozza, André Mazabraud and Muriel Perron
Laboratory of Neurobiology & Development, UPR CNRS 3294,
Université Paris-Sud Orsay,
France

1. Introduction

The human retina is a complex, layered tissue responsible for the perception of the visual stimuli coming from the external environment. Since the visual inputs account for about 30% of our sensory stimulations, it is not surprising that partial or complete blindness results in a strong decrease of life quality. Several diseases affect the retina, often leading to degeneration of one or several cell types. The damage induced by these diseases is often irreversible, thus leading to a permanent loss of the visual ability.

Over the last decades significant progress has been made to elucidate the molecular basis of retinal degenerative diseases. This knowledge is necessary in order to design valid approaches for the treatment of retinopathies. However, the extreme genetic heterogeneity of these diseases, combined with the complex molecular nature of the underlying degenerative processes, has hampered the design of definitive therapies. Indeed, pharmacological treatments are only partially effective and tend to treat the effect of the disease, not the underlying cause. Furthermore, despite the recent technical advances, drug delivery into the retina may reveal challenging (reviewed in Yasukawa et al., 2011). On the other hand, gene therapy reveals as a promising alternative approach for inherited retinopathies. However, although significant progress has been achieved in treating recessive forms of retinopathies, dominant or multi-factorial forms are more difficult to treat, revealing the current limits of gene-based therapies (Farrar et al., 2010).

Thus, other strategies are needed to treat these diseases. One interesting approach, which will not be covered in this review, is the stem cell-based therapy, where stem cells can be differentiated *in vitro* and then transplanted to replace dead cells (for a review see Bull and Martin, 2011; Dahlmann-Noor et al., 2010; Locker et al., 2010). Another promising strategy is to promote the regeneration *in vivo* of the tissues (or organs) affected by the disease. Although the field of regenerative medicine is relatively young, the continuous findings on the mechanisms regulating regeneration in model organisms is improving our knowledge on the regenerative phenomena, and in some cases this information can be used to induce or increase regeneration in particular contexts (reviewed in Stoick-Cooper et al., 2007).

In the first part of this review, we will introduce some of the most common retina degenerative diseases that affect people worldwide, namely retinitis pigmentosa (RP) and age-related macular degeneration (AMD), providing the reader with a general description of these retinopathies. In the second part, we will discuss some of the animal models used to mimic particular features of retinal diseases, useful to understand the molecular mechanisms underlying retina degeneration. The information gathered from these models will be instrumental for the design of accurate and effective therapies. In the final part of this chapter we will present animal models, like transgenic frogs or fishes, capable of regenerating their damaged retinas. The use of these models is generating important and promising knowledge on the molecular pathways regulating neural repair, which could one day be used to stimulate retinal regeneration in humans. Thus, we will also review the cellular sources and molecular components playing a fundamental role in *in vivo* regeneration. Recent studies, aimed at stimulating regeneration processes *in vivo* by genetic or pharmacological manipulation of signaling pathways, will also be covered in this part.

2. Retinal diseases

Retinopathy is a generic term regarding non-inflammatory diseases affecting the retina. The damage usually affects severely the vision ability, especially when the macula (the region of human retina where cone photoreceptors concentrate) or the optic nerve are concerned. Retinopathies can arise as a consequence of other disorders, for example diabetes, or as a consequence of aging. The degeneration process is generally progressive, resulting in severe sight impairment in older patients. Furthermore, although the pathology affects primarily a given cell type, secondary events can arise extending the degeneration to other cell types. In some cases, genetic mutations are the underlying causes of the pathology, which can manifest alone or as a part of a syndromic disease. Due to their prevalence and to the highly debilitating loss of sight they cause, RP and AMD gained much attention in medical and clinical research.

2.1 Retinitis pigmentosa

RP is a group of inherited genetic diseases with a worldwide prevalence of 1 on 4000 individuals in the industrialized world. RP was first described around 1855 by the Dutch ophthalmologist Franciscus C. Donders, while observing pigmented speckles in a patient's degenerating retina. In the most typical form of RP, rod photoreceptors undergo a gradual degeneration, eventually leading to their death which occurs either by apoptotic or non-apoptotic mechanisms (reviewed in Berson, 1993; Chang et al., 1993; Hartong et al., 2006; Sancho-Pelluz et al., 2008). As a result, the loss of rods compromises the peripheral and night vision, as evidenced by the restriction of visual field (tunnel vision) and night blindness in RP patients.

As the disease progresses, the cones residing in the macula are also affected, degenerating after the rod photoreceptor death. As a consequence, other symptoms are manifested like color-blindness and loss of the central vision, resulting in total visual impairment (Berson, 1993; Hartong et al., 2006). The onset of RP symptoms is variable: some patients develop visual impairments during childhood, while others remain asymptomatic until mid-adulthood. There is no treatment to cure patients with RP, who usually become legally blind by the 40th year of age, because of a severe constriction of the visual field. Retinas of patients affected by RP undergo dramatic changes, as can be observed by fundus examination. Besides accumulation of pigmented deposits, retina atrophy, attenuation of retinal vessels and reactive gliosis are also associated with RP (reviewed in Hamel, 2006).

The pattern of inheritance of RP is complex and depends on the gene being affected. Generally, RP presents as a monogenic disease that can be inherited as autosomal dominant (24% of the cases), recessive (41%), or X-linked (22%). However, the remaining of the cases are considered to result from non-Mendelian inheritance or environmental factors (reviewed in Wright et al., 2010). So far, more than 45 genes have been shown to be involved in RP (http://www.sph.uth.tmc.edu/retnet/), which highlights the high genetic heterogeneity of the disease. Nonetheless, the significance of gene mutations in RP, as well as the molecular mechanisms leading to photoreceptors degeneration, is still poorly understood.

2.2 Age-related macular degeneration

AMD is a group of conditions that affect the central retina, particularly the macula. AMD represents the leading cause of vision loss in individuals above the 60th year of age, with a prevalence of 30% among people older than 75. Among these patients, about 10% show signs of advanced AMD, which is characterized by an extensive deterioration of central vision, with a tremendous impact on normal daily activities such as reading or deambulating (reviewed in Prasad et al., 2010). Age is one of the risk factors for this disease. It is estimated that, since the life expectancy is increasing, with an escalation of people reaching the age of 70 or greater, the prevalence of AMD in 2020 will double (for reviews, see Liutkeviciene et al., 2010; Prasad et al., 2010). Besides age, other factors such as oxidative stress and genetics are known to play a major role in the onset of the disease. However, it is important to note that the precise causes of AMD remain still unclear. Although AMD affects mainly elderly people, young may also be affected in the context of genetic (monogenic) disorders characterized by early onset, affecting people younger than 40 (reviewed in Rattner and Nathans, 2006).

Initially, AMD is characterized by the presence of large and poorly demarcated "drusen", aggregates of lipids and proteins which accumulate between the retinal pigmented epithelium (RPE) and the Bruch's membrane. RPE underlying these drusen displays some abnormalities such as hypo-pigmentation. Although at this stage there is no significant impairment on visual acuity, there is a high risk of progression to more severe forms of AMD, or late AMD. Late AMD is characterized by localized degeneration of RPE (geographic atrophy or dry - non exudative – AMD). Eventually, new blood vessels from the choroid above the RPE can grow into the macula, leading to the so-called exudative (or wet) AMD. These blood vessels are very fragile and may result in local bleeding, leading to blurry vision and visual distortion. Ultimately, these lesions can form a fibrous scar, which typically result in irreversible loss of central vision (Prasad et al., 2010; Rattner and Nathans, 2006). It is important to note that, although RPE cells are primarily affected by AMD, it is the subsequent loss of photoreceptors that accounts for vision loss.

As stated above, genetics play also a role in macular degeneration. In particular, some inherited forms are known, characterized by early (or juvenile) onset. For example, Best Vitelliform Macular Degeneration (BVMD) and Stargardt-like disease are both dominant inherited juvenile forms of macular degeneration (reviewed in Vasireddy et al., 2010; Xiao et al., 2010). These diseases are characterized by extensive accumulation of lipofuscin, yellow granules containing lipids, atrophy of the RPE and loss of the central vision, all clinical features shared by several forms of age-related macular degeneration. However, clinical profiles of inherited macular degeneration may vary among patients, indicating the intervention of other factors in modelling the disease. One of such factors could be oxidative stress. Interestingly, susceptibility to the onset of macular degeneration is increased by oxidative damage that leads to cumulative mutations in mitochondrial DNA of retina cells (reviewed in Jarrett et al., 2010).

3. Animal models for retinopathies

Because animals develop retinal diseases with similar traits as in humans, they are commonly used to improve our understanding of the molecular pathogenesis of hereditary retinal degeneration. Animal models can also serve to screen compounds for therapeutic use. Different species such as mice, rats, dogs, cats and even pigs have contributed to the study of retinal degeneration (reviewed in Fletcher et al., 2011; Rivas and Vecino, 2009). Describing all these different models is beyond the aims of this review. Thus, in this section we will focus on some models mimicking retina degeneration, which greatly contributed to the elucidation of molecular pathways underlying the different retinopathies.

3.1 Mouse models for retina degeneration

Natural mutant mice or knock-out animals allowed to identify several important genetic factors involved in retina degeneration. Importantly, these genes share analogous mutations in human. In addition, the effects on retinal structure and function of gene mutations associated with retinal diseases can also be tested by means of transgenesis. In many cases the proteins encoded by these genetic factors were shown to be sensible targets for pharmacological treatments or gene therapy, giving important results in terms of clinical and medical investigation. Below, we will give a brief overview of the different mouse models used to mimic RP or AMD, describing the degenerative mechanisms they contributed to elucidate (see also **Table 1**).

3.1.1 Mouse models for RP: Mutations in rhodopsin

The rod visual pigment rhodopsin, a G-protein coupled receptor that initiates the visual transduction cascade, was the first protein found to be mutated in RP. Rhodopsin mutations are responsible for 30–40% of autosomal dominant forms of RP (reviewed in Fletcher et al., 2011; Wang et al., 2005). One of the most common mutation in human rhodopsin is the substitution of proline for histidine in position 23 (P23H) (Dryja et al., 1990). Mice expressing such mutated rhodopsin develop significant photoreceptor degeneration, with a consequent decrease in visual function (Olsson et al., 1992). Importantly, photoreceptor degeneration was also observed in control transgenic mice overexpressing wild type rhodopsin. Together, these data suggested that the expression levels of either normal or aberrant rhodopsin may play a role in triggering cell death (Olsson et al., 1992). Based on cell culture and transgenic mice, it was shown that most rhodopsin mutants interfere with the folding, the stability and/or the transport to the membrane of rhodopsin protein (Kaushal et al., 1994; Roof et al., 1994; Wang et al., 2005). However, since different results are obtained in various animal models, the exact mechanisms underlying rhodopsin-dependant photoreceptor cell degeneration is still not well understood and remains controversial. Therefore, generating additional, more appropriate, animal models is still an issue. Along this line, a knock-in mice model expressing P23H rhodopsin was recently reported and highlighted defective glycosilation and rapid degradation of the mutant protein, in addition to disorganization of the rod outer segments, which may account for photoreceptor degeneration (Sakami et al., 2011).

3.1.2 Mouse models for RP: Mutations affecting phototransduction pathway

Among the genes involved in phototransduction cascade, *phosphodiesterase 6* (*PDE6*) genes are the most studied. Phosphodiesterase 6 is a hetero-tetrameric enzyme consisting of one α, one ß and two γ subunits, involved in cGMP (cyclic guanosine monophosphate) breakdown. Mutations in any of the three subunits give rise to recessive forms of RP. Most of the knowledge on the molecular mechanisms concerning this specific form of RP comes from studies on one mouse strain, known as *Retinal degeneration 1* (*rd1*), where the *PDE6B* gene, encoding the ß subunit, is mutated. The corresponding mutation in humans accounts for about 5% of the recessive forms of RP (http://www.sph.uth.tmc.edu/retnet/). These animals are natural loss of function mutants closely mimicking the human disorder. In these mice, non-functional PDE6B protein leads to accumulation of high levels of cGMP in the rods (Paquet-Durand et al., 2009). Thus, the sustained levels of cations contribute to constantly depolarize rod photoreceptors, eventually causing these cells to die. More recently, it has been shown that cyclic nucleotide-gated ion channels (CNGs) play a pivotal role in the PDE6B-dependent cell death mechanism (Paquet-Durand et al., 2011). These channels allow the influx of Na^+ and Ca^{2+} inside the cells in a cGMP-dependent manner, and increased levels of cGMP are correlated with excessive intracellular concentration of Ca^{2+} ions. Indeed, as a confirmation of these results, knock-out mice for *Cngb1* gene (lacking rod CNGs) in a *rd1* mutant background, show a substantial rescue of rod and cone cell-morphology and function, compared to single *rd1* mutants. These double mutant mice show also an increased rate of photoreceptor survival compared to the single mutant *rd1* (Paquet-Durand et al., 2011). Thus, the results reported here argue for a major role of CNGs and calcium in triggering rod degeneration.

3.1.3 Mouse models for RP: Mutations affecting photoreceptor structural proteins

Photoreceptor outer segments are continuously renewed to replace proteins and lipids that are damaged by light and highly oxidative environment. Maintenance of the integrity and renewal of the outer segment disk membranes is thus critical for correct photoreceptor function. Peripherin is an integral membrane protein that localizes to the rim of photoreceptor outer segments and was proposed to act as a structural element for the outer segment membranes (Connell and Molday, 1990; Molday et al., 1987). Mutations in *Peripherin* (*PRPH*) are a common cause for autosomal dominant RP, accounting for about 5% of the cases (reviewed in Boon et al., 2008). Retina degeneration slow (rds) mice are naturally occurring homozygous mutants, carrying a mutation in the *Peripherin* gene (Connell et al., 1991; Travis et al., 1989). These mice are characterized by photoreceptor degeneration and abnormal morphological features in the outer segment discs. Transgenic mice, carrying an aminoacid substitution (P216L) in the Peripherin protein were generated (Kedzierski et al., 1997). These mice exhibited photoreceptor loss proportional to the level of transgene expression. Importantly, photoreceptor outer segments displayed abnormal shortening and central lacunae, confirming the hypothesized role for Peripherin in the stabilization of outer segment architecture and thereby in the viability of photoreceptors. The underlying mechanisms remain however to be further explored for the future development of therapeutic strategies.

3.1.4 Mouse models for RP: Mutations affecting ciliary proteins

The photoreceptor connecting cilium, a structure that connects the inner (having the metabolic machinery) with the outer segment (containing phototransduction proteins), mediates trafficking of factors involved in phototransduction. It is therefore not surprising that mutations in photoreceptor cilium proteins have been reported to cause RP, and in particular X-linked RP, which is the most severe form in terms of age onset and progression (reviewed in Hosch et al., 2011; Murga-Zamalloa et al., 2009). Mutations in *retinitis pigmentosa GTPase regulator* (*RPGR*) and *retinitis pigmentosa GTPase regulator interacting protein* (*RPGRIP*) genes account for about 70% of X-linked RP (Vervoort et al., 2000). RPGR protein localizes to the connecting cilium, and mice knock-out for this protein exhibit progressive photoreceptor degeneration. Notably, basal disc membranes were strongly disorganized in these mice, and opsin mislocalization or reduction was observed (Hong et al., 2000). Even though RPGR is a ciliary protein, the cilium structure seems to be well preserved in knock-out mice. This may indicate that RPGR regulates activities like protein transport along the cilium, more than its biogenesis/maintenance. Consistent with this, it was recently reported that RPGR interacts with RAB8A, which is a critical GTPase involved in photoreceptor protein trafficking, and that this interaction may account for photoreceptor degeneration in patients with RPGR mutations (Murga-Zamalloa et al., 2010). *RPGRIP* knock-out mice also show photoreceptor degeneration and outer segment discs appear oversized and disorganized (Zhao et al., 2003). Importantly, in these mice RPGR protein appears mislocalized, indicating that RPGRIP is involved in tethering RPGR to the connecting cilium. RP1 is another ciliary factor, a photoreceptor-specific microtubule-associated protein, whose mutation gives rise to common forms of autosomal dominant RP (5-10% of the cases). Mice expressing C-terminal truncation of this protein (resembling human disease-associated mutations) were created (Liu et al., 2003). These mice developed significant photoreceptor degeneration. Analysis of the phenotype indicated that RP1 could play a fundamental role in the organization of the outer segment discs. Given the important phenotype heterogeneity observed in patients with cilium associated RP, it is critical to further characterize animal models carrying mutations in cilium proteins to better understand the pathogenesis of these RP diseases.

3.1.5 Mouse models for age-related macular degeneration

Due to its complex genetic components and to the role played by environmental factors, AMD is more difficult to be modelled in mice. Furthermore, mice do not possess a well-defined macula, which is particularly affected in human AMD. Notwithstanding, it is possible to mimic some forms of AMD, as well as juvenile inherited forms of macular degeneration, using knock-out or transgenic mouse models (reviewed in Elizabeth Rakoczy et al., 2006; Fletcher et al., 2011). Stargardt's disease is a juvenile inherited form of macular degeneration, where defects in lipid biosynthesis and transport occur (reviewed in Molday and Zhang, 2010). *ABCR* (a member of ABC transporter family) null mice are used to mimic recessive form of Stargardt's disease while transgenic mice expressing the human mutant form of *ELOVL4* (an enzyme that catalyzes the elongation of very long-chain fatty acids) mimic the autosomal dominant Stargardt-like disease (Karan et al., 2005; Weng et al., 1999). In both the mutant mice, accumulation of lipofuscin granules in RPE is observed and is correlated with photoreceptor loss. Indeed, lipofuscin accumulation is one of the hallmarks of several forms of macular degeneration. The studies of these mice allowed to hypothesize that accumulation of lipofuscin (and its main component, the fluorophore A2E) into RPE cells would result in toxicity and death of these cells, with deleterious secondary consequences on photoreceptors.

Knock-in mice expressing a mutated form (W93C) of Bestrophin anion channel revealed to be a good model of the inherited form of macular degeneration, BVMD (Best vitelliform macular dystrophy, Zhang et al., 2010). The aminoacid substitution present in this mutated protein exerts a dominant negative effect on the channel activity. Indeed, these mice developed several features typical of BVMD, like retinal detachment in central region of the eye, accumulation of debris like shed outer segment, lipofuscin and lipid droplets. Furthermore, calcium signalling was suppressed in the transgenic mice. Such findings suggest that calcium may play a pivotal role in the pathogenesis of BVMD. Furthermore, since BVMD is the second cause of juvenile inherited macular degeneration (following Stargardt's disease), their use is thought to generate important data for therapeutical approaches. Other models, like mice lacking monocyte chemoattractant protein-1 (Ccl-2) or its cognate C-C chemokine receptor-2 (Ccr-2) also develop characteristic features of AMD, like drusen, lipofuscin accumulation and choroidal neo-vascularization (CNV), as well as photoreceptor degeneration (Ambati et al., 2003). Importantly, choroidal macrophage recruitment was impaired in Ccl-2$^{-/-}$ and Ccr-2$^{-/-}$, leading to accumulation of complement C5a and IgG proteins. This accumulation, in turn, leads to overexpression of vascular endothelial growth factor (VEGF) by RPE, likely mediating the CNV phenotype. Other interesting factors have been shown to play a role in RPE degeneration and development of particular forms of AMD. Mice lacking functional Dicer1 in the RPE, for example, develop

Mouse models for RP

Affected proteins	Affected genes	Inheritance	Phenotype
Rod visual pigment	*Rhodopsin*	Autosomal dominant	Defects in Rhodopsin sorting, glycosilation and/or folding, resulting in cell toxicity
Phototransduction proteins	*PDE6B*	Recessive	High cGMP and calcium levels, determining cell death
Structural proteins	*PRPH*	Autosomal dominant	Rod outer discs abnormalities, leading to photoreceptor death
Ciliary proteins	*RPGR, RPGRIP* *RP1*	X-linked Autosomal dominant	Defects in disc membrane organization and protein sorting/localization, leading to cell death

Mouse models for AMD

Disease category	Affected genes	Phenotype
Juvenile inherited forms of macular degeneration	*ABCR, ELOVL4, Bestrophin*	Lipofuscin accumulation, resulting in RPE degeneration
Dry AMD (geographic atrophy)	*Dicer1*	Accumulation of Alu RNAs, which results in RPE cell dismorphology and death
AMD (general features)	*Ccl-2, Ccr-2*	Accumulation of drusen and lipofuscin, choroidal neo-vascularization, leading to cell death

Table 1. Main mouse models for RP or AMD, discussed in the present manuscript.

RPE dysmorphology similar to those observed in humans affected by geographic atrophy (or dry AMD) and RPE degeneration (Kaneko et al., 2011). Importantly, Dicer1 depletion in RPE results in increased levels of Alu RNAs, which are responsible for the cell death. Thus, a variety of different mechanisms appear to be involved in different steps, or forms, of AMD.

3.1.6 Mouse models and therapies for retina degeneration

The plethora of data obtained from the studies on mouse models has provided potential targets of pharmaceutical interest. Some Ca^{2+} channel blockers were found to efficiently protect photoreceptor of *rd1* mice from degeneration. For example D-cis-diltiazem, a cGMP-gated channel blocker used to treat cardiac disfunctions, was shown to slow down photoreceptor degeneration and to preserve visual function in *rd1* mice (Frasson et al., 1999). More recently, similar observations were extended to other calcium channel blockers, able to preserve photoreceptor viability in mouse models for retinal degeneration (Takano et al., 2004). These results confirm cGMP-gated channels as important pharmacological targets for treating some forms of RP, like *PDE6B* recessive RP. Other factors play an important role in preserving photoreceptor viability. For example, cytotoxicity in P23H mutant mice was exacerbated by lack of 11-*cis*-retinal chromophore, thus indicating an important role for this molecule in protecting from P23H-induced cell death(Sakami et al., 2011). Indeed, pharmacological replacement of 11-*cis*-retinal (as well as other retinoids) showed beneficial effects in the treatment of several forms of retina degeneration (Palczewski, 2010). Proteins may be also used to treat retina degeneration. It was shown that injection of rod-derived cone viability factor (RdCVF), a protein secreted by rod cells, is able to protect cone cells from death in *rd1* mice (Leveillard et al., 2004). This observation is particularly important, since cones affection in late stages of RP is responsible for legal blindness. Thus, due to its ability to maintain cones viable and to preserve their function, future therapeutic strategies in humans should consider the possibility to deliver RdCVF protein in the patients' retina, or to promote its expression from unaffected cells (Leveillard and Sahel, 2010). Besides RdCVF, systemic administration of Insulin in *rd1* mice increased the rate of cone survival (Punzo et al., 2009). It is evident that different therapeutic strategies are becoming available to promote photoreceptor maintenance.

However, drug or molecule treatment is not the only strategy that can be used for this aim. Gene therapy, for example, showed promising results in treating some forms of RP in mouse models. Indeed, adeno-associated virus serotype 5 (AAV5) has been recently used to deliver mouse rhodopsin in P23H transgenic mice (Mao et al., 2011). These mice (which mimic autosomal dominant RP) showed significant rescue of photoreceptor function (assessed by electroretinogram recording) as well as photoreceptor viability. Although it is still not known how wild type rhodopsin rescues the dominant effect of mutated rhodopsin, this result is encouraging in the view of gene therapy treatment of human RP. Several clinical trials have shown to be effective in improving visual function in patients affected by Leber's congenital amaurosis (a recessive retinal degeneration) (Bainbridge et al., 2008; Simonelli et al., 2010). Thus, delivery of functional rhodopsin could represent a valid option for treating RP. However, since high rhodopsin levels can result in toxicity for rod cells, much work is still required to optimize these therapies in humans. Recently, adeno-associated vectors were used to deliver human RPGRIP1 gene in mouse models of Leber's congenital amaurosis (Pawlyk et al., 2010). Expression of RPGRIP1 was able to improve rod and cone survival in the retina of these mice.

Different approaches have been designed to treat also AMD. Importantly, many of these approaches have been designed to target wet (exudative) AMD (reviewed in Querques et al., 2011). As we have seen, one of the hallmarks of wet AMD is the process of neovascularization, which is correlated to severe visual loss. To this regard, several strategies aim to block the process of blood vessel formation, to minimize the effects of macular degeneration. Indeed, antibodies against the angiogenic factor VEGF, like Ranibizumab, have demonstrated successful in blocking VEGF-dependent angiogenesis in patients affected by AMD, improving their visual functions (for a review, see Lien and Lowman, 2008). Not all the antibodies against VEGF are able to efficiently penetrate the retina, and some of them show important side-effects. Thus, AAV-assisted gene therapy seems to be a valid alternative to this approach. In particular, AAV2-mediated intravitreal delivery of chimeric forms of VEGF receptor Flt-1 showed promising results in mouse models (Pechan et al., 2009). In these oxygen-induced retinopathy models, the AAV-delivered chimeric proteins were able to inhibit efficiently angiogenesis, showing no detectable signs of toxicity. Thus, these results show promising candidate molecules for the treatment of human AMD by gene therapy.

Despite the abundance of strategies to target wet AMD, no therapies are available for treating dry or atrophic AMD (Querques et al., 2011). However, a recent study shows a potential and important therapeutic target. As seen above, Alu RNAs were shown to induce cytotoxicity in murine RPE cells, which developed a degeneration phenotype similar to atrophic AMD (Kaneko et al., 2011). Notably, subretinal administration of antisense oligonucleotides, which targeted and reduced Alu RNAs, inhibited efficiently the degeneration phenotype *in vivo* (Kaneko et al., 2011). Thus, this study reveals a promising target for blocking RPE degeneration typical of AMD. Inherited forms of macular degeneration are also current targets of therapeutic treatment. For example, administration of vitamin A to murine models of Stargardt's disease was shown to efficiently decrease the accumulation of lipofuscin, and to improve visual function (Ma et al., 2010). In a different approach, lentiviral-mediated delivery of human *ABCA4* gene in *Abca4*$^{-/-}$ mice showed effective reduction in the accumulation of the lipofuscin pigment A2E (Kong et al., 2008). Thus, gene therapy could also be used to treat genetic forms of macular degeneration.

3.2 Alternative models for retina degeneration

Besides the mouse models reported above, other mammalian animal models are used to get insights into the pathogenesis of inherited retina degeneration, such as rat, dog or pig. However, because the development of the retina and its morphology are remarkably conserved in all vertebrates, non-mammalian species have also emerged during the last decade as alternative valuable models. In particular, zebrafish, which is well suited for large-scale genetic screens, revealed to be very informative in the elucidation of the processes underlying retina degeneration (reviewed in Bibliowicz et al., 2011; Brockerhoff and Fadool, 2011; Li et al., 2010). Its importance is strengthened by the fact that not all the murine models are suitable to mimic a particular pathological feature or for the development of pharmacological therapies. In zebrafish, a dozen different genetic lesions have been identified leading to photoreceptor cell death, belonging to different functional classes. For example, mutation in the gene encoding the cone specific phosphodiesterase c 6 (pde6c) was shown to induce cone cell death, similarly to *pde6b*-induced rod cell death in mouse models (Stearns et al., 2007). Importantly, some rod cells also died in these mutant fish, primarily in areas where rod density was low. This result suggests an important correlation between secondary cell death and cell density. Another recent study taking advantage of zebrafish as a model system provided novel insights into the pathogenic potential of disease-associated mutations in RPGR (Ghosh et al., 2010).

The frog *Xenopus laevis* also recently revealed important in deciphering the mechanisms underlying photoreceptor degeneration (Tam and Moritz, 2006; Tam and Moritz, 2007; Tam and Moritz, 2009; Tam et al., 2006; Zhang et al., 2008). For example, transgenic frogs expressing different glycosylation-deficient forms of rhodopsin showed rod degeneration when rhodopsin glycosylation was inhibited in position N15 (Tam and Moritz, 2009). Strikingly, the photoreceptor degeneration was exacerbated when the frogs were exposed to light, while was attenuated in dark-reared animals. This result suggests that glycosylation is involved in rhodopsin stabilization, particularly after its activation by light. Studies on *Xenopus* also showed that the localization of rhodopsin is important for photoreceptor viability. In fact, in transgenic frogs expressing a rhodopsin bearing C-terminal truncation (Q350Ter), mutant rhodopsin (but not wild type endogenous rhodopsin) failed to correctly localize and induced a severe form of photoreceptor degeneration (Tam et al., 2006). Importantly, in this case degeneration occurred even in the dark, indicating that the toxic effects of mislocalized rhodopsin were independent from its activation. Additionally, since *Xenopus* retina contains numerous cone cells (differently from mice) this system would allow to examine more easily the secondary effects of rod death on cone viability.

4. Studying retina regeneration *in vivo*

The animal models described so far have greatly contributed to the understanding of mechanistic processes of retina degeneration. Furthermore, as we have seen, this knowledge has revealed instrumental in the excogitation of valid medical approaches. However, even though the available therapies lead to amelioration of the degeneration, these treatments cannot compensate for the cells that have been already lost, often limiting their efficacy to the early steps of the pathology. In order to compensate for this limitation, regenerative medicine aims at developing treatments that improve, or even induce, regeneration of lost cells, a process that involves formation of newly born cells, which then differentiate to replace dead cells.

In order to identify mechanisms and molecules that mediate retina regeneration, it is opportune to work with suitable animal models. The mouse models reported in the last section are not appropriate for deep examination of regenerative processes. One reason is that higher vertebrates show little regenerative potential. Moreover, the transgenic, knock-out or natural mutant models presented above do not offer the possibility to control experimentally the duration of degeneration. Such a requisite is essential to observe the mechanisms acting during regeneration: first, one would induce degeneration of a particular retinal cell type; then, by blocking the degenerative processes, it would be possible to allow the regeneration to occur. This system would offer the possibility to study genes and pathways regulating regeneration.

Thus, the first problematic is the choice of the animal model. Amphibians such as frogs and salamanders, and fishes show a strong regenerative potential, maintained for a great part of their life. Importantly, it seems that part of the regenerative program active in lower vertebrates is still present in higher vertebrates like birds and mammalian, which show a limited and short regenerative response upon injuries (reviewed in Karl and Reh, 2010).

Thus, the complete understanding of the regenerative program in fish and frogs would provide a basis in designing effective strategies able to stimulate or enhance this process in humans affected by degenerative retinopathies.
The second problematic is how to induce and to reverse the process of retina degeneration. Several methods can be employed to this aim: for example, it is possible to induce the death of photoreceptor (both rods and cones) by exposing dark-adapted animals to strong white light (Reme et al., 1995; Vihtelic and Hyde, 2000), or by using laser to mediate cell ablation (Wu et al., 2001). Small molecules can also be used to destroy retina cell types: mature photoreceptors are particular sensitive to N-Methyl-N-nitrosourea (MNU) (Tsubura et al., 2011), while exposure to high level of glutamate can induce death of retinal ganglion cells (RGCs), a finding that has been linked to several cases of glaucoma (Dreyer et al., 1996). Neurotoxins like ouabain can also be used to damage RGCs and cells of the inner nuclear layer (Fimbel et al., 2007; Sherpa et al., 2008). Logically, removing the drug or switching off the light source would block the degenerative event, allowing the system to recover and to regenerate. Finally, surgical methods can also be employed to remove small part of the retina, for example by cutting a portion of retina with a surgical blade or by pricking it with a needle (Cameron, 2000; Fausett and Goldman, 2006; Goldsmith and Harris, 2003; Vergara and Del Rio-Tsonis, 2009; Yoshii et al., 2007).
Although these strategies do not resemble generally a particular pathway occurring during human retina degenerative diseases, they offer the important advantage of controlling the timing and duration of degenerative treatment. However, the systems here described present also a limitation: for example light can be used to kill only photoreceptors, while chemicals or physical damage may lack enough specificity, affecting more than a particular cell type. Thus, in the recent years more elegant and fine tuned systems have been developed, like transgenic animals expressing an inducible and reversible cytotoxic gene under control of a retina specific promoter. Below, we will give examples of transgenic animal models which are thought to bring great contribution to the study of regeneration.

4.1 Transgenic models for regeneration studies

Transgenic animals offer the unique possibility to selectively induce cell death (provided that a cell or tissue specific promoter is available) through the use of a cytotoxic gene whose activity is inducible and reversible. One powerful and elegant system to induce specific cell ablation in fish and frogs, in an inducible and reversible fashion, is the nitroreductase/metronidazole (NTR/mtz) system **(Figure 1)**. Bacterial NTR is a flavoprotein enzyme able to reduce nitro group-containing substrates. When a non-toxic molecule like mtz (prodrug), is reduced by NTR, it is converted into a powerful DNA cross-linker agent leading to cell death by apoptotic mechanisms. Thus, mtz can be added into the water of animals expressing NTR under the regulation of a cell-specific promoter to induce cell death only in those cells expressing the enzyme. Cells not expressing NTR are not directly affected by the prodrug mtz, validating the high specificity of the system. Subsequently, mtz can be removed by washing the transgenic animals in mtz-free water, allowing the study of the cellular and molecular processes of regeneration (Curado et al., 2007; Curado et al., 2008; Davison et al., 2007; Pisharath, 2007; Pisharath et al., 2007).
In the context of the retina, NTR/mtz system has been used to induce cell death successfully in different retina cells of zebrafish and *Xenopus laevis*, using enhancer traps or specific promoters (Choi et al., 2011; Montgomery et al., 2010; Zhao et al., 2009). In particular,

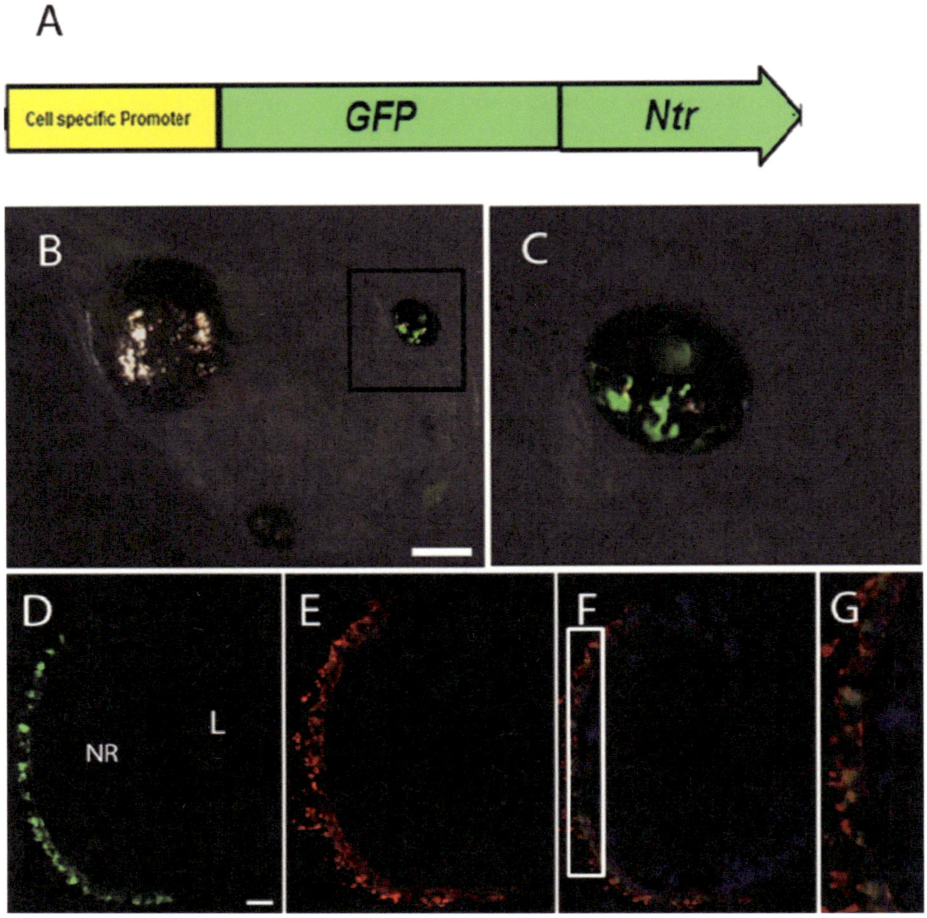

A: Schematic diagram representing a chimeric GFP-tagged NTR under the control of a specific promoter. This construct is used to create transgenic frogs. B: Ventral view of a stage 41 transgenic tadpole carrying a GFP-NTR transgene under the control of the rod specific *Xop* promoter. C: Magnification of the onset shown in B, showing GFP fluorescence in the eye. D-F: Transversal retinal section of such a transgenic tadpole, following anti-GFP (green, D) and anti-rhodopsin (red, E) immunostaining. The merge image (F) confirms the rod specificity of expression. G: magnification of the onset shown in F. l = lens; nr = neural retina. Scale bar in B=2 mm, scale bar in D=20 μm.

Fig. 1. The NTR/mtz transgenic system in frogs.

transgenic *Xenopus* frogs were obtained, expressing NTR under the control of rhodopsin promoter (*Xop::NTR*). This model is of particular importance, since it mimics human RP. Indeed, when tadpoles at stage 50 were exposed to mtz, their rods started to die soon after the treatment. Furthermore, when mtz treatment was prolonged, cone degeneration was also observed following rod cell death, a characteristic of RP (Choi et al., 2011). Importantly, upon mtz removal, newly born rods were observed in a time-range of thirty days, indicating successful regeneration of death cells. Regeneration studies were conducted also in zebrafish retina, where rod or bipolar cells were targeted for ablation. Again, treatment with mtz induced specific ablation of NTR-expressing cells, while removal of the prodrug allowed regeneration of rods and bipolar interneurons in a short period (Montgomery et al., 2010; Zhao et al., 2009). Altogether, these data show that the NTR/mtz system is a powerful tool to induce cell-death in a specific and reversible fashion, which can be applied to animals with regenerative capacity to investigate the molecular pathways governing the phenomenon of regeneration.

Other cytotoxic genes can also be used to achieve retinal cell death. For example, an inducible form of caspase-9 (iCasp9) was expressed in *Xenopus* rod photoreceptors by using rhodopsin promoter (Hamm et al., 2009). This inducible caspase is activated by the drug AP20187, added to the tadpoles' medium, thus leading to death of rod cells. Similarly to mtz, AP20187 can be removed at any time, thus switching off the stimulus for rod death. These frogs developed abnormalities in the function of cone cells, assessed by electroretinogram. Importantly, cone functionality was restored 6 months after the treatment, confirming the reversibility of this model. However, differently from the *Xop::NTR Xenopus* model, no clear signs of cone apoptosis were observed, although the reasons of this difference are not known. In conclusion, different systems can be employed to induce death of specific cell types, enabling the study of their regeneration and the secondary consequences related to their death on neighbouring tissues.

4.2 Sources of regeneration

In order to understand the retina regenerative process, the first question to answer is which cells are able to set off the regeneration? Although mammals are devoid of a regenerative capacity, they possess cells with stemness features in the ciliary body, a structure derived from the optic cup, lying between the retina and the iris. Although these cells remain mitotically quiescent *in vivo*, they are able to clonally expand *in vitro* (Ahmad et al., 2000; Coles et al., 2004; Tropepe et al., 2000). Although the proliferative and differentiation potential of such ciliary body-derived cells towards the photoreceptor lineage is controversial (Cicero et al., 2009; reviewed in Locker et al., 2010), this raises the question whether and how it is possible to stimulate the proliferative and differentiation ability of these cells in *vivo*, to regenerate injured cells. Lower vertebrates on the other hand possess a region, adjacent to ciliary body, called ciliary marginal zone (CMZ), where active proliferating retinal stem and progenitor cells reside. These cells, capable of generating all types of retinal neuron and glial cells, contribute to the perpetually expanding retina of these animals, which grow throughout life (reviewed in Otteson and Hitchcock, 2003; Wetts and Fraser, 1988; Wetts et al., 1989). CMZ cells were shown to contribute to the replacement of retinal cells after injury (reviewed in Locker et al., 2009; Locker et al., 2010; Moshiri et al., 2004). However, it seems that the origin of regenerating cells can be species-specific and changes according to the developmental stage.

4.2.1 CMZ and RPE as a source of retina regeneration

If retina is surgically removed in urodele amphibians, like salamanders, RPE cells respond to injury by re-entering the cell cycle. These cells undergo a process of transdifferentiation, loosing features of differentiated RPE like their pigmentation, and form a new (inner) layer which will give rise to a complete retina, with normal lamination and apparently functional (Araki, 2007). In the anuran amphibian *Xenopus*, both stem cells of the CMZ and cells constituting the RPE are able to regenerate the injured retina. The contribution of CMZ in *Xenopus* retina regeneration was already known since 1982 (Mitashov and Maliovanova,

1982). When neural retina was removed, partial regeneration of neural retina was observed by proliferating stem cells normally present in the CMZ. However, it was also shown that, upon surgical retinectomy, cells of RPE delaminated and migrate toward the retinal vascular membrane (RVM), forming a new layer (Yoshii et al., 2007). The cells of this layer transdifferentiate, as assessed by expression of neural retina markers. Importantly, the pigmented cells were intermingled with other cells deriving from the CMZ, which also contributed to the newly formed neural retina. This study suggested also a role for RVM in the process of regeneration, which could be fundamental. Although the molecular nature of RVM contribution in regeneration remains still elusive, it could act likely through the secretion of particular growth factors into extracellular matrix. Indeed a possible role for this tissue in retina regeneration was observed already in 1987 in a study on the amphibian *Rana catesbienna*, where close contact between RVM and delaminated RPE was observed (Reh and Nagy, 1987). Its importance is also underlined by the fact that, when absent, the neural retina is not able to regenerate. When neural retina was surgically removed, together with the RVM, no regeneration occurred in tadpoles, unless specific growth factors were added to the operated eyes (Vergara and Del Rio-Tsonis, 2009). Further work is required in order to fully elucidate the modalities through which RVM act during regeneration. It is important to note, at this point, that the mechanisms of regeneration observed in amphibians are also present in higher vertebrates such as chicken, although this potential is limited to a short temporal window. The hatched chicken retina indeed also possesses a CMZ, although with a more limited potential compared to that of fish and amphibians (Fischer and Reh, 2000). Thus, both RPE and stem cells residing in the ciliary body/CMZ are able to contribute to retinal regeneration, until approximately embryonic day 4.5 (E4.5) (reviewed in Spence et al., 2007b).

4.2.2 Müller glia as a source of retina regeneration

RPE and CMZ are not the only origin of regeneration. Müller glia were also shown to display neurogenic capacity and to contribute to the regenerative process following retinal injury in diverse vertebrate retinas (Bernardos et al., 2007; Das et al., 2006; Fausett and Goldman, 2006; Fischer and Reh, 2001; Fischer and Reh, 2003; Monnin et al., 2007; Yurco and Cameron, 2005). For instance, upon photoreceptor cell death in transgenic zebrafish expressing NTR in rod cells, Müller cells were observed to re-enter the cell cycle. This yielded clusters of neural progenitor cells, which later on differentiated in rod cells (Montgomery et al., 2010). It is important to note, however, that Müller glia was mobilized only when the damage to rod cells was extensive: when mosaic transgenic lines, expressing NTR only in a small subset of rods, were treated with mtz to induce cell death, only specialized Müller glia-derived progenitors committed to rod lineage (known as rod precursor cells) were activated for regeneration. This result demonstrates that factors, like the extent of damage, can influence the source of regeneration. The role of Müller glia in regeneration has been observed in other studies. Upon light damage in zebrafish, Müller glia cells re-enter the cell cycle and de-differentiate, as judged by the loss of specific differentiation markers such as GFAP (glial fibrillary acidic protein) and Glutamine Synthetase. Successively, these cells start to express neural marker such as *Pax6*, and form neurogenic clusters that migrate toward the outer nuclear layer, where photoreceptors reside (Thummel et al., 2008a). In particular, ability of glial cells to re-enter the cell cycle is fundamental for the regeneration. Indeed, inhibition of Müller glia division, by knock-down of PCNA (proliferating cell nuclear antigen), inhibited proliferation and expression of neural markers such as *Pax6* (Thummel et al., 2008b). This resulted in failure to regenerate photoreceptors in zebrafish light-damaged retinas confirming that Müller glia, upon proliferation, can serve as a major source of retinal regeneration. Importantly, lineage tracing experiments indicate that Müller glia can regenerate all retinal cell types, demonstrating the high plasticity of these cells (Ramachandran et al., 2010b). However, the neurogenic potential of activated Müller glia seems to be limited, as particular mature retinal cell types are rarely or never regenerated following injury (Fischer and Reh, 2001; Ooto et al., 2004). Nonetheless, this can be experimentally overcome by treatment with specific soluble factors (Ooto et al., 2004; Osakada et al., 2007) raising the possibility that Müller glia could be stimulated to produce specific retinal cell types, provided exogenous supply of endogenously limiting factors.

What initiates the regenerative response of Müller glia? This question remains still largely unanswered, however a fascinating hypothesis points to its role in phagocytosis of dead photoreceptors. Interestingly, it was observed that Müller cells were able to engulf apoptotic bodies derived from light-damaged photoreceptors. Upon phagocytosis of dead cells, it was reported that Müller glia started to proliferate, as assessed by PCNA immunolabelling, and successively regenerated dead cells (Bailey et al., 2010). On the other hand, when phagocytosis was blocked by means of inhibitors, the process of regeneration failed to occur. In mammalian retina, Müller glia can be activated in response to injury, becoming hypertrophic. However, very few cells enter the cell cycle (Bringmann et al., 2009). Thus, understanding the signals that activate these cells toward a regeneration program is a necessary step to successfully induce retinal regeneration in mammals.

4.3 Candidate genes and signaling pathways involved in regeneration

In the previous section, we have discussed about the sources of regeneration in lower vertebrates, identifying the RPE, the CMZ and the Müller glia as the main generator of regeneration. What are the molecular signals that activate these cells in response to retina injury? To date, a network of different genes and signaling has been shown to act during regeneration. An important goal of regenerative medicine is to characterize the network in fine details, so that this knowledge can be used to promote retinal regeneration in patients affected by retinopathies.

4.3.1 Transcription factors promoting regeneration

Several transcription factors involved in early retinogenesis were shown to be essential for regeneration to occur. It has been recently reported in *Xenopus* tadpoles that the transcription factor *Rx1*, a paired like homeobox gene fundamental for early eye field specification, plays a central role in retinal regeneration (Martinez-De Luna et al., 2011). When retina was physically damaged, retinal progenitor cells were found to migrate near the site of wound and to initiate regeneration of the entire damaged retina. However, when *Rx1* was knocked-down by means of small hairpin RNA (shRNA), defects in morphology of progenitor cells were observed, together with a reduced expression of progenitor markers. Examples of transcription factors acting during Müller glia-dependant regeneration come from studies in zebrafish. Indeed, the proneural transcription factor Ascl1 is essential for the regeneration process in the fish (Ramachandran et al., 2010a). *Ascl1* mRNA is rapidly induced in dedifferentiating Müller glia, and it is necessary for the transcription of the pluripotency RNA binding protein *Lin-28*. Lin-28, on its turn, inhibits the biogenesis of *Let-7*, a miRNA associated with the repression of regeneration-associated genes including *Ascl1* itself, *hspd1* and *Pax6b*. Thus, while normally *Let-7* is implicated in maintaining the differentiation status of Müller glia, its *Ascl1/Lin28*-mediated downregulation in response to injury is a pre-requisite for the re-entry in the cell cycle and dedifferentiation of the glial cells. The activity of *Pax6a* and *Pax6b* transcription factors is also essential for the division of the glia-derived neuronal progenitors. Both genes are expressed in Müller glia soon after the retinal damage (Raymond et al., 2006). Their knockdown in zebrafish reduced the number of inner nuclear layer proliferating neuronal progenitors, resulting in a corresponding loss of regenerated photoreceptors (Thummel et al., 2010). *Pax6* was also shown to be sufficient, upon ectopic expression, to promote transdifferentiation of the RPE in the chick retina both during development (Azuma et al., 2005) and during the regenerative process (Spence et al., 2007c). Due to their involvement in early retinal development, where such transcription factors regulate proliferation and specification of early retinal progenitors, it is plausible that they play an analogous role on the regenerating cells during the process of regeneration. However, such hypothesis remains to be tested and the details by which these factors act remain to be identified.

4.3.2 Signaling molecules involved in regeneration

To discover novel factors involved in retina regeneration, transcriptional profiles of injury-activated Müller glial cells were generated from adult zebrafish, at the early stages of photoreceptor regeneration (Qin et al., 2009). The chaperone hspd1 and the mitotic checkpoint kinase mps1 were shown to be up-regulated in these cells. Subsequent functional analyses indicated that both factors are required for cone regeneration, at different steps of the process (Qin et al., 2009). Although the mutation in these genes was shown to inhibit neurogenic cluster formation and neuronal progenitor proliferation, the exact mechanisms underlying their activities remain unknown. Such genes, activated during regeneration play an obvious important role, yet they represent a downstream response to the initial event that damages the retina. This implies that there must be one or more signals that initiate the regeneration process. So far, several signaling molecules have been identified, which might play such a role. It is long known that the FGF (fibroblast growth factor) pathway is a critical inductive cue for retinal regeneration. Basic FGF-2 was indeed shown to induce RPE transdifferentiation in chick (Park and Hollenberg, 1989; Park and Hollenberg, 1991) and *Xenopus* (Sakaguchi et al., 1997). The importance of FGF-2 in *Xenopus* regeneration was confirmed also recently. Indeed, administration of FGF-2 was able to stimulate regeneration of a new entire retina in *Xenopus* tadpoles, where complete neural retina (and vascular membrane) was surgically removed (Vergara and Del Rio-Tsonis, 2009). Inhibition of the MAPK pathway reduced significantly the regeneration, highlighting its importance in retina regeneration. MAPK may regulate retina regeneration also in birds. For instance, over-expression of a constitutively active form of MAPK (Mek-1) in the chick embryonic retina induced RPE transdifferentiation into neural retina, an event correlated with down-regulation of the RPE-specific marker *Mitf*, and overexpression of *Pax6* (Galy et al., 2002; Spence et al., 2007c). Among other critical factors, CNTF (ciliary neurotrophic tactor) was observed to stimulate Müller glia proliferation in zebrafish (Kassen et al., 2009) and BMP signaling was shown to regulate regeneration from the ciliary margin in chick, through Smad activation and upregulation of FGF/MAPK pathway (Haynes et al., 2007). Besides, Sonic hedgehog (Shh) also plays a key role in chick retina regeneration. In particular, exogenous activation of Shh in cooperation with FGF activation induces regeneration from the ciliary body, by stimulating the proliferation of the cells residing in this region (Spence et al., 2007a; Spence et al., 2004). More importantly, it was observed that activation of analogous pathways were able to initiate retina regeneration even in mammals. The mammalian optic nerve does not regenerate when injured. However, significant regeneration can be achieved through exogenous administration of various growth factors. For example, intravitreous injections of CTNF strongly induced optic nerve regeneration in rats where the optic nerve was cut (axotomy). This effect was shown to strongly rely on MAPK activity, since inhibition of MAPK compromised the effects of CNTF on neurite outgrowth (Muller et al., 2009). Similarly, simultaneous administration of CNTF and inhibition of Rho kinase (Rock) promoted retinal ganglion cell survival and regeneration upon axotomy (Lingor et al., 2008). Hedgehog signaling could also represent a potential candidate pathway to stimulate mouse retina regeneration. Activation of Shh pathway in *patched*$^{+/-}$ (*ptc*$^{+/-}$) mice results in persistent proliferating progenitors at the ciliary margin, resembling the CMZ of lower vertebrates. Importantly, such patched mutation in P23H mice was able to promote a partial rescue of the retina degeneration (Moshiri et al., 2004).

The factors shown here are only some of the many known to be involved or stimulate retina regeneration. Although the list of these factors is in continuous expansion, our knowledge about their *in vivo* function modalities is still modest. The most important point emerging from the present data is that the mechanisms regulating regeneration seem to be conserved among vertebrates. In particular, there are some evidences that factors normally regulating regeneration in lower vertebrates can be successfully used to improve regeneration in higher vertebrates like birds and mammals. Thus, an interesting scenario emerges from these results: once a particular factor is identified to have a role during retina regeneration in animals showing regeneration, then it is important to test whether this factor is conserved in mammals and whether it has any ability to stimulate retina regeneration.

5. Conclusion

The promise of regenerative medicine is to induce regeneration of damaged tissues or entire complex structures *in situ*. Although mammalian possess a natural regenerative capacity, this is very limited and restricted to a few tissues, such as the liver and the skin. On the other hand, amphibians and fishes have extensive regenerative capacity. Concerning the retina, they are able to regenerate all the neurons and the glia of this fundamental tissue. Understanding precisely all the molecules and genes involved in this process, as well as the network of interactions that these form *in vivo* will be fundamental for a rationale screening of pharmacological drugs, able to trigger the precise regenerative process. Furthermore, the lower vertebrates adopt mechanisms, such as transdifferentiation and dedifferentiation, which mammalian does not seem to possess. In the next future, it will be important to determine exactly how Müller glia or RPE exploit these processes to regenerate damaged tissues, so that we could use them to initiate regeneration directly *in vivo*, in patients affected by degenerative retinopathies.

6. Acknowledgments

This work is supported by grants from the CNRS, University Paris Sud, ARC, Retina France, ANR and Région Île de France (AMBRe network). G.C. is a AMBRe network fellow. We thank A. Chesneau for his technical support to make *Xenopus* transgenics shown in Figure 1.

7. References

Ahmad, I., Tang, L. and Pham, H. (2000). Identification of neural progenitors in the adult mammalian eye. *Biochem Biophys Res Commun* 270, 517-21.

Ambati, J., Anand, A., Fernandez, S., Sakurai, E., Lynn, B. C., Kuziel, W. A., Rollins, B. J. and Ambati, B. K. (2003). An animal model of age-related macular degeneration in senescent Ccl-2- or Ccr-2-deficient mice. *Nat Med* 9, 1390-7.

Araki, M. (2007). Regeneration of the amphibian retina: role of tissue interaction and related signaling molecules on RPE transdifferentiation. *Dev Growth Differ* 49, 109-20.

Azuma, N., Tadokoro, K., Asaka, A., Yamada, M., Yamaguchi, Y., Handa, H., Matsushima, S., Watanabe, T., Kida, Y., Ogura, T. et al. (2005). Transdifferentiation of the retinal pigment epithelia to the neural retina by transfer of the Pax6 transcriptional factor. *Hum Mol Genet* 14, 1059-68.

Bailey, T. J., Fossum, S. L., Fimbel, S. M., Montgomery, J. E. and Hyde, D. R. (2010). The inhibitor of phagocytosis, O-phospho-L-serine, suppresses Muller glia proliferation and cone cell regeneration in the light-damaged zebrafish retina. *Exp Eye Res* 91, 601-12.

Bainbridge, J. W., Smith, A. J., Barker, S. S., Robbie, S., Henderson, R., Balaggan, K., Viswanathan, A., Holder, G. E., Stockman, A., Tyler, N. et al. (2008). Effect of gene therapy on visual function in Leber's congenital amaurosis. *N Engl J Med* 358, 2231-9.

Bernardos, R. L., Barthel, L. K., Meyers, J. R. and Raymond, P. A. (2007). Late-stage neuronal progenitors in the retina are radial Muller glia that function as retinal stem cells. *J Neurosci* 27, 7028-40.

Berson, E. L. (1993). Retinitis pigmentosa. The Friedenwald Lecture. *Invest Ophthalmol Vis Sci* 34, 1659-76.

Bibliowicz, J., Tittle, R. K. and Gross, J. M. (2011). Toward a better understanding of human eye disease insights from the zebrafish, Danio rerio. *Prog Mol Biol Transl Sci* 100, 287-330.

Boon, C. J., den Hollander, A. I., Hoyng, C. B., Cremers, F. P., Klevering, B. J. and Keunen, J. E. (2008). The spectrum of retinal dystrophies caused by mutations in the peripherin/RDS gene. *Prog Retin Eye Res* 27, 213-35.

Bringmann, A., Iandiev, I., Pannicke, T., Wurm, A., Hollborn, M., Wiedemann, P., Osborne, N. N. and Reichenbach, A. (2009). Cellular signaling and factors involved in Muller cell gliosis: neuroprotective and detrimental effects. *Prog Retin Eye Res* 28, 423-51.

Brockerhoff, S. E. and Fadool, J. M. (2011). Genetics of photoreceptor degeneration and regeneration in zebrafish. *Cell Mol Life Sci* 68, 651-9.

Bull, N. D. and Martin, K. R. (2011). Towards Stem Cell-Based Therapies for Retinal Neurodegenerative Diseases. *Stem Cells.*

Cameron, D. A. (2000). Cellular proliferation and neurogenesis in the injured retina of adult zebrafish. *Vis Neurosci* 17, 789-97.

Chang, G. Q., Hao, Y. and Wong, F. (1993). Apoptosis: final common pathway of photoreceptor death in rd, rds, and rhodopsin mutant mice. *Neuron* 11, 595-605.

Choi, R. Y., Engbretson, G. A., Solessio, E. C., Jones, G. A., Coughlin, A., Aleksic, I. and Zuber, M. E. (2011). Cone degeneration following rod ablation in a reversible model of retinal degeneration. *Invest Ophthalmol Vis Sci* 52, 364-73.

Cicero, S. A., Johnson, D., Reyntjens, S., Frase, S., Connell, S., Chow, L. M., Baker, S. J., Sorrentino, B. P. and Dyer, M. A. (2009). Cells previously identified as retinal stem cells are pigmented ciliary epithelial cells. *Proc Natl Acad Sci U S A* 106, 6685-90.

Coles, B. L., Angenieux, B., Inoue, T., Del Rio-Tsonis, K., Spence, J. R., McInnes, R. R., Arsenijevic, Y. and van der Kooy, D. (2004). Facile isolation and the characterization of human retinal stem cells. *Proc Natl Acad Sci U S A* 101, 15772-7.

Connell, G., Bascom, R., Molday, L., Reid, D., McInnes, R. R. and Molday, R. S. (1991). Photoreceptor peripherin is the normal product of the gene responsible for retinal degeneration in the rds mouse. *Proc Natl Acad Sci U S A* 88, 723-6.

Connell, G. J. and Molday, R. S. (1990). Molecular cloning, primary structure, and orientation of the vertebrate photoreceptor cell protein peripherin in the rod outer segment disk membrane. *Biochemistry* 29, 4691-8.

Curado, S., Anderson, R. M., Jungblut, B., Mumm, J., Schroeter, E. and Stainier, D. Y. (2007). Conditional targeted cell ablation in zebrafish: a new tool for regeneration studies. *Dev Dyn* 236, 1025-35.

Curado, S., Stainier, D. Y. and Anderson, R. M. (2008). Nitroreductase-mediated cell/tissue ablation in zebrafish: a spatially and temporally controlled ablation method with applications in developmental and regeneration studies. *Nat Protoc* 3, 948-54.

Dahlmann-Noor, A., Vijay, S., Jayaram, H., Limb, A. and Khaw, P. T. (2010). Current approaches and future prospects for stem cell rescue and regeneration of the retina and optic nerve. *Can J Ophthalmol* 45, 333-41.

Das, A. V., Mallya, K. B., Zhao, X., Ahmad, F., Bhattacharya, S., Thoreson, W. B., Hegde, G. V. and Ahmad, I. (2006). Neural stem cell properties of Muller glia in the mammalian retina: regulation by Notch and Wnt signaling. *Dev Biol* 299, 283-302.

Davison, J. M., Akitake, C. M., Goll, M. G., Rhee, J. M., Gosse, N., Baier, H., Halpern, M. E., Leach, S. D. and Parsons, M. J. (2007). Transactivation from Gal4-VP16 transgenic insertions for tissue-specific cell labeling and ablation in zebrafish. *Dev Biol* 304, 811-24.

Dreyer, E. B., Zurakowski, D., Schumer, R. A., Podos, S. M. and Lipton, S. A. (1996). Elevated glutamate levels in the vitreous body of humans and monkeys with glaucoma. *Arch Ophthalmol* 114, 299-305.

Dryja, T. P., McGee, T. L., Reichel, E., Hahn, L. B., Cowley, G. S., Yandell, D. W., Sandberg, M. A. and Berson, E. L. (1990). A point mutation of the rhodopsin gene in one form of retinitis pigmentosa. *Nature* 343, 364-6.

Elizabeth Rakoczy, P., Yu, M. J., Nusinowitz, S., Chang, B. and Heckenlively, J. R. (2006). Mouse models of age-related macular degeneration. *Exp Eye Res* 82, 741-52.

Farrar, G. J., Palfi, A. and O'Reilly, M. (2010). Gene therapeutic approaches for dominant retinopathies. *Curr Gene Ther* 10, 381-8.

Fausett, B. V. and Goldman, D. (2006). A role for alpha1 tubulin-expressing Muller glia in regeneration of the injured zebrafish retina. *J Neurosci* 26, 6303-13.

Fimbel, S. M., Montgomery, J. E., Burket, C. T. and Hyde, D. R. (2007). Regeneration of inner retinal neurons after intravitreal injection of ouabain in zebrafish. *J Neurosci* 27, 1712-24.

Fischer, A. J. and Reh, T. A. (2000). Identification of a proliferating marginal zone of retinal progenitors in postnatal chickens. *Dev Biol* 220, 197-210.

Fischer, A. J. and Reh, T. A. (2001). Muller glia are a potential source of neural regeneration in the postnatal chicken retina. *Nat Neurosci* 4, 247-52.

Fischer, A. J. and Reh, T. A. (2003). Potential of Muller glia to become neurogenic retinal progenitor cells. *Glia* 43, 70-6.

Fletcher, E. L., Jobling, A. I., Vessey, K. A., Luu, C., Guymer, R. H. and Baird, P. N. (2011). Animal models of retinal disease. *Prog Mol Biol Transl Sci* 100, 211-86.

Frasson, M., Sahel, J. A., Fabre, M., Simonutti, M., Dreyfus, H. and Picaud, S. (1999). Retinitis pigmentosa: rod photoreceptor rescue by a calcium-channel blocker in the rd mouse. *Nat Med* 5, 1183-7.

Galy, A., Neron, B., Planque, N., Saule, S. and Eychene, A. (2002). Activated MAPK/ERK kinase (MEK-1) induces transdifferentiation of pigmented epithelium into neural retina. *Dev Biol* 248, 251-64.

Ghosh, A. K., Murga-Zamalloa, C. A., Chan, L., Hitchcock, P. F., Swaroop, A. and Khanna, H. (2010). Human retinopathy-associated ciliary protein retinitis pigmentosa GTPase regulator mediates cilia-dependent vertebrate development. *Hum Mol Genet* 19, 90-8.

Goldsmith, P. and Harris, W. A. (2003). The zebrafish as a tool for understanding the biology of visual disorders. *Semin Cell Dev Biol* 14, 11-8.

Hamel, C. (2006). Retinitis pigmentosa. *Orphanet J Rare Dis* 1, 40.

Hamm, L. M., Tam, B. M. and Moritz, O. L. (2009). Controlled rod cell ablation in transgenic Xenopus laevis. *Invest Ophthalmol Vis Sci* 50, 885-92.

Hartong, D. T., Berson, E. L. and Dryja, T. P. (2006). Retinitis pigmentosa. *Lancet* 368, 1795-809.

Haynes, T., Gutierrez, C., Aycinena, J. C., Tsonis, P. A. and Del Rio-Tsonis, K. (2007). BMP signaling mediates stem/progenitor cell-induced retina regeneration. *Proc Natl Acad Sci U S A* 104, 20380-5.

Hong, D. H., Pawlyk, B. S., Shang, J., Sandberg, M. A., Berson, E. L. and Li, T. (2000). A retinitis pigmentosa GTPase regulator (RPGR)-deficient mouse model for X-linked retinitis pigmentosa (RP3). *Proc Natl Acad Sci U S A* 97, 3649-54.

Hosch, J., Lorenz, B. and Stieger, K. (2011). RPGR: role in the photoreceptor cilium, human retinal disease, and gene therapy. *Ophthalmic Genet* 32, 1-11.

Jarrett, S. G., Lewin, A. S. and Boulton, M. E. (2010). The importance of mitochondria in age-related and inherited eye disorders. *Ophthalmic Res* 44, 179-90.

Kaneko, H., Dridi, S., Tarallo, V., Gelfand, B. D., Fowler, B. J., Cho, W. G., Kleinman, M. E., Ponicsan, S. L., Hauswirth, W. W., Chiodo, V. A. et al. (2011). DICER1 deficit induces Alu RNA toxicity in age-related macular degeneration. *Nature* 471, 325-30.

Karan, G., Lillo, C., Yang, Z., Cameron, D. J., Locke, K. G., Zhao, Y., Thirumalaichary, S., Li, C., Birch, D. G., Vollmer-Snarr, H. R. et al. (2005). Lipofuscin accumulation, abnormal electrophysiology, and photoreceptor degeneration in mutant ELOVL4 transgenic mice: a model for macular degeneration. *Proc Natl Acad Sci U S A* 102, 4164-9.

Karl, M. O. and Reh, T. A. (2010). Regenerative medicine for retinal diseases: activating endogenous repair mechanisms. *Trends Mol Med* 16, 193-202.

Kassen, S. C., Thummel, R., Campochiaro, L. A., Harding, M. J., Bennett, N. A. and Hyde, D. R. (2009). CNTF induces photoreceptor neuroprotection and Muller glial cell proliferation through two different signaling pathways in the adult zebrafish retina. *Exp Eye Res* 88, 1051-64.

Kaushal, S., Ridge, K. D. and Khorana, H. G. (1994). Structure and function in rhodopsin: the role of asparagine-linked glycosylation. *Proc Natl Acad Sci U S A* 91, 4024-8.

Kedzierski, W., Lloyd, M., Birch, D. G., Bok, D. and Travis, G. H. (1997). Generation and analysis of transgenic mice expressing P216L-substituted rds/peripherin in rod photoreceptors. *Invest Ophthalmol Vis Sci* 38, 498-509.

Kong, J., Kim, S. R., Binley, K., Pata, I., Doi, K., Mannik, J., Zernant-Rajang, J., Kan, O., Iqball, S., Naylor, S. et al. (2008). Correction of the disease phenotype in the mouse model of Stargardt disease by lentiviral gene therapy. *Gene Ther* 15, 1311-20.

Leveillard, T., Mohand-Said, S., Lorentz, O., Hicks, D., Fintz, A. C., Clerin, E., Simonutti, M., Forster, V., Cavusoglu, N., Chalmel, F. et al. (2004). Identification and characterization of rod-derived cone viability factor. *Nat Genet* 36, 755-9.

Leveillard, T. and Sahel, J. A. (2010). Rod-derived cone viability factor for treating blinding diseases: from clinic to redox signaling. *Sci Transl Med* 2, 26ps16.

Li, L., Li, Y., Chen, D., Shao, J., Li, X. and Xu, C. (2010). Fishing for age-related visual system mutants: behavioral screening of retinal degeneration genes in zebrafish. *Curr Aging Sci* 3, 43-5.

Lien, S. and Lowman, H. B. (2008). Therapeutic anti-VEGF antibodies. *Handb Exp Pharmacol*, 131-50.

Lingor, P., Tonges, L., Pieper, N., Bermel, C., Barski, E., Planchamp, V. and Bahr, M. (2008). ROCK inhibition and CNTF interact on intrinsic signalling pathways and differentially regulate survival and regeneration in retinal ganglion cells. *Brain* 131, 250-63.

Liu, Q., Lyubarsky, A., Skalet, J. H., Pugh, E. N., Jr. and Pierce, E. A. (2003). RP1 is required for the correct stacking of outer segment discs. *Invest Ophthalmol Vis Sci* 44, 4171-83.

Liutkeviciene, R., Lesauskaite, V., Asmoniene, V., Zaliuniene, D. and Jasinskas, V. (2010). Factors determining age-related macular degeneration: a current view. *Medicina (Kaunas)* 46, 89-94.

Locker, M., Borday, C. and Perron, M. (2009). Stemness or not stemness? Current status and perspectives of adult retinal stem cells. *Curr Stem Cell Res Ther* 4, 118-30.

Locker, M., El Yakoubi, W., Mazurier, N., Dullin, J. P. and Perron, M. (2010). A decade of mammalian retinal stem cell research. *Arch Ital Biol* 148, 59-72.

Ma, L., Kaufman, Y., Zhang, J. and Washington, I. (2010). C20-D3-vitamin A slows lipofuscin accumulation and electrophysiological retinal degeneration in a mouse model of Stargardt disease. *J Biol Chem* 286, 7966-74.

Mao, H., James, T., Jr., Schwein, A., Shabashvili, A. E., Hauswirth, W. W., Gorbatyuk, M. S. and Lewin, A. S. (2011). AAV delivery of wild-type rhodopsin preserves retinal function in a mouse model of autosomal dominant retinitis pigmentosa. *Hum Gene Ther* 22, 567-75.

Martinez-De Luna, R. I., Kelly, L. E. and El-Hodiri, H. M. (2011). The Retinal Homeobox (Rx) gene is necessary for retinal regeneration. *Dev Biol* 353, 10-8.

Mitashov, V. I. and Maliovanova, S. D. (1982). [Cellular proliferative potentials of the pigment and ciliated epithelium of the eye in clawed toads normally and during regeneration]. *Ontogenez* 13, 228-34.

Molday, R. S., Hicks, D. and Molday, L. (1987). Peripherin. A rim-specific membrane protein of rod outer segment discs. *Invest Ophthalmol Vis Sci* 28, 50-61.

Molday, R. S. and Zhang, K. (2010). Defective lipid transport and biosynthesis in recessive and dominant Stargardt macular degeneration. *Prog Lipid Res* 49, 476-92.

Monnin, J., Morand-Villeneuve, N., Michel, G., Hicks, D. and Versaux-Botteri, C. (2007). Production of neurospheres from mammalian Muller cells in culture. *Neurosci Lett* 421, 22-6.

Montgomery, J. E., Parsons, M. J. and Hyde, D. R. (2010). A novel model of retinal ablation demonstrates that the extent of rod cell death regulates the origin of the regenerated zebrafish rod photoreceptors. *J Comp Neurol* 518, 800-14.

Moshiri, A., Close, J. and Reh, T. A. (2004). Retinal stem cells and regeneration. *Int J Dev Biol* 48, 1003-14.

Muller, A., Hauk, T. G., Leibinger, M., Marienfeld, R. and Fischer, D. (2009). Exogenous CNTF stimulates axon regeneration of retinal ganglion cells partially via endogenous CNTF. *Mol Cell Neurosci* 41, 233-46.

Murga-Zamalloa, C. A., Atkins, S. J., Peranen, J., Swaroop, A. and Khanna, H. (2010). Interaction of retinitis pigmentosa GTPase regulator (RPGR) with RAB8A GTPase: implications for cilia dysfunction and photoreceptor degeneration. *Hum Mol Genet* 19, 3591-8.

Murga-Zamalloa, C. A., Swaroop, A. and Khanna, H. (2009). RPGR-containing protein complexes in syndromic and non-syndromic retinal degeneration due to ciliary dysfunction. *J Genet* 88, 399-407.

Olsson, J. E., Gordon, J. W., Pawlyk, B. S., Roof, D., Hayes, A., Molday, R. S., Mukai, S., Cowley, G. S., Berson, E. L. and Dryja, T. P. (1992). Transgenic mice with a rhodopsin mutation (Pro23His): a mouse model of autosomal dominant retinitis pigmentosa. *Neuron* 9, 815-30.

Ooto, S., Akagi, T., Kageyama, R., Akita, J., Mandai, M., Honda, Y. and Takahashi, M. (2004). Potential for neural regeneration after neurotoxic injury in the adult mammalian retina. *Proc Natl Acad Sci U S A* 101, 13654-9.

Osakada, F., Ooto, S., Akagi, T., Mandai, M., Akaike, A. and Takahashi, M. (2007). Wnt signaling promotes regeneration in the retina of adult mammals. *J Neurosci* 27, 4210-9.

Otteson, D. C. and Hitchcock, P. F. (2003). Stem cells in the teleost retina: persistent neurogenesis and injury-induced regeneration. *Vision Res* 43, 927-36.

Palczewski, K. (2010). Retinoids for treatment of retinal diseases. *Trends Pharmacol Sci* 31, 284-95.

Paquet-Durand, F., Beck, S., Michalakis, S., Goldmann, T., Huber, G., Muhlfriedel, R., Trifunovic, D., Fischer, M. D., Fahl, E., Duetsch, G. et al. (2011). A key role for cyclic nucleotide gated (CNG) channels in cGMP-related retinitis pigmentosa. *Hum Mol Genet* 20, 941-7.

Paquet-Durand, F., Hauck, S. M., van Veen, T., Ueffing, M. and Ekstrom, P. (2009). PKG activity causes photoreceptor cell death in two retinitis pigmentosa models. *J Neurochem* 108, 796-810.

Park, C.M., Hollenberg, M.J. (1989). Basic fibroblast growth factor induces retinal regeneration in vivo. *Dev. Biol.* 134, 201-5.

Park, C.M., Hollenberg, M.J. (1991). Induction of retinal regeneration in vivo by growth factors. *Dev. Biol.* 148, 322-33.

Pawlyk, B. S., Bulgakov, O. V., Liu, X., Xu, X., Adamian, M., Sun, X., Khani, S. C., Berson, E. L., Sandberg, M. A. and Li, T. (2010). Replacement gene therapy with a human

RPGRIP1 sequence slows photoreceptor degeneration in a murine model of Leber congenital amaurosis. *Hum Gene Ther* 21, 993-1004.

Pechan, P., Rubin, H., Lukason, M., Ardinger, J., DuFresne, E., Hauswirth, W. W., Wadsworth, S. C. and Scaria, A. (2009). Novel anti-VEGF chimeric molecules delivered by AAV vectors for inhibition of retinal neovascularization. *Gene Ther* 16, 10-6.

Pisharath, H. (2007). Validation of nitroreductase, a prodrug-activating enzyme, mediated cell death in embryonic zebrafish (Danio rerio). *Comp Med* 57, 241-6.

Pisharath, H., Rhee, J. M., Swanson, M. A., Leach, S. D. and Parsons, M. J. (2007). Targeted ablation of beta cells in the embryonic zebrafish pancreas using E. coli nitroreductase. *Mech Dev* 124, 218-29.

Prasad, P. S., Schwartz, S. D. and Hubschman, J. P. (2010). Age-related macular degeneration: current and novel therapies. *Maturitas* 66, 46-50.

Punzo, C., Kornacker, K. and Cepko, C. L. (2009). Stimulation of the insulin/mTOR pathway delays cone death in a mouse model of retinitis pigmentosa. *Nat Neurosci* 12, 44-52.

Qin, Z., Barthel, L. K. and Raymond, P. A. (2009). Genetic evidence for shared mechanisms of epimorphic regeneration in zebrafish. *Proc Natl Acad Sci U S A* 106, 9310-5.

Querques, G., Avellis, F. O., Querques, L., Bandello, F. and Souied, E. H. (2011). Age-related macular degeneration. *Clin Ophthalmol* 5, 593-601.

Ramachandran, R., Fausett, B. V. and Goldman, D. (2010a). Ascl1a regulates Muller glia dedifferentiation and retinal regeneration through a Lin-28-dependent, let-7 microRNA signalling pathway. *Nat Cell Biol* 12, 1101-7.

Ramachandran, R., Reifler, A., Parent, J. M. and Goldman, D. (2010b). Conditional gene expression and lineage tracing of tuba1a expressing cells during zebrafish development and retina regeneration. *J Comp Neurol* 518, 4196-212.

Rattner, A. and Nathans, J. (2006). Macular degeneration: recent advances and therapeutic opportunities. *Nat Rev Neurosci* 7, 860-72.

Raymond, P. A., Barthel, L. K., Bernardos, R. L. and Perkowski, J. J. (2006). Molecular characterization of retinal stem cells and their niches in adult zebrafish. *BMC Dev Biol* 6, 36.

Reh, T. A. and Nagy, T. (1987). A possible role for the vascular membrane in retinal regeneration in Rana catesbienna tadpoles. *Dev Biol* 122, 471-82.

Reme, C. E., Weller M., Szczesny P., Munz K., Hafezi F., Reinboth J. and M., C. (1995). in Degenerative Diseases of the Retina. *ed Anderson R. E.(Plenum Press, New York)*, pp 19–25.

Rivas, M. A. and Vecino, E. (2009). Animal models and different therapies for treatment of retinitis pigmentosa. *Histol Histopathol* 24, 1295-322.

Roof, D. J., Adamian, M. and Hayes, A. (1994). Rhodopsin accumulation at abnormal sites in retinas of mice with a human P23H rhodopsin transgene. *Invest Ophthalmol Vis Sci* 35, 4049-62.

Sakaguchi, D. S., Janick, L. M. and Reh, T. A. (1997). Basic fibroblast growth factor (FGF-2) induced transdifferentiation of retinal pigment epithelium: generation of retinal neurons and glia. *Dev Dyn* 209, 387-98.

Sakami, S., Maeda, T., Bereta, G., Okano, K., Golczak, M., Sumaroka, A., Roman, A. J., Cideciyan, A. V., Jacobson, S. G. and Palczewski, K. (2011). Probing mechanisms of photoreceptor degeneration in a new mouse model of the common form of autosomal dominant retinitis pigmentosa due to P23H opsin mutations. *J Biol Chem* 286, 10551-67.

Sancho-Pelluz, J., Arango-Gonzalez, B., Kustermann, S., Romero, F. J., van Veen, T., Zrenner, E., Ekstrom, P. and Paquet-Durand, F. (2008). Photoreceptor cell death mechanisms in inherited retinal degeneration. *Mol Neurobiol* 38, 253-69.

Sherpa, T., Fimbel, S. M., Mallory, D. E., Maaswinkel, H., Spritzer, S. D., Sand, J. A., Li, L., Hyde, D. R. and Stenkamp, D. L. (2008). Ganglion cell regeneration following whole-retina destruction in zebrafish. *Dev Neurobiol* 68, 166-81.

Simonelli, F., Maguire, A. M., Testa, F., Pierce, E. A., Mingozzi, F., Bennicelli, J. L., Rossi, S., Marshall, K., Banfi, S., Surace, E. M. et al. (2010). Gene therapy for Leber's congenital amaurosis is safe and effective through 1.5 years after vector administration. *Mol Ther* 18, 643-50.

Spence, J. R., Aycinena, J. C. and Del Rio-Tsonis, K. (2007a). Fibroblast growth factor-hedgehog interdependence during retina regeneration. *Dev Dyn* 236, 1161-74.

Spence, J. R., Gutierrez, C. and Del Rio-Tsonis, K. (2007b). Retina regeneration in the embryonic chick In *Strategies for Retinal Tissue Repair and Regeneration in Vertebrates: From Fish to Human*.

Spence, J. R., Madhavan, M., Aycinena, J. C. and Del Rio-Tsonis, K. (2007c). Retina regeneration in the chick embryo is not induced by spontaneous Mitf downregulation but requires FGF/FGFR/MEK/Erk dependent upregulation of Pax6. *Mol Vis* 13, 57-65.

Spence, J. R., Madhavan, M., Ewing, J. D., Jones, D. K., Lehman, B. M. and Del Rio-Tsonis, K. (2004). The hedgehog pathway is a modulator of retina regeneration. *Development* 131, 4607-21.

Stearns, G., Evangelista, M., Fadool, J. M. and Brockerhoff, S. E. (2007). A mutation in the cone-specific pde6 gene causes rapid cone photoreceptor degeneration in zebrafish. *J Neurosci* 27, 13866-74.

Stoick-Cooper, C. L., Moon, R. T. and Weidinger, G. (2007). Advances in signaling in vertebrate regeneration as a prelude to regenerative medicine. *Genes Dev* 21, 1292-315.

Takano, Y., Ohguro, H., Dezawa, M., Ishikawa, H., Yamazaki, H., Ohguro, I., Mamiya, K., Metoki, T., Ishikawa, F. and Nakazawa, M. (2004). Study of drug effects of calcium channel blockers on retinal degeneration of rd mouse. *Biochem Biophys Res Commun* 313, 1015-22.

Tam, B. M. and Moritz, O. L. (2006). Characterization of rhodopsin P23H-induced retinal degeneration in a Xenopus laevis model of retinitis pigmentosa. *Invest Ophthalmol Vis Sci* 47, 3234-41.

Tam, B. M. and Moritz, O. L. (2007). Dark rearing rescues P23H rhodopsin-induced retinal degeneration in a transgenic Xenopus laevis model of retinitis pigmentosa: a chromophore-dependent mechanism characterized by production of N-terminally truncated mutant rhodopsin. *J Neurosci* 27, 9043-53.

Tam, B. M. and Moritz, O. L. (2009). The role of rhodopsin glycosylation in protein folding, trafficking, and light-sensitive retinal degeneration. *J Neurosci* 29, 15145-54.

Tam, B. M., Xie, G., Oprian, D. D. and Moritz, O. L. (2006). Mislocalized rhodopsin does not require activation to cause retinal degeneration and neurite outgrowth in Xenopus laevis. *J Neurosci* 26, 203-9.

Thummel, R., Enright, J. M., Kassen, S. C., Montgomery, J. E., Bailey, T. J. and Hyde, D. R. (2010). Pax6a and Pax6b are required at different points in neuronal progenitor cell proliferation during zebrafish photoreceptor regeneration. *Exp Eye Res* 90, 572-82.

Thummel, R., Kassen, S. C., Enright, J. M., Nelson, C. M., Montgomery, J. E. and Hyde, D. R. (2008a). Characterization of Muller glia and neuronal progenitors during adult zebrafish retinal regeneration. *Exp Eye Res* 87, 433-44.

Thummel, R., Kassen, S. C., Montgomery, J. E., Enright, J. M. and Hyde, D. R. (2008b). Inhibition of Muller glial cell division blocks regeneration of the light-damaged zebrafish retina. *Dev Neurobiol* 68, 392-408.

Travis, G. H., Brennan, M. B., Danielson, P. E., Kozak, C. A. and Sutcliffe, J. G. (1989). Identification of a photoreceptor-specific mRNA encoded by the gene responsible for retinal degeneration slow (rds). *Nature* 338, 70-3.

Tropepe, V., Coles, B. L., Chiasson, B. J., Horsford, D. J., Elia, A. J., McInnes, R. R. and van der Kooy, D. (2000). Retinal stem cells in the adult mammalian eye. *Science* 287, 2032-6.

Tsubura, A., Lai, Y. C., Miki, H., Sasaki, T., Uehara, N., Yuri, T. and Yoshizawa, K. (2011). Review: Animal models of N-Methyl-N-nitrosourea-induced mammary cancer and retinal degeneration with special emphasis on therapeutic trials. *In Vivo* 25, 11-22.

Vasireddy, V., Wong, P. and Ayyagari, R. (2010). Genetics and molecular pathology of Stargardt-like macular degeneration. *Prog Retin Eye Res* 29, 191-207.

Vergara, M. N. and Del Rio-Tsonis, K. (2009). Retinal regeneration in the Xenopus laevis tadpole: a new model system. *Mol Vis* 15, 1000-13.

Vervoort, R., Lennon, A., Bird, A. C., Tulloch, B., Axton, R., Miano, M. G., Meindl, A., Meitinger, T., Ciccodicola, A. and Wright, A. F. (2000). Mutational hot spot within a new RPGR exon in X-linked retinitis pigmentosa. *Nat Genet* 25, 462-6.

Vihtelic, T. S. and Hyde, D. R. (2000). Light-induced rod and cone cell death and regeneration in the adult albino zebrafish (Danio rerio) retina. *J Neurobiol* 44, 289-307.

Wang, D. Y., Chan, W. M., Tam, P. O., Baum, L., Lam, D. S., Chong, K. K., Fan, B. J. and Pang, C. P. (2005). Gene mutations in retinitis pigmentosa and their clinical implications. *Clin Chim Acta* 351, 5-16.

Weng, J., Mata, N. L., Azarian, S. M., Tzekov, R. T., Birch, D. G. and Travis, G. H. (1999). Insights into the function of Rim protein in photoreceptors and etiology of Stargardt's disease from the phenotype in abcr knockout mice. *Cell* 98, 13-23.

Wetts, R. and Fraser, S. E. (1988). Multipotent precursors can give rise to all major cell types of the frog retina. *Science* 239, 1142-5.

Wetts, R., Serbedzija, G. N. and Fraser, S. E. (1989). Cell lineage analysis reveals multipotent precursors in the ciliary margin of the frog retina. *Dev Biol* 136, 254-63.

Wright, A. F., Chakarova, C. F., Abd El-Aziz, M. M. and Bhattacharya, S. S. (2010). Photoreceptor degeneration: genetic and mechanistic dissection of a complex trait. *Nat Rev Genet* 11, 273-84.

Wu, D. M., Schneiderman, T., Burgett, J., Gokhale, P., Barthel, L. and Raymond, P. A. (2001). Cones regenerate from retinal stem cells sequestered in the inner nuclear layer of adult goldfish retina. *Invest Ophthalmol Vis Sci* 42, 2115-24.

Xiao, Q., Hartzell, H. C. and Yu, K. (2010). Bestrophins and retinopathies. *Pflugers Arch* 460, 559-69.

Yasukawa, T., Tabata, Y., Kimura, H. and Ogura, Y. (2011). Recent advances in intraocular drug delivery systems. *Recent Pat Drug Deliv Formul* 5, 1-10.

Yoshii, C., Ueda, Y., Okamoto, M. and Araki, M. (2007). Neural retinal regeneration in the anuran amphibian Xenopus laevis post-metamorphosis: transdifferentiation of retinal pigmented epithelium regenerates the neural retina. *Dev Biol* 303, 45-56.

Yurco, P. and Cameron, D. A. (2005). Responses of Muller glia to retinal injury in adult zebrafish. *Vision Res* 45, 991-1002.

Zhang, R., Oglesby, E. and Marsh-Armstrong, N. (2008). Xenopus laevis P23H rhodopsin transgene causes rod photoreceptor degeneration that is more severe in the ventral retina and is modulated by light. *Exp Eye Res* 86, 612-21.

Zhang, Y., Stanton, J. B., Wu, J., Yu, K., Hartzell, H. C., Peachey, N. S., Marmorstein, L. Y. and Marmorstein, A. D. (2010). Suppression of Ca2+ signaling in a mouse model of Best disease. *Hum Mol Genet* 19, 1108-18.

Zhao, X. F., Ellingsen, S. and Fjose, A. (2009). Labelling and targeted ablation of specific bipolar cell types in the zebrafish retina. *BMC Neurosci* 10, 107.

Zhao, Y., Hong, D. H., Pawlyk, B., Yue, G., Adamian, M., Grynberg, M., Godzik, A. and Li, T. (2003). The retinitis pigmentosa GTPase regulator (RPGR)- interacting protein: subserving RPGR function and participating in disk morphogenesis. *Proc Natl Acad Sci U S A* 100, 3965-70.

7

A Strategy Using Pluripotent Stem Cell-Derived Hepatocytes for Stem Cell-Based Therapies

Daihachiro Tomotsune et al.*
Department of Histology and Embryology, Shinshu University School of Medicine 3-1-1 Asahi, Matsumoto, Nagano 390-8621, Japan

1. Introduction

There are millions of patients suffering from fatal liver disease in the world. Whole-organ transplantation of the liver, such as orthotopic liver transplantation (OLT), improves the survival rate of these patients (Benten et al., 2009). However, the traditional OLT is still limited due to the serious want of viable livers available for organ transplantation. Therefore, cell therapies for the treatment of end-stage hepatic diseases are currently under investigation all over the world (Ito et al., 2009). The simplest method for cell therapies is by transplantation of primary hepatocytes isolated from the donor. Experimental studies on animals have shown that transplanted primary hepatocytes into the spleen or portal vein of host animals repopulated in the liver, suggesting that the primary hepatocyte transplantation may be successful as an alternative to organ transplantation for patients with liver failure (Sutherland et al., 1977; Makowka L et al., 1980; Demetrious et al., 1986; Arkadopoulos et al., 1998; Ribeiro et al., 1992; Ito et al., 2007; Nagata et al., 2003; Kobayashi et al., 2000). However, success in clinical use is limited (Fox et al., 1998; Platt, 1998) and donor human livers to isolate the hepatocytes for hepatocyte transplantations are also limited, since these organs are needed to use in organ transplantation. Furthermore, the primary hepatocytes are cultured in vitro with very limited success due to their slow growth and instability of hepatic phenotype tending to lose differentiation character (Clayton & Darnell, 1983). Thus, although the human primary hepatocytes are ideal cells for cell therapies, an unlimited source of hepatocytes is required. There are two ways to prepare an unlimited supply of hepatocytes: 1) induction of differentiation from pluripotent stem cells, such as embryonic stem cells (ESCs) and induced pluripotent stem cells (iPSCs) and 2) production of hepatocytes from somatic cells by reactivation of mature hepatocytes themselves or reprogramming the somatic cells other than hepatocytes into hepatocytes.

ESC and iPSC are expected to be a promising alternative resources for cell therapy because they are pluripotent, making it possible to produce any type of tissue from a single resource and they are also an infinite resource expanding continuously in the undifferentiated state

*Fumi Sato[1], Susumu Yoshie[1], Sakiko Shirasawa[2], Tadayuki Yokoyama[2], Yoshiya Kanoh[1], Hinako Ichikawa[1], Akimi Mogi, Fengming Yue[1] and Katsunori Sasaki[1]
[1] *Department of Histology and Embryology, Shinshu University School of Medicine 3-1-1 Asahi, Matsumoto, Nagano 390-8621, Japan*
[2] *Bourbon Corporation. 4-2-14 Matsunami, Kashiwazaki, Niigata 945-8611, Japan*

under appropriate conditions (Hanna et al., 2010). A lot of studies have reported the successful induction of hepatic differentiation from ESCs and iPSCs (reviewed in Andersson & Lendahl, 2009; Behbahan et al., 2011; Greenbaum, 2010; Kung & Forbes, 2009). However, undifferentiated pluripotent stem cells possess intrinsic property of teratoma formation in the host after transplantation (Fong et al., 2010; Blum & Benvenisty, 2009; Knoepfler, 2009) and their induction efficiency of hepatocellular differentiation was still not enough so that various types of cells, including undifferentiated pluripotent stem cells, remained at different stages of differentiation (Sasaki et al., 2009; Shiraki et al., 2008; Teratani et al., 2005; Yoshie et al., 2010). Thus, improvement of the induction method for the hepatocellular differentiation to the efficiency of almost 100% and/or development of an efficient selection method for differentiated hepatocytes are needed.

Several strategies for isolation of ES cell-derived hepatic cells have been proposed in the last decade (Basma et al., 2009; Duan et al., 2007; Gouon-Evans et al., 2006; Heo et al., 2006; Li et al., 2010; Soto-Gutierrez et al., 2006; Suzuki et al., 2000; Yin et al., 2002). The method of isolation is based on the sorting system, such as fluorescence-activated cell sorting (FACS) or magnetic-activated cell sorting (MACS) devices and utilizes fluorescent markers or antibodies specific for hepatic cells. To realize enough quality and quantity, we propose in this paper a dual selection method; first the proliferative immature hepatocytes are isolated with a cell surface marker Liv2 antigen and after growing fully in the culture, differentiated mature hepatocytes are purified with indocyanine green (ICG) which is an organic anion that is specifically taken by mature hepatocytes. We suggest that in surface markers only Liv2 antigen is used in the specific isolation of immature hepatocytes from pluripotent stem cells (Takashimizu et al., 2009). Purification with ICG has the two major benefits of being both safe and inexpensive, since the ICG system that we are developing is non-immune based and a non-genetically engineered method for selection of differentiated hepatocytes by using flow cytometry with a newly developed laser beam suitable for excitation of ICG.

Except for cell therapy, human hepatocytes are also useful and necessary for drug development (Greenhough et al., 2010). The liver is the central site of drug metabolism and detoxication and thus liver based toxicological tests for developing drugs are necessary. However, preclinical studies of model animals are inadequate to fully evaluate toxicity due to species variation of hepatic functions, such as cytochrome P450 induction (Lake, 2009). This remark is also true for food development. Therefore, the strategy described here is applicable to a wide area of health science.

2. How to obtain hepatocytes

2.1 Hepatocyte induction by differentiating from pluripotent stem cells (ESCs or iPSCs)

Induction of differentiation from pluripotent stem cells into hepatocytes was performed by mimicking the hepatocyte differentiation during embryonic development. During embryonic development, hepatocytes are differentiated from the definitive endoderm and the definitive endoderm is generated from the mesendoderm, which is induced under the influence of Nodal, a transforming growth factor beta (TGF-beta) family gene product (Zaret, 2000; Zaret & Grompe, 2008). The hepatocyte differentiation from the definitive endoderm or mesendoderm is induced by FGF signalling from the cardiac mesoderm and BMP-4 signalling from the septum transversum mesenchyme (Duncan, 2003; Si-Tayeb et al., 2010; Zaret, 2001). Thus, the procedure for induction of hepatocyte differentiation from pluripotent stem cells consists of two steps: the first is definitive endoderm induction by activin, which acts in a similar way to Nodal, the second is hepatocyte induction by FGF and BMP-4. However, because the cells induced with this procedure are not fully maturated, we further induce maturation with serum-free Lanford medium, which is developed for maintaining human hepatocytes in vitro (Sasaki et al., 2009; Yoshie et al., 2010).

2.1.1 Induction of hepatocyte differentiation from mouse ESCs

Although, in most studies, the differentiation-inducing experiments were performed under serum-containing conditions (e.g. Ishii et al., 2005), the hepatocytes induced using these conditions have inevitable risks, such as viral infection advising against use in possible future clinical applications. Therefore, we propose a simple two-step induction method under serum-free conditions as below (Yoshie et al., 2010).

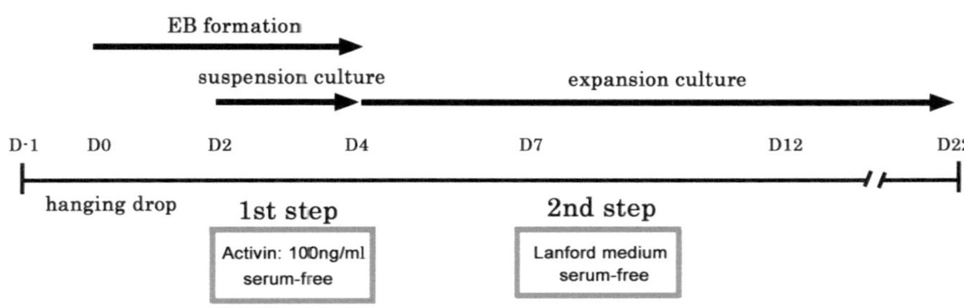

Fig. 1. Experimental design for the induction of hepatocytes from mouse ESCs.

As a first step, the mesendoderm was induced by activin as shown in Fig. 1. The expression of the mesendodermal marker goosecoid and the primary endoderm marker Foxa2 were increased approximately twofold under serum-free conditions than under serum-containing conditions at 100 ng/ml of activin which mimics the role of Nodal signals (Kubo et al., 2004), suggesting that unknown serum factors inhibited hepatic differentiation.
Next, as a second step, hepatocytes were induced from the mesendoderm with serum-free Lanford medium, which has been developed for maintaining human hepatocytes (Lanford et al., 1989). Once immersed in the Lanford medium, the embryoid bodies (EBs) began to show typical hepatic features by day 17, including the expression of hepatic lineage markers albumin (ALB), alpha fetoprotein (AFP), transthyretin (TTR) and α1-anti-trypsin (AAT) detected by RT-PCR, and ALB, AFP and cytokeratin-18 (CK18) detected by immunostaining. On day 22, these cells seemed to have become mature, functional hepatocytes characterized by the expression of metabolizing enzymes, including DPPIV/UDP-glucuronosyl transferase (Ugt1a1), Slcola4, cyp3a11, cyp2b10 and cyp7a1 detected by real-time PCR, a 50-fold greater cyp3A11 response than the control with 100uM dexamethasone stimulation. These results indicate that this simple two-step induction method under serum-free conditions induces high quality hepatic lineage cells directly from mouse embryonic stem (ES) cell-derived mesendoderm.

2.1.2 Induction of hepatocyte differentiation from human ESCs or iPSCs

From embryonic events we learn how to differentiate hepatocytes from pluripotent stem cells. Put simply, FGF produced from cardiac mesoderm triggers hepatic buds from the endoderm-originated tube, BMP-4 from septum transversum promotes differentiation and proliferation, cytokine, such as OSM, leads them to the final stage (Duncan, 2003). Many methods to differentiate have been developed and practiced (Cai et al., 2007; Hu et al., 2003; Ishii et al., 2008; Pei et al., 2009; Soto-Gutierrez et al., 2006; Teratani et al., 2005). Curiously, however, they deleted BMP-4 produced from septum transversum, which is an indispensible factor for hepatic development. So we have developed a new protocol to differentiate EB prepared from H1,khES1,2,3 ES lines into hepatocytes (Sasaki et al., 2009): 25 ng/ml acidic fibroblast growth factor (aFGF) and 25ng/ml basic fibroblast growth factor (bFGF for the first 3 days; 20 ng/ml BMP-4and 20 ng/ml hepatocyte growth factor (HGF) for the next 3 days, and 20 ng/ml oncostatin M (OSM) and 20 ng/ml vascular endothelial

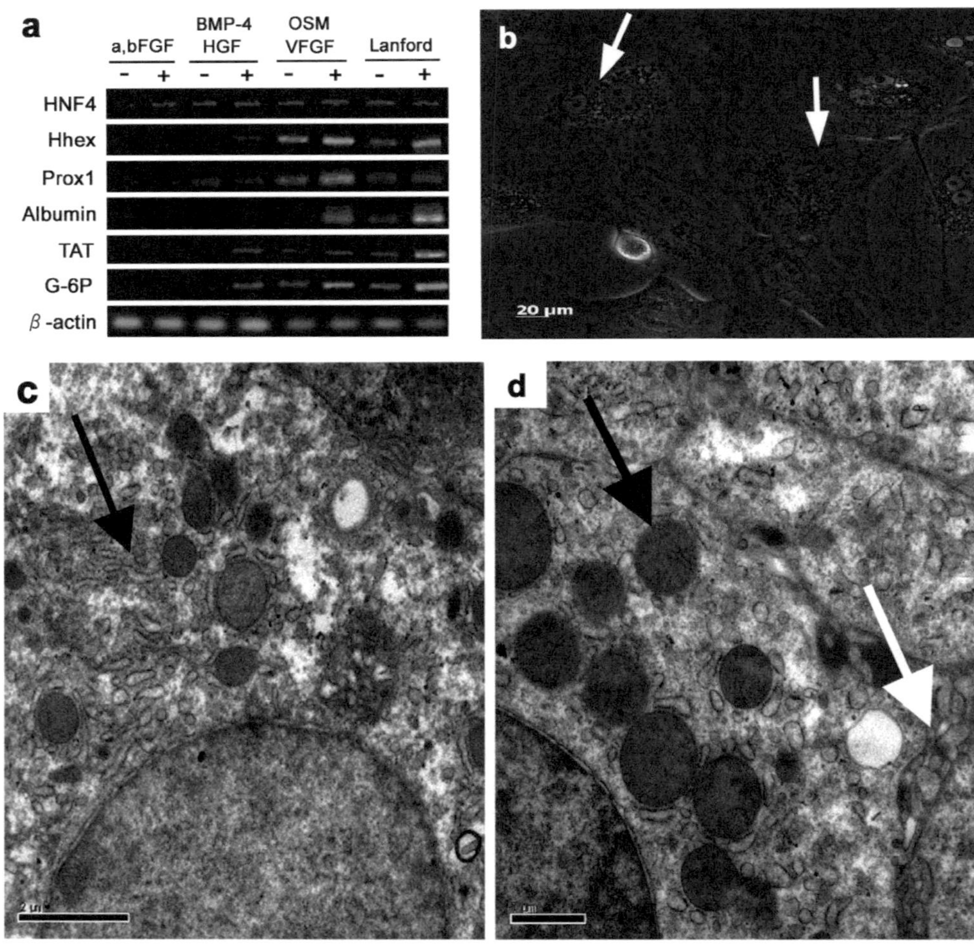

Fig. 2. Induction of hepatocyte differentiation from human ESCs . a. RT-PCR analysis of gene expressions associated with hepatic differentiation. Prox1 is definitely recognized in the OSM stage. Albumin is definite in the Lanford stage. G6P appears from the BMP-4 stage. b. The cluster of large cells with binucleate (arrows). c. Fine structures of the cytoplasm of differentiated hepatocytes. rER are layered (arrow). d. The junctional areas show bile canaliculi with microvilli (white arrow). Round mitochondria are rich in the cytoplasm (black arrow).

growth factor (VEGF) for the third set of 3 days; the modified Lanford medium for an additional 2 weeks. RT-PCR showed switch-on to hepatocyte maturation in the BMP-4 stage (Fig.2a), but it was quickly recognized that Lanford medium was a powerful promoter for cellular maturation. Therefore, Lanford medium was developed to maintain primary hepatocytes without serum. The contents contain growth factors and hormone, including EGF, LCGF, prolactin insulin and glucagon (Lanford et al., 1989). It remains unknown which factors are more effective, but it is true that some of them, or all, progress maturation powerfully (Sasaki et al., 2009; Yoshie et al., 2010). Drastic morphological changes had not been recognized during the periods from FGF stage to OMS, though gene expression showed signs of hepatocyte differentiation. However, several days later in the Lanford medium, the cells enlarged, took polyhedral forms and formed hepatocyte-like colonies with two-nuclei and large cytoplasm (Fig.2b). As described in our paper, they showed many gene expressions seen in mature hepatocytes, ICG uptake and albumin production. However, it was the TEM images that made us confirm that they were hepatocytes. They contained all the features with which morphologists determine that they are hepatocytes (Fig.2c, d). But this method, as well as other methods, has fatal defects. It is complicated, has no ability to produce a large number of hepatocytes at the final differentiation stage and requires high cost to use many growth factors. For clinical application, a simpler method has to be developed.

2.2 Hepatocyte reactivation or transdifferentiation from somatic cells

Although the liver is well known to have remarkable regenerative potential, the pathological liver has a limited ability to regenerate and the hepatocytes isolated from the living liver do not proliferate with ease in vitro (Clayton et al., 1983). However, by using gene transfer techniques similar to that used in the production of iPSC, reactivation of the hepatocytes or reprogramming of the somatic cells to hepatocytes is possible. It has been reported that rat primary hepatocytes, immortalized with oncogenic simian virus 40 T antigen (SV40Tag) could grow in vitro and maintain differentiated hepatic phenotype (Cai et al., 2000, 2002; Schumacher et al., 1996; Tada et al., 1998). Additionally, following transplantation into host rodents with liver failure, these immortalized hepatocytes function as well as primary hepatocytes and improve survival. However, although the SV40 T antigen can be excised by genetic engineering (e.g. Cre/lox recombination), the reprogramming to proliferative state by SV40Tag is thought not to be physiological as SV40 T antigen is a virus gene. We found some endogenous genes which when expressed in the proliferative hepatic cells could stimulate the growth of rat primary hepatocytes, following gene transfer, in vitro (unpublished data), so we expect that hepatocytes can be reactivated physiologically by these methods. However, to prevent immune rejection in cell therapy, these hepatocyte reactivation strategies have to utilize primary hepatocytes from the patient with end-stage liver pathologies or transfer the gene(s) into the patient's liver. Recently, it was reported that functional hepatocyte-like (iHep) cells could be induced directly from mouse tail-tip fibroblasts by gene transfer of Gata4, Hnf1a and Foxa3 (Huang et al., 2011). The transplanted iHep cells repopulated the livers of mice deficient in fumarylacetoacetate hydrolase (Fah) gene, restored liver functions and rescued almost half of recipients. Moreover, recently, we showed gene transfer unmediated transdifferentiation between hepatocytes and pancreatic cells (Kanoh et al., 2011). Thus, in addition to the induction from pluripotent stem cells, the strategy of direct induction from somatic cells will provide promising methods to prepare functional hepatocytes for regenerative medicine and toxicological tests.

3. Applicative aspects of the prepared hepatocytes

3.1 Research for realization of the cell replacement therapy

The major problems concerning realization of the hepatic cell therapy are: (i) existence of undifferentiated cells in the induced hepatocyte sample and (ii) difficulty in the establishment and functioning of transplanted hepatocytes. To solve the first problem, we are trying to establish two selection techniques: 1) isolation of immature hepatocytes with a cell surface marker Liv2 antigen and 2) purification of differentiated hepatocytes with indocyanine green (ICG). The immature hepatocytes isolated with Liv2 antigen retain proliferative capacity and therefore the collection of immature hepatocytes can be amplified in this step. Although the maturation to functional hepatocyte is usually difficult, the flow cytometry with laser beam for ICG could specifically purify functional hepatocytes, because uptake of ICG is hepatocyte-specific function and ICG is incorporated by only functional hepatocytes. The second problem could be partly solved by improving the transplant method. Because the liver with cirrhosis cannot accept transplanted cells, we are examining the transplant method that plants hepatocytes into the submucosa of the duodenum as mentioned in the last section of this chapter. In this way, the bile produced by the hepatocyte may be flowed into the lumen of the duodenum appropriately.

3.1.2 Isolation of immature hepatocytes with Liv2 antibody

Although various cytoplasmic markers, such as ALB, AFP and glucose-6-phosphatase (G6P), have been reported as useful for identifying immature and mature hepatocytes, these are not surface markers, so it is impossible to isolate only living hepatocytes using the fluorescence-activated cell sorting (FACS) or magnetic-activated cell sorting (MACS) method. An isolating method using several surface antigens of hematopoietic markers to select hepatocytes has been reported (Suzuki et al., 2000), but the method is insufficient due to inaccurate cell counts and contamination of other types of cells in addition to hematopoietic cells and hepatocytes. Liv2, which was first reported by Watanabe et al. (Watanabe et al., 2002) is selectively detected in the cell surface of fetal mouse liver cells of embryonic day E9.5 to E12, so it can be used as a surface marker to isolate immature hepatocytes.

We have performed immunocytochemical analysis to verify the presence of Liv2 in immature hepatocytes derived from ES cells (Takashimizu et al., 2009) and we found clusters of Liv2-positive cells in EB outgrowth with the same patterns as the immunoreactivity seen in the E9.5 immature hepatocytes. In addition, we analyzed microlocalization of the Liv2 antigen by immuno-TEM (ImmunoGold) and confirmed definitely that Liv2 is a surface antigen. Next, we analyzed the property of Liv2-positive cells isolated by MACS. Cell counting and MTT assay revealed that Liv2-positive cells isolated by MACS from the primary hepatocytes obtained from the fetal livers of E12 mouse embryos have moderate proliferative potential (Figure 3).

Double-Immunostaining of the isolated cells of 1 day after Liv2-MACS showed that all of Liv2-positive cells analyzed express Prox1, AFP and ALB (Figure 4). Both Prox1 and AFP are markers of immature hepatocytes and not express in the adult liver cells. Thus, it was demonstrated that in the cells obtained from E12 fetal livers, only immature hepatocytes were positive for Liv2.

Further cultivation of the isolated cells confirmed that these Liv2-positive cells were progenitors of mature hepatocytes and could differentiate into mature hepatic cells. Immunocytochemical analysis showed that although population of Liv2-positive cells was decreased 5 days after culture, positive cells for AFP and ALB increased (Figure 4).

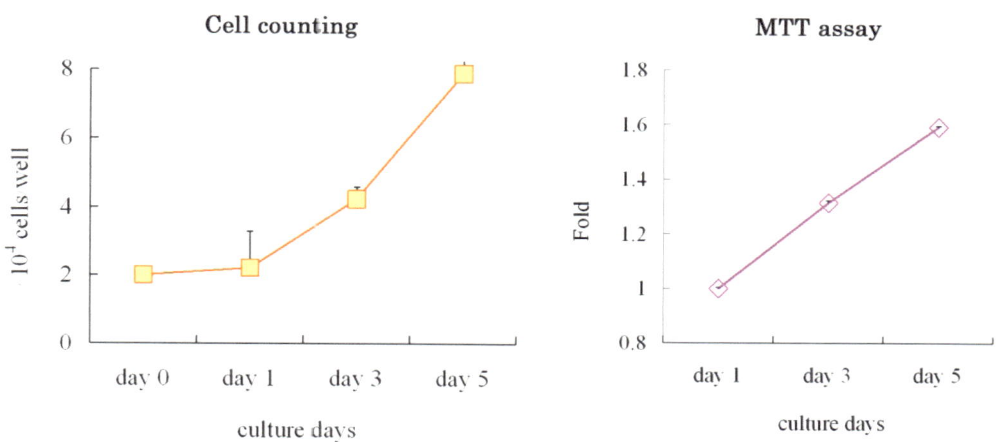

Fig. 3. Proliferation of Liv2-positive cells isolated by MACS. a: cell counting. b: MTT assay.

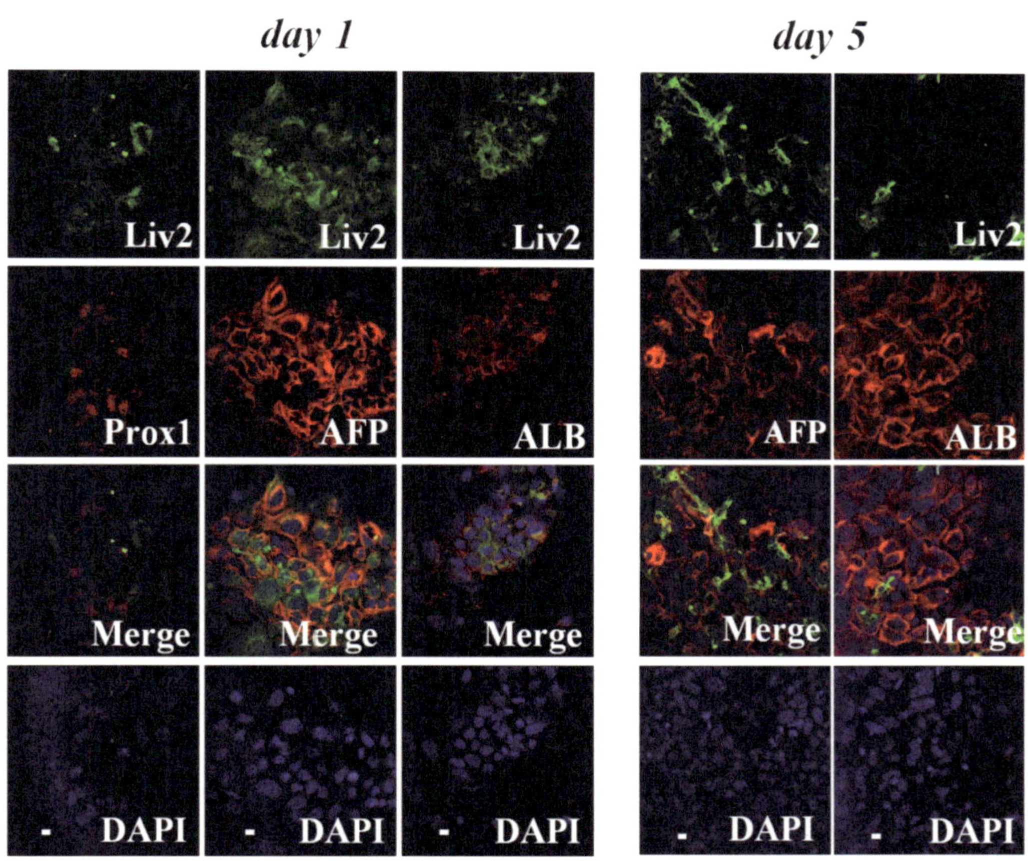

Fig. 4. Immunocytochemical analysis of cultured cells isolated by MACS with Liv2 antibody.

RT-PCR analysis revealed that the expression of mature hepatocyte markers CYP7A1, G6P, TAT and TTR was initially very low or absent and increased gradually afterward (Figure 5A). Moreover, the expression of CK7, a maker of bile duct cells, was extremely low on the first day and drastically up-regulated on day 5 (Figure 5b), suggesting that the Liv2-positive cells were bipotential progenitors capable of producing both hepatocytes and bile duct cells.

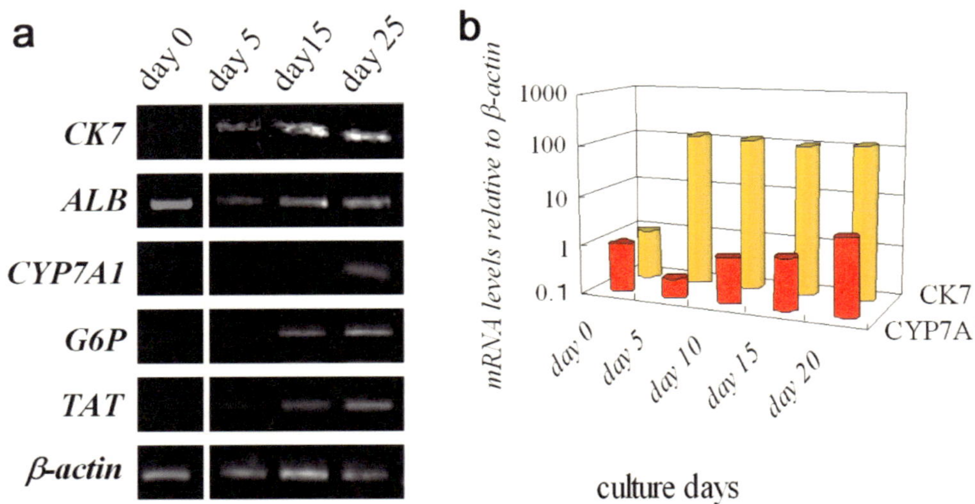

Fig. 5. Analysis of gene expression of mature hepatocyte makers in Liv2 positive cells cultured with Lanford medium. a; RT-PCR , b; Real-Time RT-PCR.

Taken together, these data strongly suggest that Liv2-positive cells are immature hepatocytes normally contributing to embryonic development and that anti-Liv2 antibody can be used to isolate immature hepatocytes derived from ESCs.

There is only one report adopting a similar strategy by purifying hepatic progenitor cells rather than mature hepatic cells with surface antigens (Li et al., 2010). They successfully isolated hepatic progenitors by FACS with c-Kit and EpCAM antibody, and demonstrated that these c-Kit-negative, EpCAM-positive cells could undergo long-term expansion with sustained hepatoblast-like characteristics. Moreover, they cloned the hepatoblast-like cells and showed these clones repopulated in the host livers without inducing tumorigenesis. Although EpCAM expressed in hepatic progenitor cells rather than adult hepatocytes, EpCAM also expressed in many other cells such as undifferentiated ESCs and definitive endoderm cells. Therefore, they complemented EpCAM in positive selection with c-Kit in negative selection since c-Kit is only expressed on ESCs, definitive endoderm cells and some mesoderm cells, but not hepatic cells. However, there may be other cells with c-Kit-negative and EpCAM-positive other than hepatic progenitor and the success of their strategy is thought to depend on their efficient induction of definitive endoderm from mouse ESCs. On the other hand, to isolate immature hepatocytes, we use single antibody (Liv2) highly specific to immature hepatocytes. Although anti-Liv2 antibody specifically recognizes rodent Liv2 antigen, but not human and gene encoding, its Liv2 antigen is still unclear, however, a human counterpart must exist considering the evolutionary relationship between rodent and human. Developing tools for biochemical analysis, such as mass spectrometry, are expected tc reveal the Liv2 gene. Thus, we expect the development of an isolating system for human immature hepatocytes with anti-Liv2 antibody recognizing human Liv2 in the future.

3.1.3 Isolation of mature hepatocytes with ICG

There are a lot of protocols for generating hepatocytes from ESCs or iPSCs in vitro, but purifying mature hepatocytes is still difficult and many different cell types remain. Several studies have used ESC lines transfected with the green fluorescent protein reporter gene controlled by promoters of hepatic genes, such as Foxa2 (Gouon-Evans et al., 2006), Afp (Yin et al., 2002), ALB (Heo et al., 2006; Soto-Gutierrez et al., 2006) and a1-antitrypsin (Aat) (Duan et al., 2007). However, these strategies need genetic manipulation that is not suited to the future therapeutic applications and GFP is known to have slight toxicity to cells. Moreover, mature hepatocytes rather than immature hepatocytes are known to be unstable in vitro. Thus, a particularly safe method is required for selecting mature hepatocytes derived from ESCs or iPSCs.
Indocyanine green (ICG) is a nontoxic organic anion that is eliminated exclusively by hepatocytes and is clinically used as a test substance to evaluate liver function (Berk & Stremmel, 1986). The uptake and release of ICG are frequently used to identify and/or evaluate ESC-derived hepatocytes (Agarwal et al., 2008; Farzaneh et al., 2010; Yamada et al., 2002). The fluorescence of ICG can be observed at 800-900 nm with appropriate excitation of near 780 nm. Therefore, we developed a flow cytometer equipped with an excitation laser of 785 nm to detect ICG-positive cells (Yoshie et al., manuscript in preparation). We determined optimal concentration and incubation time for detection of ICG uptake and showed that rat primary hepatocytes and ESC-derived hepatocytes selectively took up ICG. Although irradiation of ICG with a laser at 100 J/cm^2 was reported to generate singlet oxygen ($^1O_2$), which is damaging to cells (Hirano et al., 2007), the power of our laser (FISHMAN-R) was 5 mJ/cm^2 and much less likely to generate singlet oxygen. In fact, in analysis with trypan blue exclusion, the analyzed cells were not damaged by FISHMAN-R laser flow cytometry.
Similar sorting methods of hepatocytes or hepatocyte-like cells by use of antibodies are frequently reported, e.g. isolation of mature hepatocyte-like cells by FACS with sialoglycoprotein-receptor (Basma et al., 2009). However, because the purification of mature hepatocytes is thought to be a final step and the mature hepatocytes have limited proliferative potential, this purification step must be massive i.e. sorting a lot of hepatocytes. Therefore, use of antibodies in this step is inadequate in the cost performance, as the antibodies are expensive. Our detection system which relies on the selective uptake by hepatocytes of ICG, a stable and inexpensive fluorescent chemical compound, overcomes this problem and will lead to the development of an effective system for purifying hepatocytes derived from ESCs, iPSCs or somatic cells.

3.2 Use of hepatocytes for toxicology and drug development

The liver is centre of metabolism and the hepatocytes are responsible for metabolizing most compounds in vivo. Therefore, the cultured hepatocyte can be used to predict how drugs are metabolized and to what extent they may be toxic. These tests include use of primary rodent hepatocytes and human immortalised hepatocyte cell lines, such as HepG2 (Rudzok et al., 2010). However, the rodent hepatocytes have species-specific differences in metabolizing function from human (Lake, 2009) and immortalised hepatocytes have poor hepatic function and sometimes have an abnormal karyotype (Wong et al., 2000). Thus, human primary hepatocytes have become a standard tool for evaluating hepatic drug metabolism in vitro. However, to do so demands a lot of functional hepatocytes and the utility of the human primary hepatocyte is seriously limited. Our induction and selection methods will be helpful to meet the demand of these examination systems as an alternative to primary human hepatocytes. Particularly, dual selection strategies mentioned above are suited for scalable production of human hepatocytes. Immature hepatocytes isolated with Liv2 were suggested to have unlimited proliferative potential (Figure 3) and could serve the scalable production. Although expanded immature hepatocytes have to differentiate into mature hepatocytes, our protocol could efficiently induce maturation with Lanford medium and the functional, mature hepatocytes that must be in large amounts could be purified inexpensively by FACS with ICG.

4. Further problems

4.1 In vitro tissue formation using induced hepatocytes

Besides immature hepatocytes being acceptable in some cases of hepatocyte transplantation due to hepatic maturation after transplantation, immature hepatocytes are absolutely unacceptable for use in evaluating hepatic drug metabolism in vitro. The mature functional hepatocytes are required in drug tests. However, in vitro induction of hepatic maturation or terminal differentiation has not been achieved enough. Realization of fully functional hepatocytes may require tissue organization similar to that of the liver in the living body. There is a dilemma in this problem: although the purification of hepatocytes is necessary for the improvement of the quality of induced hepatocytes, fully functional liver tissue requires various types of cells other than hepatic parenchymal cells. To solve this problem, we have three options: (i) selecting several types of cells and organizing them into tissue structure, (ii) selecting the precursor cells for histogenesis of liver, inducing differentiation and prompting self-organization and (iii) constructing artificial culture microenvironments mimicking native environments of liver. The culture microenvironments, such as microwell architecture (Mori et al., 2008), are well known to improve the function of cells, including hepatocytes (Sharma et al., 2010).

Previously, we showed the induction of retinal cells by co-culturing with ESC-derived RPE cells (Yue et al., 2010). Now, we try to apply a combination of these co-culture and microwell methods to induce mature function of hepatocytes.

4.2 Hepatocytes transplantation: Where or how are hepatocytes transplanted?

Purified pluripotent stem cell-derived hepatocytes are transplanted into patients with fatal hepatic disease, in whom hepatic tissue structures are newly developed and function. Thus, a new therapy for incurable hepatic disease substituting liver transplantation is established, which is a final goal of regenerative medicine.

The problems to overcome for successful regenerative medicine are to determine the appropriate sites for effective cellular survival and function, and the effective way, i.e. which should be selected for transplantation, cell type or tissue type. The best way still remains unknown. Most experimental studies show injection of pluripotent stem cell-derived hepatocytes through the vein or directly into the liver as cell type. Mouse ESC-derived hepatocytes were injected into the liver through the portal vein (Yamada et al., 2002). HGF-treated bone marrow mesenchymal cells were transplanted through the caudal vein into the liver of CCl4-injured rats and this was effective against liver damage (Oyagi et al., 2006). After multipotent progenitor cells derived from human umbilical cord blood were transplanted into the left lateral lobe of the liver of the CCl4-injured rats or through the portal vein, they differentiated into the hepatocyte-like cells and were useful in treating the injury (Moon et al., 2009).

But there is extreme pessimism about these ideas, methods and practices because fatally damaged livers provide poor and severe environments, such as extreme fibrosis in cirrhosis, rapid cellular necrosis in liver failure etc. for transplanted fresh hepatocytes. It is more reasonable to transplant hepatocytes into good conditions with a rich blood supply ectopically. In fact, the first transplanted site performed in a human patient with cirrhosis was the spleen (Mito et al., 1992). According to their report, a part of the cirrhotic liver was resected, from which intact hepatocytes were isolated and were injected into the spleen; afterwards, the survival was confirmed. Their experimental reports using rats indicated that 6 months later splenic tissues were substituted by the hepatic tissues consisting of hepatic cords and hepatic sinus (Kusano & Mito, 1982; Mito et al., 1979). Except the spleen, fat tissues are good sites for breeding hepatocytes and their survival has been shown in fat pads (Jirtle & Chalopoulos, 1965). One of the authors transplanted rat primary hepatocytes into the rat mesenteric fat tissues and confirmed their survival and tissue structure (Sasaki et al., 1983) (Figure 6a). But ectopic transplantation has a significant problem: it has no way and route to discharge bile juice. Any definite answers have not been given yet, but we have some hints that might help to resolve the problem. For example, glands e.g. duodenal glands or pancreatic tissues are found in the submucosa of the intestine (Fig.6b, c). Clinically ectopic tissues e.g. pancreatic or stomach tissues are recognized. In evolution, primitive hepatic tissues are incorporated into the midintestine in the earth worm as chloragocytes with varied functions, including a storing function of endogenous substances, such as glycogen and lipids, and metals (Ireland & Richards, 1977) and intensive DAB reactivity (Fischer & Horvath,

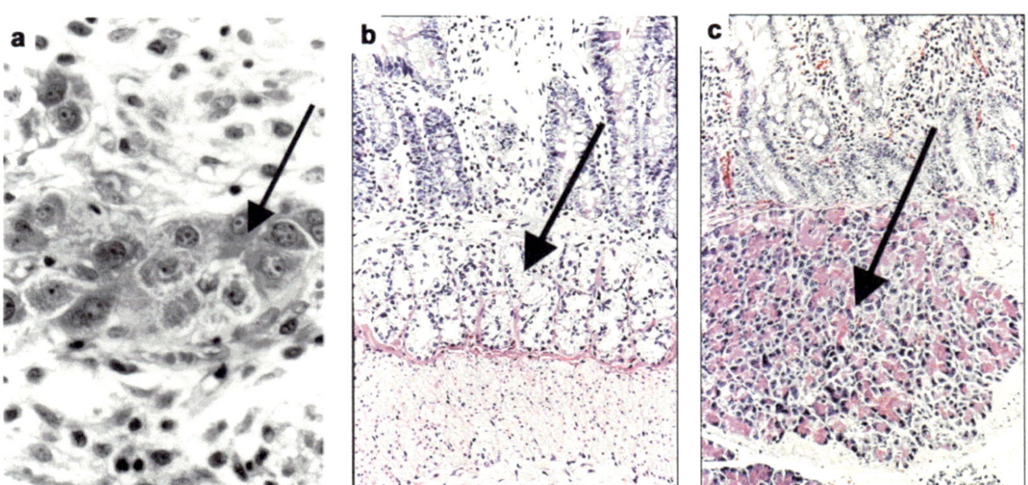

Fig. 6. Strategy of ectopic transplantation. a. The survival hepatocyte cluster in the rat intra-mesenterium (arrow). b. The duodenal gland in the submucosa of the rat duodenum (arrow). c. The pancreatic tissue in the submucosa of the small intestine (arrow).

1978). The next function in shipworms is midgut glands consisting of large cells with binucleate (Strunk, 1959) and enhanced hepatic function in crabs and lobsters as hepatopancreas (James, 1989). In mammalian embryonic development, it is well-known that hepatic buds grow from the duodenum described above. Therefore, our idea is to transplant hepatocytes into the submucosa and return them into the mother site, the intestine.

Our first trial was to confirm whether human ES cell-derived embryo bodies, which were treated with the previous cocktail (Takashimizu et al., 2009) and containing ICG-positive cells (Fig.7a), survived in the submucosa or in the intestine. Injection was insecure in that it was difficult to confirm where it was injected or whether hepatocytes or EBs remained there. Then, we have established a new technique for transplantation. The peritoneum and muscle layer of the nude rat was cut by an electronic scalpel and separated (Fig.7b), in whose shallow furrow EBs were arranged in order (Fig.7c). After the wound was closed (Fig.7d), one or two weeks later, the transplanted site was prepared for histological investigation. In the submucosa, unique tissues including duct like structures were found (fig.7e). The vascular

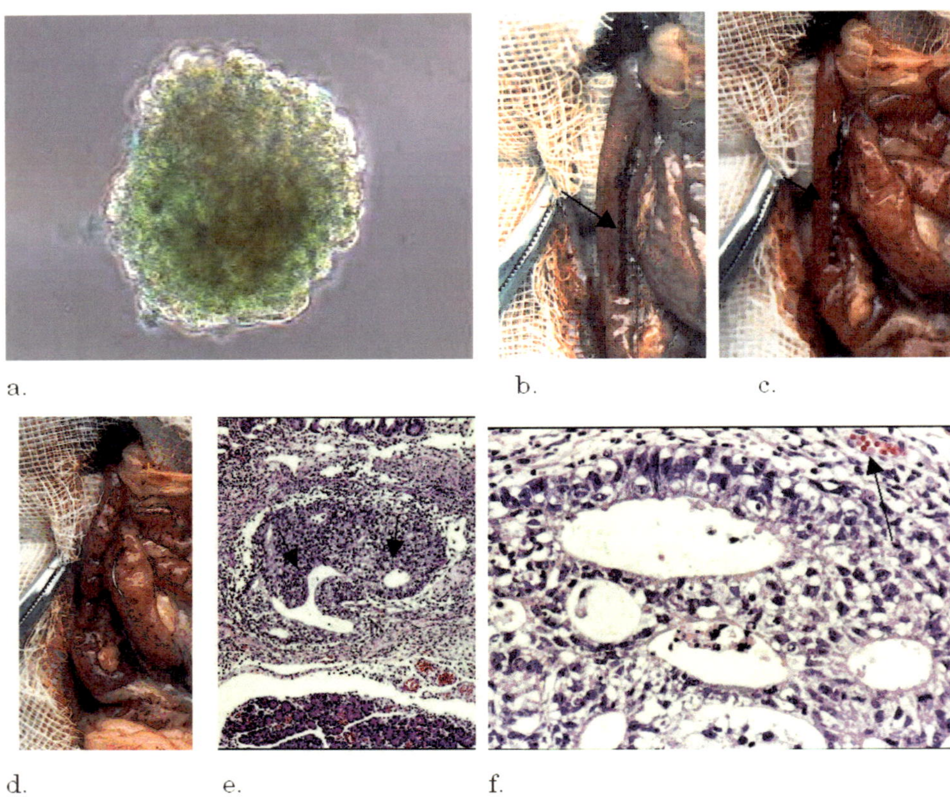

Fig. 7. Transplantation of human ESC-derived hepatocyte-like cells into a rat duodenum. a. Human EB containing ICG positive cells (green). b. The linear cut line performed on the surface of the nude rat duodenum (arrow). c. EBs (arrow) are arranged in the shallow furrow. d. The wound was closed with 7-0 nylon. e. Histology of the treated duodenum. EB survives in the submucosa along with duct formation (arrows). f. Vascular supply is confirmed (arrow), but is not sinusoid-like.

supply did not penetrate the cell cluster, but rounded the periphery (Fig.7f). Hepatic sinus did not appear to be differentiated yet. We recognized transplanted cell clusters survived in this method, but did not confirm that they differentiated into functional hepatocytes from EBs containing ICG-positive cells. At that time, we did not have the method to purify pluripotent stem cell-derived hepatocytes from mixed varied cell types yet. As the ICG-selection method has been established, as described above, purified hepatocytes will be transplanted and be confirmed to function and discharge bile juice into the intestine.

Further, for clinical trials, we provide a new technique, fibre scope transplantation. To cut the muscular layer is to subject patients to opening the abdomen, whereas if hepatocytes were injected into the submucosa, patients' discomfort would significantly decrease. But it remains unknown how hepatocytes are injected into the submucosa with the fibre scope, as fibre scope has not yet been developed for rats. The scope with the smallest diameter for use on a dog was tried through the stomach to attain the duodenum (Fig.8a). It was very difficult to inject into the mucosa on the same side, but was successful in swelling the mucosa on the opposite side (Fig.8b). Fibre scope transplantation will be an effective tool for endoderm-derived differentiated cells in addition to hepatocytes.

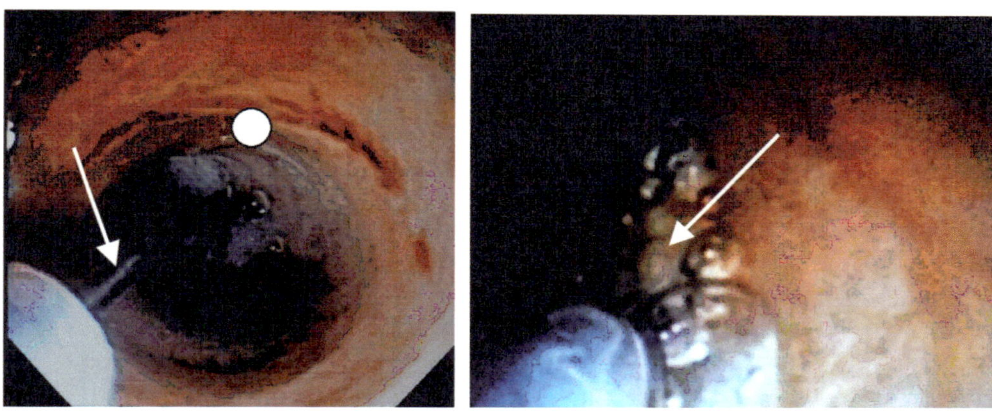

a. b.

Fig. 8. Fibre scope transplantation. a. The interior of the nude rat duodenum. The red mucosa and the needle for injection of the fibre scope (arrow) are recognized. White circle is the mark for the needle to insert. b. The needle is inserted into the mucosa. The mucosa swells after fluid injection (arrow).

On the other hand, another problem is whether cell transplantation or tissue transplantation is better, more effective or reasonable. It will take a long time to determine this, or each method may be used according to varied conditions, such as kinds of disease, its progress and permitted transplanted site. But in conclusion we prefer tissue transplantation, because effective function is due to cell-to-cell interaction and tissue organization. Kikuchi et al. have developed a unique in site cell micropatterning system (Kikuchi et al., 2009). A photoresponsive cell culture surface was geometrically processed in situ with the UV irradiation to increase the cell adhesiveness of HepG2. After confirmation of HepG2 adhesion to the first dot pattern, fibroblast, which adhered easily to the plate without the UV irradiation were introduced to fill gaps among the HepG2 dot pattern. This simple

geometric pattern caused a 50-fold increase of CYP3A than conventionally cultured HepG2. Cell transplantation breaks once-established cell-to-cell interaction and requires reorganization in vivo, which creates double the work.
But three methods to complete tissue transplantation have to be established, the in vitro method of tissue organization, which geometric patterning described above helps, the technique to transport in vitro established tissue into in vivo without destruction and the method to acquire rapid blood supply after transplantation.
Scaffold-engineering may be essential to tissue transplantation, because it is easily transported in intact conditions and is processed for drug delivery systems e.g. slow releasing angiogenic factors (Hou et al., 2011). Collagen or modified collagen sponges were conventionally used as cell carriers in many labs, as well as in ours (Imamura et al., 2004; Kanematsu et al., 2004; Takimoto et al., 2003). The above collagen-based scaffold is not complete, because cells did not expand rapidly within the sponge in vitro. After transplantation, the carrier was bounded on the periphery by the non-penetrating host tissue. Moreover, function, such as urea synthetic rate, decreased below the control level (Hasirci et al., 2001). In addition to collagen-based materials, many other materials and methods are proposed (Kano et al., 2008; Katsuda et al., 2010; Torok et al., 2011), which will steadily improve tissue transplantation techniques. In the future it is expected that simpler, easier, more effective and more practical transplantation techniques will be developed.

5. Conclusion

In this chapter, we described the production and selection of hepatocytes and show a strategy for regenerative medicine using pluripotent-derived hepatocytes. There are two sources to obtain hepatocytes: (i) somatic cells, such as mature hepatocytes themselves and (ii) multi- or pluripotent stem cells, such as embryonic stem cells (ESCs) and induced pluripotent stem cells (iPSCs). We have reported the methods for induction of differentiation from pluripotent stem cells. Our protocols for induction of differentiation from pluripotent stem cells mimic the hepatocellular differentiation during embryonic development.
One important problem of the differentiation-mediated preparation of hepatocytes is how to exclude undifferentiated cells, while a lot of hepatocytes are produced. Additionally, it is impossible to improve the efficiency of hepatocellular differentiation to 100% and the remaining undifferentiated cells have the potential to cause teratoma after being transplanted into the host. To isolate hepatocytes, we proposed a dual selection method. The first is isolation of immature hepatocytes by cell sorting with Liv2, which specifically detects immature hepatocytes. The second is purification of mature hepatocytes by low cost sorting with ICG, following massive expansion of immature hepatocytes and induction of differentiation into mature hepatocytes with Lanford medium. The flow of our method is as below:

i. Induction of immature hepatocytes from pluripotent stem cells
ii. Isolation of proliferative immature hepatocytes with anti-Liv2 antibody
iii. Large scale culture expanding the immature hepatocytes and induction of differentiation into mature hepatocytes
iv. Low cost and safe purification of functional mature hepatocytes by FACS with ICG

The next step is in vitro and in vivo tissue formation. The culture condition mimicking tissue organization improves the function of cells, including hepatocytes. Moreover, to realize the therapy for patients with incurable hepatic disease, it is necessary to promote tissue organization artificially in vivo. We propose that for the artificial organization of hepatic tissue, heterotropic transplantation into submucosa of the duodenum is most suitable, since the liver with incurable disease does not accept the transplanted hepatocytes easily.

6. Acknowledgments

We gratefully thank Ms. Kayo Suzuki and Dr. Kiyokazu Kametani (Research Center for Instrumental Analysis, Shinshu University) for their excellent technical assistance.

7. References

Andersson, ER. & Lendahl, U. J (2009). Regenerative medicine: a 2009 overview. *Intern Med.* 266(4):303-310

Agarwal, S.; Holton, K.L. & Lanza, R. (2008). Efficient differentiation of functional hepatocytes from human embryonic stem cells. *Stem Cells* 26: 1117-1127

Arkadopoulos, N.; Chen, SC.; Khalili, T.M.; Detry O.; Hewitt, WR.; Lilja, H.; Kamachi, H.; Petrovic, L.; Mullon, CJ.; Demetriou, AA. & Rozga, J. (1998). Transplantation of hepatocytes for prevention of intracranial hypertension in pigs with ischemic liver failure. *Cell Transplant.*7:357-363

Basma, H.; Soto-Gutierrez, A.; Yannam, GR.; Liu, L.; Ito, R.; Yamamoto, T.; Ellis, E.; Carson, SD.; Sato, S.; Chen, Y.; Muirhead, D.; Navarro-Alvarez, N.; Wong, RJ.; Roy-Chowdhury, J.; Platt, JL.; Mercer, DF.; Miller, JD.; Strom, SC.; Kobayashi, N.; Fox, IJ. (2009). Differentiation and transplantation of human embryonic stem cell-derived hepatocytes *Gastroenterology* 136:990-999

Behbahan, IS.; Duan, Y.; Lam, A.; Khoobyari, S.; Ma, X.; Ahuja, TP. & Zern, MA. (2011). New approaches in the differentiation of human embryonic stem cells and induced pluripotent stem cells toward hepatocytes. *Stem Cell Rev.* Feb 19. [Epub ahead of print]

Berk, PD. & Stremmel, W. (1986). Hepatocellular uptake of organic anions. *Prog. Liver Dis.* 8:125-144

Benten, D.; Staufer, K. & Sterneck, M. (2009). Orthotopic liver transplantation and what to do during follow-up: recommendations for the practitioner. *Nat. Clin. Pract. Gastroenterol Hepatol.* 2009 6(1):23-36

Blum, B.& Benvenisty, N. (2009). The tumorigenicity of diploid and aneuploid human pluripotent stem cells. *Cell Cycle* 8(23):3822-3830

Cai, J.; Ito, M.; Westerman, K.; Kobayashi, N.; Leboulch, P. & Fox, IJ. (2000). Construction of a non-tumorigenic rat hepatocyte cell line for transplantation: reversal of hepatocyte immortalization by site-specific excision of the SV40 T antigen. *J. Hepatol.* 33:701-708

Cai, J.; Ito, M.; Nagata, H.; Westerman, KA.; Lafleur, D.; Chowdhury, JR.; Leboulch, P. & Fox, IJ. (2002). Treatment of liver failure in rats with end-stage cirrhosis by transplantation of immortalized hepatocytes. *Hepatology.* 36(2):386-394

Cai, J.; Zhao, Y.; Liu, Y.; Ye, F.; Song, Z.; Qin, H.; Meng, S.; Chen, Y.; Zhou, R.; Song, X.; Guo, Y.; Ding, M. & Deng, H. (2007). Directed differentiation of human embryonic stem cells into functional hepatic cells. *Hepatology* 45,1229-1239

Clayton, DF. & Darnell, JE. Jr. (1983). Changes in liver specific compared to common gene transcription during primary culture of mouse hepatocytes. *Molecular and Cellular Biology* 3 (9).; 1552-1561

Demetrious, AA.; Levenson, SM.; Novikoff, PM.; Novikovv, AB.; Chowdhury, NR.; Whiting, J.; Reisner, A. & Chowdhury, JR. (1986). Survival, organization and function of microcarrier-attached hepatocytes transplanted in rats. *Proc. Natl. Acad. Sci. USA* 83:7475-7479

Duncan, SA.; (2003). Mechanisms controlling early development of the liver. *Mech.Dev.*120 19-33

Duan, Y.; Catana, A.; Meng, Y.; Yamamoto, N.; He, S.; Gupta, S.; Gambhir, SS.; Zern, MA. (2007). Differentiation and enrichment of hepatocyte-like cells from human embryonic stem cells in vitro and in vivo. *Stem Cells* 25:3058-3068

Farzaneh, Z.; Pournasr, B.; Ebrahimi, M.; Aghdami, N. & Baharvand, H. (2010). Enhanced functions of human embryonic stem cell-derived hepatocyte-like cells on three-dimensional nanofibrillar surfaces. *Stem Cells Rev.* 6: 601-610

Fischer, E. & Horvath, I. (1978). Evidence of the presence of extraperoxisomal catalase in chloragogen cells of the earthworm, Lumbricus terrestris L. *Histochemistry* 56(2):165-171

Fong, CY.; Gauthaman, K. & Bongso, A. (2010). Teratomas from pluripotent stem cells: a clinical hurdle. *J. Cell Biochem.* 111(4):769-781

Fox, IJ.; Chowdhury, JR.; Kaufman, SS.; Goertzen, TC.; Chowdhury, NR.; Warkentin, PI.; Dorko, K.; Sauter, BV. & Strom, SC. (1998). Treatment of Crigler-Najjar syndrome type I with hepatocyte transplantation. *N. Engl. J. Med.* 338:1422-1426

Greenbaum LE. (2010). From skin cells to hepatocytes: advances in application of iPS cell technology. *J. Clin. Invest.* 120(9):3102-3105

Gouon-Evans, V.; Boussemart, L.; Gadue, P.; Nierhoff, D.; Koehler, CI.; Kubo, A.; Shafritz, DA.; Keller, G. (2006). BMP-4 is required for hepatic specification of mouse embryonic stem cell-derived definitive endoderm. *Nat Biotechnol* 24:1402-1411

Greenhough, S.; Medine, CN. & Hay DC. (2010). Pluripotent stem cell-derived hepatocyte-like cells and their potential in toxicity screening. *Toxicology* 278(3):250-255

Hanna, JH.; Saha K. & Jaenisch R. (2010) Pluripotency and cellular reprogramming: facts, hypotheses, unresolved issues. *Cell.* 143(4):508-525

Hasirci, V.; Berthiaume, F.; Bondre, SP.; Gresser, JD.; Trantolo, DJ.; Toner, M. & Wise DL. (2001). Expression of liver-specific functions by rat hepatocytes seeded in treated poly (lactic-co-glycolic) acid biodegradable foams. *Tissue Eng.* 7(4):385-394

Heo, J.; Factor, VM.; Uren, T.; Takahama, Y.; Lee, JS.; Major, M.; Feinstone, SM.; Thorgeirsson, SS. (2006). Hepatic precursors derived from murine embryonic stem cells contribute to regeneration of injured liver. *Hepatology* 44:1478-1486

Hirano, T.; Kohno, E.; Gohto, Y. & Obana, A. (2007). Singlet oxygen generation by irradiation of Indocyanina Green (ICG) and its effect to tissues. *Nippon Laser Igakkaishi* 28: 122-128

Hou, YT.; Ijima, H.; Takei, T.& Kawakami, K. (2011).Growth factor/heparin-immobilized collagen gel system enhances viability of transplanted hepatocytes and induces angiogenesis. *J. Biosci. Bioeng.* Jun 1. [Epub ahead of print]

Hu, A.; Cai, J.; Zheng, Q.; He, X.; Pan, Y. & Li, L. (2003) Hepatic differentiation from embryonic stem cells in vitro. *Chin. Med. J.* 116, 1893-1897

Huang, P.; He, Z.; Ji, S.; Sun, H.; Xiang, D.; Liu, C.; Hu, Y.; Wang, X. & Hui, L. (2011). Induction of functional hepatocyte-like cells from mouse fibroblasts by defined factors. *Nature* May 11. [Epub ahead of print]

Imamura, T.; Cui, L.; Teng, R.; Johkura, K.; Okouchi, Y.; Asanuma, K.; Ogiwara, N. & Sasaki, K. (2004). Embryonic stem cell-derived embryoid bodies in three-dimensional culture system form hepatocyte-like cells in vitro and in vivo. *Tissue Eng.* 10(11-12):1716-1724

Ireland, MP. & Richards, KS. (1977). The occurrence and localisation of heavy metals and glycogen in the earthworms Lumbricus rubellus and Dendrobaena rubida from a heavy metal site. *Histochemistry* 51(2-3):153-166

Ishii, T.; Yasuchika, K.; Fujii, H.; Hoppo, T.; Baba, S.; Naito, M.; Machimoto, T.; Kamo, N.; Suemori, H.; Nakatsuji, N. & Ikai I. (2005). In vitro differentiation and maturation of mouse embryonic stem cells into hepatocytes. *Exp. Cell Res.* 309(1):68-77

Ishii, T.; Fukumitsu, K.; Yasuchika, K.; Adachi, K.; Kawase, E.; Suemori, H.; Nakatsuji, N.; Ikai, I. & Uemoto, S.(2008). Effects of extracellular matrixes and growth factors on the hepatic differentiation of human embryonic stem cells. *Am. J. Physiol. Gastrointest. Liver Physiol.* 295, G313-321

Ito, M.; Nagata, H.; Yamamoto, T.; Yoshihara, D.; Fox, IJ. & Miyakawa, S. (2007). Intrasplenic hepatocyte transplantation prolonged the survival in Nagase analbuminemic rats with liver failure induced by common bile duct ligation. *Cell Transplant.* 16(5):547-553

Ito, M.; Nagata, H.; Miyakawa, S. & Fox IJ. (2009). Review of hepatocyte transplantation. *J. Hepatobiliary Pancreat Surg.* 16(2):97-100

James MO. (1989). Cytochrome P450 monooxygenases in crustaceans. *Xenobiotica* 19(10):1063-1076

Jirtle, RJ. & Chalopoulos, GM. (1964). Effects of partial hepatectomy on transplanted hepatocytes. *Cancer Res* 24:1611-1625

Kanematsu, A.; Yamamoto, S.; Ozeki, M.; Noguchi, T.; Kanatani, I.; Ogawa, O. & Tabata, Y. (2004). Collagenous matrices as release carriers of exogenous growth factors. *Biomaterials* 25(18):4513-4520

Kano, K.; Yamato, M. & Okano, T. (2008). Ectopic transplantation of hepatocyte sheets fabricated with temperature-responsive culture dishes. *Hepatol Res.* 38(11):1140-1147

Kanoh, Y.; Tomotsune, D.; Shirasawa, S.; Yoshie, S.; Ichikawa, H.; Yokoyama, T.; Mae, S.; Ito, J.; Mizuguchi, M.; Matsumoto, K.; Yue, F. & Sasaki, K. (2011). In vitro transdifferentiation of HepG2 cells to pancreatic-like cells by CCl4, D-Galactosamine, and ZnCl2. *Pancreas* [Epub ahead of print]

Katsuda, T.; Teratani, T.; Ochiya, T. & Sakai Y. (2010). Transplantation of a fetal liver cell-loaded hyaluronic acid sponge onto the mesentery recovers a Wilson's disease model rat. *J. Biochem* 148(3):281-288

Kikuchi, K.; Sumaru, K.; Edahiro, J.; Ooshima, Y.; Sugiura, S.; Takagi, T. & Kanamori, T. (2009). Stepwise assembly of micropatterned co-cultures using photoresponsive culture surfaces and its application to hepatic tissue arrays. *Biotechnol Bioeng.* 103(3):552-561

Kusano, M. & Mito M. (1982). Observations on the fine structure of long-survived isolated hepatocytes inoculated into rat spleen. *Gastroenterology.* 82(4):616-628

Knoepfler, PS. (2009). Deconstructing stem cell tumorigenicity: a roadmap to safe regenerative medicine. *Stem Cells* 27(5):1050-1056

Kobayashi, N.; Ito, M.; Nakamura, J.; Cai J.; Hammel, JM.& Fox, IJ. (2000). Treatment of carbon tetrachloride and phenobarbital-induced chronic liver failure with intrasplenic hepatocyte transplantation. *Cell Transplant* 9(5):671-673

Kubo, A.; Shinozaki, K.; Shannon, JM.; Kouskoff, V.; Kennedy, M.; Woo, S,; Fehling, HJ. & Keller, G. (2004). Development of definitive endoderm from embryonic stem cells in culture. *Development* 131(7): 1651-1662

Kung, JW. & Forbes, SJ. (2009). Stem cells and liver repair. *Curr Opin Biotechnol*. 20(5):568-74.

Lake, BG.; (2009). Species' differences in the hepatic effects of inducers of CYP2B and CYP4A subfamily forms: relationship to rodent liver tumour formation. *Xenobiotica* 39:582-596

Lanford, RE.; Carey, KD.; Estlack, LE.; Smith, GC. & Hay, RV. (1989). Analysis of plasma protein and lipoprotein synthesis in long-term primary cultures of baboon hepatocytes maintained in serum-free medium. *In Vitro Cell Dev Biol*. 25(2):174-182

Li, F.; Liu, P.; Liu, C.; Xiang, D.; Deng, L.; Li, W.; Wangensteen, K.; Song, J.; Ma, Y.; Hui, L.; Wei, L.; Li, L.; Ding, X.; Hu, Y.; He, Z.; Wang, X. (2010). Hepatoblast-like progenitor cells derived from embryonic stem cells can repopulate livers of mice. *Gastroenterology* 139(6):2158-2169

Makowka, L.; Rotstein. LE.; Falk, RE.; Falk, J. A.; Langer, B.; Nossal, NA.; Blendis, LM. & Phillips, MJ. (1980). Reversal of toxic and anoxic induced hepatic failure by syngeneic, allogeneic and xenogeneic hepatocyte transplantation. *Surgery* 88:244-253

Mito, M.; Kusano, M. & Kawaura, Y. (1992). Hepatocyte transplantation in man. *Transplant Proc*. 24(6):3052-3053

Mito, M.; Ebata, H.; Kusano, M.; Onishi, T.; Saito, T.; Sakamoto, S. (1979). Morphology and function of isolated hepatocytes transplanted into rat spleen. *Transplantation* 28(6):499-505

Moon, YJ.; Yoon, HH.; Lee, MW.; Jang, IK.; Lee, DH.; Lee, JH.; Lee, SK.; Lee, KH.; Kim, YJ.; Eom, YW. (2009). Multipotent progenitor cells derived from human umbilical cord blood can differentiate into hepatocyte-like cells in a liver injury rat model. *Transplant Proc*. 41(10):4357-4360

Mori, R.; Sakai, Y. & Nakazawa, K. (2008). Micropatterned organoid culture of rat hepatocytes and HepG2 cells. *J. Biosci. Bioeng*. 106(3):237-242

Nagata, H.; Ito M.; Shirota, C.; Edge, A.; McCowan, TC.& Fox, IJ. (2003) Route of hepatocyte delivery affects hepatocyte engraftment in the spleen. Transplantation.76(4):732-734

Oyagi, S.; Hirose, M.; Kojima, M.; Okuyama, M.; Kawase, M.; Nakamura, T.; Ohgushi, H.; Yagi, K. (2006). Therapeutic effect of transplanting HGF-treated bone marrow mesenchymal cells into CCl4-injured rats. *J. Hepatol*. 44(4):742-748

Pei, H.; Yang, Y.; Xi, J.; Bai, Z.; Yue, W.; Nan, X.; Bai, C.; Wang, Y. & Pei, X. (2009). Lineage restriction and differentiation of human embryonic stem cells into hepatic progenitors and zone 1 hepatocytes. *Tissue Eng. Part C Methods* 15, 95-104

Platt, J. (1998). New directions for organ transplantation. *Nature*. 392:11-17

Ribeiro, J.; Nordlinger, B.; Ballet, F.; Cynober, L.; Coudray-Lucas, C.; Baudrimont, M.; Legendre, C.; Delelo, R. & Panis, Y. (1992). Intrasplenic hepatocellular transplantation corrects hepatic encephalopathy in portacaval-shunted rats. *Hepatology*. 15:12-18

Rudzok, S.; Schlink, U.; Herbarth, O. & Bauer, M.; (2010). Measuring and modeling of binary mixture effects of pharmaceuticals and nickel on cell viability/cytotoxicity in the human hepatoma derived cell line HepG2. *Toxicol. Pharmacol*. 244:336-343

Sasaki, K.; Miyano, T.; Suruga, K.; Nozawa, R.; Yokota, K. (1983). Isografts of rat hepatocytes into intramesenterium. *Jap J. Pediatr Surg*. 15:1450-1451

Sasaki, K., Ichikawa, H., Takei, S., No, H., Tomotsune, D., Kano, Y., Yokoyama, T., Shirasawa, S., Mogi, A., Yoshie, S., Sasaki, S., Yamada, S., Matsumoto, K., Mizuguchi, M., Yue, F. & Tanaka, Y. (2009). Hepatocyte differentiation from human ES cells using the simple embryoid body formation method and the staged additional cocktail. *ScientificWorldJournal* 9:884-890

Schumacher, IK.; Okamoto, T.; Kim, BH.; Chowdhury, NR.; Chowdhury, JR. & Fox, IJ. (1996). Transplantation of conditionally immortalized hepatocytes to treat hepatic encephalopathy. *Hepatology* 24:337-343

Sharma, R.; Greenhough, S.; Medine, CN. & Hay, DC. (2010). Three-dimensional culture of human embryonic stem cell derived hepatic endoderm and its role in bioartificial liver construction. *J. Biomed. Biotechnol*. 2010:236147

Shiraki, N., Umeda, K., Sakashita, N., Takeya M., Kume, K. & Kume, S. (2008). Differentiation of mouse and human embryonic stem cells into hepatic lineages. *Genes Cells* 13, 731-746

Si-Tayeb, K.; Lemaigre, FP. & Duncan, SA.; (2010). Organogenesis and development of the liver. *Dev Cell*. 18(2):175-189

Soto-Gutierrez A.; Kobayashi N.; Rivas-Carrillo, JD.; Navarro-Alvarez, N.; Zhao, D.; Okitsu, T.; Noguchi, H.; Basma, H.; Tabata, Y.; Chen Y.; Tanaka K.; Narushima M.; Miki, A.; Ueda, T.; Jun, HS.; Yoon, JW.; Lebkowski, J.; Tanaka, N.; Fox, IJ. (2006). Reversal of mouse hepatic failure using an implanted liver-assist device containing ES cell-derived hepatocytes. *Nat. Biotechnol*. 24:1412-1419

Strunk SW. (1959). The formation of intracellular crystals in midgut glands of Limnoria lignorum. *J. Biophys. Biochem. Cytol*. 5(3):385-392

Sutherland, D. E. R.; Numate M, Matas AJ, Simmons RL, Najaran JS. (1977). Hepatocellular transplantation in acute liver failure. *Surgery* 82:124-132

Suzuki, A., Zheng, Y.W., Kondo, R., Kusakabe, M., Takada, Y., Fukao, K., Nakauchi, H., and Taniguchi, H. (2000). Flow-cytometric separation and enrichment of hepatic progenitor cells in the developing mouse liver. *Hepatology* 32,1230-1239

Tada, K.; Roy-Chowdhury, N.; Prasad, V.; Kim, BH.; Manchikalapudi, P.; Fox, IJ.; van Duijvendijk, P.; Bosma, PJ.; Roy-Chowdhury, J.; (1998). Long-term amelioration of bilirubin glucuronidation defect in Gunn rats by transplanting genetically modified immortalized autologous hepatocytes. *Cell Transplant* 7:607-616

Takashimizu, I.; Tanaka, Y.; Yoshie, S.; Kano, Y.; Ichikawa, H.; Cui, L.; Ogiwara, N.; Johkura. K. & Sasaki, K. (2009). Localization of Liv2 in EB Outgrowth. *The Scientific World Journal* 9:190-199

Takimoto, Y.; Dixit, V.; Arthur, M. & Gitnick G. (2003). De novo liver tissue formation in rats using a novel collagen-polypropylene scaffold. *Cell Transplant* 12(4):413-421

Teratani, T., Yamamoto H., Aoyagi, K., Sasaki, H., Asari, A., Quinn, G., Sasaki, H., Terada, M. & Ochiya, T. (2005). Direct hepatic fate specification from mouse embryonic stem cells. *Hepatology* 41, 836-846

Torok, E.; Lutgehetmann, M.; Bierwolf, J.; Melbeck, S.; Dullmann, J.; Nashan, B.; Ma, PX. & Pollok, JM. (2011). Primary human hepatocytes on biodegradable poly (l-lactic acid) matrices: a promising model for improving transplantation efficiency with tissue engineering. *Liver Transpl.* 17(2):104-114

Watanabe, T.; Nakagawa, K.; Ohata, S.; Kitagawa, D.; Nishitai, G.; Seo, J.; Tanemura, S.; Shimizu, N.; Kishimoto, H.; Wada, T.; Aoki, J.; Arai, H.; Iwatsubo, T.; Mochita, M.; Watanabe, T.; Satake, M.; Ito, Y.; Matsuyama, T.; Mak, TW.; Penninger, JM.; Nishina, H. & Katada, T. (2002) SEK1/MKK4-mediated SAPK/JNK signaling participates in embryonic hepatoblast proliferation via a pathway different from NF-_B-induced anti-apoptosis. *Dev. Biol.* 250.; 332-347

Wong, N.; Lai, P.; Pang, E.; Leung, TW.; Lau, JW. & Johnson, PJ. (2000). A comprehensive karyotypic study on human hepatocellular carcinoma by spectral karyotyping. *Hepatology* 32:1060-1068

Yamada, T.; Yoshikawa, M.; Kanda, S.; Kato, Y.; Nakajima, Y.; Ishizaka, S. & Tsunoda, Y. (2002). In vitro differentiation of embryonic stem cells into hepatocyte-like cells identified by cellular uptake of indocyanine green. *Stem Cells* 20(2):146-154

Yin, Y.; Lim, YK.; Salto-Tellez, Ng SC.; Lin, CS.; Lim, SK. (2002). AFP(+).; ESC-derived cells engraft and differentiate into hepatocytes in vivo. *Stem Cells* ;20:338-346

Yoshie, S., Shirasawa, S., Yokoyama, T., Kano, Y., Takei, S., Mizuguchi, M., Matsumoto, K., Tomotsune, D., and Sasaki, K. (2010). Lanford medium induces high quality hepatic lineage cell differentiation directly from mouse embryonic stem cell-derived mesendoderm. *Biochem Biophys Res Commun.* 391, 1477-1482

Yue, F.; Johkura, K.; Shirasawa, S.; Yokoyama, T.; Inoue, Y.; Tomotsune, D. & Sasaki, K. (2010). Differentiation of primate ES cells into retinal cells induced by ES cell-derived pigmented cells. *Biochem Biophys Res Commun.* 394: 877-883

Zaret, KS.; (2000). Liver specification and early morphogenesis. *Mech. Dev.* 92 83-88.

Zaret, KS.; (2001). Hepatocyte differentiation from the endoderm and beyond. *Curr. Opin. Genet. Dev.*11 568-574

Zaret, KS. & Grompe, M,; (2008). Generation and regeneration of cells of the liver and pancreas. *Science* 322(5907):1490-1494

8

Amniotic Fluid Progenitor Cells and Their Use in Regenerative Medicine

Stefano Da Sacco, Roger E. De Filippo and Laura Perin
The Saban Research Institute, Children's Hospital Los Angeles USA

1. Introduction

In recent past, the potential use of stem cells and the advancement in stem cell research for Regenerative Medicine is considered as an alternative therapeutic strategy for a broad range of genetic and acquired diseases.
The interest in stem cells has been increasing over the past years, since their discovery in the early '90s, and they might represent a promising tool for regenerative purposes because of their capability to become almost any cell of an adult organism.
Despite the discoveries and the promising results, many are the controversies raised by stem cells. Feasibility of their use for human therapeutic purposes is regulated by many requirements such as safety, accessibility to a source that can provide an adequate amount of cells for in vitro expansion, absence of ethical issues and repeatability of the results.
Different lines of stem cells are investigated for understanding the basic mechanism of cellular differentiation and the potential for regenerative medicine purposes. However, to overcome safety and ethical issues, scientists are still looking for alternative sources that may provide easy and safe access to a cell population that may be used for cellular therapy.
Amniotic fluid, due to its contact with the fetus, has been considered an interesting source for undifferentiated or partially differentiated cells.
More recently, interest has been rising on more committed cell lines that may possibly provide new, more specific, tools for tissue regeneration. In particular, the isolation of cells already committed to a specific fate has been performed for kidney, pancreas and other organs and the study of these novel cell populations may give us an insight on cellular development and provide a more precise way of driving cell differentiation into a mature cell type.
Nevertheless, our knowledge of amniotic fluid cellular composition is still incomplete and only in the last few years some studies have been published describing the different cell types that can be retrieved. As long as new discoveries are shared and new insights are given on amniotic fluid cellular composition and cell differentiation we can gain a better understanding of the mechanism underlying development. The main goal of this chapter is to provide the readers with a broad knowledge regarding the work that has been done until now to undisclose the heterogeneous amniotic fluid cellular composition and their use for regenerative medicine purposes.

2. Amniotic fluid

Amniotic fluid is a clear, fluid that fills the amniotic cavity. It provides an ideal and protective environment in which floats the developing embryo and later on the fetus. It also helps to regulate the temperature of the fetus during the pregnancy.

2.1 Origin and molecular composition

During embryogenesis, maternal plasma is the main protagonist of amniotic fluid volume increase and water flows osmotically through fetal membranes, and, later on, through the placental membrane. The volume and the composition change during pregnancy following the physiological variations of the developing fetus (Fig. 1).

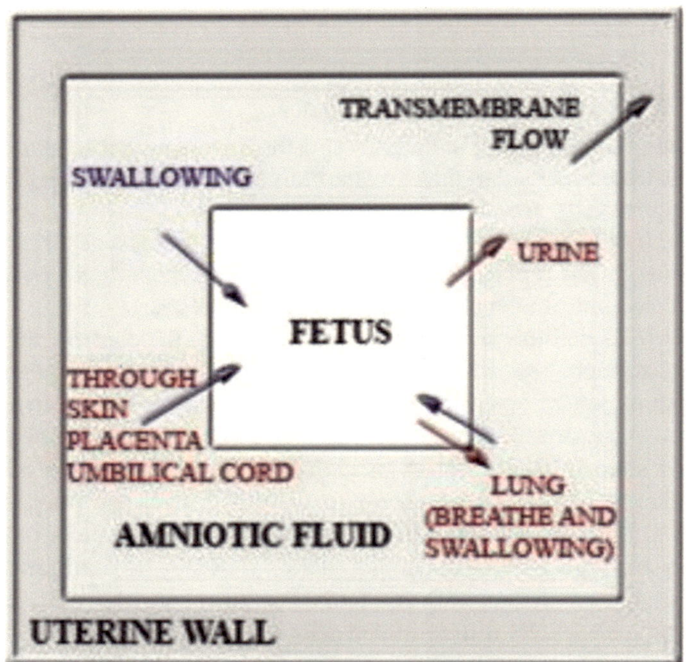

Amniotic fluid composition and volume are the result of exchanges and interaction with many different sources, either fetal or maternal. the above figure shows a schematic representation of the most important overall contributions to amniotic fluid.

Fig. 1. Amniotic fluid origin and composition

During the first weeks of gestation, the composition is comparable to the fetal plasma and its volume increases from 25 ml at 10 weeks to about 400 ml at 20 weeks (Underwood et al., 2005). By 8 weeks of gestation the fetal kidney begins fluid production that rapidly increases in volume during the second trimester. The exchange of fluids through the skin is present until keratinisation that occurs between 20 and 24 weeks of gestation. The molecular composition and the presence of nutritive substances have been shown to play a key role, in animals, in the proliferation and differentiation of various intestinal cell types such as epithelial and mucosa cells (Underwood et al., 2005).

2.2 Amniotic fluid in diagnostic

The use of amniotic fluid to determine the status of health of the fetus has been an important diagnostic tool for many years. Back in the 60s, it was considered an invaluable source of information for the diagnosis of fetal distress, haemolytic disease and fetal maturity (Horger et al., 1969), neural tube defects and lung maturity (Underwood et al., 2005). Over the years, the diagnostic techniques have been greatly improved and new fields of investigation have tried to tie various conditions with preterm labour, infective processes and embryo diseases. In particular, it has been used as a safe and reliable screening tool for genetic and congenital diseases in the fetus.

2.3 Isolation, expansion and characterization of amniotic fluid cells

The possibility of using amniotic fluid derived pluripotent and multipotent stem cells has been found appealing due to the relative easiness and safe procedure required to retrieve the cells from its source. Furthermore, the use of multipotent progenitors has been considered an attractive alternative to the use of pluripotent cells due to their already committed phenotype. Cells can be isolated from the liquid collected by amniocentesis. Briefly, prior to amniocentesis an ultrasound is performed to confirm fetal viability, gestational age, number of fetuses, placental location, volume, fetal anatomical survey, uterine cavity abnormalities and to evaluate the best needle insertion site.

A 20 cc syringe is used to aspirate the liquid. The first 2 cc collected should be discharged and then using another syringe and then, using another syringe, additional 15 to 20cc are aspirated.. Removal of the fluid generally takes less than 1 minute. After collection the cells are seeded with specific culture media and the adherent fraction is expanded.

Contact between amniotic fluid and compartments of the developing fetus, such as lung and gastrointestinal tract can explain the presence of different types of cells (Fig. 2). Moreover, cells detaching from the forming kidney or exfoliating from the fetal skin may contribute significantly to cellular composition. In particular, the presence of mature cell lines derived from all three germ layers has been identified (Hoehn et al., 1982; Gosden et al., 1983). Mesenchymal and hematopoetic progenitor cells have also been shown to exist before the 12th week of gestation in humans (Torricelli et al., 1993) together with cells expressing proteins and various genetic markers from specific tissue types including brain, heart, and pancreas have all been discovered (Tsangaris et al., 2004, Bossolasco et al., 2006; McLaughlin et al., 2006; Da Sacco et al., 2010).

Fauza et al. reported the successful isolation and expansion of unfractioned mesenchymal stem cells (AFMC) from human samples between 20 and 37 weeks of gestation, confirming the presence of a multipotent mesenchymal cell types over the progression of gestation (Kunisaki et al., 2007). A fully characterization of amniotic fluid pluripotent cell population has first been reported by Atala in 2007 (De Coppi et al., 2007). This newly isolated stem cell population (AFSC) is characterized by expression of c-kit, a surface marker expressed by stem cells of mesenchymal origin. AFSC express some surface markers and transcription factors distinctive of ESC such us OCT-4 and SSEA-4 indicating they can actually posses some important characteristics that also ESC have, showing their pluripotential capability. They stained positively for a number of surface markers characteristic of mesenchymal and/or neural stem cells, including CD29, CD44 (hyaluronan receptor), CD73, CD90 and CD105 (De Coppi et al., 2007).

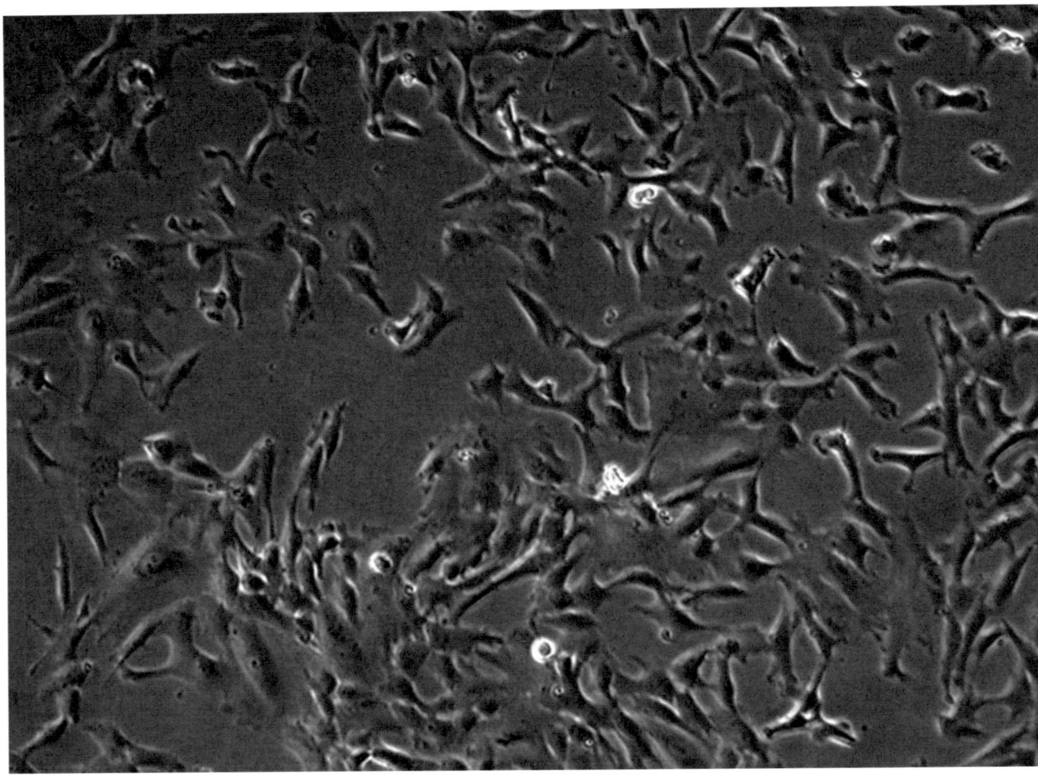

Amniotic fluid cells can be easily collected and expanded in vitro and exhibit a heterogeneous morphology with a preponderance of fibroblastoid, mesenchymal like cell shape (Unpublished Picture, Da Sacco et al.)

Fig. 2. Amniotic fluid cells morphology

However, a comprehensive analysis of amniotic fluid cellular composition has been missing and only in 2010 Da Sacco et al. for the first time demonstrated that the cellular composition varies in a timely fashion (Fig.3). Expression of markers for cells belonging to early endodermal and mesodermal germ layer differentiation pathway is predominant in the earlier weeks of gestation and constantly decreases over time to disappear at 17-18 weeks of gestation. Ectodermal markers, probably because of the exfoliating fetal skin, maintain a stable expression in all the samples analyzed. Interesting, it is shown that concurrent with the decrease of germ layers markers there is a noticeable increase of organ specific progenitor cell marker expression. Proteins expressed during lung, liver, heart and kidney differentiation are highly expressed starting form 17-18 weeks while expression of pluripotent markers such as OCT-4 and c-kit was found stable over time in all the samples analyzed, suggesting that, at least in the time range analyzed, the pluripotent cells are undergoing self renewal. Furthermore, mesenchymal marker CD90 is present in all the samples analyzed while hematopoietic marker CD34 decreased its expression over time.

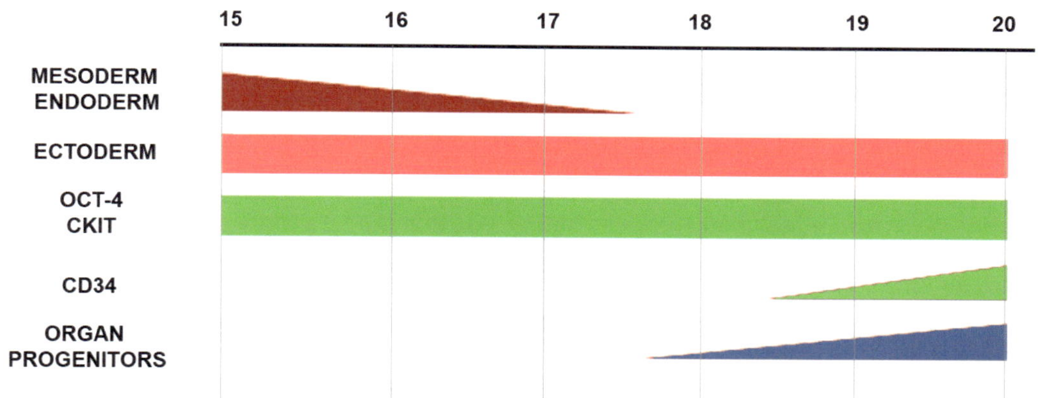

Mesodermal and endodermal cellular markers decrease over time and are not detectable after 17-18 weeks of gestation while various organ progenitor cell markers concentration rise after 17-18 weeks. Pluripotent markers OCT-4 and c-kit remain unchanged, as well as ectodermal markers, probably because of skin exfoliation. CD34 cellular marker rises after 18 weeks of gestation.

Fig. 3. Schematic representation of changes in composition within amniotic fluid between 15 and 20 weeks of gestation. (Da Sacco et al., 2010)

3. Amniotic fluid cells and organ specific regenerative medicine

Due to an easy and safe collection procedure, amniotic fluid has quickly gained interest as a potential source of pluripotent/multipotent cells for regenerative medicine purposes. Amniotic fluid stem cells have been shown to be easily cultured and expanded upon collection and isolation De Coppi et al., 2007, Perin et al., 2007, Da Sacco et al., 2010, Kunisaki et al., 2007). De Coppi et al. (De Coppi et al., 2007) and Arnhold et al. (Arnhold et al. 2011) proved that after c-kit selection these cells are still exhibiting optimal growth rate. Their potential for differentiation has been proved in many published works and cells can be retrieved from different species like humansDe Coppi et al. 2007, Perin et al., 2007, Da Sacco et al., 2010), pigs (Zheng et al., 2010), goat (He et al., 2011; Zheng et al., 2011), mouse (De Coppi et al., 2007) and buffalo (Yadav et al., 2010).

A recent report following a comparative analysis of AFSC and BM-MSC cells on proliferative potential and immunogenicity analysis showed the AFSC are less immunogenic and harbor a higher proliferation rate than BM-MSC (Mirebella et al., 2011).

3.1 Amniotic fluid cells and kidney

The complexity of the kidney and the multiple functions of the renal compartment are a great challenge to a successful therapeutic approach for its recovery and the regeneration. Beside the use of endogenous stem cells and other traditional and advanced therapies, the administration of exogenous stem cells, including AFSC has been proposed (Perin et al., 2011).

In the recent past, Perin et al. showed the capability of AFSC to participate in vitro to the development of embryonic kidneys. In particular, cells labeled with the surface marker CM-Dil were shown able to integrate within the structures of the developing kidney. Integration into the metanephric structures was additionally confirmed by the migration of the injected

cells to the periphery of the embryonic kidney. This data strongly correlates to the centrifugal pattern of induction, morphogenesis and differentiation of the metanephros, proceeding from the center to the periphery of the embryonic organ (Perin et al., 2007).
Moving into an in vivo model, the same group for the first time proved the potential of human AFSC to participate to the regeneration of kidneys undergoing acute tubular necrosis (Perin et al., 2010). After intra renal injection, cells were showed to survive, integrate into renal structures, and differentiate into tubular cells expressing proximal as well as distal epithelial tubular markers and persist over the long term.
However, as the Authors highlight in their study, the main mechanism of action seems to lie into the ability of AFSC to modulate the immune response by lowering pro-inflammatory cytokines while stimulating the expression of anti inflammatory molecules, and by lowering apoptosis and increasing endogenous proliferation.
On a different model of acute renal injury, Camussi's research group confirmed the positive results and the comparable efficacy between BM-MSC and AFSC (Hauser et al., 2010).
Beside the use of pluripotent cells, in 2010, we reported the isolation and characterization of more committed Amniotic Fluid derived Kidney Progenitor Cells (AFKPC) (Da Sacco et al., 2010). Cells expressing both CD24 and OB-Cadherin were sorted and characterized for a wide range of kidney markers such as PAX-2, LIM-1, GDNF, ZO-1. Additional selections were performed on the CD24+OB-cadherin+ cells to isolate cells committed to mesangial differentiation, podocyte differentiation, mesenchymal to epithelial transition cells and vascular progenitors. Characterization of marker expression for these subpopulations showed significant differences in gene expression, confirming their different commitment to renal fate.

3.2 Amniotic fluid cells and lung

In uterus, the developing lungs of the fetus are filled with fetal lung liquid which is actively secreted into the amniotic fluid. In the late gestational period, surfactant produced by the fetal lungs contributes to the composition of amniotic fluid and can be measured to determine the developmental stage of the surfactant system within the organ. Contact between the developing lung and the fluid make it a possible important reservoir for cells to be used in lung regenerative medicine. In fact, AFSC were shown able to integrate and proliferate into mouse embryonic lung and express human lung epithelial cell markers (Carraro et al., 2008).
Following hyperoxia injury, a tail vein injection of cells into nude mice showed localization in the distal lung with expression of both TTF1 and type II pneumocyte marker surfactant protein C. In the same work, specific Clara cells damage through naphthalene injury was followed by integration and differentiation of AFSC at the bronchioalveolar and bronchial positions with expression of specific Clara cell 10-kDa protein (Carraro et al., 2008). The positive results obtained by Warburton's research group were the first to prove the use of AFSC for in vivo organ regeneration. However, as underlined by the author, the number of cells homing and integrating within the lung was considerably low and the effects on tissue regeneration may be due on mechanisms different from integration and proliferation. However, our knowledge on this field is still lacking and more studies should be performed to clarify molecular pathways and suggest a plausible mechanism of action.

3.3 Amniotic fluid cells and heart

Heart failure remains one of the major causes of mortality in the United States (Honold et al., 2004). Stem cells have been proposed as an alternative, innovative approach for the

treatment of heart disease and cardiac differentiation. AFSC have been tested in the past years for their potential of becoming functional cardiomyocytes.
Hoerstrup's research group used amniotic fluid derived cells to successfully repopulate heart valves. After isolation, CD133- and CD133+ cells were isolated, characterized and subsequently seeded onto tissue engineered scaffolds. Feasible heart valve leaflets were obtained in vitro with the use of both fibroblast-like and endothelial like cells (Schmidt et al., 2007). However, Chiavegato et al., in 2007 showed that injections of human AFSC into a rat normal or ischemic myocardium was ineffective and cells were targeted by the immune response with consequent rejection of the xenotransplanted cells. On the other hand, the use of a xenotransplantation model, even when cells were injected in immunodeficient animals, may not be ideal for immunogenicity studies.
New insights on the cardiomyogenic potential of amniotic fluid cells have been published in 2010 by Soker et al. showing the in vitro capability of AFSC to be differentiated into cardiac cells when co-cultured with rat cardiomyocytes (Guan et al., 2010). Along with this work, Sung's research group reported the differentiation of AFMC into cardiomyocytes and endothelial cells (Yeh et al., 2010). Bollini et al. in two different works demonstrated the potential of AFSC to differentiate into cardiomyocytes both in vitro (Bollini et al., June 2011) and in vivo showing their cardioprotective effect following acute myocardial infarction (Bollini et al., May 2011).
In summary, the results obtained with amniotic fluid derived stem cells for cardiomyocyte differentiation are contrasting, mostly due to the lack of a specific model and the use of different species and differentiation protocols. More studies should be performed in order to truly confirm their capability to provide an effective tool for cardiovascular regenerative medicine.

3.4 Amniotic fluid cells and hematopoietic system

C-kit positive/ Lin – cells derived from both human and mouse, have been shown to have hematopoietic potential (Ditadi et al., 2009). These cells were capable of differentiating into erythroid, myeloid, and lymphoid lineages in vitro as well as in vivo, in the peripheral blood of irradiated mice. Furthermore, single cells analysis was able to assess the expression of several genes important during different stages of hematopoietic differentiation.

3.5 Amniotic fluid cells and pancreas

The occurrence of pancreatic damage and diabetes has dramatically increased in the last years. The rise of this emergency has strongly encouraged physicians and scientist to search for alternative therapeutic approaches. In 2009 was suggested that stem cells derived from amniotic fluid could be of use for pancreatic regeneration (Furth et al., 2009).
However, the first attempts to differentiate amniotic fluid cells into functional pancreatic cells were unsuccessful. In fact, the use of obestatin, a molecule proven to efficiently increase expression of pancreatic beta cell genes, was unable to stimulate pdx-1 expression these cells (Trovato et al., 2009).
A better knowledge of developmental pathways and gene cascades involved in pancreatic specification brought, a year later, to a growing number of successes. In fact, differentiation into pancreatic cells was proven using a variety of different procedures. In particular, transfection with the PDX-1 gene was able to induce pancreatic features on cells from AFMC (Gage et al., 2010).

With an interesting approach, Li et al. were able to prove differentiation into insulin producing cells by silencing several neuronal genes by use of small interference NRSF RNA. This was shown as crucial for pancreatic differentiation and for the expression of pancreatic markers including Pdx1, Hnf4α, Isl-1, Nkx6.1, Insulin, and Glut (Li et al., 2010). A different approach was taken by Zou et al. Knowing that the expression of particular surface markers can identify cell populations with specific traits and defined commitment, a CD44+/CD105+ population was isolated and successfully differentiated into pancreatic cells expressing PDX-1 (Zou et al., 2011). The increasing number of studies reported in the last two years suggests that the interest for amniotic fluid cells for beta cell differentiation is a growing research subject. Moreover, differentiation into pancreatic beta cells is been proven as possible in vitro settings. However, no in vivo studies have been published reporting their potential in acute and chronic pancreatic diseases.

3.6 Amniotic fluid cells and brain

The differentiation of pluripotent and multipotent cells into neural cells has been considered fundamental for understanding brain differentiation and for the establishment of innovative approaches for the healing of brain injuries. Many different studies have been performed on amniotic fluid cells and their expression of neuronal markers. However, their ability to differentiate into functional brain cells has being highly debated.

First reports on amniotic fluid progenitor cells commitment to neuronal cell lineage were published in 2006. In fact, McLaughlin's research group reported the ability to isolate and expand them in culture. Their studies showed that these novel progenitors are committed to mesencephalic dopaminergic neurons (McLaughlin et al., 2006). AFMC were shown to be able to differentiate into brain cells both in vitro (Prusa et al., 2004, Tsai et al., 2004; Tsai et al., 2007; Jiang et al., 2010, Mareschi et al., 2009) and in vivo (Cheng et al., 2010). Selection of specific cell population based on specific surface marker expression didn't show to really improve the neuronal potential of amniotic fluid cells. Cells isolated by use of different surface markers like c-kit, (De Coppi et al., 2007), sox-2 (Jezierski et al., 2010) were shown able to differentiate into neuronal like cells. However, in 2009 was reported the inability of AFSC to differentiate into dopamine neurons both in vitro and in vivo assays (Donaldson et al., 2009). An interesting recent study, investigated the impact of extracellular signals on neural differentiation, where it was confirmed that extracellular matrix has an essential role on neurogenic differentiation and therefore regulates its efficiency (Orciani et al., 2011).

While many studies seems to prove that differentiation of amniotic fluid stem cells, either AFSC or AFMC, into neural cell types, there are still too many open questions about functionality of the differentiation, ideal cell population and best differentiation cocktail. While the current status of the research gives great hope for the future, to confirm of deny the possible use of amniotic fluid cells for brain regeneration more in vitro and in vivo data are certainly required.

3.7 Amniotic fluid cells and liver

Only a few studies have been reported that investigate the potential of amniotic fluid derived cells for hepatocyte differentiation. Zheng et al. in their work, claimed that AFSC had a better response to the differentiation when compared with BM-MSC under the same conditions (Zheng et al., 2008). Later on, differentiation into the hepatic lineage was

successfully obtained by Gasbarrini research group (Saulnier et al., 2009) that showed the equal potential of adult and fetal derived cells, including AFSC, for liver regeneration.
However, beside these encouraging results, more studies are required prior to confirm the suitability of amniotic fluid stem cells for liver therapy.

3.8 Amniotic fluid cells and bone

In 2010, it was reported a positive effect of transient ethanol exposure during early differentiation of AFSC into osteoblasts (Hipp et al., 2010).
Papaccio's research group showed the ability of AFMC to differentiate into bone cells when co-cultured with dental pulp cells proving potential for bone engineering (De Rosa et al., 2010).
Osteogenic progenitors have been found within amniotic fluid (Antonucci et al., February 2009). In this work, they were able to obtain calcium mineralization and osteogenic differentiation of AFMC. Expression of various osteogenic markers after 30 days in culture was demonstrated. Similar results were obtained by two other research groups (Antonucci et al., October 2009, Steigman et al., 2009 and Sun et al., 2010).
Peister in 2011 showed that AFSC were capable of a greater differentiation potential compared to mesenchymal stem cells although the latter response to the differentiative cocktail was occurring at earlier times (Peister et al., 2011). However, in vivo data are still lacking and the osteogenic potential of amniotic fluid cells in a complex environment should be undisclosed.

3.9 Amniotic fluid cells and chondrocytes, adipose tissue and skeletal muscular cellular differentiation

3.9.1 Chondrocytes

The regenerative capacity of the cartilage is limited. The ability to differentiate stem cells into cartilage may provide a better alternative to primary culture of chondrocytes that in vitro dedifferentiate losing their characteristics (Kramer et al., 2008).
Fauza's research group demonstrated the ability of ovine AFMC to successfully differentiate into chondrocytes on 3D scaffolds expressing several markers of cartilage (Kunisaki et al., 2006). Atala's group showed the ability of AFSC to differentiate into chondrocytes (De Coppi et al., 2007). However, no functional studies were performed to confirm the possible use of these amniotic fluid derived cartilage cells.

3.9.2 Adipocytes

Adipocyte differentiation was proven in 2007 for AFSC when these cells were first characterized and tested for their pluripotentiality (De Coppi et al., 2007).
In addition, adipogenic differentiation for goat derived AFMC was shown in 2011 (He et al., 2011) proving their differentiative potential.

3.9.3 Myocytes

Muscular tissue is well known to harbour endogenous stem cells that help recovering the tissue after an injury. However, the differentiation potential of these pluripotent stem cells and when the extent of the injury, due to an acute or chronic insult, is too heavy, muscular degeneration occurs with loss of motility and impaired function. The study of cells feasible

for muscular differentiation and regeneration has been considered essential for a successful therapeutic approach.
Amniotic fluid cells have been studied for their capability to differentiate into functional muscular cells. In particular, Streubel reported using non-hematopoetic AFMC for the conversion of amniocytes into myocytes. (Streubel et al., 1996). De Coppi showed the ability of AFSC to differentiate into myocites in vitro by expression of markers expressed by the differentiating and mature muscle fibers (De Coppi et al., 2007) and the results were later confirmed by studies both in vitro and in vivo on scid mice (Gekas et al., 2010)

4. Amniotic fluid derived cells and their role as cytokine modulators

In the last years new evidences have been found that correlates the administration of stem cells with the modulation of inflammatory and fibrotic processes through cytokine mediated cross-talk between the pluripotent cells and the surrounding environment. New studies have highlighted the possibility that the same mechanism of action can be used to explain the effect of amniotic fluid stem cells in many diseases. In particular, Perin showed that in a murine model of acute tubular necrosis, the expression of inflammatory cytokines is strongly regulated after injection of AFSC (Perin et al., 2010). Down regulation of pro-inflammatory molecules and up-regulation of pro-regenerative and anti-flogistic cytokines resulted in a faster regeneration of the damaged tissue with higher proliferation rate, lower apoptosis and an overall better physiological profile of different renal parameters. A broad study performed by Yoon (Yoon et al., 2010) investigated the in vitro production of cytokines by AFMC in the cultured media. Presence of several inflammatory molecules was reported such as IL-8, IL-6, TGF-β, TNFRI, VEGF, and EGF and other molecules involved in the TGFB/SMAD2 pathway. The conditioned culture media proved to be useful for enhancing wound healing in an in vivo murine model. While studying the angiogenic potential of AFSC, Teodolinda et al. (Teodolinda et al., 2011) reported the ability of the cells to produce and release several cytokines and chemoattractant molecules that are able to modulate not only the vessel growth but also the activity of macrophages/monocytes and other cells involved in inflammation and immunoresponse.

5. Conclusions

In the last few years, an increasing number of studies have been performed on amniotic fluid derived stem cells and progenitor cells. Exciting results have been reported on amniotic fluid cell population characterization of composition, growth kinetics and potential for specific organ regeneration. However, further investigation is still required to completely categorize cells according to origin and function. Improving the efficiency and specificity of differentiation into various mature and functional cell types to prevent their attrition towards unrelated cell types would be an important factor to control in regenerative medicine applications. In this very same direction, the establishment of protocols and differentiative media will better allow us to compare the different populations and understand their mechanism of action. In addition, knowledge about how the different compartments of the developing fetus are contributing to the cellular composition may undisclose important information about the development and the amniotic fluid composition.

6. References

Antonucci I, Stuppia L, Kaneko Y, Yu S, Tajiri N, Bae EC, Chheda SH, Weinbren NL, Borlongan CV. (2009 Feb) Isolation of osteogenic progenitors from human amniotic fluid using a single step culture protocol. *BMC Biotechnol.* 16;9:9.

Antonucci I, Pantalone A, De Amicis D, D'Onofrio S, Stuppia L, Palka G, Salini V. (2009 Oct-Dec) Human amniotic fluid stem cells culture onto titanium screws: a new perspective for bone engineering. *J Biol Regul Homeost Agents.*;23(4):277-9.

Arnhold S, Glüer S, Hartmann K, Raabe O, Addicks K, Wenisch S, Hoopmann M. (2011) Amniotic-Fluid Stem Cells: Growth Dynamics and Differentiation Potential after a CD-117-Based Selection Procedure. *Stem Cells Int.* 23;715341.

Bollini S, Cheung KK, Riegler J, Dong X, Smart N, Ghionzoli M, Loukogeorgakis SP, Maghsoudlou P, Dubé KN, Riley PR, Lythgoe MF, De Coppi P. (2011 May 3.) Amniotic fluid stem cells are cardioprotective following acute myocardial infarction. *Stem Cells Dev.*

Bollini S, Pozzobon M, Nobles M, Riegler J, Dong X, Piccoli M, Chiavegato A, Price AN, Ghionzoli M, Cheung KK, Cabrelle A, O'Mahoney PR, Cozzi E, Sartore S, Tinker A, Lythgoe MF, De Coppi P. (2011 Jun) In vitro and in vivo cardiomyogenic differentiation of amniotic fluid stem cells. *Stem Cell Rev.*;7(2):364-80.

Bossolasco P, Montemurro T, Cova L, Zangrossi S, Calzarossa C, Buiatiotis S, Soligo D, Bosari S, Silani V, Deliliers GL, Rebulla P, Lazzari L., (2006 Apr) Molecular and phenotypic characterization of human amniotic fluid cells and their differentiation potential., *Cell Res.*;16(4):329-36.

Carraro G, Perin L, Sedrakyan S, Giuliani S, Tiozzo C, Lee J, Turcatel G, De Langhe SP, Driscoll B, Bellusci S, Minoo P, Atala A, De Filippo RE, Warburton D.. (2008) Human amniotic fluid stem cells can integrate and differentiate into epithelial lung lineages. *Stem Cells.*; 26(11):2902-11.

Cheng FC, Tai MH, Sheu ML, Chen CJ, Yang DY, Su HL, Ho SP, Lai SZ, Pan HC.. (2010 Apr) Enhancement of regeneration with glia cell line-derived neurotrophic factor-transduced human amniotic fluid mesenchymal stem cells after sciatic nerve crush injury. *J Neurosurg.*;112(4):868-79.

Chiavegato A, Bollini S, Pozzobon M, Callegari A, Gasparotto L, Taiani J, Piccoli M, Lenzini E, Gerosa G, Vendramin I, Cozzi E, Angelini A, Iop L, Zanon GF, Atala A, De Coppi P, Sartore S.. (2007 Apr) Human amniotic fluid-derived stem cells are rejected after transplantation in the myocardium of normal, ischemic, immuno-suppressed or immuno-deficient rat. *J Mol Cell Cardiol.*;42(4):746-59.

Da Sacco S, Sedrakyan S, Boldrin F, Giuliani S, Parnigotto P, Habibian R, Warburton D, De Filippo RE, Perin L. (2010 Mar) Human amniotic fluid as a potential new source of organ specific precursor cells for future regenerative medicine applications. *J Urol.*;183(3):1193-200.

De Coppi P, Bartsch G, Siddiqui MM, Xu T, Santos TX, Perin L, Mostoslavsky G, Serre AC, Snyder EY, Yoo JJ, Furth ME, Soker S, Atala A. (2007) Isolation of amniotic stem cell lines with potential for therapy. *Nat Biotechnol* 25:100-6.

De Rosa A, Tirino V, Paino F, Tartaglione A, Mitsiadis T, Feki A, d'Aquino R, Laino L, Colacurci N, Papaccio G. (2010 Oct 5.) Amniotic fluid-derived MSCs lead to bone differentiation when co-cultured with dental pulp stem cells. *Tissue Eng Part A.*

Ditadi A, de Coppi P, Picone O, Gautreau L, Smati R, Six E, Bonhomme D, Ezine S, Frydman R, Cavazzana-Calvo M, André-Schmutz I. (2009 Apr) Human and murine amniotic fluid c-Kit+Lin- cells display hematopoietic activity. *Blood.* 23;113(17):3953-60.

Donaldson AE, Cai J, Yang M, Iacovitti L. (2009 Sep) Human amniotic fluid stem cells do not differentiate into dopamine neurons in vitro or after transplantation in vivo. *Stem Cells Dev.*;18(7):1003-12.

Furth ME, Atala A. (2009 Mar) Stem cell sources to treat diabetes. *J Cell Biochem.* 1;106(4):507-11. Review.

Gage BK, Riedel MJ, Karanu F, Rezania A, Fujita Y, Webber TD, Baker RK, Wideman RD, Kieffer TJ. (2010 Sep-Oct) Cellular reprogramming of human amniotic fluid cells to express insulin. *Differentiation.*;80(2-3):130-9.

Gekas J, Walther G, Skuk D, Bujold E, Harvey I, Bertrand OF.. (2010 Mar) In vitro and in vivo study of human amniotic fluid-derived stem cell differentiation into myogenic lineage. *Clin Exp Med.*;10(1):1-6.

Gosden CM., (1983 Oct) Amniotic fluid cell types and culture., *Br Med Bull.*;39(4):348-54

Guan X, Delo DM, Atala A, Soker S. (2010 Aug 4.) In vitro cardiomyogenic potential of human amniotic fluid stem cells. *J Tissue Eng Regen Med.*

Hauser PV, De Fazio R, Bruno S, Sdei S, Grange C, Bussolati B, Benedetto C, Camussi G. (2010 Oct) Stem cells derived from human amniotic fluid contribute to acute kidney injury recovery. *Am J Pathol.*;177(4):2011-21.

He X, Zheng YM, Qiu S, Qi YP, Zhang Y. (2011 Jan 19.) Adipogenic differentiation and EGFP gene transfection of amniotic fluid derived stem cells from goat fetus at terminal gestational age. *Cell Biol Int.*

Hipp JA, Hipp JD, Atala A, Soker S. (2010 Oct) Ethanol alters the osteogenic differentiation of amniotic fluid-derived stem cells. *Alcohol Clin Exp Res.*;34(10):1714-22.

Hoehn H, Salk D., (1982) Morphological and biochemical heterogeneity of amniotic fluid cells in culture., *Methods Cell Biol.*;26:11-34.

Honold J, Assmus B, Lehman R, Zeiher AM, Dimmeler S. (2004 Jul) Stem cell therapy of cardiac disease: an update. *Nephrol Dial Transplant.*;19(7):1673-7.

Horger EO 3rd, Hutchinson DL. (1969) Diagnostic use of amniotic fluid. *J Pediatr.* Sep;75(3):503-8.

Jezierski A, Gruslin A, Tremblay R, Ly D, Smith C, Turksen K, Sikorska M, Bani-Yaghoub M. (2010 Jun) Probing stemness and neural commitment in human amniotic fluid cells. *Stem Cell Rev.*;6(2):199-214.

Jiang TM, Yang ZJ, Kong CZ, Zhang HT. (2010 Oct) Schwann-like cells can be induction from human nestin-positive amniotic fluid mesenchymal stem cells. In Vitro *Cell Dev Biol Anim* ;46(9):793-800.

Kramer J, Böhrnsen F, Schlenke P, Rohwedel J. (2006 Apr) Stem cell-derived chondrocytes for regenerative medicine. *Transplant Proc.*;38(3):762-5.

Kunisaki SM, Jennings RW, Fauza DO. (2006 Apr) Fetal cartilage engineering from amniotic mesenchymal progenitor cells. *Stem Cells Dev.*;15(2):245-53.

Kunisaki SM, Armant M, Kao GS, Stevenson K, Kim H, Fauza DO. (2007 Jun) Tissue engineering from human mesenchymal amniocytes: a prelude to clinical trials. *J Pediatr Surg*. 42(6):974-9; discussion 979-80.

Li B, Wang S, Liu H, Liu D, Zhang J, Zhang B, Yao H, Lv Y, Wang R, Chen L, Yue W, Li Y, Pei X. (2010 Nov) Neuronal Restrictive Silencing Factor Silencing Induces Human Amniotic Fluid-Derived Stem Cells Differentiation into Insulin-Producing Cells. *Stem Cells Dev.*

Manuelpillai U, Moodley Y, Borlongan CV, Parolini O. (2011 May) Amniotic membrane and amniotic cells: Potential therapeutic tools to combat tissue inflammation and fibrosis? *Placenta.*

Mareschi K, Rustichelli D, Comunanza V, De Fazio R, Cravero C, Morterra G, Martinoglio B, Medico E, Carbone E, Benedetto C, Fagioli F. (2009) Multipotent mesenchymal stem cells from amniotic fluid originate neural precursors with functional voltage-gated sodium channels. *Cytotherapy.*;11(5):534-47.

McLaughlin D, Tsirimonaki E, Vallianatos G, Sakellaridis N, Chatzistamatiou T, Stavropoulos-Gioka C, Tsezou A, Messinis I, Mangoura D., (2006 May) Stable expression of a neuronal dopaminergic progenitor phenotype in cell lines derived from human amniotic fluid cells., *J Neurosci Res.* 83(7):1190-200

Mirebella T, Poggi A, Scaranari M, Mogni M, Lituania M, Baldo C, Cancedda R, Gentili C. (2011 Jun)Recruitment of host's progenitor cells to sites of human amniotic fluid stem cells implantation. *Biomaterials.*;32(18):4218-27.

Orciani M, Morabito C, Emanuelli M, Guarnieri S, Sartini D, Giannubilo SR, Di Primio R, Tranquilli AL, Mariggiò MA. (2011 Jan-Mar) Neurogenic potential of mesenchymal-like stem cells from human amniotic fluid: the influence of extracellular growth factors. *J Biol Regul Homeost Agents.*;25(1):115-30.

Peister A, Woodruff MA, Prince JJ, Gray DP, Hutmacher DW, Guldberg RE. (2011 Jul) Cell sourcing for bone tissue engineering: Amniotic fluid stem cells have a delayed, robust differentiation compared to mesenchymal stem cells. *Stem Cell Res.*;7(1):17-27.

Perin L, S Giuliani, D Jin. (2007) Renal differentiation of amniotic fluid stem cells. *Cell Prolif* 40:936-48.

Perin L, Sedrakyan S, Giuliani S. (2010 Feb) Protective effect of human amniotic fluid stem cells in an immunodeficient mouse model of acute tubular necrosis. *PLoS One* 24;5(2):e9357.

Perin L, Da Sacco S, De Filippo RE. (2011 Apr) Regenerative medicine of the kidney. *Adv Drug Deliv Rev.* 63(4-5):379-87.

Prusa AR, Marton E, Rosner M, Bettelheim D, Lubec G, Pollak A, Bernaschek G, Hengstschläger M (2004) Neurogenic cells in human amniotic fluid. *Am J Obstet Gynecol* 191:309–314

Rosner M, Mikula M, Preitschopf A, Feichtinger M, Schipany K, Hengstschläger M. (2011 May) Neurogenic differentiation of amniotic fluid stem cells. *Amino Acids.* 15.

Roubelakis MG, Bitsika V, Zagoura D, Trohatou O, Pappa KI, Makridakis M, Antsaklis A, Vlahou A, Anagnou NP. (2010 Sep) In vitro and in vivo properties of distinct populations of amniotic fluid mesenchymal progenitor cells. *J Cell Mol Med.*

Saulnier N, Lattanzi W, Puglisi MA, Pani G, Barba M, Piscaglia AC, Giachelia M, Alfieri S, Neri G, Gasbarrini G, Gasbarrini A. (2009 Mar) Mesenchymal stromal cells multipotency and plasticity: induction toward the hepatic lineage. *Eur Rev Med Pharmacol Sci.*;13 Suppl 1:71-8.

Schmidt D, Achermann J, Odermatt B, Breymann C, Mol A, Genoni M, Zund G, Hoerstrup SP. (2007 Sep) Prenatally fabricated autologous human living heart valves based on amniotic fluid derived progenitor cells as single cell source. *Circulation.* 11;116(11 Suppl):I64-70.

Siegel N, Rosner M, Unbekandt M. (2010 Sep) Contribution of human amniotic fluid stem cells to renal tissue formation depends on mTOR. *Hum Mol Genet.* 1;19(17):3320-31.

Steigman SA, Ahmed A, Shanti RM, Tuan RS, Valim C, Fauza DO. (2009 Jun) Sternal repair with bone grafts engineered from amniotic mesenchymal stem cells. *J Pediatr Surg.*;44(6):1120-6; discussion 1126.

Streubel B, Martucci-Ivessa G, Fleck T, Bittner RE. (1996) In vitro transformation of amniotic cells to muscle cells--background and outlook. *Wien Med Wochenschr.*;146(9-10):216-7.

Sun H, Feng K, Hu J, Soker S, Atala A, Ma PX. (2010 Feb) Osteogenic differentiation of human amniotic fluid-derived stem cells induced by bone morphogenetic protein-7 and enhanced by nanofibrous scaffolds. *Biomaterials.*;31(6):1133-9.

Teodelinda M, Michele C, Sebastiano C, Ranieri C, Chiara G. (2011 May) Amniotic liquid derived stem cells as reservoir of secreted angiogenic factors capable of stimulating neo-arteriogenesis in an ischemic model. *Biomaterials.*;32(15):3689-99.

Torricelli F, Brizzi L, Bernabei PA, Gheri G, Di Lollo S, Nutini L, Lisi E, Di Tommaso M, Cariati E. (1993 Apr-Jun) Identification of hematopoietic progenitor cells in human amniotic fluid before the 12th week of gestation. *Ital J Anat Embryol.*;98(2):119-26. Erratum in: Arch Ital Anat Embriol 1993 Jul-Sep;98(3):215.

Trovato L, De Fazio R, Annunziata M, Sdei S, Favaro E, Ponti R, Marozio L, Ghigo E, Benedetto C, Granata R. (2009 Dec) Pluripotent stem cells isolated from human amniotic fluid and differentiation into pancreatic beta-cells. *J Endocrinol Invest.*;32(11):873-6.

Tsai M-S, Lee J-L, Chang Y-J, Hwang S-M (2004) Isolation of human multipotent mesenchymal stem cells from second-trimester amniotic fluid using a novel two-stage culture protocol. *Hum Reprod* 19:1450–1456

Tsai M-S, Hwang S-M, Tsai Y-L, Cheng F-C, Lee J-L, Chang Y-J (2006) Clonal amniotic fluid-derived stem cells express characteristics of both mesenchymal and neural stem cells. *Biol Reprod* 74:545–551

Tsangaris, Rachel Weitzdörfer, Daniela Pollak, Gert Lubec, Michael Fountoulakis, (2004) The amniotic fluid cells proteome, *Electrophoresis*, 25, 1168–1173

Underwood MA, WM Gilbert and MP Sherman. (2005) Amniotic fluid: not just fetal urine anymore. *J Perinatol* 25:341-8.

Yadav P, Mann A, Singh V, Yashveer S, Sharma R, Singh I. (2010 Dec) Expression of Pluripotency Genes in Buffalo (Bubalus bubalis) Amniotic Fluid Cells. *Reprod Domest Anim.*.

Yeh YC, Wei HJ, Lee WY, Yu CL, Chang Y, Hsu LW, Chung MF, Tsai MS, Hwang SM, Sung HW. (2010 Jun) Cellular cardiomyoplasty with human amniotic fluid stem cells: in vitro and in vivo studies *Tissue Eng Part A.*;16(6):1925-36.

Yoon BS, Moon JH, Jun EK, Kim J, Maeng I, Kim JS, Lee JH, Baik CS, Kim A, Cho KS, Lee JH, Lee HH, Whang KY, You S. (2010 Jun) Secretory profiles and wound healing effects of human amniotic fluid-derived mesenchymal stem cells. *Stem Cells Dev.*;19(6):887-902.

Zheng YB, Gao ZL, Xie C, Zhu HP, Peng L, Chen JH, Chong YT. (2008 Nov) Characterization and hepatogenic differentiation of mesenchymal stem cells from human amniotic fluid and human bone marrow: a comparative study. *Cell Biol Int.*;32(11):1439-48.

Zheng YM, Dang YH, Xu YP, Sai WJ, An ZX. (2010 Nov) Differentiation of AFS cells derived from the EGFP gene transgenic porcine fetuses. *Cell Biol Int.*

Zheng YM, Zheng YL, He XY, He XN, Zhao X, Sai WJ. (2011 May) Multipotent differentiation of the EGFP gene transgenic stem cells derived from amniotic fluid of goat at terminal gestational age. *Cell Biol Int.*

Zou G, Liu T, Zhang L, Liu Y, Li M, Du X, Xu F, Guo L, Liu Z.(2011 May) Induction of Pancreatic β-Cell-Like Cells from CD44(+)/CD105(+) Human Amniotic Fluids via Epigenetic Regulation of the Pancreatic and Duodenal Homeobox Factor 1 Promoter. *DNA Cell Biol.*

Part 2

Cell Communicators

9

Inflammation-Angiogenesis Cross-Talk and Endothelial Progenitor Cells: A Crucial Axis in Regenerating Vessels

Michele M. Ciulla, Paola Nicolini, Gianluca L. Perrucci,
Chiara Benfenati and Fabio Magrini
University of Milan
Italy

1. Introduction

The normal cell is confined by its genetic program to a rather narrow range of morphofunctional characteristics within which it is nevertheless able to handle normal physiological demands (Kumar et al., 2011). A single cell, placed in an appropriate environment, contains, enclosed in its genome, sufficient information to generate a variety of differentiated cell types, whose spatial and temporal dynamics interact to form detailed morphological patterns (Geard and Wiles, 2005). Similarly, **in multicellular organisms, it is the genetic program that establishes networks basics -including differentiation, functional hierarchy and interaction patterns- with each level in the system representing an increase in organizational complexity.** Indeed the emergence from a single or few stem cells of multicellular organisms with a complex organization must be seen as a necessity in evolution since it alone can ensure an optimal use of resources (Furusawa and Kaneko, 2000). From such evolutionary perspective, **the development of a cardiovascular system in complex organisms fulfills the need for transport of nourishment, oxygen and metabolic waste as well as the need for communication between distant districts.**

All closed biological systems facing the external environment, whatever their level of complexity, are characterized by the ability to maintain a stable, constant condition by regulating themselves and the internal environment. In the early 19th century Cannon (Cannon, 1926) defined this property "homeostasis", from the Greek: ὅμοιος, *hómoios*, "similar" and στάσις, *stásis*, "standing still". This condition of relative equilibrium grants the biological system a certain degree of independence from the environment and enables what we call life.

Any change, either decremental or incremental, in the internal/external environment of the cell/organism such that it requires an active response from the cell/organism can be termed demand. In the case of complex organisms demands are met by the functional reserve of the cell/tissue/organ. The functional reserve is involved in the process of adaptation by which a cell/organism acquires morphofunctional features that allow it to better fit in with the changed environment and thus attain a new equilibrium (Figure 1). Adaptation is driven by the genetic program, via the activation of appropriate subsets of genes, and is influenced by the qualitative and quantitative characteristics of the demand such as its nature, intensity and duration. Adaptation can be considered to include both plasticity and adaptability. Plasticity is the ability to express a broad variety of phenotypes in response to environmental changes and is at a maximum during embryogenesis and early extrauterine life. The term adaptability could instead be applied to the (more limited) process of adaptation which occurs in adult life and varies between organs and species. In complex organisms two points deserve particular mention. First, it should be remarked that, since complex organisms possess cognitive functions, demands can be not only physical but also mental or emotional, and that no line can be drawn between body and mind. Second, the functional hierarchy of complex organisms is paramount in allocating functional reserve where it is needed so that priorities can be established in the distribution of the available resources.

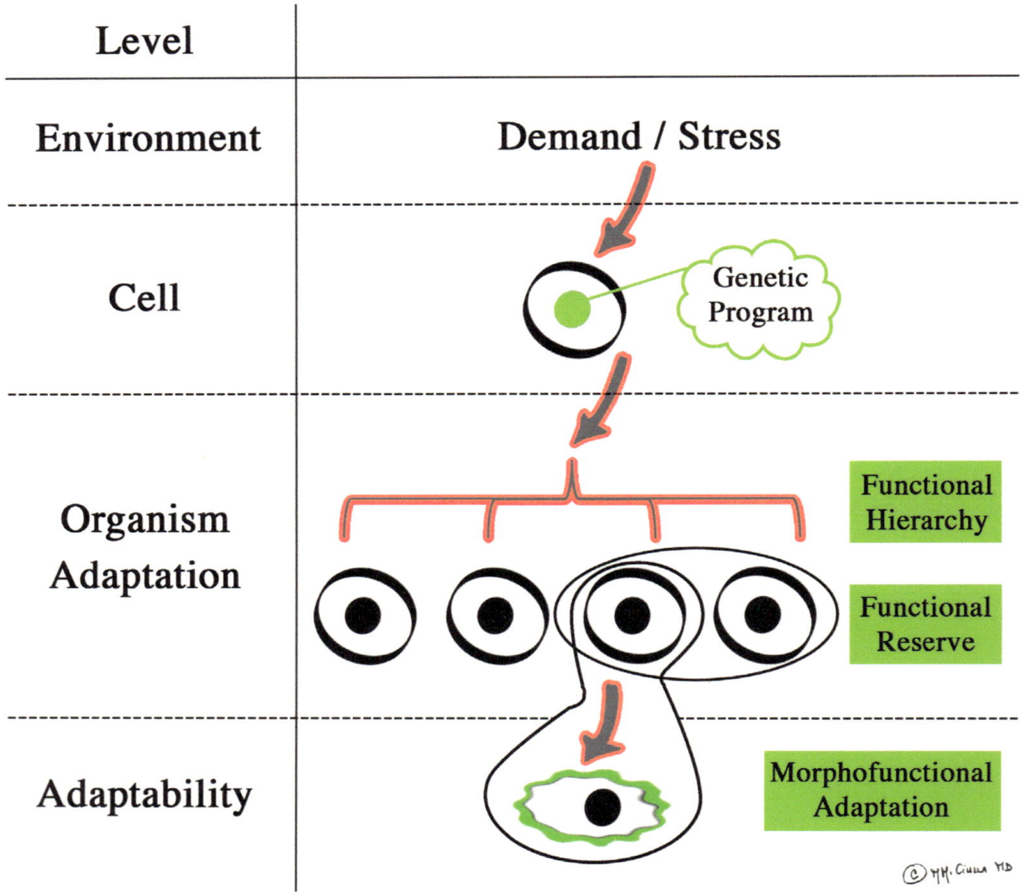

Fig. 1. Genetic programming and functional hierarchy
Schematic representation of the interplay between environmental demands and the genetic-driven adaptation process in adult life. In complex organisms genetic programming is paramount in allocating functional reserve where it is needed. Adaptation is the response to environmental demands and involves the acquisition by a cell/organism of morphofunctional features that allow it to better fit in with the changed environment. Adaptation includes plasticity and adaptability. During development plasticity is progressively replaced by adaptability.

Thus, the adaptation process -whether be it grounded in plasticity or adaptability- is determined by the genetic program. Since the genetic program is finite so is also the functional reserve, and there is necessarily an intrinsic limit to adaptation. However, this limit is extremely variable from subject to subject because each organism possesses a peculiar genotype and thus a unique functional reserve.
It must be underscored that there is a lifelong interplay between genome and environment, i.e. the genome-driven adaptation process is continuously called upon to buffer environmental changes throughout the life of an individual. What happens then when the functional reserve is inadequate per se and/or in relation to the characteristics of the demand? There arises a condition which is currently classified as stress and appears to be more common than is generally believed, although it is not formally considered a disease. In fact stress may cause cell injury but a disease can be recognized as such only when cell injury leads to functional deficits in the organism and to the occurrence of signs and/or symptoms. Hence, the notion of disease has very much to do with communication. It can be conceived as communication at multiple levels of complexity: from cell to neighbouring cells, from organs to organism and, finally from organism to organism. The latter is the apex of the communication process and corresponds to what is normally witnessed in everyday life when the patient attends the doctor's office.
In summary, it is the balance between genetically-determined functional reserve and environmental changes that delineates the fuzzy boundary between physiology and pathology, between what we define a current and an extra demand and, ultimately, between what we define health and disease. **In complex organisms with an evolved circulatory system, the role of the circulation as a dynamic interface between blood and tissues implies that it is continuously challenged by noxious stimuli of different kinds (mechanical, chemical, biological). Thus the circulatory system plays a key role in communicating stress at a distance and in responding to it through the vascular changes that characterize inflammation, even by replacing its own damaged endothelial cells.** This is the conceptual basis of vascular homeostasis and the reason why in organisms with a circulatory system **inflammation acts as a bridge between environment (demand) and genome (response).**

2. The circulation of blood and its constituents: an evolved system of transport and communication

There are about 100 trillion cells in the human body, a number that is about ten thousand times greater than the number of human beings on Earth. Even though they are an awesome multitude all these cells can communicate directly or indirectly with each other. Every single cell needs to receive nourishment, replenish oxygen and remove waste in order to survive and work properly. It must also be able to inform other cells about its fate, the changes it experiences, the answers it provides to stimuli and, in general, the actions it intends to take. **Successful cell functioning therefore requires the existence of an excellent "communication route" that enables cells to exchange information with the external environment, and with distant cells via humoral signaling.** This route is the circulatory system, which is a perfect model of a dynamic and well-organized highway. Consistently with its function, it is composed of an extensive network of vessels: De Witt (2005) has calculated that blood vessels from an adult would circle the earth twice if placed end to end (DeWitt, 2005). Busy but efficient, the circulatory system consists of two basic elements: the vessels (composed by endothelium and connective tissue) and the blood. It is a highly dynamic system which is always "turned on" and ready to respond to the changing demands of the organism. **Given the dynamic nature of the vascular system, vascular injury requires rapid repair and this means that "spare parts" must be readily available.** Later in this chapter we will see which are the main players in vascular turnover and which are the cellular and molecular processes involved in the repair of vascular damage.

2.1 Open and closed circulatory systems

With very few exceptions, all coelomates (by the broadest definition, every organism with a fluid-filled cavity) possess a circulatory system (Ruppert and Carle, 1983). The walls of the blood vessels are made of a layer of epithelial cells which in most invertebrates forms part of the coelomic lining (called mesothelium or peritoneum); the mesothelium lining the blood vessels contains myofibrils (myo-epithelium) whose contractility determines the flow of blood. In vertebrates and some invertebrates, blood vessels are lined by a non-muscular endothelium and are surrounded by a separate muscle cell layer. Blood vessels can form two different types of circulatory systems: closed, e.g. in chordates and annelids, or open at both ends, e.g. in arthropods such as insects, and mollusks. In the former case, the body fluid is contained within two separate compartments, one inside the lumen of the circulatory system (blood), the other inside the coelom (coelomic fluid). In the latter case, there is a single compartment filled with a body fluid called hemolymph. Blood, coelomic fluid and hemolymph contain dispersed cells, generally termed blood cells or hemocytes (Ratcliffe and Rowley, 1979). Contractile blood vessels, such as the dorsal vessel/heart in insects, are suspended within the hemolymphatic space and cause a streaming motion of the hemolymph by means of rhythmic contractions. Evolution well illustrates how an increase in complexity is always paralleled by an increase in needs and thus requires the establishment of appropriate structures to support them. **A closed vascular system, under neural control, is consistent with the functional hierarchy of complex organisms, since it allows a fine-tuned modulation of the blood flow to allocate resources (nutrients, oxygen) to the body districts that most require them.**

2.2 Embryo vasculogenesis

The development of the circulatory system is a key event during embryogenesis in many animal species. The process of vasculogenesis involves the differentiation of local mesodermal precursors into vascular and endothelial cells (ECs). These are clustered in blood islands whose growth and fusion leads to the formation of a primary vascular plexus (Carmeliet, 2000; Risau and Flamme, 1995; Timmermans et al., 2009). After the onset of the blood circulation, the yolk sac capillary network differentiates into an arteriovenous vascular system (Risau et al., 1988). The integral relationship between the elements that circulate in the vascular system (the blood cells) and the cells that are mainly responsible for the vessels themselves (the endothelial cells) is reflected by the composition of the embryonic blood islands. In fact the cells which are destined to generate hematopoietic cells are located in the center of the blood islands and are termed hematopoietic stem cells (HSCs) while endothelial progenitor cells (EPCs), or angioblasts, can be found at the periphery of the blood islands. In addition to this spatial association, HSCs and EPCs share certain antigenic markers, including Flk-1, Tie-2, Sca-1 and CD34, which will be discussed later in greater detail. On the basis of such embryological and molecular data there is evidence for a close developmental relationship between these two types of progenitor cells,

thus EPCs and HSCs have been considered to derive from a putative common precursor, termed hemangioblast (Choi et al., 1998; Flamme and Risau, 1992; Weiss and Orkin, 1996).
For a long time the existence of a common precursor for both hematopoietic and endothelial cells was believed to be restricted to embryonic development but recent studies have identified a postnatal hemangioblast. Over recent years, much attention has focused on bone marrow (BM) stem cell potentiality. It has been reported that the bone marrow contains cells termed multipotent adult progenitor cells (MAPCs), which, at a single-cell level, can differentiate into a large number of cell types, including endothelial cells (Reyes et al., 2002). Bone marrow–derived EPCs originate from CD34+ stem cells, which are able to differentiate via separate pathways not only in endothelial cells but also in erythrocytes, thrombocytes and various lineages of leukocytes.
It is possible that circulating EPCs mobilized from the bone marrow are in fact progenitor or stem cells with a broader differentiation potential which is directed towards the formation of endothelial cells at sites of neovascularization, where the microenvironment is adequate for such process to occur (Asahara et al., 1997; Peichev et al., 2000). Indeed, a decade ago, two groups reported that human CD34+ cells isolated from circulating peripheral blood (PB), umbilical cord blood (UCB) and BM, could differentiate into ECs in vitro and in vivo in mouse models, thereby contributing to neoendothelialization and neovascularization in the adult organism (Asahara et al., 1997; Shi et al., 1998).

2.3 Stemness and the hemangioblast

As we have seen, **stem cells have a key role in the setting of both embryonic and postnatal vasculogenesis.** The two main players on stage are the hemangioblast and the EPC. The hemoangioblast is a pluripotent stem cell that is able to differentiate into cells belonging to the hemopoietic lineage and into EPCs. The EPC is a multipotent stem cell that can differentiate into various cell types including vessel-lining endothelial cells. Hemoangioblast and EPC are distinguished on the basis of a number of characteristics, including the developmental stage in which they are identified, their tissue localization, their molecular profile and their properties in vitro. Still, as new information is acquired and as experimental techniques are refined, the classification of stem cells is becoming less clear-cut and much more complex. Indeed a review by Blau and colleagues (Blau et al., 2001) endorses the pioneering view that a stem cell should not be considered a specific cellular entity but rather a biological function which can be taken on by numerous cell types expressing different genes. **Traditionally, stem cells have been regarded as undifferentiated cells capable of self-renewal and production of a large number of differentiated progeny, with a sharp distinction being made between embryonic and adult stem cells.** According to a large body of in vitro and in vivo evidence only embryonic stem cells (ES) were believed to be pluripotent –i.e. capable of differentiating into a wide range of cell types. Their high degree of plasticity fitted in well with the notion that maximum "flexibility" is required during the early stages of development, when a variety of body tissues are in the process of being formed. Adult stem cells were thought to exhibit a limited differentiative potential- i.e. to be able to differentiate exclusively into cells of the specific tissue in which they resided. This idea was supported by several examples of tissue-specific regeneration: the proliferation of liver cells after surgical removal of part of the liver, the production of blood elements by transplanted hematopoietic stem cells after chemo- or radio-induced BM ablation, the existence of satellite cells that repair damaged skeletal muscles, the participation of keratinocyte precursors in the healing of skin wounds.

However, it is increasingly recognized that **adult stem cells have a broader regenerative potential than was originally supposed**, and that it is probably involved not only in the response to tissue injury but also in physiological tissue homeostasis. BM-derived cells have been reported to yield not only blood cells but also cells belonging to the brain, liver and heart. Stromal cells in the BM, other than HSC, have been found to give rise to a number of different cell types. Going one step further, even highly specialized cells such as muscle cells and central nervous system cells have been shown to be able to contribute to the replenishment of the blood. It is therefore speculated that adult stem cells possess a significant amount of plasticity which can be channelled along diverse differentiation pathways by the interactions they establish with the environment. Indeed, resorting once again to the "highway" metaphor, the circulatory system can be envisioned as the route through which stem cells gain access to all districts of the body. It has been documented experimentally that BM-derived cells enter different organs such as the heart, brain, skeletal muscle, and liver (Bittner et al., 1999; Brazelton et al., 2000; Ciulla et al., 2007c; Gussoni et al., 1999; Jackson et al., 1999; Krause et al., 2001; Lagasse et al., 2000; Mezey et al., 2000). The recruitment of stem cells from the circulation into a specific organ would be brought about by homing signals, as has been described for the homing of leukocytes (Butcher, 1991). Likewise, the behavior of the stem cells - in terms of generative potential- would then be determined by the surrounding microenvironment, including the extracellular matrix (Hay, 1991), the local milieu (Studer et al., 2000) as well as growth and differentiation factors. Stem cells would thus be stimulated to assume morphologies and functions typical of the new environment in which they have migrated. Therefore, **stem cell plasticity coupled with the availability of a route connecting different compartments of the organism, explains how stem cells residing in one tissue would be able to contribute to the regeneration of distal tissues.**
In summary, the concept of stemness should be conceived as a highly dynamic one, with no fixed tissue-specific boundary either in the localization or in the differentiation of stem cells.

2.3.1 The hemangioblast

It has already been mentioned that the formation of blood islands in the extraembyronic yolk sac marks the onset of hematopoiesis and vasculogenesis in the developing embryo. These blood islands derive from aggregates of mesodermal cells that colonize the presumptive yolk sac. In about 12 hours, the central cells within these clusters give rise to embryonic hematopoietic cells while the peripheral cells differentiate into endothelial cells which form the first vascular structures that surround the inner blood cells (Wagner, 1980). The close developmental association of the hematopoietic and endothelial lineages within the blood islands has led to speculate that they arise from a common precursor, the hemangioblast (Murray, 1932; Sabin, 1920; Wagner, 1980). This hypothesis has gained most support from the observation that the hematopoietic and endothelial lineages share the expression of a number of different genes (Anagnostou et al., 1994; Asahara et al., 1997; Fina et al., 1990; Kabrun et al., 1997; Kallianpur et al., 1994; Millauer et al., 1993; Yamaguchi et al., 1993; Young et al., 1995). Moreover, recent gene-targeting experiments demonstrating that a functional Flk-1 receptor tyrosine kinase is required for the development of the blood islands are consistent with the notion that these lineages derive from a common precursor (Shalaby et al., 1997; Shalaby et al., 1995). However, there is still no definite proof of the existence of the hemoangioblast. Past studies aimed at identifying and characterizing the putative hemangioblast have been hampered by difficulties in accessing the embryo prior to the establishment of the blood islands and by the limited number of cells present at this stage of development. Differentiation of embryonic stem (ES) cells to hematopoietic and endothelial cells in culture provides an alternative approach for investigating these early commitment steps since both cellular and molecular analyses have documented that the sequence of events giving rise to these lineages is similar in vitro and in the normal mouse embyo (Keller et al., 1993; Keller, 1995; Nakano et al., 1994; Risau et al., 1988; Vittet et al., 1996; Wiles and Keller, 1991). Indeed, using this in vitro model, it has recently been shown that embryoid bodies (EBs) generated from ES cells allowed to differentiate for 3-3.5 days contain a unique precursor population with both primitive and definitive hematopoietic potential (Kennedy et al., 1997). When cultured in the presence of vascular endothelial growth factor (VEGF), *c-kit* ligand (KL) and conditioned medium from an endothelial cell line, D4T, these precursors form colonies consisting of immature or blast-like cells that express a number of genes common to both the hematopoietic and endothelial lineages, including *tal-1*/SCL, CD34 and the VEGF receptor, *flk-1* (Kennedy et al., 1997). The responsiveness to VEGF of these embryonic precursors together with the gene expression pattern of their blast cell progeny suggests that this population could have the potential to generate cells of the endothelial lineage in addition to hematopoietic precursors (Choi et al., 1998).

3. Cellular identity, a dynamic concept

Until a few decades ago, many dogmas of cell biology had not yet been dismantled; among them was the idea that cells might not have an alternative fate after maturation according to their specific identity.

The effects of nuclear transfer to enucleated oocytes in adult cells (Gurdon, 1960) have experimentally shown that the mechanisms that regulate cell identity follow rules, bound to processes of genetic reprogramming, which are still unclear today. There are two well-known processes by which cells are able to turn into other cell types: transdifferentiation, i.e. the direct conversion from one cell type to another, and dedifferentiation, i.e. the reversion to a less-differentiated cell type and the subsequent maturation to a different lineage.

After Dolly the sheep (Campbell et al., 1996), cell biology moved under the spotlight and there arose within the scientific community questions which have still not found exhaustive answers. Cell identity, which had so far been considered a rigid and durable characteristic involving a one-way process from precursor to mature cell, was shown to exhibit not only intrinsic plasticity (Scadden, 2007) but also a large degree of adaptation depending on the interplay between genome and microenvironment.

In fact it was demonstrated that mature cells are able to switch not only their functional phenotype but also their gene expression profile into that of stem cells, thereby acquiring pluripotent plasticity. Such findings yielded several questions: can all somatic cells dedifferentiate? What kind of epigenetic events are able to reverse the fate of a cell which is already committed and in what order can they do so? Factors that can reverse the fate of the cells are only endogenous or can they be exogenous too? As far as regenerative medicine is concerned, the latter question is the most important, since the application of potential drug treatments that can revert a pool of cells into stem cells capable of regenerating damaged tissues in the setting of injury or disease holds great promise.

In the field of regenerating vessels the central issue is: what kind of cells are able to repair vascular damage? The answer to this apparently straightforward question – i.e. EPCs - leads us into a topic fraught with controversy. In fact the identity of EPCs from both a molecular and technical standpoint is still open to discussion and experimentation. It would therefore be helpful to find biological markers of progenitor cells in order to validate a reproducible method for the enumeration and distinction of EPCs from circulating endothelial cells (CECs). Different approaches have been used to detect these cells, including in vitro culture, magnetic bead separation and fluorescence microscopy, immunocytochemistry and flow cytometry (Bull et al., 2003; Del Papa et al., 2004; George et al., 1993; Goon et al., 2006; Mancuso et al., 2001; Mutunga et al., 2001; Nakatani et al., 2003), each having specific strengths and limitations.

3.1 EPC Identity

Numerous markers of EPC lineage have been proposed in the literature, subcategorized into stem cell makers (such as CD34, CD133, CD45, and c-kit) and endothelial-like markers (such as VEGF receptor (VEGFR), CD31, CD146, and von Willebrand factor) (Hristov et al., 2003a; Timmermans et al., 2009). However, the precise definition of what constitutes an EPC is the subject of an extensive debate (Asosingh et al., 2009; Diller et al., 2008). At present, the only EPC phenotype based on surface antigenic markers that provides strong and reproducible correlations across multiple studies on vascular damage and cardiovascular risk is CD34+/VEGFR+ (Fadini et al., 2008). An additional phenotype that has recently been employed in the literature involves the inclusion of CD133 as a secondary stem cell marker (Friedrich et al., 2006); however, Timmermans et al. have recently questioned its utility as an EPC marker (Timmermans et al., 2009; Timmermans et al., 2007). Notably, the intersection of the CD34+/CD133+ and CD34+/ VEGFR+ cell phenotypes (i.e., CD34+/CD133+/ VEGFR+ cells) is known to be extremely rare (Fadini et al., 2008).

Many investigators have identified or designated putative circulating EPCs (CEPCs) with flow cytometry using a single surface marker such as CD34 or CD133, or various combinations of surface markers, which has actually resulted in a complicated list of putative CEPC immunophenotypes both in humans and mice. Most of the surface marker combinations used in flow cytometry studies included the marker CD34 and VEGFR-2, because initial studies in the field had reported that CD34+ and VEGFR-2+ cells purified from various sources like UCB, PB and BM were able to generate ECs in vitro, suggesting that CD34+ cells contain CEPCs (Asahara et al., 1997; Bompais et al., 2004; Shi et al., 1998). One specific subset of CD34+ cells, designated as CD34+VEGFR-2+CD133+ cells, is widely accepted to correspond to 'true CEPCs' in humans but yet these cells were never directly tested for their ability to generate new ECs in vitro or in vivo, which is essential to validate CD34+VEGFR-2+CD133+ cells as true CEPCs (Bertolini et al., 2006; Kondo et al., 2004; Peichev et al., 2000). Recently, however, Case et al. , using in vitro hematopoiesis assays, reported for the first time that isolated human UCB or mobilized adult PB CD34+VEGFR-2+CD133+ cells in fact represent an enriched population of CD45+ hematopoietic precursors, but that CD34+VEGFR-2+CD133+ did not contribute to the formation of ECs in vitro (Case et al., 2007). Similarly, CD34+CD45+CD146+ cells previously considered to be CEPCs, were not directly assayed in vitro, nor in vivo for their ability to contribute to newly formed endothelium, and thus, it is difficult to know whether this cell type acts as a true CEPC (Delorme et al., 2005). Therefore, the scientific foundation for using CD34+ plus variable combinations of surface markers remains elusive. Moreover, the use of these diverse combinations to define a singular entity (the CEPC), makes the significance of flow cytometry studies difficult to interpret, creates obstacles to the direct comparison of data between laboratories, and may result in discrepancies in the interpretations of study results among different laboratories. Therefore, investigators should strongly consider that any 'putative' CEPC, whatever its phenotype, be carefully assessed by validating its postnatal endothelial differentiation capacity in vitro and in vivo. An effective way of doing so is to test the ability of CEPCs to form endothelial colonies in vitro (Ciulla et al., 2006) or actual vessels in vivo (Ciulla et al., 2007b; Ciulla et al., 2007c). However, even the latter approaches pose a series of challenges because they involve a number of variables such as the exact combination of growth factors to which the cells are exposed in vitro and, in vivo, the type of animal model used, its genetic background and the nature of the angiogenic stimulus applied. Furthermore, a potential limitation of single-cell clonal assays is that a cell may behave differently according to whether it is isolated or in the presence of other cells, and thus transplanting or culturing a single cell may not be an appropriate EPC assay, and the presence of supportive cells with a different phenotype or function may be required.

In summary, **one of the greatest problems to date in the field of EPC biology is the lack of a unique marker, or combination of markers, that solely identifies the rare CEPCs in humans (between 0.01% and 0.0001% of peripheral mononuclear cells)** (Shaffer et al., 2006). As Blau and colleagues (Blau et al., 2001) very appropriately point out, this obstacle may be technical or indeed purely biological- i.e. specific molecular markers for stem cells are not available either because they have not yet been discovered or because they simply do not exist.

4. The endothelium: More than a lining

The endothelium is a key component of the vascular system, being strategically located to provide a physical interface between blood and tissues. Far from being an inert layer of "nucleated cellophane" (Chen et al., 2008; Cines et al., 1998; Galley and Webster, 2004) it participates in a number of physiological processes such as the transfer of biologically active molecules to the underlying interstitium and cells, the selective transmigration of cells in and out of the bloodstream, the modulation of vasomotor tone and the regulation of hemostasis/coagulation (Chen et al., 2008; Cines et al., 1998; Galley and Webster, 2004; Hunting et al., 2005; Mikirova et al., 2009). In an average adult it contains ten trillion (10^{13}) cells, weighs about 1 kg and covers a surface area of approximately 7 m^2 (Cines et al., 1998; Galley and Webster, 2004; Lin et al., 2000). It has traditionally been considered a rather static structure with little turnover (0.1% replications per day) (Cines et al., 1998), but over the last two decades its highly dynamic nature is being increasingly recognized (Dignat-George and Sampol, 2000; Mikirova et al., 2009). Indeed the endothelium is continuously exposed to different kinds of noxious stimuli (mechanical, chemical, biological) that damage endothelial cells and injured endothelial cells need to be replaced in order to guarantee structural/functional vascular integrity and prevent atherosclerotic disease. Initially endothelial repair was believed to occur exclusively through the proliferation and migration of mature endothelial cells surrounding the lesion. However, these are terminally differentiated cells with low proliferative potential (Caplan and Schwartz, 1973; Haudenschild and Studer, 1971; Haudenschild and Schwartz, 1979; Hristov et al., 2003a; Jujo et al., 2008; Kunz et al., 1978; Malczak and Buck, 1977; Schwartz et al., 1980; Schwartz et al., 1981; Schwartz et al., 1975; Taylor and Lewis, 1986; Tongers et al., 2010) and it was reasonable to suppose that an alternative mechanism would be required for efficient endothelial regeneration. Research conducted along this line provided growing evidence that a crucial role in endothelial turnover is played by EPCs (Asahara et al., 1999; Asahara et al., 1997; George et al., 2011; Hristov et al., 2003a; Hunting et al., 2005; Jujo et al., 2008; Mikirova et al., 2009; Peichev et al., 2000; Tongers et al., 2010). These are BM-derived cells that can be found in the peripheral blood and have properties similar to those of embryonic hemangioblasts, being able to circulate, proliferate and differentiate into mature endothelial cells (Asahara et al., 1999; Asahara et al., 1997; Dignat-George and Sampol, 2000; George et al., 2011; Hristov et al., 2003b; Jujo et al., 2008; Peichev et al., 2000). In particular, they are mobilized from the BM into the circulation, travel to the site of vessel damage (a process known as recruitment or homing) and integrate into the endothelial monolayer to substitute damaged cells (Asahara et al., 1999; Asahara et al., 1997; George et al., 2011; Hristov et al., 2003b; Hunting et al., 2005; Jujo et al., 2008; Tongers et al., 2010). It is also becoming clear that EPCs do not merely have a direct, structural function –producing the "hardware" of the vessel –but also an important indirect, paracrine role, behaving as "cytokine factories"- i.e. they secrete proteins (growth factors, chemokines, cytokines) that have a variety of proangiogenic actions including the recruitment of additional EPCs and the suppression of apoptosis as well as the activation of "resident" EPCs (George et al., 2011; Jujo et al., 2008; Rehman et al., 2003; Tongers et al., 2010). As far as the latter are concerned, emerging data points to the presence of *in situ* EPCs, embedded within the vessel wall, that may aid in vascular repair, although their existence is still a matter of controversy (Ingram et al., 2005; Jujo et al., 2008; Torsney and Xu, 2011).

Animal studies have extensively demonstrated the involvement of BM cells in tissue regeneration after vascular damage. In a rat model of myocardial cryodamage it has been shown that labelled donor rat BM mononuclear cells (BM-MC) injected peripherally through the femoral vein were found after one week in the injured myocardium (Ciulla et al., 2003) in a number proportional to the size of the infarcted area (Ciulla et al., 2004) and were located within small early-stage vessels (Ciulla et al., 2007b). Similarly, in a rat model of myocardial ischemia induced by coronary artery ligation, human CD34+ cells injected intravenously were reported to infiltrate the infarcted zone within 48 hours and form capillaries of human origin after two weeks, improving echocardiographically-assessed left ventricular function (Kocher et al., 2001). Asahara and colleagues described the incorporation of BM-derived EPCs into sites of active angiogenesis one week after surgically-induced hindlimb and myocardial ischemia in a seminal work on BM transplant (BMT) murine models (Asahara et al., 1999; Asahara et al., 1997). Likewise, in a mouse model of hindlimb ischemia intravenous human EPCs were integrated into limb vessels at two weeks, significantly increasing blood flow at a Doppler evaluation. "Urbich C, Heeschen C, Aicher A, Dernbach E, Zeiher AM, Dimmeler S (2003) Relevance of monocytic features for neovascularization capacity of circulating endothelial progenitor cells. *Circulation* 2003 18;108(20):2511-6." In murine models of wire-induced arterial injury (carotid and femoral artery denudation), EPCs from donor mice were detected within the neoendothelium after two weeks (Ii et al., 2006; Urao et al., 2006). In a canine BMT model it was reported that a synthetic graft of the descending thoracic aorta was coated with endothelial cells derived from the donor three months post-angioplasty (Shi et al., 1998). With regard to humans, Lin and colleagues (Lin et al., 2000) investigated BMT recipients who had received gender-mismatched transplants and identified in the peripheral blood a small population of highly proliferative donor-derived cells, expanding more than a 1000-fold after one month in culture.

5. Inflammation-angiogenesis cross-talk

Inflammation is the body's protective response to noxious stimuli of different kinds (mechanical, chemical, biological). It has the purpose to dilute, circumscribe or destroy the injurious agent, and to promote the healing process. **In complex organisms with a circulatory system, the distinctive feature of inflammation is a vascular reaction characterized by changes in the caliber and permeability of blood vessels** (Kumar et al., 2011). Angiogenesis is closely related to inflammation and the endothelial cell lies at the very core of this association (Rajashekhar et al., 2006). In fact, activated inflammatory cells (macrophages, T cells) secrete Interleukin-1 (IL-1) and Tumoral Necrosis Factor α and β (TNF) which stimulate the production of VEGF from the endothelium (Naldini and Carraro, 2005). VEGF, in turn, is a major proangiogenic factor with a number of effects on endothelial cells. Originally discovered as a potent and rapid inducer of vascular permeability (Vascular Permeability Factor, VPF) (Senger et al., 1983), it was then found to have a modest mitogenic effect on mature endothelial cells (Keck et al., 1989; Leung et al., 1989; Plouet et al., 1989) and finally it was shown to be an important participant in the regulation of EPC kinetics (Asahara et al., 1999; Olsson et al., 2006). Specifically, VEGF is believed to promote the mobilization of EPCs from the BM into the circulation by a) stimulating BM EPC proliferation b) providing a chemoattractive gradient for EPC migration towards the site of injury and c) modulating the BM-blood barrier, by increasing its permeability and modifying the expression of adhesion molecules (Asahara et al., 1999). VEGF is a dimeric glycoprotein of approximately 40 kDa whose actions on vascular and progenitor endothelial cells are mainly mediated by the VEGFR-2 (Homsi and Daud, 2007; Olsson et al., 2006; Schabbauer et al., 2007). This is a tyrosine kinase containing in its cytoplasmic domain 19 tyrosine residues which are in part phosphorylated upon activation by receptor ligands and function as specific docking sites for molecules that initiate cytoplasmic signaling cascades (Olsson et al., 2006; Schabbauer et al., 2007). In particular, it has been shown that both Early Growth Response Protein-1 (EGR-1) and Nuclear Factor of Activated T cells (NFAT) are transcription factors involved in VEGF-mediated gene induction in endothelial cells (Schabbauer et al., 2007; Schweighofer et al., 2007). Matrix metalloproteinases (MMPs), a group of proteolytic enzymes which degrade the extracellular matrix (ECM), also appear to be involved in the mobilization of EPCs. Recently, a cross-talk between VEGF and MMPs has been suggested, with VEGF activating MMPs to favor the release of EPCs from the BM stroma and MMPs rendering available VEFG sequestered in the ECM (Ebrahem et al., 2010). VEGF is likewise produced by inflammatory cells (macrophages, neutrophils, T cells), consistently with the known redundancy of the cytokine system. It has been proposed that VEGF from different sources (endothelial and non-endothelial) may have different functional significance: endothelial VEGF seems to take part in an autocrine loop that conveys survival signals to the endothelium while paracrine VEGF is considered the main contributor to the angiogenic cascade (Lee et al., 2007).

Finally, it is important to remark that VEGF also acts on inflammatory cells: it is capable of recruiting macrophages, drive T cell differentiation towards a proinflammatory Th1 phenotype, and stimulate the production of IL-1 and TNF by peripheral blood mononuclear cells (Angelo and Kurzrock, 2007; Mor et al., 2004; Naldini and Carraro, 2005; Noonan et al., 2008). Thus **the inflammatory process is marked by an extensive bidirectional communication between the endothelium and inflammatory cells, via cytokines and growth factors, which establishes an inflammation-angiogenesis cross-talk**.

5.1 Effects of C reactive protein on EPCs

Within the scenario of inflammation-angiogenesis the role of C Reactive Protein (CRP) on EPCs remains as yet uncertain and deserves a brief discussion. CRP has long been recognized as an acute phase protein whose plasma levels increase significantly (up to 1000-fold) during systemic inflammation (Calabro et al., 2009). In such context it is synthesized by the liver in response to IL-6, IL-1 and TNF, and is a key component of innate immunity (Calabro et al., 2009; Li and Fang, 2004; Slevin and Krupinski, 2009). However, it is now acknowledged that CRP is also produced within the vascular wall at the site of inflammation. In vitro models have demonstrated that CRP is secreted by human coronary artery muscle cells after stimulation with cytokines (Calabro et al., 2003), by human aortic endothelial cells after stimulation with macrophage-conditioned medium (Venugopal et al., 2005) and by human brain endothelial microvessel cells following oxygen-glucose deprivation (Slevin et al., 2010). An in vivo study in patients with coronary artery disease (CAD) undergoing percutaneous coronary intervention (PCI) showed that CRP levels distal to the target lesion were higher than proximal levels and that CRP levels in coronary sinus blood increased in a time-dependent manner after stenting, consistently with the hypothesis of a local production of CRP (Inoue et al., 2005).

Moreover, it was found that CRP levels were higher in patients with unstable than stable angina and that the transcardiac CRP gradient at 48 hours after PCI was predictive of late-lumen loss due to neointimal thickening (Inoue et al., 2005). Similarly, elevated preprocedural plasma CRP predicted restenosis after stenting (Hong et al., 2005) in one study and ergovine-induced coronary vasospasm during coronary angiography in another (Hung et al., 2005). Taken together these data provide evidence that CRP is a useful index of the "activity" of the atherosclerotic process. In fact atherosclerosis is no longer considered a mere lipid storage disease, stemming from the passive accumulation of lipids within the artery wall, but a chronic inflammatory disease which involves inflammation at all stages, from initiation to progression and, eventually, plaque rupture (Calabro et al., 2009; Paoletti et al., 2004). This is the reason why high-sensitivity CRP (hs-CRP)- i.e. the low CRP concentrations found in the peripheral blood in the absence of overt inflammation - has gained ample consensus in the evaluation of cardiovascular risk. Indeed hs-CRP is a stronger predictor of cardiovascular events - such as CAD, stroke, PAD and sudden cardiac death- than LDL cholesterol, and it adds prognostic value to the Framingham score (Ridker, 2001; Slevin and Krupinski, 2009).

It seems therefore established that there is a relationship between CRP and atherogenesis, but the role of CRP as an innocent bystander or active participant is still an unsettled and highly debated question (Calabro et al., 2009; Genest, 2010; Kushner et al., 2010; Lee et al., 2007; Slevin and Krupinski, 2009; Verma et al., 2004). According to several authors, CRP has a direct proatherogenic effect by exercising a wide spectrum of detrimental actions on both vessel wall cells and EPCs (Calabro et al., 2009; Li and Fang, 2004; Slevin and Krupinski, 2009; Verma et al., 2004). As far as mature endothelial cells are concerned it would a) decrease the expression and activity of nitric oxide synthase, causing endothelial dysfunction, b) increase the expression of surface adhesion molecules such as Vascular Cell Adhesion Molecule-1 (VCAM-1), Intercellular Adhesion Molecule-1 (ICAM-1) and selectins, favouring the recruitment of mononuclear cells, and c) induce the expression of oxidized LDL receptor, promoting lipid accumulation (Calabro et al., 2009; Verma et al., 2004). Macrophages and vascular smooth muscle cells (VSMCs) would also be involved, the former being stimulated to phagocyte oxidized low density lipoproteins (LDLs), the latter being induced to proliferate and migrate towards the intima (Calabro et al., 2009; Verma et al., 2004). With regard to EPCs, CRP has been shown to reduce their survival, differentiation and migration, thus impairing a crucial component of vascular integrity and regeneration (Verma et al., 2004).

However, when healthy subjects with no cardiovascular risk factors are considered the picture seems to change. We found evidence of a possible CRP-enhanced functioning of EPCs, by demonstrating that endothelial colony-forming capacity is related to CRP levels in healthy subjects (Ciulla et al., 2006). In order to investigate the burden of angiogenesis in physiological conditions we enrolled 37 subjects who were carefully selected for the lack of cardiovascular risk factors, had low hs-CRP (< 1 mg/L), normal Echocardiogram and carotid Doppler examinations and were not on pharmacological treatment. We found that 15 subjects (40.5% of the sample) formed endothelial cell colonies from peripheral blood and that, relative to the non-colony formers, they had higher levels of hs-CRP as well as of VEGF. These results suggest that **CRP may directly increase the clonogenic potential of EPCs, yet it may also be that higher hs-CRP is merely a marker of more active microinflammation - due to minimal endothelial stress- and greater VEGF production.** Although the association between hs-CRP and EPC clonogenic capacity appears in contrast with reports from other investigators (Verma et al., 2004), it must be emphasized that we focused on a physiological setting and that the effect of CRP on EPCs may be influenced by different factors such as the concentration and conformation of CRP. For instance, the hs-CRP levels in our study were markedly lower (<0.1 mg/L) than those used for the in vitro demonstration of EPC inhibition (>10 mg/L) (Verma et al., 2004). Indeed Turu and colleagues subsequently showed that CRP at concentrations of 1-5 mg/L had a strong proangiogenic effect on bovine aortic and human coronary artery endothelial cells and that this was in part mediated by the expression of the VEGF receptor (Turu et al., 2008). Also, it is increasingly acknowledged that CRP exists in two distinct conformations, native circulating CRP and monomeric tissue-associated CRP, and that it is the latter which is responsible for the proangiogenic actions of CRP (Schwedler et al., 2006; Slevin and Krupinski, 2009; Slevin et al., 2010). Whatever the interpretation, the finding of colony-forming potential in healthy individuals underscores the notion that **vascular integrity is a "round-the-clock" process characterized by a continuous balance between endothelial injury and regeneration-** i.e. between environmental challenges and genetically-determined functional reserve. In this perspective cardiovascular disease can be viewed as the result of a disruption of this balance, with the threshold between physiology and pathology being specific for each single individual.

6. Current and extra demands on the endothelium, the tip of the iceberg

The endothelial cells lining the blood vessels are continuously exposed to different kinds of stimuli consisting in changes in the physicochemical characteristics of blood. These reflect the interplay between the internal and external environments, are transmitted by the pulsatile blood flow generated by the pressure gradient produced by the heart, and can trigger inflammation. We have already seen that **environmental changes requiring an active response from the cell can be called demands**. Demands can be further classified in current or extra according to the efficiency with which they can be handled by the homeostatic machinery. **Current demands can be adequately dealt with in a physiological context, even though some degree of cell injury can still occur. Extra demands cannot be**

sufficiently buffered and lead to functional impairment and, eventually, disease. Since the circulation of blood is a highly dynamic phenomenon it can be argued that current demands represent the bulk of all demands while extra demands are more often occasional. If we use the iceberg metaphor, current demands represent the huge part of the iceberg below the water while extra demands causing disease are only the tip of the iceberg. The water line is the boundary between health and overt disease, which varies from individual to individual according to the genetic program. Between overt disease and health we must conceive a grey zone (subclinical disease) in which disease may be unrecognized due to lack of knowledge or appropriate communication (Figure 2).
In the following paragraphs we will discuss the effect on the endothelium of blood pressure (BP) and high altitude (real or simulated) - taken as instances of current demands- as well as of ischemia, exemplifying an extra demand.

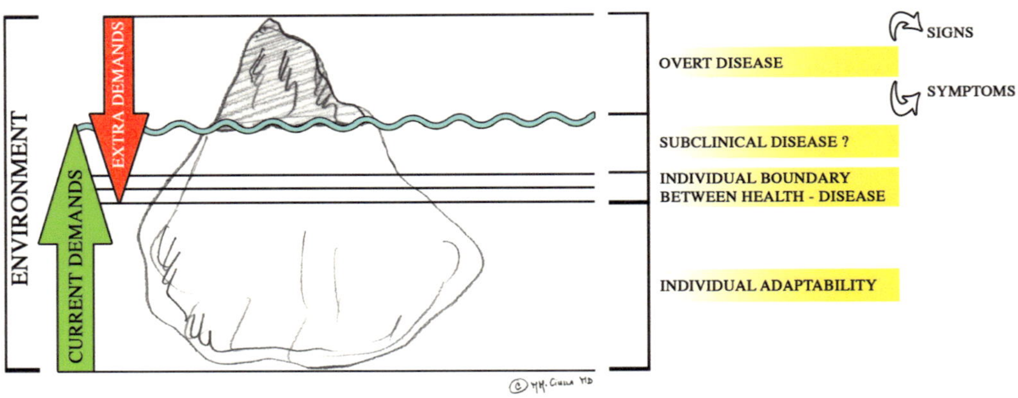

Fig. 2. The iceberg of disease
Pictorial representation of the iceberg metaphor, illustrating the boundary between health (as a result of individual adaptability) and disease. The tip of the iceberg corresponds to overt disease; the huge part below the water line is where individual adaptability successfully buffers environmental demands; just below the surface is the grey zone of subclinical disease. Demands are defined as current or extra according to how efficiently they can be handled by the single organism.

6.1 Blood pressure as a mechanical stimulus

BP is a highly dynamic phenomenon, whose profile is characterized by a series of peaks and nadirs, and it is one of the most important mechanical stimuli acting on vessels. Thus the endothelium is constantly exposed to enormous shear stress (Lu and Kassab, 2011), defined as the component of stress coplanar with a material cross-section and tending to cause a deformation of the material by slippage along this plane. Although it has been shown that exercise can increase circulating EPCs in mice (Cheng et al., 2010; Laufs et al., 2004) and humans (Rehman et al., 2004) and that shear stress promotes endothelial differentiation of EPCs in vitro (Ye et al., 2008) there is scant information on the specific effect of BP on endothelial turnover in healthy subjects. In order to address this issue we enrolled 12 normotensive volunteers who underwent a cycloergometer stress test (Bruce protocol) with BP monitoring (Ciulla et al., 2009). Samples of peripheral blood were obtained immediately before and after the test to determine the number of EPCs and circulating endothelial cells (CECs). The latter are mature endothelial cells that are dislodged from the endothelium by injurious stimuli via different mechanisms of cell detachment, including apoptosis and cytokine-mediated proteolysis of the ECM (Dignat-George and Sampol, 2000; Hunting et al., 2005; Mikirova et al., 2009). They can be considered a marker of vascular injury and their levels are generally proportional to those of EPCs (Hunting et al., 2005). We found that the number of CECs increased significantly (1.5-fold) with exercise and that such increase was directly associated with the value of peak systolic BP. The number of EPCs also increased, albeit not in a statistically significant way, probably because their mobilization from the BM requires a longer time. It can therefore be speculated that in physiological conditions BP fluctuations contribute to the renewal of the endothelium by a "mechanical clean-up" process, with systolic BP peaks removing damaged endothelial cells, just like the wind blows away the withered leaves from a tree. On the contrary, in hypertensive subjects, in whom BP levels are constantly elevated with frequent peaks, BP and shear stress can cause damage to the endothelium (Figure 3).

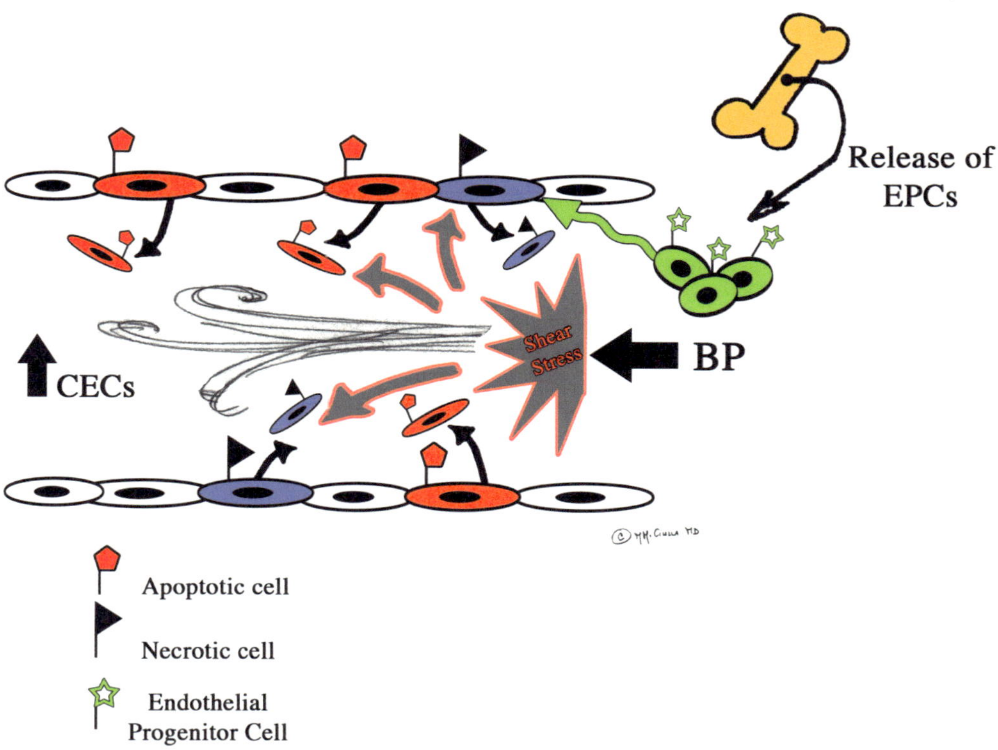

Fig. 3. Effect of blood pressure on the endothelium
Schematic illustration of the effects of shear stress on endothelial turnover. In healthy subjects systolic blood pressure peaks are responsible for a "mechanical clean up" process by dislodging damaged cells, thereby increasing the number of CECs. Concurrently, EPCs are moblized from the bone marrow into the circulation to replace detached cells. In hypertensives, BP may cause direct damage to the endothelium.

6.2 Hypoxia: a common cause of cell damage

Hypoxia is defined as a decrease in the normal level of tissue oxygen tension and can be classified as hypoxic or ischemic based on the underlying pathophysiological mechanism. Hypoxic hypoxia is characterized by a low partial pressure of oxygen in arterial blood (PaO_2) while in ischemic hypoxia PaO_2 is normal, and so is the oxygen content of blood, but oxygen delivery to tissues is impaired because of a reduction in blood flow (Kumar et al., 2011).

It is widely acknowledged that hypoxia and inflammation are closely intertwined: hypoxia stimulates the production of proinflammatory cytokines and inflammation is associated with tissue hypoxia due to the increased metabolic demands of cells (Eltzschig and Carmeliet, 2011; Frede et al., 2007). The link to angiogenesis is established via VEGF which is expressed in response to both hypoxia and proinflammatory cytokines (IL-1, TNF) (Homsi and Daud, 2007; Naldini and Carraro, 2005). At a molecular level, hypoxia and cytokines mainly rely on two distinct signaling pathways in which upregulation of the VEGF gene is brought about by Hypoxia Inducible Factor-1 (HIF-1) (Dehne and Brune, 2009; Eltzschig and Carmeliet, 2011; Harris, 2002; Jewell et al., 2001) and Nuclear Factor kappa-B (NFkB) respectively (Hofer and Schweighofer, 2007; Huang et al., 2000; Sprague and Khalil, 2009; Winsauer and de Martin, 2007). HIF is a heterodimeric transcription factor composed of a PaO_2-sensitive subunit (HIF-1 α) and a constitutive subunit (HIF-1 β). The enzyme prolyl hydroxylase (PHD) acts as an oxygen sensor within the cell since in normoxic conditions it is activated and hydroxylates proline residues on the HIF-1 α subunit, thus creating a binding site for the Von Hippel Lindau protein (VHL). The interaction with VHL then causes HIF-1 α to become ubiquitylated and targeted to the proteosome for degradation. Conversely, in hypoxic conditions PHD is inactive, HIF-1 is stabilized and translocates to the nucleus where it binds to hypoxia-responsive elements (HRE) in oxygen-responsive genes such as VEGF. NFkB is a heterodimeric transcription factor consisting of the p50 and p65 subunits. Normally it is associated to an inhibitory protein (IκBα) which retains it in the cytoplasm. Upon stimulation by IL-1 and TNF, IκBα is phosphorylated on two serine residues by a specific kinase (IKK) and undergoes ubiquitylation and proteosomal degradation. This allows translocation of NFkB to the nucleus where it binds to the promoter regions of genes encoding a number of biologically active molecules such as VEGF and proinflammatory cytokines. It must however be emphasized that the hypoxia and cytokine signaling pathways intersect at multiple points, consistently with the notion of an extensive cross-talk between hypoxia and inflammation. For instance, IL-1 and TNF are both able to activate HIF-1 (Dehne and Brune, 2009; Frede et al., 2007); HIF can stimulate the transcription of NFkB and NFkB can do the same with HIF (Angelo and Kurzrock, 2007; Eltzschig and Carmeliet, 2011); hypoxia leads to the formation of hydrogen peroxide which directly activates NFkB (Angelo and Kurzrock, 2007).

Although VEGF is the key player in angiogenesis, other proangiogenic factors deserve a brief mention, namely Hepatocte Growth Factor (HGF), Erythropoietin (Epo) and Stromal Cell Derived Factor-1 (SDF-1). HFG, also known as Scatter Factor (SF), is a large multidomain protein similar to plasminogen which is produced by cells of mesenchymal origin (such as vascular smooth muscle cells, pericytes and fibroblasts) and acts in a paracrine way on neighbouring endothelial cells via its receptor, a transmembrane tyrosine kinase encoded by the Met proto-oncogene (Abounader and Laterra, 2005; Desiderio, 2007; Ding et al., 2003; Schroder et al., 2010). It has been shown to be a powerful stimulator of angiogenesis, mobilizing EPCs from the BM into the circulation, either directly or by promoting the expression of VEGF (Desiderio, 2007; Schroder et al., 2010; Zhang et al., 2005; Zhu et al., 2010). Epo, better known for its effects on the hematopoietic system, has also gained recognition as an enhancer of endothelial cell functions, especially for its ability to mobilize EPCs from the BM (Bahlmann et al., 2004; Heeschen et al., 2003; Schroder et al., 2010). The chemokine SDF-1/CXCL12 and its receptor CXCR4 have been found to regulate the mobilization of EPCs from the BM and their homing to sites of tissue injury (Shen et al., 2011; Walter et al., 2005). Remarkably, also HGF, Epo and SDF-1 are involved in the HIF-1 signaling pathway: the former can induce HIF-1 α (Desiderio, 2007), the latter (and CRXR4) are activated by HIF-1 (Dehne and Brune, 2009; Harris, 2002).

Several studies have investigated the effect of hypoxic hypoxia on EPCs, both in vitro and in vivo. In vitro, it has been shown that rat EPCs exposed for 48 hours to 5% O_2 improved their ability to proliferate, migrate and express VEGF (Wang et al., 2010). Similarly, human peripheral blood mononuclear cells (PB-MC) cultured for one week in hypoxic conditions were stimulated to differentiate into EPCs and express VEGF (Akita et al., 2003), and human EPCs cultured in 1% O_2 were found to accumulate HIF-1 and upregulate VEGF (Abaci et al., 2010). In vivo, hypoxic hypoxia can be obtained either by exposure to a low fraction of inspired oxygen (FiO_2) - i.e. normobaric hypoxia- or to a low atmospheric pressure - i.e. hypobaric hypoxia. In a mouse model of normobaric hypoxia a significant increase in circulating EPCs was observed after exposure to a 10% FiO_2 for four days (Schroder et al., 2009). In healthy subjects we demonstrated that breathing through a hypoxicator an oxygen mixture at 11.2 % (corresponding to a simulated altitude of 4850 m above sea level) for one hour resulted in a marked increase in the number of PB-EPCs, which then normalized after one week, and that the percentage increase in EPCs from baseline was inversely correlated with the baseline PaO_2 (Ciulla et al., 2007a). In keeping with the underlying molecular mechanisms, we also found that such brief standardized hypoxia was associated with an increase in the serum levels of HGF and Epo as well in the expression of HIF-1 within EPCs (Ciulla et al., 2007a). In a healthy individual undergoing a trek in the Himalayas at a mean altitude of 3900 m we described a significant increase in the number and clonogenic activity of EPCs one day after the trek, returning to normal after 45 days at sea level (Ciulla et al., 2005). A recent study conducted on healthy volunteers participating in a one-week hiking program on the Austrian Alps at 1700 m reported an increase in circulating EPCs (Schobersberger et al., 2010). It must however be remarked that in the case of these high- and moderate-altitude studies it is not possible to dissociate the effects of hypoxia from those of physical activity which is also a powerful stimulus for EPC upregulation (Cheng et al., 2010; Laufs et al., 2004).

With regard to ischemic hypoxia there is a wealth of evidence from animal and human studies that acute ischemic conditions are marked by an increase in plasma levels of VEGF and in the number of EPCs in the PB. In rabbit and murine models of hindlimb ischemia (Takahashi et al., 1999) an increase in the frequency of circulating EPCs, reaching a maximum 1 week after the onset of ischemia, has been described. In a murine model of soft tissue ischemia (Tepper et al., 2005) VEGF levels were found to be increased by 2.5-fold after one week and correlated with the levels of circulating EPCs. As far as human studies are concerned, researchers have mainly investigated vascular diseases such as myocardial infarction (MI), ischemic stroke (IS) and peripheral artery disease (PAD). In the first study carried out on patients with acute MI (Shintani et al., 2001), the levels of VEGF and EPCs in the peripheral blood were significantly increased relative to controls the first day after the event and peaked at one week. Moreover, they were closely correlated, supporting the notion that VEGF induces EPC mobilization from the BM. Subsequent studies demonstrated an increase in the number of EPCs, peaking on admission, in patients with recent ST-elevation MI (<12 hours) (Wojakowski et al., 2004), and elevated levels of EPCs in patients with acute MI as early as 3 hours from the onset of symptoms (Massa et al., 2009). Over the last years several studies on patients with acute IS stroke have also been conducted, showing that EPCs are mobilized in response to cerebral ischemia and that the extent of such mobilization appears to have a prognostic value in terms of the likelihood of recovery. In patients with acute IS a two-fold mean increase in peripheral blood mononuclear CD34+ cells has been found when compared to healthy controls and it has been demonstrated that neurological and functional outcomes at one and three months were significantly improved in those defined as "high mobilizers"(count > 15000 cells/ml) (Dunac et al., 2007). In a similar study (Sobrino et al., 2007), patients with a greater EPC rise during the first week after a first-ever nonlacunar IS had better outcomes at three months according to the National Institutes of Health Stroke Scale (NIHSS) and the Rankin Scale (RS). In a larger sample of patients with IS (Yip et al., 2008), circulating levels of EPCs were about twice higher in subjects with IS than in controls matched for risk factors, and low ECP levels were predictive of both severe concurrent neurological impairment at the NIHSS and combined major adverse outcomes at three months. Finally, Sandri and colleagues (Sandri et al., 2005; Sandri et al., 2011) investigated the acute and chronic effects of exercise-induced peripheral limb ischemia in subjects with PAD. After four weeks of ischemic treadmill training PAD patients exhibited a 3-fold increase in VEGF levels and a four-fold increase in EPCs (Sandri et al., 2005). After a single maximal treadmill test, VEGF levels increased more than four-fold after one day and EPCs rose three-fold at one to two days (Sandri et al., 2011).

7. Enumerating ECPs: a tricky question

CD34+VEGFR-2+CD133+ cells are more similar to hematopoietic-derived cells, which may contribute to vascular repair and homeostasis in an indirect manner. In fact cells belonging to the hematopoietic lineage can be recruited to injured or angiogenic sites and secrete regulatory cytokines that promote vessel homeostasis and repair by local cells, including local vessel wall ECs. It is possible that cardiovascular risk factors and established cardiovascular disease decrease the circulating number and properties of the hematopoietic-derived cells. Hence, low levels of these circulating cells might correlate with adverse cardiovascular outcomes. However, this concept is entirely different from the notion of EPC which was proposed 10 years ago, when EPCs were suggested to function as a structural backup from the BM (Asahara et al., 1997). Mature hematopoietic cells include red blood cells, platelets, myeloid cells such as monocytes/macrophages and granulocytes, dendritic cells and lymphoid cells including B cells, T cells, NK cells and NKT cells. Hematopoietic cells derive from HSCs and hematopoietic progenitor cells (HPCs) that reside within the BM. Importantly, hematopoietic-derived cells such as monocytes, granulocytes, platelets and even HSCs/HPCs have been shown to be involved in vascular repair (Heil et al., 2006; Langer et al., 2006). However, because both endothelial lineage cells and hematopoietic cells are present at sites of neovascularization and co-express a host of similar surface markers, it can be difficult to discriminate them from each other at sites of vascular repair, and appreciate their individual contribution to the healing or regenerative process. Therefore, the diverse cell types now known to be recruited at sites of neo- vascularization are highly likely to have previously been lumped into the single term 'EPC' in many early studies of postnatal vasculogenesis, explaining some of the apparent controversy in the field.

One of the models that has been used to directly investigate donor cell differentiation into ECs in vivo has been a transgenic mouse whose cells express a fluorescent marker (e.g. green fluorescent protein (GFP)) only in the presence of an endothelium-specific gene, such as Tie-2 (Schlaeger et al., 1997). Therefore, transplantation of BM cells from transgenic mice into wild-type mice allows the tracking of the cells of interest, and of their fate during mobilization from the BM into sites of vascular injury, and discriminates them from other cell types and from the host cells involved in vascular repair and regeneration. However, even these sophisticated approaches have often yielded contradictory results, probably because the expression of Tie-2 is not entirely restricted to the endothelial lineage, but is also found in pericytes and hematopoietic (derived) cells such as monocytes, that also migrate to sites of vascular repair (De Palma et al., 2003).

On the other hand, it has been argued that the failure to retrieve genetically labelled BM-derived Tie-2+ ECs in the work by De Palma et al. might be due to the fact that the exogenous Tie-2 promoter used may not mark all mature ECs, or that during the random genetic manipulation of cells the CEPC population might not have been targeted with the viral Tie-2 vector (Nolan et al., 2007). Also, it is possible that the failure to detect few, if any, BM-derived ECs in the neovasculature of experimental models may be related to poor engraftment of the EPC compartment following its transplantation (De Palma et al., 2003; Gothert et al., 2005). Although the latter pitfall can be circumvented by the use of a parabiosis model (where two mice are surgically connected and share a common circulatory system allowing exchange of circulating cells), conflicting results have also been reported in the parabiosis model (Purhonen et al., 2008).

In order to definitely demonstrate in vivo that the EPC-derived progeny in the newly formed vasculature is truly endothelial in nature, a more direct or convincing approach would be to extract the putative EPC progeny from tissues (by means of a genetic tracer such as GFP) and FACS-analyze these cells (with or without previous culturing) using a wide panel of antigens (CD11, CD45, VE-Cadherin, CD146, CD31, CD13, CD105, etc.) so as to be able to discriminate them from other cell lineages, especially the hematopoietic one. In addition, these extracted cells could also be tested functionally by evaluating their proliferative or tube-forming capacity. So far, only one group has used such a FACS-based strategy to identify EPC progeny in vivo (Nolan et al., 2007), but the combination of markers used (CD31+Lectin+GFP+) does not allow a clear discrimination from hematopoietic (derived) cells that also display the same functional abilities (Nolan et al., 2007; Yoder et al., 2007).

Other methodological issues - such as the number of tissue sites sampled, the time frame of the study intervals, and the microscopic technique employed in tissue analysis - may explain why in different animal models different findings were obtained with regard to the extent of EPC contribution to neovascularization. For instance, while the use of confocal microscopy may permit a complete volumetric 3-dimensional rendering of donor cell contribution to new blood vessels in a damaged tissue, the use of light microscopy can make it difficult to discriminate whether cells are are integrated within the endothelial layer, or merely located at periluminal sites, just beneath the endothelial layer (De Palma et al., 2003; Galimi et al., 2005). Thus, differences in the microscopic technique, and/or in the method of image analysis, may also account for the huge variability in the rate of EPC incorporation into repairing vessels that has been reported in the literature, ranging from 0 to 90%. Furthermore, it remains uncertain how many of the 'luminal integrated cells' (e.g. monocytes, that phenotypically overlap with ECs) are in fact passenger cells participating in an inflammatory reaction to vascular injury, or cells in the process of transmigrating deeper into the vessel wall and interstitial tissues. Thus, whenever possible, the use of 3-dimensional imaging, coupled with antibodies which are specific and validated for the identification of cells and subcellular organelles as well as extracellular structures, provides the most sensitive and specific information on the location and contribution of donor cells to tissues under repair.

As it is increasingly recognized that cell fate changes are a property of stem cells (Blau et al., 2001) there arises the need to adopt more stringent criteria to determine if a cell fate transition has taken place. Blau and colleagues (Blau et al., 2001) propose a set of such criteria. The first criterion is the demonstration that the cell expresses a previously silent gene specific for the new cell type. This can be accomplished by investigating protein expression via antibodies and FACS analysis. Of course, in vivo evaluation must be considered superior to in vitro evaluation. The second criterion is the demonstration that the cell is well integrated into the tissue and has acquired morphological characteristics identical to those of neighbouring resident cells. The third -and most important- criterion is the demonstration, by means of a functional assay, that the cell is able to perform tissue-specific functions. For instance, using appropriate genetically-modified animal models it can be shown that it produces de novo an organ-specific molecule which is essential for survival or improvement of disease-associated deficits. With reference to the cardiovascular domain, convincing evidence of a cell fate transition in BM-derived cells could be obtained by verifying that these cells not only produce myocardium-specific proteins but are also able to contract in synchrony with the cardiac syncytia.

8. Conclusion

The circulatory system is involved in the transport of a wide variety of biological molecules and cells and can be considered the body's basic communication system. It consists of an intricate network of vessels lined by endothelium. Although the endothelium has long been viewed as an inert layer of "nucleated cellophane" its highly dynamic nature is being increasingly recognized. **Endothelial cells are exposed round the clock to proinflammatory stimuli of different kinds (mechanical, chemical, biological), even in healthy individuals, and vascular homeostasis is ensured by a continuous balance between injury and repair- i.e. between environmental challenges and genetically-determined functional reserve.** A disruption of this balance leads to endothelial dysfunction and atherosclerotic disease, with the threshold between physiology (current demands) and pathology (extra demands) being different and specific for each individual, as unique as is the genome.

There is growing evidence that EPCs are a major player in the process of vessel regeneration, being mobilized from the BM into the PB and homing to sites of injury where they are incorporated into the neoendothelium to serve a structural and paracrine role. Indeed we demonstrated, in an animal model of myocardial cryodamage, that peripherally injected BM mononuclear cells are able to target the site of damage and form early-stage small vessels.

Since 1997, when Asahara and colleagues isolated EPCs from human PB by means of magnetic beads, the identity and characterization of EPCs have been the subject of heated debate. However, it is generally accepted that progenitor cells similar to embryonic hemangioblasts participate in adult angiogenesis, thus establishing the notion of postnatal vasculogenesis.

In complex organisms inflammation and angiogenesis are closely related and the endothelial cell lies at the very core of this association. Inflammatory cells secrete cytokines (IL-1 and TNF) which elicit VEGF production from the endothelium. VEGF, in turn, is a powerful proangiogenic factor which regulates EPC kinetics and stimulates inflammatory cells. Such extensive communication between endothelial and inflammatory cells is the basis of the inflammation-angiogenesis cross-talk.

There is a wealth of evidence from animal and human studies that ischemia enhances the number and functions of circulating EPCs. Less information is available on the effect of physiological stimuli on EPCs or CECs in healthy subjects. Our group showed that blood pressure-induced shear stress increases the number of CECs, that hypoxia (normo- and hypobaric) increases the number and/or clonogenic capacity of EPCs, and that hs-CRP levels are higher in colony-formers.

9. References

Abaci, HE, Truitt, R, Luong, E, Drazer, G, Gerecht, S (2010) Adaptation to oxygen deprivation in cultures of human pluripotent stem cells, endothelial progenitor cells, and umbilical vein endothelial cells. In: Am J Physiol Cell Physiol: United States, pp. C1527-1537.

Abounader, R, Laterra, J (2005) Scatter factor/hepatocyte growth factor in brain tumor growth and angiogenesis. *Neuro Oncol* 7: 436-451.

Akita, T, Murohara, T, Ikeda, H, Sasaki, K, Shimada, T, Egami, K, Imaizumi, T (2003) Hypoxic preconditioning augments efficacy of human endothelial progenitor cells for therapeutic neovascularization. *Lab Invest* 83: 65-73.

Anagnostou, A, Liu, Z, Steiner, M, Chin, K, Lee, ES, Kessimian, N, Noguchi, CT (1994) Erythropoietin receptor mRNA expression in human endothelial cells. *Proc Natl Acad Sci U S A* 91: 3974-3978.

Angelo, LS, Kurzrock, R (2007) Vascular endothelial growth factor and its relationship to inflammatory mediators. In: Clin Cancer Res: United States, pp. 2825-2830.

Asahara, T, Masuda, H, Takahashi, T, Kalka, C, Pastore, C, Silver, M, Kearne, M, Magner, M, Isner, JM (1999) Bone marrow origin of endothelial progenitor cells responsible for postnatal vasculogenesis in physiological and pathological neovascularization. *Circ Res* 85: 221-228.

Asahara, T, Murohara, T, Sullivan, A, Silver, M, van der Zee, R, Li, T, Witzenbichler, B, Schatteman, G, Isner, JM (1997) Isolation of putative progenitor endothelial cells for angiogenesis. *Science* 275: 964-967.

Asosingh, K, Erzurum, SC, Yoder, MC, Tuder, RM (2009) Letter by asosingh et Al regarding article, "circulating endothelial progenitor cells in patients with eisenmenger syndrome and idiopathic pulmonary arterial hypertension". In: Circulation: United States, p. e230; author reply e231.

Bahlmann, FH, De Groot, K, Spandau, JM, Landry, AL, Hertel, B, Duckert, T, Boehm, SM, Menne, J, Haller, H, Fliser, D (2004) Erythropoietin regulates endothelial progenitor cells. In: Blood: United States, pp. 921-926.

Bertolini, F, Shaked, Y, Mancuso, P, Kerbel, RS (2006) The multifaceted circulating endothelial cell in cancer: towards marker and target identification. In: Nat Rev Cancer: England, pp. 835-845.

Bittner, RE, Schofer, C, Weipoltshammer, K, Ivanova, S, Streubel, B, Hauser, E, Freilinger, M, Hoger, H, Elbe-Burger, A, Wachtler, F (1999) Recruitment of bone-marrow-derived cells by skeletal and cardiac muscle in adult dystrophic mdx mice. *Anat Embryol (Berl)* 199: 391-396.

Blau, HM, Brazelton, TR, Weimann, JM (2001) The evolving concept of a stem cell: entity or function? In: Cell: United States, pp. 829-841.

Bompais, H, Chagraoui, J, Canron, X, Crisan, M, Liu, XH, Anjo, A, Tolla-Le Port, C, Leboeuf, M, Charbord, P, Bikfalvi, A, Uzan, G (2004) Human endothelial cells derived from circulating progenitors display specific functional properties compared with mature vessel wall endothelial cells. In: Blood: United States, pp. 2577-2584.

Brazelton, TR, Rossi, FM, Keshet, GI, Blau, HM (2000) From marrow to brain: expression of neuronal phenotypes in adult mice. In: Science: United States, pp. 1775-1779.

Bull, TM, Golpon, H, Hebbel, RP, Solovey, A, Cool, CD, Tuder, RM, Geraci, MW, Voelkel, NF (2003) Circulating endothelial cells in pulmonary hypertension. In: Thromb Haemost: Germany, pp. 698-703.

Butcher, EC (1991) Leukocyte-endothelial cell recognition: three (or more) steps to specificity and diversity. In: Cell: United States, pp. 1033-1036.

Calabro, P, Golia, E, Yeh, ET (2009) CRP and the risk of atherosclerotic events. *Semin Immunopathol* 31: 79-94.

Calabro, P, Willerson, JT, Yeh, ET (2003) Inflammatory cytokines stimulated C-reactive protein production by human coronary artery smooth muscle cells. In: Circulation: United States, pp. 1930-1932.

Campbell, KH, McWhir, J, Ritchie, WA, Wilmut, I (1996) Sheep cloned by nuclear transfer from a cultured cell line. *Nature* 380: 64-66.

Cannon, W (1926) Physiological regulation of normal states: some tentative postulates concerning biological homeostatics. Éditions Médicales.

Caplan, BA, Schwartz, CJ (1973) Increased endothelial cell turnover in areas of in vivo Evans Blue uptake in the pig aorta. *Atherosclerosis* 17: 401-417.

Carmeliet, P (2000) Mechanisms of angiogenesis and arteriogenesis. *Nat Med* 6: 389-395.

Case, J, Mead, LE, Bessler, WK, Prater, D, White, HA, Saadatzadeh, MR, Bhavsar, JR, Yoder, MC, Haneline, LS, Ingram, DA (2007) Human CD34+AC133+VEGFR-2+ cells are not endothelial progenitor cells but distinct, primitive hematopoietic progenitors. In: Exp Hematol: Netherlands, pp. 1109-1118.

Chen, X, Andresen, BT, Hill, M, Zhang, J, Booth, F, Zhang, C (2008) Role of Reactive Oxygen Species in Tumor Necrosis Factor-alpha Induced Endothelial Dysfunction. *Curr Hypertens Rev* 4: 245-255.

Cheng, XW, Kuzuya, M, Kim, W, Song, H, Hu, L, Inoue, A, Nakamura, K, Di, Q, Sasaki, T, Tsuzuki, M, Shi, GP, Okumura, K, Murohara, T (2010) Exercise training stimulates ischemia-induced neovascularization via phosphatidylinositol 3-kinase/Akt-dependent hypoxia-induced factor-1 alpha reactivation in mice of advanced age. In: Circulation: United States, pp. 707-716.

Choi, K, Kennedy, M, Kazarov, A, Papadimitriou, JC, Keller, G (1998) A common precursor for hematopoietic and endothelial cells. *Development* 125: 725-732.

Cines, DB, Pollak, ES, Buck, CA, Loscalzo, J, Zimmerman, GA, McEver, RP, Pober, JS, Wick, TM, Konkle, BA, Schwartz, BS, Barnathan, ES, McCrae, KR, Hug, BA, Schmidt, AM, Stern, DM (1998) Endothelial cells in physiology and in the pathophysiology of vascular disorders. *Blood* 91: 3527-3561.

Ciulla, MM, Cortiana, M, Silvestris, I, Matteucci, E, Ridolfi, E, Giofre, F, Zanardelli, M, Paliotti, R, Cortelezzi, A, Pierini, A, Magrini, F, Desiderio, MA (2007a) Effects of simulated altitude (normobaric hypoxia) on cardiorespiratory parameters and circulating endothelial precursors in healthy subjects. In: Respir Res: England, p. 58.

Ciulla, MM, Ferrero, S, Gianelli, U, Paliotti, R, Magrini, F, Braidotti, P (2007b) Direct visualization of neo-vessel formation following peripheral injection of bone marrow derived CD34+ cells in experimental myocardial damage. In: Micron: England, pp. 321-322.

Ciulla, MM, Ferrero, S, Lazzari, L, Pacchiana, R, Paliotti, R, Gianelli, U, Busca, G, Esposito, A, Bosari, S, Magrini, F, Rebulla, P (2004) The translocation of marrow MNCs after experimental myocardial cryoinjury is proportional to the infarcted area. In: Transfusion: United States, pp. 239-244.

Ciulla, MM, Gianni, C, Broglia, P, Lonati, S, Silvestris, I, Paliotti, R, Giofre, F, Rampoldi, E, Cortelezzi, A, Magrini, F (2009) Systolic blood pressure peak during maximal exercise testing: a possible determinant of endothelial turnover in healthy subjects. *Am J Hematol* 84: 449-450.

Ciulla, MM, Giorgetti, A, Giordano, R, Silvestris, I, Cortiana, M, Paliotti, R, Lazzari, L (2007c) Circulating endothelial progenitor cell colony-forming capacity in healthy subjects: how does an endothelial colony look like? In: Am J Cardiol: United States, pp. 559-560.

Ciulla, MM, Giorgetti, A, Lazzari, L, Cortiana, M, Silvestris, I, Annoni, G, De Asmundis, C, Fiore, AV, Montelatici, E, Paliotti, R, Magrini, F, Rebulla, P, Cortelezzi, A (2005) High-altitude trekking in the Himalayas increases the activity of circulating endothelial cells. *Am J Hematol* 79: 76-78.

Ciulla, MM, Giorgetti, A, Silvestris, I, Cortiana, M, Montelatici, E, Paliotti, R, Annoni, GA, Fiore, AV, Giordano, R, De Marco, F, Magrini, F, Rebulla, P, Cortelezzi, A, Lazzari, L (2006) Endothelial colony forming capacity is related to C-reactive protein levels in healthy subjects. *Curr Neurovasc Res* 3: 99-106.

Ciulla, MM, Lazzari, L, Pacchiana, R, Esposito, A, Bosari, S, Ferrero, S, Gianelli, U, Paliotti, R, Busca, G, Giorgetti, A, Magrini, F, Rebulla, P (2003) Homing of peripherally injected bone marrow cells in rat after experimental myocardial injury. *Haematologica* 88: 614-621.

De Palma, M, Venneri, MA, Roca, C, Naldini, L (2003) Targeting exogenous genes to tumor angiogenesis by transplantation of genetically modified hematopoietic stem cells. In: Nat Med: United States, pp. 789-795.

Dehne, N, Brune, B (2009) HIF-1 in the inflammatory microenvironment. In: Exp Cell Res: United States, pp. 1791-1797.

Del Papa, N, Colombo, G, Fracchiolla, N, Moronetti, LM, Ingegnoli, F, Maglione, W, Comina, DP, Vitali, C, Fantini, F, Cortelezzi, A (2004) Circulating endothelial cells as a marker of ongoing vascular disease in systemic sclerosis. *Arthritis Rheum* 50: 1296-1304.

Delorme, B, Basire, A, Gentile, C, Sabatier, F, Monsonis, F, Desouches, C, Blot-Chabaud, M, Uzan, G, Sampol, J, Dignat-George, F (2005) Presence of endothelial progenitor cells, distinct from mature endothelial cells, within human CD146+ blood cells. In: Thromb Haemost: Germany, pp. 1270-1279.

Desiderio, MA (2007) Hepatocyte growth factor in invasive growth of carcinomas. *Cell Mol Life Sci* 64: 1341-1354.

DeWitt, N (2005) Angiogenesis. Nature, p. 931.

Dignat-George, F, Sampol, J (2000) Circulating endothelial cells in vascular disorders: new insights into an old concept. *Eur J Haematol* 65: 215-220.

Diller, GP, van Eijl, S, Okonko, DO, Howard, LS, Ali, O, Thum, T, Wort, SJ, Bedard, E, Gibbs, JS, Bauersachs, J, Hobbs, AJ, Wilkins, MR, Gatzoulis, MA, Wharton, J (2008) Circulating endothelial progenitor cells in patients with Eisenmenger syndrome and idiopathic pulmonary arterial hypertension. In: Circulation: United States, pp. 3020-3030.

Ding, S, Merkulova-Rainon, T, Han, ZC, Tobelem, G (2003) HGF receptor up-regulation contributes to the angiogenic phenotype of human endothelial cells and promotes angiogenesis in vitro. In: Blood: United States, pp. 4816-4822.

Dunac, A, Frelin, C, Popolo-Blondeau, M, Chatel, M, Mahagne, MH, Philip, PJ (2007) Neurological and functional recovery in human stroke are associated with peripheral blood CD34+ cell mobilization. *J Neurol* 254: 327-332.

Ebrahem, Q, Chaurasia, SS, Vasanji, A, Qi, JH, Klenotic, PA, Cutler, A, Asosingh, K, Erzurum, S, Anand-Apte, B (2010) Cross-talk between vascular endothelial growth factor and matrix metalloproteinases in the induction of neovascularization in vivo. In: Am J Pathol: United States, pp. 496-503.

Eltzschig, HK, Carmeliet, P (2011) Hypoxia and inflammation. *N Engl J Med* 364: 656-665.

Fadini, GP, Baesso, I, Albiero, M, Sartore, S, Agostini, C, Avogaro, A (2008) Technical notes on endothelial progenitor cells: ways to escape from the knowledge plateau. In: Atherosclerosis: Ireland, pp. 496-503.

Fina, L, Molgaard, HV, Robertson, D, Bradley, NJ, Monaghan, P, Delia, D, Sutherland, DR, Baker, MA, Greaves, MF (1990) Expression of the CD34 gene in vascular endothelial cells. *Blood* 75: 2417-2426.

Flamme, I, Risau, W (1992) Induction of vasculogenesis and hematopoiesis in vitro. *Development* 116: 435-439.

Frede, S, Berchner-Pfannschmidt, U, Fandrey, J (2007) Regulation of hypoxia-inducible factors during inflammation. In: Methods Enzymol: United States, pp. 405-419.

Friedrich, EB, Walenta, K, Scharlau, J, Nickenig, G, Werner, N (2006) CD34-/CD133+/VEGFR-2+ endothelial progenitor cell subpopulation with potent vasoregenerative capacities. In: Circ Res: United States, pp. e20-25.

Furusawa, C, Kaneko, K (2000) Complex organization in multicellularity as a necessity in evolution. *Artif Life* 6: 265-281.

Galimi, F, Summers, RG, van Praag, H, Verma, IM, Gage, FH (2005) A role for bone marrow-derived cells in the vasculature of noninjured CNS. In: Blood: United States, pp. 2400-2402.

Galley, HF, Webster, NR (2004) Physiology of the endothelium. In: Br J Anaesth: England, pp. 105-113.

Geard, N, Wiles, J (2005) A gene network model for developing cell lineages. *Artif Life* 11: 249-267.

Genest, J (2010) C-reactive protein: risk factor, biomarker and/or therapeutic target? *Can J Cardiol* 26 Suppl A: 41A-44A.

George, AL, Bangalore-Prakash, P, Rajoria, S, Suriano, R, Shanmugam, A, Mittelman, A, Tiwari, RK (2011) Endothelial progenitor cell biology in disease and tissue regeneration. In: J Hematol Oncol: England, p. 24.

George, F, Brouqui, P, Boffa, MC, Mutin, M, Drancourt, M, Brisson, C, Raoult, D, Sampol, J (1993) Demonstration of Rickettsia conorii-induced endothelial injury in vivo by measuring circulating endothelial cells, thrombomodulin, and von Willebrand factor in patients with Mediterranean spotted fever. *Blood* 82: 2109-2116.

Goon, PK, Boos, CJ, Stonelake, PS, Blann, AD, Lip, GY (2006) Detection and quantification of mature circulating endothelial cells using flow cytometry and immunomagnetic beads: a methodological comparison. In: Thromb Haemost: Germany, pp. 45-52.

Gothert, JR, Gustin, SE, Hall, MA, Green, AR, Gottgens, B, Izon, DJ, Begley, CG (2005) In vivo fate-tracing studies using the Scl stem cell enhancer: embryonic hematopoietic stem cells significantly contribute to adult hematopoiesis. In: Blood: United States, pp. 2724-2732.

Gurdon, JB (1960) The developmental capacity of nuclei taken from differentiating endoderm cells of Xenopus laevis. *J Embryol Exp Morphol* 8: 505-526.

Gussoni, E, Soneoka, Y, Strickland, CD, Buzney, EA, Khan, MK, Flint, AF, Kunkel, LM, Mulligan, RC (1999) Dystrophin expression in the mdx mouse restored by stem cell transplantation. *Nature* 401: 390-394.

Harris, AL (2002) Hypoxia--a key regulatory factor in tumour growth. *Nat Rev Cancer* 2: 38-47.

Haudenschild, C, Studer, A (1971) Early interactions between blood cells and severely damaged rabbit aorta. *Eur J Clin Invest* 2: 1-7.

Haudenschild, CC, Schwartz, SM (1979) Endothelial regeneration. II. Restitution of endothelial continuity. *Lab Invest* 41: 407-418.

Hay, ED (1991) Collagen and other matrix glycoproteins in embryogenesis. In: Cell biology of extracellular matrix. Press, NYP, Ed., pp. 419-462.

Heeschen, C, Aicher, A, Lehmann, R, Fichtlscherer, S, Vasa, M, Urbich, C, Mildner-Rihm, C, Martin, H, Zeiher, AM, Dimmeler, S (2003) Erythropoietin is a potent physiologic stimulus for endothelial progenitor cell mobilization. In: Blood: United States, pp. 1340-1346.

Heil, M, Eitenmuller, I, Schmitz-Rixen, T, Schaper, W (2006) Arteriogenesis versus angiogenesis: similarities and differences. In: J Cell Mol Med: Romania, pp. 45-55.

Hofer, E, Schweighofer, B (2007) Signal transduction induced in endothelial cells by growth factor receptors involved in angiogenesis. In: Thromb Haemost: Germany, pp. 355-363.

Homsi, J, Daud, AI (2007) Spectrum of activity and mechanism of action of VEGF/PDGF inhibitors. *Cancer Control* 14: 285-294.

Hong, YJ, Jeong, MH, Lim, SY, Lee, SR, Kim, KH, Sohn, IS, Park, HW, Kim, JH, Kim, W, Ahn, Y, Cho, JG, Park, JC, Kang, JC (2005) Elevated preprocedural high-sensitivity C-reactive protein levels are associated with neointimal hyperplasia and restenosis development after successful coronary artery stenting. In: Circ J: Japan, pp. 1477-1483.

Hristov, M, Erl, W, Weber, PC (2003a) Endothelial progenitor cells: isolation and characterization. In: Trends Cardiovasc Med: United States, pp. 201-206.

Hristov, M, Erl, W, Weber, PC (2003b) Endothelial progenitor cells: mobilization, differentiation, and homing. In: Arterioscler Thromb Vasc Biol: United States, pp. 1185-1189.

Huang, S, Robinson, JB, Deguzman, A, Bucana, CD, Fidler, IJ (2000) Blockade of nuclear factor-kappaB signaling inhibits angiogenesis and tumorigenicity of human ovarian cancer cells by suppressing expression of vascular endothelial growth factor and interleukin 8. *Cancer Res* 60: 5334-5339.

Hung, MJ, Cherng, WJ, Yang, NI, Cheng, CW, Li, LF (2005) Relation of high-sensitivity C-reactive protein level with coronary vasospastic angina pectoris in patients without hemodynamically significant coronary artery disease. In: Am J Cardiol: United States, pp. 1484-1490.

Hunting, CB, Noort, WA, Zwaginga, JJ (2005) Circulating endothelial (progenitor) cells reflect the state of the endothelium: vascular injury, repair and neovascularization. In: Vox Sang: England, pp. 1-9.

Ii, M, Takenaka, H, Asai, J, Ibusuki, K, Mizukami, Y, Maruyama, K, Yoon, YS, Wecker, A, Luedemann, C, Eaton, E, Silver, M, Thorne, T, Losordo, DW (2006) Endothelial progenitor thrombospondin-1 mediates diabetes-induced delay in reendothelialization following arterial injury. In: Circ Res: United States, pp. 697-704.

Ingram, DA, Mead, LE, Moore, DB, Woodard, W, Fenoglio, A, Yoder, MC (2005) Vessel wall-derived endothelial cells rapidly proliferate because they contain a complete hierarchy of endothelial progenitor cells. In: Blood: United States, pp. 2783-2786.

Inoue, T, Kato, T, Uchida, T, Sakuma, M, Nakajima, A, Shibazaki, M, Imoto, Y, Saito, M, Hashimoto, S, Hikichi, Y, Node, K (2005) Local release of C-reactive protein from vulnerable plaque or coronary arterial wall injured by stenting. In: J Am Coll Cardiol: United States, pp. 239-245.

Jackson, KA, Mi, T, Goodell, MA (1999) Hematopoietic potential of stem cells isolated from murine skeletal muscle. *Proc Natl Acad Sci U S A* 96: 14482-14486.

Jewell, UR, Kvietikova, I, Scheid, A, Bauer, C, Wenger, RH, Gassmann, M (2001) Induction of HIF-1alpha in response to hypoxia is instantaneous. *FASEB J* 15: 1312-1314.

Jujo, K, Ii, M, Losordo, DW (2008) Endothelial progenitor cells in neovascularization of infarcted myocardium. In: J Mol Cell Cardiol: England, pp. 530-544.

Kabrun, N, Buhring, HJ, Choi, K, Ullrich, A, Risau, W, Keller, G (1997) Flk-1 expression defines a population of early embryonic hematopoietic precursors. *Development* 124: 2039-2048.

Kallianpur, AR, Jordan, JE, Brandt, SJ (1994) The SCL/TAL-1 gene is expressed in progenitors of both the hematopoietic and vascular systems during embryogenesis. *Blood* 83: 1200-1208.

Keck, PJ, Hauser, SD, Krivi, G, Sanzo, K, Warren, T, Feder, J, Connolly, DT (1989) Vascular permeability factor, an endothelial cell mitogen related to PDGF. *Science* 246: 1309-1312.

Keller, G, Kennedy, M, Papayannopoulou, T, Wiles, MV (1993) Hematopoietic commitment during embryonic stem cell differentiation in culture. *Mol Cell Biol* 13: 473-486.

Keller, GM (1995) In vitro differentiation of embryonic stem cells. In: Curr Opin Cell Biol: United States, pp. 862-869.

Kennedy, M, Firpo, M, Choi, K, Wall, C, Robertson, S, Kabrun, N, Keller, G (1997) A common precursor for primitive erythropoiesis and definitive haematopoiesis. *Nature* 386: 488-493.

Kocher, AA, Schuster, MD, Szabolcs, MJ, Takuma, S, Burkhoff, D, Wang, J, Homma, S, Edwards, NM, Itescu, S (2001) Neovascularization of ischemic myocardium by human bone-marrow-derived angioblasts prevents cardiomyocyte apoptosis, reduces remodeling and improves cardiac function. In: Nat Med: United States, pp. 430-436.

Kondo, T, Hayashi, M, Takeshita, K, Numaguchi, Y, Kobayashi, K, Iino, S, Inden, Y, Murohara, T (2004) Smoking cessation rapidly increases circulating progenitor cells in peripheral blood in chronic smokers. In: Arterioscler Thromb Vasc Biol: United States, pp. 1442-1447.

Krause, DS, Theise, ND, Collector, MI, Henegariu, O, Hwang, S, Gardner, R, Neutzel, S, Sharkis, SJ (2001) Multi-organ, multi-lineage engraftment by a single bone marrow-derived stem cell. In: Cell: United States, pp. 369-377.

Kumar, V, Abbas, AK, Fausto, N, Aster, J (2011) Robbins and Cotran Pathologic Basis of Disease.

Kunz, J, Schreiter, B, Schubert, B, Voss, K, Krieg, K (1978) [Experimental investigations on the regeneration of aortic endothelial cells. Automatic and visual evaluation of autoradiograms (author's transl)]. *Acta Histochem* 61: 53-63.

Kushner, I, Samols, D, Magrey, M (2010) A unifying biologic explanation for "high-sensitivity" C-reactive protein and "low-grade" inflammation. *Arthritis Care Res (Hoboken)* 62: 442-446.

Lagasse, E, Connors, H, Al-Dhalimy, M, Reitsma, M, Dohse, M, Osborne, L, Wang, X, Finegold, M, Weissman, IL, Grompe, M (2000) Purified hematopoietic stem cells can differentiate into hepatocytes in vivo. *Nat Med* 6: 1229-1234.

Langer, H, May, AE, Daub, K, Heinzmann, U, Lang, P, Schumm, M, Vestweber, D, Massberg, S, Schonberger, T, Pfisterer, I, Hatzopoulos, AK, Gawaz, M (2006) Adherent platelets recruit and induce differentiation of murine embryonic endothelial progenitor cells to mature endothelial cells in vitro. In: Circ Res: United States, pp. e2-10.

Laufs, U, Werner, N, Link, A, Endres, M, Wassmann, S, Jurgens, K, Miche, E, Bohm, M, Nickenig, G (2004) Physical training increases endothelial progenitor cells, inhibits neointima formation, and enhances angiogenesis. In: Circulation: United States, pp. 220-226.

Lee, S, Chen, TT, Barber, CL, Jordan, MC, Murdock, J, Desai, S, Ferrara, N, Nagy, A, Roos, KP, Iruela-Arispe, ML (2007) Autocrine VEGF signaling is required for vascular homeostasis. In: Cell: United States, pp. 691-703.

Leung, DW, Cachianes, G, Kuang, WJ, Goeddel, DV, Ferrara, N (1989) Vascular endothelial growth factor is a secreted angiogenic mitogen. *Science* 246: 1306-1309.

Li, JJ, Fang, CH (2004) C-reactive protein is not only an inflammatory marker but also a direct cause of cardiovascular diseases. In: Med Hypotheses: Scotland, pp. 499-506.

Lin, Y, Weisdorf, DJ, Solovey, A, Hebbel, RP (2000) Origins of circulating endothelial cells and endothelial outgrowth from blood. *J Clin Invest* 105: 71-77.

Lu, D, Kassab, GS (2011) Role of shear stress and stretch in vascular mechanobiology. In: J R Soc Interface.

Malczak, HT, Buck, RC (1977) Regeneration of endothelium in rat aorta after local freezing. A scanning electron microscopic study. *Am J Pathol* 86: 133-148.

Mancuso, P, Burlini, A, Pruneri, G, Goldhirsch, A, Martinelli, G, Bertolini, F (2001) Resting and activated endothelial cells are increased in the peripheral blood of cancer patients. *Blood* 97: 3658-3661.

Massa, M, Campanelli, R, Bonetti, E, Ferrario, M, Marinoni, B, Rosti, V (2009) Rapid and large increase of the frequency of circulating endothelial colony-forming cells (ECFCs) generating late outgrowth endothelial cells in patients with acute myocardial infarction. In: Exp Hematol: Netherlands, pp. 8-9.

Mezey, E, Chandross, KJ, Harta, G, Maki, RA, McKercher, SR (2000) Turning blood into brain: cells bearing neuronal antigens generated in vivo from bone marrow. In: Science: United States, pp. 1779-1782.

Mikirova, NA, Jackson, JA, Hunninghake, R, Kenyon, J, Chan, KW, Swindlehurst, CA, Minev, B, Patel, AN, Murphy, MP, Smith, L, Alexandrescu, DT, Ichim, TE, Riordan, NH (2009) Circulating endothelial progenitor cells: a new approach to anti-aging medicine? In: J Transl Med: England, p. 106.

Millauer, B, Wizigmann-Voos, S, Schnurch, H, Martinez, R, Moller, NP, Risau, W, Ullrich, A (1993) High affinity VEGF binding and developmental expression suggest Flk-1 as a major regulator of vasculogenesis and angiogenesis. In: Cell: United States, pp. 835-846.

Mor, F, Quintana, FJ, Cohen, IR (2004) Angiogenesis-inflammation cross-talk: vascular endothelial growth factor is secreted by activated T cells and induces Th1 polarization. *J Immunol* 172: 4618-4623.

Murray, PDF (1932) The Development in vitro of the Blood of the Early Chick Embryo. Proceedings of the Royal Society of London, pp. 497-521.

Mutunga, M, Fulton, B, Bullock, R, Batchelor, A, Gascoigne, A, Gillespie, JI, Baudouin, SV (2001) Circulating endothelial cells in patients with septic shock. *Am J Respir Crit Care Med* 163: 195-200.

Nakano, T, Kodama, H, Honjo, T (1994) Generation of lymphohematopoietic cells from embryonic stem cells in culture. *Science* 265: 1098-1101.

Nakatani, K, Takeshita, S, Tsujimoto, H, Kawamura, Y, Tokutomi, T, Sekine, I (2003) Circulating endothelial cells in Kawasaki disease. In: Clin Exp Immunol: England, pp. 536-540.

Naldini, A, Carraro, F (2005) Role of inflammatory mediators in angiogenesis. *Curr Drug Targets Inflamm Allergy* 4: 3-8.

Nolan, DJ, Ciarrocchi, A, Mellick, AS, Jaggi, JS, Bambino, K, Gupta, S, Heikamp, E, McDevitt, MR, Scheinberg, DA, Benezra, R, Mittal, V (2007) Bone marrow-derived endothelial progenitor cells are a major determinant of nascent tumor neovascularization. In: Genes Dev: United States, pp. 1546-1558.

Noonan, DM, De Lerma Barbaro, A, Vannini, N, Mortara, L, Albini, A (2008) Inflammation, inflammatory cells and angiogenesis: decisions and indecisions. *Cancer Metastasis Rev* 27: 31-40.

Olsson, AK, Dimberg, A, Kreuger, J, Claesson-Welsh, L (2006) VEGF receptor signalling - in control of vascular function. In: Nat Rev Mol Cell Biol: England, pp. 359-371.

Paoletti, R, Gotto, AM, Jr., Hajjar, DP (2004) Inflammation in atherosclerosis and implications for therapy. In: Circulation: United States, pp. III20-26.

Peichev, M, Naiyer, AJ, Pereira, D, Zhu, Z, Lane, WJ, Williams, M, Oz, MC, Hicklin, DJ, Witte, L, Moore, MA, Rafii, S (2000) Expression of VEGFR-2 and AC133 by circulating human CD34(+) cells identifies a population of functional endothelial precursors. *Blood* 95: 952-958.

Plouet, J, Schilling, J, Gospodarowicz, D (1989) Isolation and characterization of a newly identified endothelial cell mitogen produced by AtT-20 cells. *EMBO J* 8: 3801-3806.

Purhonen, S, Palm, J, Rossi, D, Kaskenpaa, N, Rajantie, I, Yla-Herttuala, S, Alitalo, K, Weissman, IL, Salven, P (2008) Bone marrow-derived circulating endothelial precursors do not contribute to vascular endothelium and are not needed for tumor growth. In: Proc Natl Acad Sci U S A: United States, pp. 6620-6625.

Rajashekhar, G, Willuweit, A, Patterson, CE, Sun, P, Hilbig, A, Breier, G, Helisch, A, Clauss, M (2006) Continuous endothelial cell activation increases angiogenesis: evidence for the direct role of endothelium linking angiogenesis and inflammation. In: J Vasc Res: Switzerland, pp. 193-204.

Ratcliffe, NA, Rowley, AF (1979) A comparative synopsis of the structure and function of the blood cells of insects and other invertebrates. *Dev Comp Immunol* 3: 189-221.

Rehman, J, Li, J, Orschell, CM, March, KL (2003) Peripheral blood "endothelial progenitor cells" are derived from monocyte/macrophages and secrete angiogenic growth factors. *Circulation* 107: 1164-1169.

Rehman, J, Li, J, Parvathaneni, L, Karlsson, G, Panchal, VR, Temm, CJ, Mahenthiran, J, March, KL (2004) Exercise acutely increases circulating endothelial progenitor cells and monocyte-/macrophage-derived angiogenic cells. In: J Am Coll Cardiol: United States, pp. 2314-2318.

Reyes, M, Dudek, A, Jahagirdar, B, Koodie, L, Marker, PH, Verfaillie, CM (2002) Origin of endothelial progenitors in human postnatal bone marrow. *J Clin Invest* 109: 337-346.

Ridker, PM (2001) High-sensitivity C-reactive protein: potential adjunct for global risk assessment in the primary prevention of cardiovascular disease. *Circulation* 103: 1813-1818.

Risau, W, Flamme, I (1995) Vasculogenesis. *Annu Rev Cell Dev Biol* 11: 73-91.

Risau, W, Sariola, H, Zerwes, HG, Sasse, J, Ekblom, P, Kemler, R, Doetschman, T (1988) Vasculogenesis and angiogenesis in embryonic-stem-cell-derived embryoid bodies. *Development* 102: 471-478.

Ruppert, EE, Carle, KJ (1983) Morphology of metazoan circulatory systems. 193-208.

Sabin, FR (1920) Studies on the origin of blood vessels and of red corpuscles as seen in the living blastoderm of the chick during the second day of incubation. Contribution to Emmbryology, pp. 231-262.

Sandri, M, Adams, V, Gielen, S, Linke, A, Lenk, K, Krankel, N, Lenz, D, Erbs, S, Scheinert, D, Mohr, FW, Schuler, G, Hambrecht, R (2005) Effects of exercise and ischemia on mobilization and functional activation of blood-derived progenitor cells in patients with ischemic syndromes: results of 3 randomized studies. In: Circulation: United States, pp. 3391-3399.

Sandri, M, Beck, EB, Adams, V, Gielen, S, Lenk, K, Hollriegel, R, Mangner, N, Linke, A, Erbs, S, Mobius-Winkler, S, Scheinert, D, Hambrecht, R, Schuler, G (2011) Maximal exercise, limb ischemia, and endothelial progenitor cells. *Eur J Cardiovasc Prev Rehabil* 18: 55-64.

Scadden, DT (2007) The weight of cell identity. *J Clin Invest* 117: 3653-3655.

Schabbauer, G, Schweighofer, B, Mechtcheriakova, D, Lucerna, M, Binder, BR, Hofer, E (2007) Nuclear factor of activated T cells and early growth response-1 cooperate to mediate tissue factor gene induction by vascular endothelial growth factor in endothelial cells. In: Thromb Haemost: Germany, pp. 988-997.

Schlaeger, TM, Bartunkova, S, Lawitts, JA, Teichmann, G, Risau, W, Deutsch, U, Sato, TN (1997) Uniform vascular-endothelial-cell-specific gene expression in both embryonic and adult transgenic mice. *Proc Natl Acad Sci U S A* 94: 3058-3063.

Schobersberger, W, Leichtfried, V, Mueck-Weymann, M, Humpeler, E (2010) Austrian Moderate Altitude Studies (AMAS): benefits of exposure to moderate altitudes (1,500-2,500 m). *Sleep Breath* 14: 201-207.

Schroder, K, Kohnen, A, Aicher, A, Liehn, EA, Buchse, T, Stein, S, Weber, C, Dimmeler, S, Brandes, RP (2009) NADPH oxidase Nox2 is required for hypoxia-induced mobilization of endothelial progenitor cells. In: Circ Res: United States, pp. 537-544.

Schroder, K, Schutz, S, Schloffel, I, Batz, S, Takac, I, Weissmann, N, Michaelis, UR, Koyanagi, M, Brandes, RP (2010) Hepatocyte growth factor induces a pro-angiogenic phenotype and mobilizes endothelial progenitor cells by activating Nox2. *Antioxid Redox Signal.*

Schwartz, SM, Gajdusek, CM, Reidy, MA, Selden, SC, 3rd, Haudenschild, CC (1980) Maintenance of integrity in aortic endothelium. *Fed Proc* 39: 2618-2625.

Schwartz, SM, Gajdusek, CM, Selden, SC, 3rd (1981) Vascular wall growth control: the role of the endothelium. *Arteriosclerosis* 1: 107-126.

Schwartz, SM, Stemerman, MB, Benditt, EP (1975) The aortic intima. II. Repair of the aortic lining after mechanical denudation. *Am J Pathol* 81: 15-42.

Schwedler, SB, Filep, JG, Galle, J, Wanner, C, Potempa, LA (2006) C-reactive protein: a family of proteins to regulate cardiovascular function. In: Am J Kidney Dis: United States, pp. 212-222.

Schweighofer, B, Schultes, J, Pomyje, J, Hofer, E (2007) Signals and genes induced by angiogenic growth factors in comparison to inflammatory cytokines in endothelial cells. *Clin Hemorheol Microcirc* 37: 57-62.

Senger, DR, Galli, SJ, Dvorak, AM, Perruzzi, CA, Harvey, VS, Dvorak, HF (1983) Tumor cells secrete a vascular permeability factor that promotes accumulation of ascites fluid. *Science* 219: 983-985.

Shaffer, RG, Greene, S, Arshi, A, Supple, G, Bantly, A, Moore, JS, Parmacek, MS, Mohler, ER, 3rd (2006) Effect of acute exercise on endothelial progenitor cells in patients with peripheral arterial disease. *Vasc Med* 11: 219-226.

Shalaby, F, Ho, J, Stanford, WL, Fischer, KD, Schuh, AC, Schwartz, L, Bernstein, A, Rossant, J (1997) A requirement for Flk1 in primitive and definitive hematopoiesis and vasculogenesis. In: Cell: United States, pp. 981-990.

Shalaby, F, Rossant, J, Yamaguchi, TP, Gertsenstein, M, Wu, XF, Breitman, ML, Schuh, AC (1995) Failure of blood-island formation and vasculogenesis in Flk-1-deficient mice. *Nature* 376: 62-66.

Shen, L, Gao, Y, Qian, J, Sun, A, Ge, J (2011) A novel mechanism for endothelial progenitor cells homing: The SDF-1/CXCR4-Rac pathway may regulate endothelial progenitor cells homing through cellular polarization. In: Med Hypotheses, A 2010 Elsevier Ltd: United States, pp. 256-258.

Shi, Q, Rafii, S, Wu, MH, Wijelath, ES, Yu, C, Ishida, A, Fujita, Y, Kothari, S, Mohle, R, Sauvage, LR, Moore, MA, Storb, RF, Hammond, WP (1998) Evidence for circulating bone marrow-derived endothelial cells. *Blood* 92: 362-367.

Shintani, S, Murohara, T, Ikeda, H, Ueno, T, Honma, T, Katoh, A, Sasaki, K, Shimada, T, Oike, Y, Imaizumi, T (2001) Mobilization of endothelial progenitor cells in patients with acute myocardial infarction. *Circulation* 103: 2776-2779.

Slevin, M, Krupinski, J (2009) A role for monomeric C-reactive protein in regulation of angiogenesis, endothelial cell inflammation and thrombus formation in cardiovascular/cerebrovascular disease? *Histol Histopathol* 24: 1473-1478.

Slevin, M, Matou-Nasri, S, Turu, M, Luque, A, Rovira, N, Badimon, L, Boluda, S, Potempa, L, Sanfeliu, C, de Vera, N, Krupinski, J (2010) Modified C-reactive protein is expressed by stroke neovessels and is a potent activator of angiogenesis in vitro. In: Brain Pathol: Switzerland, pp. 151-165.

Sobrino, T, Hurtado, O, Moro, MA, Rodriguez-Yanez, M, Castellanos, M, Brea, D, Moldes, O, Blanco, M, Arenillas, JF, Leira, R, Davalos, A, Lizasoain, I, Castillo, J (2007) The increase of circulating endothelial progenitor cells after acute ischemic stroke is associated with good outcome. In: Stroke: United States, pp. 2759-2764.

Sprague, AH, Khalil, RA (2009) Inflammatory cytokines in vascular dysfunction and vascular disease. In: Biochem Pharmacol: England, pp. 539-552.

Studer, L, Csete, M, Lee, SH, Kabbani, N, Walikonis, J, Wold, B, McKay, R (2000) Enhanced proliferation, survival, and dopaminergic differentiation of CNS precursors in lowered oxygen. *J Neurosci* 20: 7377-7383.

Takahashi, T, Kalka, C, Masuda, H, Chen, D, Silver, M, Kearney, M, Magner, M, Isner, JM, Asahara, T (1999) Ischemia- and cytokine-induced mobilization of bone marrow-derived endothelial progenitor cells for neovascularization. *Nat Med* 5: 434-438.

Taylor, RG, Lewis, JC (1986) Endothelial cell proliferation and monocyte adhesion to atherosclerotic lesions of white carneau pigeons. *Am J Pathol* 125: 152-160.

Tepper, OM, Capla, JM, Galiano, RD, Ceradini, DJ, Callaghan, MJ, Kleinman, ME, Gurtner, GC (2005) Adult vasculogenesis occurs through in situ recruitment, proliferation, and tubulization of circulating bone marrow-derived cells. In: Blood: United States, pp. 1068-1077.

Timmermans, F, Plum, J, Yoder, MC, Ingram, DA, Vandekerckhove, B, Case, J (2009) Endothelial progenitor cells: identity defined? In: J Cell Mol Med: England, pp. 87-102.

Timmermans, F, Van Hauwermeiren, F, De Smedt, M, Raedt, R, Plasschaert, F, De Buyzere, ML, Gillebert, TC, Plum, J, Vandekerckhove, B (2007) Endothelial outgrowth cells are not derived from CD133+ cells or CD45+ hematopoietic precursors. In: Arterioscler Thromb Vasc Biol: United States, pp. 1572-1579.

Tongers, J, Roncalli, JG, Losordo, DW (2010) Role of endothelial progenitor cells during ischemia-induced vasculogenesis and collateral formation. In: Microvasc Res, 2010 Elsevier Inc: United States, pp. 200-206.

Torsney, E, Xu, Q (2011) Resident vascular progenitor cells. In: J Mol Cell Cardiol, 2010 Elsevier Ltd: England, pp. 304-311.

Turu, MM, Slevin, M, Matou, S, West, D, Rodriguez, C, Luque, A, Grau-Olivares, M, Badimon, L, Martinez-Gonzalez, J, Krupinski, J (2008) C-reactive protein exerts angiogenic effects on vascular endothelial cells and modulates associated signalling pathways and gene expression. In: BMC Cell Biol: England, p. 47.

Urao, N, Okigaki, M, Yamada, H, Aadachi, Y, Matsuno, K, Matsui, A, Matsunaga, S, Tateishi, K, Nomura, T, Takahashi, T, Tatsumi, T, Matsubara, H (2006) Erythropoietin-mobilized endothelial progenitors enhance reendothelialization via Akt-endothelial nitric oxide synthase activation and prevent neointimal hyperplasia. In: Circ Res: United States, pp. 1405-1413.

Urbich C, Heeschen C, Aicher A, Dernbach E, Zeiher AM, Dimmeler S (2003) Relevance of monocytic features for neovascularization capacity of circulating endothelial progenitor cells. *Circulation* 2003 18;108(20):2511-6.

Venugopal, SK, Devaraj, S, Jialal, I (2005) Macrophage conditioned medium induces the expression of C-reactive protein in human aortic endothelial cells: potential for paracrine/autocrine effects. In: Am J Pathol: United States, pp. 1265-1271.

Verma, S, Kuliszewski, MA, Li, SH, Szmitko, PE, Zucco, L, Wang, CH, Badiwala, MV, Mickle, DA, Weisel, RD, Fedak, PW, Stewart, DJ, Kutryk, MJ (2004) C-reactive protein attenuates endothelial progenitor cell survival, differentiation, and function: further evidence of a mechanistic link between C-reactive protein and cardiovascular disease. In: Circulation: United States, pp. 2058-2067.

Vittet, D, Prandini, MH, Berthier, R, Schweitzer, A, Martin-Sisteron, H, Uzan, G, Dejana, E (1996) Embryonic stem cells differentiate in vitro to endothelial cells through successive maturation steps. *Blood* 88: 3424-3431.

Wagner, R (1980) Endothelial cell embryology and growth. Adv. Microcirc., p. 45.

Walter, DH, Haendeler, J, Reinhold, J, Rochwalsky, U, Seeger, F, Honold, J, Hoffmann, J, Urbich, C, Lehmann, R, Arenzana-Seisdesdos, F, Aicher, A, Heeschen, C, Fichtlscherer, S, Zeiher, AM, Dimmeler, S (2005) Impaired CXCR4 signaling contributes to the reduced neovascularization capacity of endothelial progenitor cells from patients with coronary artery disease. In: Circ Res: United States, pp. 1142-1151.

Wang, DH, Li, FR, Zhang, Y, Wang, YQ, Yuan, FH (2010) Conditioned medium from renal tubular epithelial cells stimulated by hypoxia influences rat bone marrow-derived endothelial progenitor cells. *Ren Fail* 32: 863-870.

Weiss, MJ, Orkin, SH (1996) In vitro differentiation of murine embryonic stem cells. New approaches to old problems. *J Clin Invest* 97: 591-595.

Wiles, MV, Keller, G (1991) Multiple hematopoietic lineages develop from embryonic stem (ES) cells in culture. *Development* 111: 259-267.

Winsauer, G, de Martin, R (2007) Resolution of inflammation: intracellular feedback loops in the endothelium. In: Thromb Haemost: Germany, pp. 364-369.

Wojakowski, W, Tendera, M, Michalowska, A, Majka, M, Kucia, M, Maslankiewicz, K, Wyderka, R, Ochala, A, Ratajczak, MZ (2004) Mobilization of CD34/CXCR4+, CD34/CD117+, c-met+ stem cells, and mononuclear cells expressing early cardiac, muscle, and endothelial markers into peripheral blood in patients with acute myocardial infarction. In: Circulation: United States, pp. 3213-3220.

Yamaguchi, TP, Dumont, DJ, Conlon, RA, Breitman, ML, Rossant, J (1993) flk-1, an flt-related receptor tyrosine kinase is an early marker for endothelial cell precursors. *Development* 118: 489-498.

Ye, C, Bai, L, Yan, ZQ, Wang, YH, Jiang, ZL (2008) Shear stress and vascular smooth muscle cells promote endothelial differentiation of endothelial progenitor cells via activation of Akt. In: Clin Biomech (Bristol, Avon): England, pp. S118-124.

Yip, HK, Chang, LT, Chang, WN, Lu, CH, Liou, CW, Lan, MY, Liu, JS, Youssef, AA, Chang, HW (2008) Level and value of circulating endothelial progenitor cells in patients after acute ischemic stroke. In: Stroke: United States, pp. 69-74.

Yoder, MC, Mead, LE, Prater, D, Krier, TR, Mroueh, KN, Li, F, Krasich, R, Temm, CJ, Prchal, JT, Ingram, DA (2007) Redefining endothelial progenitor cells via clonal analysis and hematopoietic stem/progenitor cell principals. In: Blood: United States, pp. 1801-1809.

Young, PE, Baumhueter, S, Lasky, LA (1995) The sialomucin CD34 is expressed on hematopoietic cells and blood vessels during murine development. *Blood* 85: 96-105.

Zhang, QW, Liu, HJ, Duan, HF, Ha, XQ, Wang, H, Jia, XX, Lu, ZZ, Wu, CT, Wang, LS (2005) [Hepatocyte growth factor recruits endothelial progenitor cells from bone marrow into blood circulation]. *Zhongguo Ying Yong Sheng Li Xue Za Zhi* 21: 100-103.

Zhu, G, Huang, L, Song, M, Yu, Z, Wu, X, Zhao, X, Jin, J, Zhao, G, Chen, J, Yu, S (2010) Over-expression of hepatocyte growth factor in smooth muscle cells regulates endothelial progenitor cells differentiation, migration and proliferation. In: Int J Cardiol: Netherlands, pp. 70-80.

10

Cellular Stress Responses

Irina Milisav
Institute of Pathophysiology, Faculty of Medicine, University of Ljubljana
Faculty of Health Sciences, University of Ljubljana
Slovenia

1. Introduction

Cells encounter many internal and external stimuli, some of which may induce stresses, when they are a part of a normal tissue or when they grow in a culture. These stresses trigger responses which may change cellular responses to subsequent environmental signals or even cause cell death. Exposure to stress over time may cause the accumulation of damage to DNA, proteins and lipids. If not repaired, this will enhance susceptibility to aging associated illnesses, like neurodegenerative diseases, diabetes, heart diseases, etc., and to cancer. The cellular stress responses must be taken into account when the cells are used in cell therapies and in regenerative medicine.

The cellular stress response is a reaction to changes or fluctuations of extracellular conditions, which damage the structure and function of macromolecules (Kültz, 2003). Depending on the severity and duration of stress encountered, cells either re-establish cellular homeostasis to the former state or adopt an altered state in the new environment. Therefore, different stressors and different intensities of stresses trigger different cellular responses: (1) induce cell repair mechanisms; these use considerable amounts of available resources and often result in recovery of normal cells, (2) induce cell responses that result in temporary adaptation to some stressors, (3) induce autophagy or (4) trigger cell death. Some important mechanisms of cell adaptation to stress and their inducers will be described in this chapter.

1.1 Cell repair mechanisms

1.1.1 Altered gene expression

Stressors can damage intracellular macromolecules, including proteins, DNA, RNAs and lipids. The stress responses are mediated through induction of molecular chaperones (Buchberger et al., 2010), clearance of damaged macromolecules (Kroemer et al., 2010), growth arrest and changed gene expression patterns (Spriggs et al., 2010), etc. First we shall describe some examples of altered transcription.

1.1.1.1 Modulation of transcription upon stress

The changes in gene expression are often mediated by micro ribonucleic acids (miRNAs, reviewed in Leung and Sharp, 2010), which are short noncoding RNAs of about 22 nucleotides. There were about 950 miRNA discovered by April 2010, which could target about 60% of mammalian mRNAs. The stress responses modify the synthesis of miRNAs and the activities of miRNA-protein complexes, consequently the expression of their mRNA

targets. miRNAs bind to mRNAs and either accelerate the degradation or inhibit the translation of mRNA, i.e. modulate the stability and/or translational potential of their targets.
The transcription factor, transformation-related protein 53 (p53, chapter 1.1.3.2) regulates the expression of miRNA at the levels of transcription and processing. p53 was found to induce the transcription of the primary transcripts miR-34a, miR-34b and miR-34c upon DNA damage. These repress some target genes to promote growth arrest and apoptosis. The expression of p53 itself must be precisely regulated. A miRNA (miR-125b) was found, which keeps the expression of p53 low in humans. The repression of miR-125b is reversed upon DNA damage with the activation of p53 through a protein kinase cascade (Leung and Sharp, 2010).
The level of target gene repression depends on the relative concentrations of target genes and miRNAs. For example, MICA and MICB are extracellular ligands of an immune activating receptor (NKG2D) of natural killer cells (NK) and T cells. MICA and MICB are induced by several types of cells under stress, like heat shock, viral infection, oxidative stress and DNA-damage. Their translation is inhibited by miRNAs in normal cells, however, upon stress the transcription of MICA and MICB is upregulated, while the levels of appropriate inhibiting miRNAs remain unchanged. It seems that the levels of mRNA exceed the quantity that can be inhibited by miRNA; consequently MICA and MICB are expressed. Their binding to the receptor NKG2D, which is expressed on natural killer cells and T cells helps to eliminate the virus-infected cells. Interactions in the cells are more complex, as the level of miRNA-mediated repression depends also on the expression amount of other mRNAs of the transcriptome, which are targeted by the same miRNA. The outcome of miRNA repression depends also on interactions with other stress proteins that can modulate the activity of miRNA protein complexes, e.g. by inhibiting the access to target mRNA.
The subcellular location of miRNA can change as a consequence of stress. Most of miRNA are diffused in the cytoplasm; they have to associate with the member of Argonaute protein family for activity. After a maturation process at Argonaute protein, mature miRNA guide the Argonaute-containing complexes to target sites in mRNAs that are partially complementary to the miRNA sequence, and induce repression of gene expression at the level of mRNA stability or translation. Upon stress some of miRNAs, mRNAs and Argonaute are in stress granules.
Sometimes, miRNA can time the stress response. Timing is important in acute stress responses, such as during inflammation. Then nuclear factor-κB (NF-κB, chapter 1.1.3.3) upregulates the transcription of miRNAs along with other inflammatory responsive genes through a cascade of reactions in macrophages. RNAs are synthesized as pre-RNAs; all pre-RNAs, including miRNA, are synthesized in about 2 hours. The processed mature miRNA peak about 24 hours later; therefore, the action of miRNA is delayed.

1.1.1.2 Translational regulation of gene expression during stress

Nutrient stress, temperature shock, DNA damage and hypoxia can lead to changes in gene expression patterns caused by shutdown and reprogramming of protein synthesis through selective recruitment of ribosomes to mRNAs (Spriggs et al., 2010). This is regulated by elements in 5′ and 3′ untranslated regions of mRNAs, like internal ribosome entry segments, upstream open reading frames and miRNA target sites.
In eukaryotes, the initiation of translation is inhibited often by phosphorylation of the eukaryotic initiation factor 2 (eIF2), which is a part of the so called ternary complex. The

ternary complex is composed of eIF2, initiator tRNA (tRNAi) and GTP and is loaded onto a small ribosome subunit, which binds mRNA and recognizes the start of translation, codon AUG. This triggers the hydrolysis of GTP, uncoupling of tRNAi from eIF2, release of initiation factors and start of translation (Spriggs et al., 2010). In mammalian cells, four stress-related kinases phosphorylate eIF2, which lead to reduction in initiation codon recognition. A second mechanism for nonspecifically reducing levels of protein synthesis is by preventing recruitment of the translational machinery to the mRNA. This is done by interfering with m7G cap recognition, which is a modified base at the extreme 5′ end of the mRNA. mRNAs for stress response proteins evade global repression of translation by several mechanisms. For example, a cap-dependent recruitment of mRNA may be bypassed, by the internal ribosome entry sites (IRES). IRES were originally detected in viruses and are parts of RNA, which facilitate binding of mRNAs to 40S ribosomal subunits (Spriggs et al., 2010). This involves cofactors IRES trans-acting factors (ITAF). Changes in the abundance or activity of ITAFs influences the degree of IRES mediated translation; this is often used during stress conditions.

Heat shock factors (HSF) are inducible transcriptional regulators of genes encoding stress proteins, like molecular chaperones and others. HSF1 is the most important regulator of expression of heat shock proteins (Hsp, chapter 1.1.2.1) in vertebrates (Akerfelt et al., 2010). Hsp are molecular chaperones and among others assist proteins in folding or prevent and reverse protein missfolding and aggregation. HSF1 is converted from monomer to trimer in response to temperature shock and oxidative stress. Monomeric HSF1 is a phosphorylated protein and is interacting with Hsp90. On stress, HSF1 dissociates from Hsp90, which enables its trimerization and binding to heat shock elements of Hsp genes. To enable a versatile regulation, these processes are more complex. HSF1 interacts with different HSP at different phases of its activation cycle. Trimeric HSF1 is inactive when bound to multimeric chaperon complex composed of Hsp90, co-chaperone p23 and immunophilin FK506-binding protein 5 (FKBP52, also FKBP4, Akerfelt et al., 2010). Elevated levels of both, Hsp90 and Hsp70, prevent trimer formation of HSF1. Activated HSF1 can bind to Hsp70 and Hsp40. There seem to be activation-attenuation cycles of HSF1, during which there are extensive posttranslational modifications of HSF1, including acetylation, phosphorylation and sumoylation. HSF1 is phosphorylated also under non-stress conditions, while phosphorylation-mediated sumoylation of a Lys residue of the regulatory domain occurs on exposure to heat shock. SUMO stands for small ubiquitin-related modifier. This are small proteins that are covalently attached to and detached from other proteins to modify their function. While stress-induced sumoylation is rapid after the heat shock, acetylation is delayed, as it is regulated by the balance of acetylation by p300-CBP (CREB-binding protein) and deacetylation by nicotinamide adenine dinucleotide dependent histone deacetylase sirtuin 1, SIRT1. Increased expression of SIRT1 was reported to enhance and prolong the DNA-binding activity of HSF1 at the promoter of HSP70.1. Therefore SIRT1 maintains HSF1 in a state which enables DNA binding, while the acetylated form can not. Sirtuins are a family of NAD^+ (nicotinamide adenine dinucleotide) dependent histone deacetylases, which influence gene transcription, metabolism, DNA repair and organism life span (Majmundar et al., 2010). Sirtuins are sensors of the cellular redox state, as they respond to the changes of ratios of oxidized/reduced forms of NAD^+.

Other HSF, like HSF2 also bind to the promoters of HSP genes. Upon stress-induced transcription of HSP genes, both, HSF1 and HSF2 accumulate into nuclear stress bodies.

Nuclear stress bodies are thought to participate in rapid, transient, and global reprogramming of gene expression through different types of mechanisms including chromatin remodeling and trapping of transcription and splicing factors (Biamonti & Vourc'h, 2010).

1.1.2 Protein quality control and repair

Proteins have numerous functions in cells: enzymatic, transport, structural, for molecular recognition, signal transduction, etc. They are often damaged as a consequence of stress; even their normal biogenesis is an error-prone process and may lead to stress. For example, truncated polypeptides that result from incomplete translation, misfolded intermediates, and unassembled subunits of protein complexes have exposed hydrophobic regions, which may facilitate aggregation (Buchberger et al., 2010). Failure to clear aggregated proteins leads to cell stress common to many disorders, especially to neurodegenerative diseases. The environmental stress triggers can induce nonnative posttraslational modifications and damage proteins in other ways and consequently induce cell stress as well.

Some degree of protein damage occur normally in every cell, however, the extent of protein damage increases by adverse intrinsic and environmental conditions, like unbalanced protein synthesis, oxidative stress, metabolic stress, some environmental toxins and pollutants, elevated temperature, high-energy radiation, etc. To cope with considerable extent of protein damage, the damaged proteins are either repaired by molecular chaperones or degraded by the ubiquitin proteasome system or autophagy (Buchberger et al., 2010). Ubiquitin is a protein of 76 kDa, which marks the proteins for degradation by the protease, 26S proteasome. The attachment of ubiquitin (ubiquination or ubiquitinylation) is an ATP-consuming process by the cascade of enzymes: ubiquitin-activating enzyme (E1), ubiquitin-conjugating enzyme (E2) and ubiquitin protein ligase (E3). E2 and E3 enzymes cooperate with molecular chaperones. Polyubiquinated substrates are transferred to the proteasome for degradation. Aggregated proteins, which are degraded by selective autophagy, are also ubiquinated.

Some types of protein damage are prevalent in specific organelles of eukaryotic cells (Buchberger et al., 2010). The errors in glycosylation, disulphide bond formation and membrane insertion are more frequent in the endoplasmic reticulum (ER) and secretory pathway, while these proteins are generally thermodynamically more stable, therefore less sensitive to heat stress compared to cytosolic proteins.

1.1.2.1 Stress repair and stress response

The stress repair mechanisms enable cells to survive the stress induced damage. Chaperones are a family of proteins that assist the non-covalent folding or unfolding of proteins and the assembly or disassembly of multimeric macromolecular structures, but do not occur in these structures during their normal biological functions when these are in correctly folded state. There are different classes (families) of chaperones, which serve different functions, including folding of newly made proteins, the repair of the damage caused by misfolding, membrane transport, keeping the protein precursor in translocation-competent state, assistance in protein degradation, etc. Chaperons are efficient quality control components for proteins. Many chaperons were discovered as their expression was elevated after the heat shock, thus they were named heat shock proteins (Hsp). Chaperons of Hsp70 family interact with short extended peptide stretches of hydrophobic and basic amino acid residues of unfolded, natively folded or aggregated proteins (Buchberger et al., 2010). The binding of Hsp70 prevents aggregation of these proteins and can even induce conformational changes. The rounds of Hsp70 binding and release to protein substrates can promote the disaggregation of proteins. This is enhanced by cooperation with Hsp100 family chaperones. The cells have several Hsp40 chaperones (J domain containing chaperones), which provide substrate specificity of Hsp70, stabilize Hsp70-substrate interactions and trigger ATP hydrolysis by Hsp70. Nuclectide exchange factors, like Hsp110 are also required for ADP release and rebuilding of ATP at Hsp70.

Chaperon Hsp90 is thought to bind and stabilize partially folded but inactive conformations of its substrates; some of the substrates may be recognized in extended conformations. It hydrolyzes ATP like Hsp70; it interacts with a large number of co-chaperones. Its substrates are proteins involved in signal transduction, like protein kinases and transcription factors (Buchberger et al., 2010). Chaperones Hsp90 and Hsp70 can cooperate; then they are physically coupled by co-chaperone Hop.

An Hsp60 family member in eukaryotic cytosol is TRiC (TCP-1 Ring Complex or chaperonin containing TCP-1; CCT). Like other Hsp60 family members, TRiC is a barrel shaped protein complex that encapsulates substrates into protected folding environment. Its substrats are subunits of oligomers, with beta sheet secondary structures (Buchberger et al., 2010) and possibly late folding intermediates or misfolded proteins. There are also other small heat shock protein family members in the cytosol. ATP-independent small heat shock proteins (sHsp) bind misfolded proteins to prevent their aggregation and may loosen aggregates by coagreggation therefore facilitating subsequent re-folding by Hsp70. Hsp70 and Hsp40 present aggregated proteins to Hsp100. The solubilized protein can re-enter chaperone-mediated folding cycles or degradation by ubiquitin-proteasome system; this is implied by the presence of ubiquinated proteins within aggregates, colocalization of 26S proteasome, and the increased aggregate formation and delayed removal of aggregates upon inhibition of proteasome. However, ubiquitin proteasome system is not the only pathway for aggregate removal, especially as 26S proteasome can become inhibited by the aggregate formation. Irreversibly aggregated proteins in aggresomes are degraded primarily by selective autophagy. This autophagy requires ubiquitin-binding proteins; perhaps ubiquination is important for this degradation as well.

Stressed cells attempt to repair or degrade acutely damaged proteins and may undergo adaptive responses to reduce protein damage by decreasing global protein translation and increasing the molecular chaperones and proteins of proteolytic system. The well known adaptive responses in eukaryotic cells are the heat shock response (HSR) and unfolded protein response, which is the response to an accumulation of unfolded or misfolded proteins in the endoplasmic reticulum (UPR, Buchberger et al., 2010). Expression of genes, which are upregulated by HSR, are under the control of HSF1 (chapter 1.1.1.2, Akerfelt et al., 2010). Under non stress conditions, the monomeric HSF1 is in the cytosol bound by Hsp90, perhaps also Hsp70/40. These chaperones are tugged away under heat stress, which enables trimerization of HSF1, the activation of HSF1 trimer and expression of HSR target genes.

The ability to respond to ER stress is critical for cell survival, as chronic or unresolved ER stress can lead to apoptosis. Increases of protein synthesis or protein missfolding rates that exceed the capacity of chaperones, changes in calcium concentration in the ER lumen, oxidative stress and disturbances of the redox balance in the ER lumen contribute to development of ER stress (Tabas & Ron, 2011). In eukaryotic cells, the ER stress is sensed by three upstream signaling proteins that start cascades of corrective reactions. The activities of these three pathways are collectively called unfolded protein response. All ER stress response pathways are activated when there is imbalance of unfolded proteins and chaperones. The evolutionary oldest pathway is triggered by activation of IRE1 (inositol-requiring protein-1). In mammalian cells there are two isoforms: IRE1α in all cells, and IRE1β in gastrointestinal and respiratory tracts. IRE1 is likely activated by dissociation of BiP and binding of unfolded proteins. ER stress induces the dimerization of the IRE1 luminal domains and positions the cytosolic domains close for trans-autophosphorylation of the kinase activation loop. Phosphorylated IRE1 is the specific endonuclease which cleaves the mRNA of X-box binding protein 1 (XBP-1), which is then translated into a transcription factor, which induces the expression of many genes involved in UPR. The second pathway of UPR is initiated by activation of kinase PERK (protein kinase RNA (PKR)-like ER kinase). Similarly to IRE1, PERK is autophosphorylated and undergo homomultimerization upon stress. Then it phosphorylates the α-subunit of the translation initiation factor eIF2 (eukaryotic translation initiation factor-2), which results in attenuation of global translation initiation. Translation of the gene encoding the transcription factor ATF4 (activating transcription factor-4) is favored by limiting amounts of eIF2. It enables the expression of CHOP (C/EBP-homologous protein, GADD153, gene name Ddit3), which through interaction with other transcriptional regulators induces and suppresses numerous genes, which repair the ER stress. CHOP induces also the transcription of GADD34 (growth arrest and DNA damage-inducible protein-34), which dephosphorylates phosphorylated eIF2α and restores global protein translation.

A third pathway is mediated through transcriptional factor ATF6 (activating transcription factor-6, Buchberger et al., 2010). In unstressed cells ATF6 is bound to BiP in the ER. Upon ER stress, BiP dissociates to allow transport of ATF6 to Golgi, where it is processed and its fragment is released to cytosol for the subsequent activation of genes in the nucleus. The three pathways can be activated by any type of ER stress; there is different timing of activation (Tabas & Ron, 2011). IRE1, ATF6 and PERK pathways are sequentially activated upon the prolonged ER stress. Prolonged IRE1 and CHOP activation can trigger apoptosis.

Molecular chaperones seem to have also other biological functions, which are not connected with protein folding (Henderson, 2010). Such proteins are called moonlighting proteins. Based on the extreme sequence conservation between the chaperone members within families, the researchers have long assumed that they have the same functions. However, it appears that they may have different moonlighting activities despite of close sequence similarities. Some molecular chaperones act as receptors in the plasma membrane and have signaling functions in the extracellular fluid. The latter may be important in stress response. Circulating molecular chaperones may be a danger signal, as stress is a danger to all organisms and needs to be integrated into the homeostatic regulation.

1.1.3 Signaling pathways

1.1.3.1 mTOR pathway

Protein mTOR (mammalian target of rapamycin, also FK506 binding protein 12-rapamycin associated protein 1; FRAP1) is a serine/threonine protein kinase, which regulates growth by maintaining the balance between anabolic processes, like macromolecular synthesis, and catabolic processes, such as autophagy. It is involved in cell cycle progression, DNA recombination, and DNA damage detection. mTOR functions in regulatory pathways that control ribosome biogenesis and cell growth when nutrient concentrations change (particularly the levels of essential amino acids), like in a hypoxic environment of solid tumors (OMIM, 2011).

mTOR is the catalytic subunit of two complexes, mTORC1 and mTORC2 (Sengupta et al., 2010). mTORC1 is a homodimer, composed of mTOR and regulatory-associated protein of mTOR (raptor), mammalian lethal with Sec13 protein 8 (mLST8), proline-rich AKT substrate 40 kDa (PRAS40), and DEP-domain-containing mTOR-interacting protein (Deptor). mTORC2 is composed of Rictor (RPTOR-independent companion of mTOR), mLST8, Deptor, mammalian stress-activated protein kinase interacting protein (mSIN1) and protein observed with Rictor-1 (Protor-1, also PRR5). Rictor is required for mTORC2 catalytic activity; it is proposed to recruit substrates to mTORC2.

The upstream signals that regulate the activity of mTORC1 are growth factors, amino acids, glucose, oxygen levels (Sengupta et al., 2010). The downstream actions of mTORC1 include protein synthesis, autophagy and many metabolic pathways. mTORC1 is a critical mediator of the cellular response to many types of stress, such as DNA damage, drops in the level of energy, oxygen, amino acids and glucose. It is therefore involved in many stress responses in physiological and pathophysiological states, possibly in aiding the resistance of tumor cells to conventional therapy.

1.1.3.2 p53

Transformation-related protein 53 (p53) is a transcription factor, which responds to many types of cellular stress, such as DNA damage, hypoxia and oncogene activation. It regulates target genes that induce cell cycle arrest, apoptosis, senescence, DNA repair or changes in metabolism (Speidel, 2010). In addition to its functions as a transcription factor in the nuclei, cytoplasmic p53 has transcription independent activities. Both, transcription dependent and transcription independent activities of p53 may lead to apoptosis. There are accounts when transcription independent mechanisms are essential for the full apoptotic response (Speidel, 2010). p53 is also a BH3 domain protein and can interact with some members of BCL2 family (chapter 1.4.1.3). The apoptotic stimuli will induce transcription dependent and transcription independent activities of p53 in most cases; the two complement each other. A few connections of the two signaling pathways have been identified by demonstrating the roles of p53 transcription targets Puma, Mdm2, IGFBP1 in transcription-independent apoptosis.

p53 is kept inactive in unstressed cells. The concentration and activity of p53 are regulated mostly on the posttranslational level. The half life of p53 protein in normal, unstressed cells, is about 6 to 20 minutes, while it can be hours under stress. Post-translational modifications are selectively activated by different stress signals. p53 is then modified by phosphorylation, methylation, monoubiquination, sumoylation, neddylation and glycosylation (Ak & Levine, 2010). p53 is polyubiquinated under non-stressed conditions by mouse double minute 2 homolog (MDM-2, E3 ubiquitin ligase, chapter 1.1.2), which speeds up its degradation on the proteasome. On the other hand, MDM-2 is transcriptionally regulated by p53, so there is an autoregulatory loop. MDM-2 functions as a heterodimer with MDM4. There is another autoregulatory loop between MDM-2 and MDM-4, as MDM-2 ubiquinates MDM-4. Further, MDM-2/MDM-4 and MDM-2/p53 complexes are regulated by protein kinases CHK-1/2 and ATM (Ak & Levine, 2010). DNA damage activates the ATM kinase (ataxia-telangiectasia mutated gene), which inhibits MDM-2 and activates p53. The members of calcium calmodulin kinase superfamily, CHK-1/2 kinases, phosphorylate MDM-2, which dissociates from MDM-4 resulting in increased levels of p53. The failure to polyubiquinate p53 results in its high intracellular levels. Ubiquitin proteases, HAUSP (herpesvirus-associated ubiquitin-specific protease, also ubiquitin-specific protease 7; USP7) and USP42 remove ubiquitin from p53 and stabilize it. Stress signal results in higher levels of p53, which leads to transcription of selected genes and cell cycle arrest, apoptosis or the cells loose the ability to divide (are senescent). Different stress signals result in different post-translational modifications of p53 and transcription of different sets of genes.

p53 regulates the expression of miRNA (chapter 1.1.1.1) at the levels of transcription and processing. Most p53 mutations found in cancers are in a domain required for miRNA processing and transcription (Leung & Sharp, 2010). p53 enhances the processing of a population of pre-miRNA in cancer cells. Therefore loss of p53 function in transcription and processing might contribute to tumor progression.

The interruption of ribosomal biogenesis during the cell replication cycle results in an increase in free ribosomal proteins; some of them bind to MDM-2 and inhibit its polyubiquination, which stabilizes and activates p53 (Ak & Levine, 2010). Mutations in some tumor suppressor genes, e.g. retinoblastoma protein, free the transcription factor E2F-1, which transcribes ARF tumor suppressor gene. ARF is transcribed from an alternative reading frame of the cyclin-dependent kinase inhibitor 2A gene (CDKN2A). ARF protein binds and inhibits MDM-2 and raises levels of p53 within the cells. Similarly, the mutations that activate oncogenes Ras and myc increase ARF levels. In this way ARF acts as tumor suppressor by initiating p53-dependent responses. Activity of p53 is ubiquitously lost in human cancer either by mutation of the p53 gene itself or by loss of cell signaling upstream or downstream of p53 (Toledo & Wahl, 2006).

p53 responds to the intrinsic stress. There is another transcription factor, nuclear factor-κB complex (NF-κB), which responds to extrinsic stress (chapter 1.1.3.3, K10). p53 and NF-κB cannot function at the same time in the same cell. On activation of one, the other is inactivated.

1.1.3.3 NF-κB

NF-κB has been detected in many cell types, which express cytokines, chemokines, growth factors, cell adhesion molecules, and some acute phase proteins in health and in various disease states. It is activated by different stimuli, like cytokines, free radicals, ultraviolet irradiation, and bacterial or viral products. Inappropriate activation of NF-κB leeds to inflammatory events associated with autoimmune arthritis, asthma, septic shock, lung fibrosis, glomerulonephritis, atherosclerosis, and AIDS, while its persistent inhibition may result in apoptosis, inappropriate immune cell development, and delayed cell growth (Chen et al., 1999).

NF-kB is a complex; a heterodimer from the protein of Rel family (p65, Rel A, Rel B, c-Rel) and from the p50/p52 set of proteins (Ak & Levine, 2010). NF-κB is in the cytosol associated with the inhibitor IκB in the absence of stress signals. IκB is phosphorylated upon stress by IκB kinase (IKK), then ubiquinated and degraded by a proteasome. NF-κB shifts to the nucleus resulting in transcription of genes with NF-κB response elements. The activation of these genes results in cellular replication, inflammatory responses mediated by tumor necrosis factor, and cell survival signals. NF-κB transcribes the gene of IκB-α, which is one of its negative regulators. The result is the autoregulatory loop of the transcribed IκB-α with NF-κB, which alternatively activates and shuts down the transcription of IκB-α.

1.2 Adaptation to stress

Acclimation and physiological adjustments that lead to tolerance to a certain degree of stress is a well known and proved concept, which enables the organisms to survive in different environmental conditions. Some experimental proofs are emerging that the same concept is true at the cellular level; i.e. that the cells can adapt to environmental stress to a certain degree and this makes them more resilient to environmental stresses. In 2010 our group has shown that a mild stress can inhibit triggering of apoptosis through the intrinsic pathway (chapter 1.4.1.1) in the primary liver cells (hepatocytes; Nipic et al., 2010). This is the first evidence to our knowledge of apoptotic pathway shut down caused as a consequence of mild stress. The state of the cells after encountering the mild stress, was named preapoptotic cell stress response. This state is temporary, since the cells revert to a normal state in a few days in the absence of further stressors (Nipic et al., 2010, Banic et al., 2011). Despite the inactivation of apoptosis triggering by caspase-9 through the intrinsic pathway, apoptosis can be triggered when the inducer is strong enough, however, the apoptosis executing enzymes caspase-3 and -7 are activated to a lesser degree than when the cells are in the normal state. This seems like a mechanism of an adaptive response in the stressed cells.

Adaptations to sublethal stress, which result in greater stress tolerance, were observed also in mouse blastocysts, where applying hydrostatic pressure improved their survival after freezing and in suboptimal culture conditions (Pribenszky et al., 2005). Similar results were obtained also in bovine blastocysts (Pribenszky & Vajta, 2011). Also, signals from DNA damage can facilitate osmotic stress adaptation (Kültz, 2005).

The phenomenon that a sublethal stress induces a resistance to mild stress, was observed from bacteria to multicellular organisms and humans. Some cells have to tolerate large changes in the environment because of their position in the organism. The cells in papilla of mammalian kidney have to tolerate varying degrees of hyperosmotic stress during urine concentration; the degree of hyperosmolarity depends also on the organism's hydration state. These cells are adapted to ever changing environment. The proteomic approach was used to compare the expression in hyperosmotic renal papilla and an adjacent iso-osmotic region, the cortex (Gabert & Kültz, 2011). Of 1877 proteins that were common to both regions, there were 212 comparably overexpressed in the cortex and 80 proteins in the papilla. In response to tonicity changes in papilla, protein expression altered significantly, mainly that of metabolic enzymes, molecular chaperones, proteins involved in redox balance, transport and transcription. During antidiuresis 15 different proteins changed significantly, while 18 different proteins changed significantly during diuresis relatively to normally hydrated controls. Proteins significantly altered by diuretic state are structure proteins (actin, tubulin), signaling (Rho GDP dissociation inhibitor, abhydrolase domain-containing protein 14B), chaperones (Hsp beta-1, αB crystallin, T complex protein-1) and those with anti-oxidant functions (α-enolase, GAPDH, LDH). Therefore, many genes, which are commonly overexpressed in stress, have been identified to be induced as a result to environmental changes in the kidney papilla. The possible pathways of adaptation remain to be determined.

Discovering the mechanisms of adaptation to stress may prove tremendously important, also because it is conceivable that the cells, during the process of malignant transformation and the resulting tumor cells, undergo adaptations, which enable them to resist the signals that would trigger cell death in their normal state.

1.3 Autophagy

Autophagy is an intracellular lysosomal (vacuolar) degradation process characterized by the formation of double-membrane vesicles, autophagosomes, which sequester cytoplasm. It is involved in growth, survival, development and death of cells. Autophagy is a term used for several processes: (1) macroautophagy, (2) microautophagy and (3) chaperone-mediated autophagy (Funderburk et al., 2010). (1) In macroautophagy, the cytoplasmic material is engulfed by a double-membrane, which fuses subsequently to the lysosome. Sequestered material is indiscriminately removed. (2) Lysosomes engulf a portion of cytosol in microautophagy. (3) Chaperone-mediated autophagy is a way for removal of selected proteins. These proteins can be modified, e.g. ubiquinated. As macroautophagy is the most prevalent form, it is often referred to as autophagy; this term shall be used in such manner below. Autophagy starts by formation of an isolation membrane (phagophore) around the portion of cytosol. The membrane elongates and seals on itself to form a double membrane vacuoles autophagosomes. These then fuse with lysosomes where the entrapped components are degraded.

Autophagy is conserved among eukaryotic cells and occurs at a basal rate in most cells. It is a mechanism for quality control to eliminate protein aggregates and damaged organelles, like mitochondria (Scarlatti et al., 2009). During cell stress, autophagy is a process through which the cells can reuse the resources. For example, the starvation-induced autophagy helps to recycle the amino acids for protein synthesis and produce the substrates for oxidative phosphorylation when the supplies of nutrients are limited. Degradation of whole regions of cytoplasm therefore generates free amino acids, which can be metabolized to meet energy demand during periods of stress. Autophagy may be the last attempt to rescue the cells from dying. As the intracellular processes are complex and intertwined, autophagy may have a role in some cases of cell death, too. While this was shown during the development of *Drosophilla* and *Dictyostelium discoideum*, there is no evidence for such process in mammals so far. However, autophagy can be involved in cell death in cultured mammalian cells and can occur upstream of, alongside to or during the final stages of apoptosis. It is often challenging to determine the role of autophagy in cell death, as observing autophagic structures is not sufficient to demonstrate the involvement of autophagy in cell death.

Apoptosis and autophagy may be co-regulated in the same directions, as the anti-apoptotic Bcl-2 and Bcl-xL proteins negatively regulate autophagy by binding to Beclin 1 (mammalian Atg6, see below), and pro-apoptotic BH3-only proteins may reverse this effect by displacing these interactions. Apoptosis can also suppress autophagy (Luo & Rubinsztein, 2010). The interplay between autophagy and apoptosis is currently a fast developing field of research.

The signaling of autophagy is mainly through serine/threonine kinase mTOR (chapter 1.1.3.1). Amino acid deprivation inhibits mTORC1, which leads to induction of autophagy (Sengupta et al., 2010). The autophagic process liberates amino acids, which reactivate mTORC1 and the replacement of lysosomes consumed during autophagy. Another main regulator of autophagy is a protein Beclin 1 (Bcl-2-interacting protein, reviewed in Scarlatti et al., 2009, Funderburk et al., 2010). Beclin 1 was discovered first as the interacting partner of the anti-apoptotic protein Bcl-2 through its BH3 domain (chapter 1.4.1.3). Subsequent studies have indicated that the endogenous Bcl-2 regulates Beclin 1. Also, it was found out that Beclin 1 interacts with many different proteins; however, the physiological roles of many of these interactions are not fully understood. In mammals, the core autophagy complex is thought to form by binding of Beclin 1 to the class III phosphatidylinositol 3-kinase VPS34 and protein VPS15 (Funderburk et al., 2010). This is based also on the similarity with the autophagy complex of its yeast counterparts Atg6 and Vps34. The association of Beclin 1 with additional sets of proteins implies that Beclin 1 may be a connection point between autophagy, endocytic and cell death pathways. Elucidating its diverse roles may uncover its involvement in heart disease, pathogen infection, development and neurodegeneration.

1.4 Cell death

There are many pathways for a cell to die; the best known are necrosis, apoptosis and in some circumstances autophagy.

1.4.1 Apoptosis

The term apoptosis was first used by Kerr and coworkers (1972) to describe a cell death with several of typical morphological manifestations (Kroemer et al., 2009): rounding up of the cell, reduction of cellular volume, chromatin condensation, nuclear fragmentation, little or no ultrastructural modifications of organelles, plasma membrane blebbing, maintenance of plasma membrane integrity until the final stages of the process, phagocytosis of remains of the cells. There is no inflammation in tissue as the consequence of apoptosis (Savill & Fadok, 2000; Kurosaka et al., 2003).
There are biochemical and functional heterogeneities, as apoptosis is triggered through different biochemical pathways. The cysteine proteases caspases are central to triggering apoptosis (caspase-dependent triggering), although the apoptosis triggering can be caspase-independent, too. These and some of the main apoptosis regulators shall be described in the following chapters.

1.4.1.1 Caspases

The term caspases (cysteine-dependent aspartate-specific protease) describes a family of proteolytic enzymes with cystein in the active site, which cleave the substrates after the aspartate residue (Denault & Salvesen, 2002). Caspases -1, - 4, -5, -11 and -12 are important in development of cytokines, like interleukines 1ß and 18. The caspases -2, -3, -6, -7, -8, -9 and -10 are important in apoptosis signaling. In normal cells, they are inactive zymogens (procaspases), which are activated through dimerization and proteolytic cleavage upon the apoptotic stimuli (Denault & Salvesen, 2002). Caspases can trigger and execute apoptosis through cascades of reactions. Caspases -2, -8, -9 and -10 are initiating caspases, i.e. activate the executioner caspases, -3, -6 and -7. These activate other proteins, whose action result in apoptosis morphology.
There are several molecular pathways of triggering apoptosis; however the central two involve proteases caspases. These are (1) the extrinsic pathway, which originates from the cell surface, and involves activation of caspase-8 and (2) the intrinsic or mitochondrial pathway originating through the activation of caspase-9 (Salvesen & Dixit, 1997). The two pathways converge to activate caspase-3, which is the best understood executioner caspase. The extrinsic pathway is important in immune responses. The intrinsic or mitochondrial pathway can be activated by many stress stimuli, including DNA damage or extensive perturbation of mitochondrial membrane potential. It is also activated through caspase-3; it then enhances apoptosis.
Although not all caspases are involved in the regulation of apoptosis, the overexpression of any one of them culminates in cell death (Norberg et al., 2010). Knockout of any of caspases results in higher cell numbers.

1.4.1.2 Caspase – independent pathways

Many apoptosis regulators are associated with mitochondrial membranes and are released into the cytosol upon apoptotic stimuli. One of the better known pathways of caspase-independent triggering of apoptosis involves apoptosis-inducing factor (AIF, Norberg et al., 2010). It is anchored with its N-terminal into the mitochondrial inner membrane. AIF is triggered by increased calcium or early lysosomal permeabilization; these are frequent in cell death signalling after ischaemia/reperfusion injury and treatment with cytotoxic drugs. AIF can be cleaved by calpain I, which has a mitochondrial localization signal and needs μM amounts of Ca^{2+} for activation. Then cleaved AIF is released from mitochondria upon the mitochondrial permeabilization. It is transported into the nucleus, where it contributes to a large scale DNA fragmentation and chromatin condensation. Several studies detected that antioxidants can inhibit AIF-induced cell death and intracellular ROS levels may regulate AIF cleavage and release.

AIF has to be cleaved from the membrane in order to be released from mitochondria. Unless it is cleaved before the permeabilization of the mitochondrial outer membrane, it would be reasonable to expect that other soluble mitochondrial proteins, e.g. cytochrome c, would be released first and would activate apoptosis through caspase-dependent mechanisms (Norberg et al., 2010). The elimination of AIF does not protect the cells from apoptosis induced by most drugs.

1.4.1.3 Apoptosis regulators - BCL2 family proteins

The proteins of BCL2 family (B cell lymphoma-2) are important regulators of apoptosis; some of them are pro-apoptotic, others are anti-apoptotic. All family members share characteristic BCL2 homology domains (BH). The anti-apoptotic members, like BCL-2 and BCL-xl (BCL-2-related gene, long isoform) have four BH domains (BH1-BH4). Some pro-apoptotic BCL2 members have three (effector domains), others one BH domains (BH3-only proteins). BAX (BCL-2-associated x protein) and BAK (BCL-2 antagonist killer 1) have three BH domains (BH1-BH3), while e.g. BID (BCL-2-interacting domain death agonist), BIM (BCL-2-interacting mediator of cell death), Puma (p53-upregulated modulator of apoptosis) and Noxa have a single BH domain (BH3). Anti-apoptotic BCL2 proteins control the integrity of the mitochondrial outer membrane in mammals by inhibiting pro-apoptotic members (Chipuk et al., 2010). Upon activation, the effector pro-apoptotic proteins homo-oligomerize into pores in the outer mitochondrial membrane and promote its permeabilization. This releases apoptosis regulators, among others cytochrome c, which binds to APAF-1 (apoptotic protease activating factor-1) and procaspase-9 in the cytosol to oligomerize and form a complex named apoptosome. This activates procaspase-9, therefore turns on the intrinsic apoptotic pathway (chapter 1.4.1.1).

The BH3-only proteins are activated in response to cellular stress. Some of them (at least BID and BIM) are called direct activators, as they can promote the oligomerization of BAX and BAK. For example, the stress activated BID may be sequestered by an anti-apoptotic member, which prevents apoptosis. In the case of the further stress, other BH3-only protein can replace the BID, so it may activate BAX or BAK. Cellular stress can cause also transcriptional regulation of the BCL-2 family. The newly synthesized BH3-only proteins can interact with anti-apoptotic proteins and lower the threshold for BAK and BAX activation. For example, if BCL-2 is associated with PUMA, any future induction of BIM is not inhibited and results in the permeabilization of mitochondrial outer membrane.

Therefore the progression of stress to apoptosis is determined through complex interactions between the BCL-2 family proteins.

1.5 Cellular stress responses in regenerative medicine

As it was described above, the cellular responses to stress result in reparation of damage, cell death or in adaptation to mild stress that prevents excessive apoptosis. The processes of cellular adaptation to stress, i.e. the acquired resilience of cells to apoptosis by mild stress may be of value in regenerative medicine. There are indications that modulation of stress mechanisms is useful to improve the outcome of transplantations. For example, cold ischaemia pretreatment correlates with increased regeneration of epithelial cells immediately after the transplantation of kidney allografts (Naesens, 2011). Also, the adaptations to increased hydrostatic pressure improve the survival of murine and bovine blastocysts after freezing or in suboptimal culture conditions (Pribenszky & Vajta, 2011, chapter 1.2). On the other hand, it was observed that deterioration of mechanisms of cellular adaptation to stress result in lesser survival of kidney grafts. Old donor age decreases the chances of successful kidney transplantation (Naesens, 2011). Namely, many processes associated with aging are general pathways involved in tissue damage and stress responses; examples are the changes in mitochondrial physiology, increased susceptibility to apoptosis, impaired regeneration and repair, replicative senescence (when the cells become senescent as the result of breaks in DNA), etc.

There are examples of pre-treatments of cells to manipulate stress response pathways that minimise cellular damage and improve the transplantation outcome. Preconditioning of model cells of retinal pigment epithelium with non-lethal oxidative stress protects these cells from cell death induced by oxidative-stress (Sharma et al., 2009). The cell line ARPE-19 was used to mimick the conditions of oxidative stress, which is encountered by retinal cells transplanted for repairing the age-related macular degeneration.

The application of moderate shear stress on liver tissue slices was better than no shear or high shear stress, with the conclusion, that perioperative flow management is needed to regulate shear stress, i.e. to avoid the excessive shear stress on liver tissue upon a massive liver resection (Torii et al., 2005). Interestingly, the absence of shear stress also resulted in destruction of sinusoidal structures, which supports the need for constant perfusion of donor liver.

Targeting specific proteins involved in stress response was shown to increase the stress resistence of grafts in several cases. Increased synthesis of heme oxygenase-1, which is known to increase the cellular resistance against oxidative injury, improved liver graft viability in Lewis rats (Uchida et al., 2003). Other stress proteins, like heat shock proteins, are protective in models of transplantation; however, there is a need to develop strategies for their upregulation in clinical practice. Overexpression of heat shock protein 90-binding agent geldanamycin and some of its analogs protected renal cells from oxidative stress and reduced kidney ischaemia-reperfusion injury in a mouse model (Harrison et al., 2008). Rat mesenchymal stem cells engineered to overexpress Hsp20 were resistant to oxidative stress; this increased their survival after transplantation into infarcted heart by about twofold (Wang et al., 2009). A further example of targeting the stress responses is the protection of β cells by inhibition of iNOS (Hynes et al., 2011). Transplanted β cells fail because of specific autoimmune reactions and also due to non-specific inflammatory reactions. Proinflamatory cytokines, like interleukin 1β, can induce β cell destruction in a nitric oxide-dependent manner, as it stimulates inducible nitric oxide synthase iNOS, which produces NO cytotoxic to β cells. Lentiviral-based strategy was used to inhibit iNOS expression through short hairpin interfering RNA; this improved protection of β cells (Hynes et al., 2011).

Pharmacological preconditioning may improve the survival of grafted cells as well. For example, trimetazidine (1-[2,3,4-trimethoxybenzyl]piperazine, TMZ) is a widely used anti-ischaemic drug for treating angina in cardiac patients. The stem cells were preconditioned with TMZ and used in an *in vivo* rat model of myocardial infarction (Wisel et al., 2009). A significant increase in the recovery of myocardial function and up-regulation of Akt and BCL-2 levels were observed in hearts transplanted with TMZ-preconditioned cells. Similarly, the treatment of ventromesencephalic grafts with the p53 inhibitor, pifithrin-α enhanced the survival of dopamine cell transplants and augmented behavioral recovery in Parkinsonian rats (Chou et al., 2011).

Clearly, using cells' ability to adapt to stress conditions or manipulating the stress response mechanisms to improve the cellular adaptation to stress betters the survival of cells and the transplantation outcome in experimental models. The challenge remaining is the finetuning of experimental techniques to suit the needs of regenerative medicine.

1.6 From triggers to consequences of stress responses

Any deviation from the ideal environment could be a stressor for cells, i.e. the stress is caused by too much or too little of an agent, stimulus or other environmental condition. Heat or cold, modification of pH, hyper- or hypo- osmolarity, the increased concentrations of reactive oxygen species, etc., all result in cellular stress.

Stressors can trigger two types of cellular responses, from within cells and by the immune system. The stress responses of non-immune cells, are described sometimes as an intrinsic stress, while the stress responses of the immune system are called extrinsic stress (Ak & Levine, 2010). The latter functions by signaling with cytokines and clonal selection of cells. The transcription of many of the genes that participate in immune response is regulated by NF-κB (chapter 1.1.3.3), which is a growth and division-promoting factor and can thus turn into an oncogene. The intrinsic stresses can be DNA damage, hypoxia, low levels of glucose and amino acids, interference with mitochondrial and ribosomal biogenesis, the action of some toxins, etc. The protein p53 responds to such stressors (chapter 1.1.3.2). p53 also regulates many genes that prevent DNA damage or help in the DNA repair. Hypoxia, glucose levels and mitochondrial and ribosomal biogenesis are regulated by the interactions of the p53 pathway genes with the insulin-like growth factor 1 (IGF-1)/mTOR pathways and the regulation of the endosomal compartment by p53-induced genes (Sengupta et al., 2010). The activation of p53 or NF-κB including pathways are mutually exclusive within the cells. The activation of p53 results in slowing glycolysis and restoring oxidative phosphorylation, while the activation of NF-κB pathway activates cell division and utilizes large amounts of glucose and predominant use of glycolysis (Ak & Levine, 2010).

Encountering the stressors is the normal consequence of living in a fluctuating environment, therefore, the cells have developed mechanisms to ameliorate the stress or to adapt to it. This is achieved through the repair of damage, adaptation, reuse of resources and a limited cell death. As living with stressors is unavoidable in the life of organisms and cells, does the stress matter? The cells have to divert at least some of their resources from other pathways, to deal with stressors, as it is described throughout this chapter. Our cells are well adapted to a mild stress for a short time, however, there are potentially serious consequences of the long term stress.

One of the recently established hallmarks of cancer is the presence of stress phenotypes (Leung & Sharp, 2010). The expression of microRNA is often aberrant in cancer; also, microRNA expression patterns often correlate with clinically relevant tumor characteristics. Several microRNAs regulate genes that control proliferation, apoptosis, differentiation, tumor invasion or tumor metastases, and are therefore directly involved in cancer initiation and progression. MicroRNAs are also important in regulating the development of immune cells and in modulating innate and adaptive immune responses (Stern-Ginossar et al., 2008).

The connection of prolonged stress to the development or deterioration of various pathologies is clearly seen in the consequences of aggregated proteins accumulation, which is a hallmark of many neurodegenerative diseases (Petrozzi et al., 2007). Cell stress and stress proteins have a profound effect in triggering/developing the cardiovascular diseases, too. It was first observed in 1970's that the patients infected with *Mycobacterium tuberculosis* or *M. leprae* have antibodies to an antigen, which was later identified as Hsp60 (chapter 1.1.2.1). It is now confirmed that Hsp60 and other chaperones, like Hsp10, Hsp70 and Hsp90 family members are strong imunogens and imunomodulators in experimental models of arthritis, diabetes and atherosclerosis (Shamaei-Tousi et al., 2007). The reason for this may be that many chaperones are potent activators of immune cells and may act as adjuvants and as immunogens, possibly as the latter in the case of Mycobacterial infection. Atherogenesis may be driven by crossreactive immunity to bacterial Hsp60 proteins. Namely, the host Hsp60 are expressed on the stressed endothelial cells. Endogenous chaperones may protect ischaemic myocardium. Hsp70 members inhibit some caspase-dependent and independent apoptosis. Hsp27 was shown also to inhibit cytochrome c-dependent activation of apoptosis and to stablize cytoskeletal structures. The major cytoskeletal lesions of myocardium may occur during ischemic injury, therefore, the stabilization of these structures could be important for cell survival (Haigs & Yankner, 2010).

Accumulation of damaged macromolecular structures over time was long recognized to lead to aging. The rate of aging and the appearance of age-related signs are modulated by stress responses (Haigs & Yankner, 2010). Central to it are protein and DNA damage repair mechanisms and mitochondrial respiratory metabolism. These are controlled by insulin/IGF-1, mTOR, sirtuin and AMPK (AMP-activated protein kinase) pathways. The coordinated action of these pathways are central to maintaining homeostasis in normal and stress conditions through the regulation of nutrient sensing and stress response pathways. There are many studies, which link the decline in the effectiveness and integration of stress responses to ageing and the development of age-related diseases. Modulation of mitochondrial and metabolic functions and mobilization of macromolecular maintenance and repair leads to life extension in modeling organisms.

Cellular stress can at least contribute to, or even trigger, many diseases and malignant transformations and has an important role in aging. To ameliorate or repair these processes the cells are being used in cell therapies and regenerative medicine. For maximal therapeutic success it is important to use the cells in the best condition possible or those adapted to stress as described in chapter 1.5. Discovering the detailed mechanisms of stress responses, some of which were described in this chapter, will improve the assessment of the condition of transplanted cells and to better their handling and transplant survival rates.

2. References

Ak P, Levine AJ. p53 and NF-κB: different strategies for responding to stress lead to a functional antagonism. FASEB J. 2010 Oct;24(10):3643-52. Epub 2010 Jun 8. Review. PubMed PMID: 20530750.

Akerfelt M, Morimoto RI, Sistonen L. Heat shock factors: integrators of cell stress, development and lifespan. Nat Rev Mol Cell Biol. 2010 Aug;11(8):545-55. Epub 2010 Jul 14. Review. PubMed PMID: 20628411.

Banič B, Nipič D, Suput D, Milisav I. DMSO modulates the pathway of apoptosis triggering. Cell Mol Biol Lett. 2011 Jun;16(2):328-41. Epub 2011 Mar 20. PubMed PMID: 21442445.

Biamonti G, Vourc'h C. Nuclear stress bodies. Cold Spring Harb Perspect Biol. 2010 Jun 1;2(6):a000695. Epub 2010 Apr 28. Review. PubMed PMID: 20516127; PubMed Central PMCID: PMC2869524.

Buchberger A, Bukau B, Sommer T. Protein quality control in the cytosol and the endoplasmic reticulum: brothers in arms. Mol Cell. 2010 Oct 22;40(2):238-52. Review. PubMed PMID: 20965419.

Chen F, Castranova V, Shi X, Demers LM. New insights into the role of nuclear factor-kappaB, a ubiquitous transcription factor in the initiation of diseases.

Clin Chem. 1999 Jan;45(1):7-17. Review. PubMed PMID: 9895331.

Chipuk JE, Moldoveanu T, Llambi F, Parsons MJ, Green DR. The BCL-2 family reunion. Mol Cell. 2010 Feb 12;37(3):299-310. Review. PubMed PMID: 20159550.

Chou J, Greig NH, Reiner D, Hoffer BJ, Wang Y. Enhanced survival of dopaminergic neuronal transplants in hemi-Parkinsonian rats by the p53 inactivator PFT-α. Cell Transplant. 2011 Feb 3. [Epub ahead of print] PubMed PMID: 21294958.

Denault JB, Salvesen GS. Caspases: keys in the ignition of cell death. Chem Rev. 2002 Dec;102(12):4489-500. Review. PubMed PMID: 12475198.

Funderburk SF, Wang QJ, Yue Z. The Beclin 1-VPS34 complex--at the crossroads of autophagy and beyond. Trends Cell Biol. 2010 Jun;20(6):355-62. Epub 2010 Mar 29. Review. PubMed PMID: 20356743.

Gabert BJ, Kültz D. Osmoprotective proteome adjustments in mouse kidney papilla. Biochim Biophys Acta. 2011 Mar;1814(3):435-48. Epub 2011 Jan 12. PubMed PMID: 21236367; PubMed Central PMCID: PMC3045564.

Haigis MC, Yankner BA. The aging stress response. Mol Cell. 2010 Oct 22;40(2):333-44. Review. PubMed PMID: 20965426; PubMed Central PMCID: PMC2987618.

Harrison EM, Sharpe E, Bellamy CO, McNally SJ, Devey L, Garden OJ, Ross JA, Wigmore SJ. Heat shock protein 90-binding agents protect renal cells from oxidative stress and reduce kidney ischemia-reperfusion injury. Am J Physiol Renal Physiol. 2008 Aug;295(2):F397-405. Epub 2008 Jun 18. PubMed PMID: 18562631.

Henderson B. Integrating the cell stress response: a new view of molecular chaperones as immunological and physiological homeostatic regulators. Cell Biochem Funct. 2010 Jan;28(1):1-14. Review. PubMed PMID: 19830685.

Hynes SO, McCabe C, O'Brien T. β cell protection by inhibition of iNOS through lentiviral vector-based strategies. Methods Mol Biol. 2011;704:153-68. PubMed PMID: 21161636.

Kerr JF, Wyllie AH, Currie AR. Apoptosis: a basic biological phenomenon with wide-ranging implications in tissue kinetics. Br J Cancer. 1972 Aug;26(4):239-57. Review. PubMed PMID: 4561027; PubMed Central PMCID: PMC2008650.

Kroemer G, Galluzzi L, Vandenabeele P, Abrams J, Alnemri ES, Baehrecke EH, Blagosklonny MV, El-Deiry WS, Golstein P, Green DR, Hengartner M, Knight RA, Kumar S, Lipton SA, Malorni W, Nuñez G, Peter ME, Tschopp J, Yuan J, Piacentini M, Zhivotovsky B, Melino G; Nomenclature Committee on Cell Death 2009. Classification of cell death: recommendations of the Nomenclature Committee on

Cell Death 2009. Cell Death Differ. 2009 Jan;16(1):3-11. Epub 2008 Oct 10. PubMed PMID: 18846107; PubMed Central PMCID: PMC2744427.

Kroemer G, Mariño G, Levine B. Autophagy and the integrated stress response. Mol Cell. 2010 Oct 22;40(2):280-93. Review. PubMed PMID: 20965422.

Kurosaka K, Takahashi M, Watanabe N, Kobayashi Y. Silent cleanup of very early apoptotic cells by macrophages. J Immunol. 2003 Nov 1;171(9):4672-9. PubMed PMID: 14568942.

Kültz D. Evolution of the cellular stress proteome: from monophyletic origin to ubiquitous function. J Exp Biol. 2003 Sep;206(Pt 18):3119-24. Review. PubMed PMID: 12909693.

Kültz D. DNA damage signals facilitate osmotic stress adaptation. Am J Physiol Renal Physiol. 2005 Sep;289(3):F504-5. Review. PubMed PMID: 16093426.

Leung AK, Sharp PA. MicroRNA functions in stress responses. Mol Cell. 2010 Oct 22;40(2):205-15. Review. PubMed PMID: 20965416; PubMed Central PMCID: PMC2996264.

Luo S, Rubinsztein DC. Apoptosis blocks Beclin 1-dependent autophagosome synthesis: an effect rescued by Bcl-xL. Cell Death Differ. 2010 Feb;17(2):268-77. Epub 2009 Aug 28. PubMed PMID: 19713971; PubMed Central PMCID: PMC2894406.

Majmundar AJ, Wong WJ, Simon MC. Hypoxia-inducible factors and the response to hypoxic stress. Mol Cell. 2010 Oct 22;40(2):294-309. Review. PubMed PMID: 20965423.

Naesens M. Replicative senescence in kidney aging, renal disease, and renal transplantation. Discov Med. 2011 Jan;11(56):65-75. Review. PubMed PMID: 21276412.

Nipic D, Pirc A, Banic B, Suput D, Milisav I. Preapoptotic cell stress response of primary hepatocytes. Hepatology. 2010 Jun;51(6):2140-51. PubMed PMID: 20513000.

Norberg E, Orrenius S, Zhivotovsky B. Mitochondrial regulation of cell death: processing of apoptosis-inducing factor (AIF). Biochem Biophys Res Commun. 2010 May 21;396(1):95-100. Review. PubMed PMID: 20494118.

In *Online Mendelian Inheritance in Man, OMIM*. Johns Hopkins University, Baltimore, MD. MIM Number: 601231: Date last edited: 2/7/2011. World Wide Web URL: http://omim.org/

Petrozzi L, Ricci G, Giglioli NJ, Siciliano G, Mancuso M. Mitochondria and neurodegeneration. Biosci Rep. 2007 Jun;27(1-3):87-104. Review. PubMed PMID: 17486441.

Pribenszky C, Molnár M, Cseh S, Solti L. Improving post-thaw survival of cryopreserved mouse blastocysts by hydrostatic pressure challenge. Anim Reprod Sci. 2005 Jun;87(1-2):143-50. Epub 2004 Dec 9. PubMed PMID: 15885447.

Pribenszky C, Vajta G. Cells under pressure: how sublethal hydrostatic pressure stress treatment increases gametes' and embryos' performance. Reprod Fertil Dev. 2011;23(1):48-55. Review. PubMed PMID: 21366980.

Salvesen GS, Dixit VM. Caspases: intracellular signaling by proteolysis. Cell. 1997 Nov 14;91(4):443-6. Review. PubMed PMID: 9390553.

Savill J, Fadok V. Corpse clearance defines the meaning of cell death. Nature. 2000 Oct 12;407(6805):784-8. Review. PubMed PMID: 11048729.

Scarlatti F, Granata R, Meijer AJ, Codogno P. Does autophagy have a license to kill mammalian cells? Cell Death Differ. 2009 Jan;16(1):12-20. Epub 2008 Jul 4. Review. PubMed PMID: 18600232.

Sengupta S, Peterson TR, Sabatini DM. Regulation of the mTOR complex 1 pathway by nutrients, growth factors, and stress. Mol Cell. 2010 Oct 22;40(2):310-22. Review. PubMed PMID: 20965424; PubMed Central PMCID: PMC2993060.

Shamaei-Tousi A, Halcox JP, Henderson B. Stressing the obvious? Cell stress and cell stress proteins in cardiovascular disease. Cardiovasc Res. 2007 Apr 1;74(1):19-28. Epub 2006 Nov 1. Review. PubMed PMID: 17141205.

Sharma RK, Netland PA, Kedrov MA, Johnson DA. Preconditioning protects the retinal pigment epithelium cells from oxidative stress-induced cell death. Acta Ophthalmol. 2009 Feb;87(1):82-8. Epub 2008 May 20. PubMed PMID: 18494742.

Speidel D. Transcription-independent p53 apoptosis: an alternative route to death. Trends Cell Biol. 2010 Jan;20(1):14-24. Epub 2009 Oct 30. Review. PubMed PMID: 19879762.

Spriggs KA, Bushell M, Willis AE. Translational regulation of gene expression during conditions of cell stress. Mol Cell. 2010 Oct 22;40(2):228-37. Review. PubMed PMID: 20965418.

Stern-Ginossar N, Gur C, Biton M, Horwitz E, Elboim M, Stanietsky N, Mandelboim M, Mandelboim O. Human microRNAs regulate stress-induced immune responses mediated by the receptor NKG2D. Nat Immunol. 2008 Sep;9(9):1065-73. PubMed PMID: 18677316.

Tabas I, Ron D. Integrating the mechanisms of apoptosis induced by endoplasmic reticulum stress. Nat Cell Biol. 2011 Mar;13(3):184-90. Review. PubMed PMID: 21364565; PubMed Central PMCID: PMC3107571.

Toledo F, Wahl GM. Regulating the p53 pathway: in vitro hypotheses, in vivo veritas. Nat Rev Cancer. 2006 Dec;6(12):909-23. Review. PubMed PMID: 17128209.

Torii T, Miyazawa M, Koyama I. Effect of continuous application of shear stress on liver tissue: continuous application of appropriate shear stress has advantage in protection of liver tissue. Transplant Proc. 2005 Dec;37(10):4575-8. PubMed PMID: 16387174.

Uchida Y, Tamaki T, Tanaka M, Kaizu T, Tsuchihashi S, Takahashi T, Kawamura A, Kakita A. Induction of specific stress response increases resistance of rat liver allografts to cold ischemia and reperfusion injury. Transpl Int. 2003 Jun;16(6):396-404. Epub 2003 Mar 26. PubMed PMID: 12819870.

Wang X, Zhao T, Huang W, Wang T, Qian J, Xu M, Kranias EG, Wang Y, Fan GC. Hsp20-engineered mesenchymal stem cells are resistant to oxidative stress via enhanced activation of Akt and increased secretion of growth factors. Stem Cells. 2009 Dec;27(12):3021-31. PubMed PMID: 19816949; PubMed Central PMCID: PMC2806498.

Wisel S, Khan M, Kuppusamy ML, Mohan IK, Chacko SM, Rivera BK, Sun BC, Hideg K, Kuppusamy P. Pharmacological preconditioning of mesenchymal stem cells with trimetazidine (1-[2,3,4-trimethoxybenzyl]piperazine) protects hypoxic cells against oxidative stress and enhances recovery of myocardial function in infarcted heart through Bcl-2 expression. J Pharmacol Exp Ther. 2009 May;329(2):543-50. Epub 2009 Feb 13. PubMed PMID: 19218529; PubMed Central PMCID: PMC2672865.

Part 3

Tissue Engineering

11

Tissue Engineering of Tubular and Solid Organs: An Industry Perspective

Joydeep Basu and John W. Ludlow
Bioprocess Research and Assay Development, Tengion Inc., North Carolina, USA

1. Introduction

Tissue engineering and regenerative medicine (TE/RM) represents a broad spectrum of cell and biomaterial based approaches aiming to repair, augment and regenerate diseased tissues and organs. The successful recent clinical implantation of tissue engineered trachea, bladder and bladder derivative (Neo-Urinary Conduit™) has highlighted the emergence of common methodological frameworks for the development of TE/RM approaches that may be broadly applicable towards the regeneration of multiple, disparate tubular organs.

Similarly, recent progress towards regeneration of heart, kidney, liver, pancreas, spleen and central nervous system is identifying shared methodologies that may underlie the development of foundational platform technologies broadly applicable towards the regeneration of multiple solid organ systems. Central themes emerging for both tubular and solid neo-organs include the application of a biodegradable scaffold to provide structural support for developing neo-organs and the role of committed or progenitor cell populations in establishing the regenerative micro-environment of key secreted growth factors and extra-cellular matrix critical for catalyzing *de novo* organogenesis.

However, aspects of these strategies currently under active development for tissue engineering of tubular and solid neo-organs may not be relevant for successful *commercialization* of neo-organs as novel TE/RM *products* for clinical application. For example, difficulties in large scale sourcing and quality control of biomaterials derived from decellularization of cadaveric organs imply that such biomaterials may be less suitable for incorporation into TE/RM products when compared to biomaterials of synthetic origin. Similarly, TE/RM technologies that attempt to leverage populations of stem and progenitor cells are less likely than platforms focused on committed cell populations or acellular biomaterials to facilitate rapid development of viable products.

In this chapter, we will present our experience in the development of the Neo-Bladder Augment™, Neo-Urinary Conduit™ and Neo-Kidney Augment™ to identify elements of this foundational organ regeneration technology platform that may be broadly applicable towards the design and development of additional tubular neo-organ *products*, including the esophagus and small intestine, the lung, the vasculature and the male or female genito-urinary tract as well as additional solid neo-organs. We will focus specifically on highlighting aspects of these neo-organ regenerative platforms conducive to the commercial viability of such technologies.

2. Tissue engineering of tubular organs

Two significant recent studies have brought the concept of *de novo* regeneration of tubular organs within humans towards proof-of-practice in the clinic. In the first example, seeding of tubular, biodegradable scaffolds with autologous urothelial and smooth muscle cells was shown to catalyze regeneration of a complete bladder with laminarly organized histology and associated urologic functionality upon implantation within seven pediatric patients (Atala et al., 2006). In the second example, a tissue engineered, functional human trachea was created using a decellularized, cadaveric tracheal segment as scaffold and seeded with autologous respiratory epithelial cells and chondrocytes generated by the directed differentiation of the patient's own bone marrow cells (Asnaghi et al., 2009; Macchiarini et al., 2008).

Both studies leveraged a scaffold seeded with autologous cells to trigger a regenerative response within the patient, ultimately leading to complementation of organ functionality with concomitant histogenesis of laminarly organized tissue structures. These approaches have focused on the regeneration of tubular organs and, taken together with additional recent developments in the regeneration of additional hollow organs, provide perspective into overlapping technology platforms and insight into key methodological differences that may impact the commercial feasibility of tubular organ engineering. In the following section (based on Basu & Ludlow, 2010), we discuss recent developments in tissue engineering of tubular organs.

2.1 A Platform technology for tubular organ regeneration

Organ regeneration technologies aim to restore the original structure and functionality of a diseased organ. Typically, healing responses within mammals are characterized by fibrosis and scar tissue formation, not regeneration. Tubular organ regeneration involves a specific, temporal sequence of cellular infiltration, vasculogenesis, neurogenesis and the defined differentiation of mucosal, stromal and parenchymal laminar tissue architectures (reviewed by Basu & Ludlow, 2010). Strategies for organ and tissue regeneration must therefore achieve the dual objectives of triggering a true regenerative response while ameliorating any tendency towards fibrotic repair. The methodology first pioneered for regeneration of the bladder (illustrated in Figure 1) may serve as a foundational platform for the regeneration of any tubular organ.

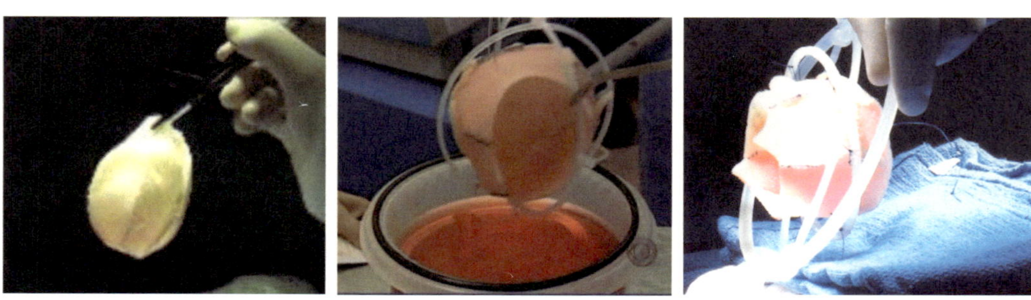

Fig. 1. Tengion Neo-Bladder Augment™ (NBA). Dome shaped biodegradable bladder scaffold (left panel) is placed in a support prior to seeding with cells and incubated in a bioreactor. The bioreactor is transported to the clinical site and the supported seeded scaffold removed (middle panel) and positioned for patient implantation (right panel).

Examination of the temporal sequence of neo-bladder regeneration in dogs illustrates the dichotomy in outcomes between implantation of acellular and cellularized scaffolds. This distinction is a principal theme of most studies describing the regeneration of tubular organs. Cell seeded scaffolds mediated a regenerative response within one month post-implantation, characterized by induction of heavily vascularized, smooth muscle-like parenchyma. In contrast, acellular scaffolds triggered a principally fibrotic, reparative outcome characterized by randomly organized collagen fibers with minimal vascularization.

Baseline urodynamics were reconstituted within four months of implantation with cell seeded scaffold, whereas the urodynamic profile of animals implanted with acellular scaffold remained abnormal throughout the trial period (Jayo et al., 2008a).

In a related dog study, restoration of tri-laminar bladder wall architecture occurred within three months post-implantation and normal compliance characteristics of a urinary bladder wall developed by 12 months. Regenerated bladders were functionally and structurally stable for up to two years post-implantation. Importantly, although the construct volume was constant at implantation within variably sized dogs, the ratio of the regenerated bladder's volume to host body weight adapted to the recipient animal's size, demonstrating that the neo-organ responds to homeostatic mechanisms regulating organ volume (Jayo et al., 2008b).

Although these canine studies utilized both bladder-derived urothelial and smooth muscle cells, urothelial cells have been shown to *not* be essential for bladder regeneration, thereby greatly facilitating process development and commercialization (Jack et al., 2009). However, use of bladder-derived smooth muscle cells is problematic in patients presenting with bladder related malignancies. Therefore, a number of alternate sources of smooth muscle cells have been investigated. Such alternate cell sources may have broad application in the regeneration of additional, laminarly organized tubular organs.

One possible alternate source of smooth muscle cells may be the directed differentiation of mesenchymal stem cells (MSC) with recombinant, inductive cytokines such as transforming growth factor β (TGF-β). Such MSC may be procured from bone marrow or adipose (Tian et al, 2010a,b; Sakuma et al., 2009). Additionally, autologous smooth muscle cells may be isolated from the vasculature of omentum or other adipose tissue as well as directly from the mononuclear fraction of peripheral blood (Basu et al., 2011a; Toyoda et al., 2009; Lin et al., 2008; Metharom et al., 2008).

Although the use of MSC or other stem cell populations (adult-derived, embryonic or iPS (induced pluripotent stem cell)) for organ regeneration is informative as proof-of-concept, from a process development perspective focused on eventual manufacturing and product release, the requirements to monitor and control stem cell potential and directed differentiation towards a smooth muscle lineage substantially complicates efforts to streamline, standardize and assure quality, as well as leading to significant increases in cost of goods. To simplify bioprocessing procedures, enhance product consistency, reduce the risk of immune rejection and to ensure a robust supply of cellular raw material, we are focused on developing classes of *committed* smooth muscle cells for tubular organ regeneration (Basu et al., 2011a).

An alternative platform technology for regeneration of bladder and related tubular organs including vas deferens and uterus is based around application of the peritoneal cavity as a living bioreactor to trigger the encapsulation of a tubular scaffold with myofibroblasts, as demonstrated in rabbit and rat models (Campbell et al., 2008). Two to three weeks after implantation, tubular constructs were observed to be heavily encapsulated with myofibroblasts. These tissue engineered composites were anastamosed to the native organ and allowed to mature *in vivo* for up to 14 months. Histological analysis of regenerated tissue was performed together with evaluation of organ functionality. Bladder function was confirmed by normal urine flow, vas deferens function by ejaculation and uterine function by the ability to sustain pregnancy.

With all three organ systems, an epithelial layer and laminarly organized musculature was observed with minimal evidence of inflammation. However, this approach may be impractical for widespread clinical application, as the cell encapsulation protocol represents an additional major surgical event. *In vitro* scaffold cell seeding is a simple and straightforward methodology sufficient to induce neo-organ regeneration *in vivo*. Use of alternate, non-bladder sources of smooth muscle-like cells will likely eliminate any requirement for peritoneal cavity based tissue engineering of bladder or bladder-like derivatives.

2.2 Trachea

Tissue engineered trachea illustrates another iteration of the foundational, bladder-based platform for tubular organ regeneration. Here, a decellularized, cadaveric tracheal segment was used as a scaffold. The recipient's own epithelial cells seeded the luminal surface, while MSC derived from the recipient's bone marrow was differentiated towards a chondrocytic lineage with recombinant cytokines prior to application on the abluminal surface. Although the patient showed reconstitution of pulmonary function within four months, detailed histological examination of the regenerated organ was not possible (Macchiarini et al., 2008).

However, in a swine model of this tissue engineered trachea, both the chondrocytes derived from differentiated MSC as well as the epithelial cells were needed for host survival (Go et al., 2010). While this methodology is valuable as proof of concept, it is clearly not amenable to large-scale process development. Decellularization of cadaveric scaffolds is an uncontrolled methodology that may not entirely alleviate immune response (Kasimir et al., 2006; Zhou et al., 2010), as well as increasing cost of goods by the requirement to stringently monitor the decellularization protocol. Broader application requires synthetic biodegradable scaffold materials, facilitating reproducible manufacturability at large scale with defined chemical and physical properties. Furthermore, the application of directed differentiation strategies on cells destined for patient implantation may trigger additional regulatory hurdles, further complicating commercialization efforts. To this end, chondrocytes derived from autologous tracheal explants may be leveraged (Komura et al., 2008 a,b).

2.3 Gastro-intestinal tract (GI)

Individual components of the GI represent locally specialized iterative variations of essentially the same laminarly organized tubular histologic architecture as the bladder. However, from a commercial perspective, the small intestine represents by far the most pressing current clinical need, with small bowel transplantation being an unsatisfactory current standard of care for pediatric small bowel syndrome (SBS) (http://digestive.niddk.nih.gov/ddiseases/pubs/shortbowel/). Esophageal replacement for esophageal cancer may also be commercially viable.

The intestinal epithelium is the most highly regenerative tissue within adult mammals and may therefore be expected to be highly amenable towards tissue engineering or regenerative medicine methodologies. In one such study, *in vivo* derived organoid units, formed from incompletely disassembled clusters of epithelial and mesenchymal cells generated through partial digestion of intestinal epithelium (and therefore likely incorporating resident intestinal stem cells) were used to seed PLLA scaffold tubes that were subsequently matured within the peritoneal cavity of pigs. Seven weeks post-implantation, this tissue-engineered small intestine recapitulated the gross overall laminar organization of native small intestine (SI) (Sala et al., 2009).

Significantly, acellular scaffolds did not result in the regeneration of tissue engineered gastrointestinal structures. These data notwithstanding, anastamosis of these tissue engineered small intestines to host SI within a large animal model remains to be demonstrated. Additionally, up to 10cm of host SI was harvested to derive donor organoids that are not readily expandable *in vitro*. Whether organoid units capable of seeding a scaffold structure may be isolated from diseased human intestine, and how much diseased donor material may be needed, remain factors yet to be elucidated. The requirement to leverage the peritoneal cavity as a bioreactor for tissue engineering may also impede widespread application of this approach.

The bladder-derived organ regeneration platform of biopolymeric scaffold seeded with smooth muscle cells may be applicable for regeneration of SI. To this end, stomach derived smooth muscle cells were used to seed a collagen-based scaffold prior to implantation within surgically isolated ileal loops of dogs for eight weeks, prior to re-anastamosis to the native intestine. Acellular collagen scaffold was used as a control. By 12 weeks post-surgery, macroscopic analysis of the cell seeded scaffold implantation site demonstrated regeneration of neo-mucosa resembling native mucosa. However, in animals containing an acellular scaffold, the implant site remained ulcerated up to 12 weeks post-implantation. Additional histological data showed significantly enhanced vascularization, epithelialization and organization of the circular muscle layer at the cell seeded scaffold defect site relative to acellular control (Nakase et al., 2006).

Increasing the number of smooth muscle cells seeded onto the scaffold increased the area of regenerated SI tissue, although no concomitant increase in the thickness of the smooth muscle layer was observed (Nakase et al., 2007). Nevertheless, these data suggest that a simple regenerative platform composed of biodegradable scaffold nucleated with smooth muscle cells may be adequate to facilitate SI regeneration. Although this approach must be reproduced using a directly anastamosed tubular scaffold and alternate sources of smooth muscle cells, if successful, this methodology represents the most straightforward, clinically relevant and commercially viable strategy for regeneration of the SI.

This organ regeneration platform technology may also be leveraged for regeneration of the esophagus. In one such example, patch defects were created in the abdominal esophagus of 27 female rats, subsequently implanted with gastric acellular matrix (GAM). Of the 24 survivors, none showed evidence of regeneration of the lamina muscularis mucosa even 18 months post-implantation (Urita et al., 2007).

In contrast, a study of a canine model of esophageal resection and replacement demonstrated that PGA tubes seeded with a mixture of keratinocytes and fibroblasts triggered regeneration of smooth muscle laminar organization similar to native esophagus within 3 weeks post-implantation, whereas acellular PGA tubes formed esophageal strictures associated with near complete obstruction within two to three weeks (Nakase et al., 2008).

In another dog study, cervical esophageal defects were patched with either small intestinal submucosa (SIS) alone, or SIS seeded with autologous oral mucosal epithelial cells. After four weeks, dogs implanted with cell seeded SIS showed almost complete re-epithelialization and minimal evidence of inflammation and, by eight weeks post surgery, regeneration of the underlying smooth muscle layer. Acellular SIS grafted animals presented only partial re-epithelialization and a more extensive inflammatory response by four weeks, and no muscular regeneration by eight weeks post-implantation. Attempts to introduce an acellular SIS tubular construct into the cervical esophagus of piglets were also unsuccessful, demonstrating scarification and a minimal regenerative response (Doede et al., 2009).

Progress has also been made in efforts to tissue engineer the stomach. Stomach derived organoid units, (analogous to the SI organoids used to tissue engineer the SI) upon seeding of a biopolymeric scaffold, triggered reconstitution of the gastric and muscularis mucosa in stomach tissue engineered within the peritoneal cavities of swine (Sala et al., 2009). In another study, circular defects were created in the stomach of seven dogs and a composite biodegradable scaffold ("New-sheet"), soaked with either autologous peripheral blood or bone marrow aspirate, was sutured over the defect. By 16 weeks post implantation, the defect site had formed regenerated stomach with evidence of re-epithelialization, formation of villi, vascularization and fibrosis within the submucosal layer. However, minimal regeneration of the smooth muscle layer was observed, as shown by expression of smooth muscle α-actin, without expression of calponin, a marker of mature smooth muscle cells (Araki et al., 2009).

Though strictly not a tubular organ, the anal sphincter is a component of the gastrointestinal tract and is critical in regulating patency of the large intestine. Recent efforts to engineer the anal sphincter leverage the same general platform used to catalyze bladder regeneration. To this end, smooth muscle cells isolated from human internal anal sphincter were seeded onto fibrin gels poured around a central mold. Cell mediated contraction of the gel around the mold resulted in the formation of a three-dimensional cylindrical tube of sphincteric smooth muscle tissue. Although *in vivo* anastamosis remains to be demonstrated, this bio-engineered anal sphincter demonstrated contractile properties and response to defined neurotransmitters consistent with the functionality of native anal sphincter (Hashish et al., 2010; Somara et al., 2009). Use of alternatively sourced smooth muscle cells may facilitate the transition of engineered sphincter towards commercial production.

2.4 Neo-vessel

Neo-blood vessels represent a major commercial opportunity for application of a tubular organ platform technology towards patients receiving bypass surgery. In a recently published clinical trial of vascular grafts composed of autologous bone marrow aspirate seeded onto a PGA/PCL scaffold and implanted into a cohort of 25 patients presenting with single ventricle physiology, all patients were asymptomatic 30 days post implantation and 24/25 patients were alive at one year post-implantation (Hibino et al., 2010).

Efforts to seed PGA tubular neo-vessel scaffolds with smooth muscle cells derived from the directed differentiation of bone marrow or adipose derived MSC with TGF-β have also been described (Wang et al., 2010; Gong and Niklason, 2008). The requirement to induce directed differentiation with TGF-β or related agents as well as the prolonged maturation period under pulsatile conditions needed to achieve a mature smooth muscle phenotype will likely make this approach impractical for commercial application. A reliable source of committed smooth muscle cells (e.g., the vascular fraction of adipose or omentum) may represent a more commercially feasible platform.

2.5 Lung

Lung may be regarded as a highly specialized tubular organ amenable to regeneration with the bladder-based platform outlined above. Evidence to this effect was provided by a study examining the regenerative potential of PGA felt sheets seeded with adipose derived stromal cells in triggering pulmonary regeneration within a rat lung lobectomy model (Shigemura et al., 2006). The cell seeded PGA sheet was sealed onto the remaining lung lobe. Alveolar and vascular regeneration was observed within 1 week of implantation, with concomitant recovery of pulmonary functionality. Paracrine action by secreted factors including VEGF, HGF and FGF was suggested as a possible mechanism of action for triggering the regenerative effect.

In another study, fetal rat lung cells were seeded onto gelfoam sponge-based scaffolds and implanted into adult rat lung. Alveolar-like structures with apparent vascular networks were observed to regenerate within degrading scaffold by 4 months post-implantation (Andrade et al., 2007). Importantly, the formation of these alvelolar like networks was observed to be strongly dependant on prior seeding of the scaffold with lung cells. As with the previous study, the authors suggest paracrine action of secreted factors from the seeded lung cells acting to facilitate regenerative in-growth of lung cells from the surrounding native lung tissue as the likely mechanism of action.

Growth of lung cells in three-dimensions is essential to induce expression of epithelial genes related to lung morphogenesis, including FGFR2 (Mondrinos et al., 2006). Appropriate combinations of exogenous fibroblast growth factors chosen to target specific receptor isoforms may facilitate appropriate lung epithelial and mesenchymal cell behavior conducive to tissue regeneration (Mondrinos et al., 2007). Finally, tissue engineered lungs created from recellularized scaffolds derived from decellularized lung have also been recently characterized (Ott et al., 2010; Petersen et al., 2010; Cortiella et al., 2010).

2.6 Genito-urinary system

The recent reports of functional regenerated neo-phallus and neo-vagina within a rabbit model illustrate how foundational principles of tubular organ regeneration pioneered for bladder may be extrapolated to facilitate organogenesis of functionally distinct tubular organs (De Filippo et al., 2008). Decellularized corpora cavernosa was used as a collagen-based scaffold matrix for seeding autologous corporal smooth muscle and endothelial cells in a rabbit model of penile replacement. Implantation of decellularized matrix alone led to formation of a non-functional, fibrotic phallus. However, cell seeded scaffolds regenerated corporal tissue organization histologically similar to native controls within 3-6 months post-implantation. Tissue engineered phallus was functional as demonstrated by the ability of recipient animals to copulate normally (Chen et al., 2010).

For the neo-vagina, autologous vaginal epithelial and smooth muscle cells were seeded onto the luminal and abluminal surfaces of PGA tubular scaffolds, preconfigured to resemble native rabbit vagina. Seeded composites were implanted in place of the native vagina of nine rabbits, with unseeded controls introduced into six other animals. As has been observed for multiple organs, unseeded scaffold failed to trigger a regenerative response, whereas cell seeded scaffolds generated stage specific histogenesis, vascularization, innervation and regeneration of a patent neo-vagina by 6 months post-implantation with a defined muscular layer and a luminal invaginated epithelium. Organ functionality was confirmed by a graded contractile response of the musculature to electrical stimulation in a manner paralleling native tissue (De Filippo et al., 2008).

2.7 Mechanism of action for tubular organ regeneration

The lineage of cell populations constituting a regenerated neo-organ remains unclear. What is the contribution of the original cell population used to seed the scaffold compared to that derived from cellular migration from the surrounding tissue and/or omentum typically used to wrap neo-organs at the time of implantation as a source of neo-vascularization? Data addressing this issue was provided for the regenerated neo-vagina, where both the epithelial and smooth muscle cell populations used to seed the neo-vagina were labeled independently with fluorochromes and tracked to three months post-surgery. Labeled cells of both types were found to compose over 85% of the regenerated neo-organ (De Filippo et al., 2008).

However, implantation of a human bone marrow derived cell seeded neo-vessel scaffold into immuno-deficient mice led to loss of all seeded human cells within days, followed by re-population of the scaffold with mouse monocytes and subsequent re-population with mouse smooth muscle cells and endothelial cells recruited via the monocyte chemoattractant protein MCP-1 (Roh et al, 2010). No direct contribution of the seeded bone marrow cells to the regenerated neo-vessel was observed.

Organ specific mechanistic differences notwithstanding, these and results from other tubular organs demonstrate that defined populations of stem and progenitor cells are not required to trigger neo-organ formation. Smooth muscle cells and other committed cell types appear capable of mechanistically substituting for stem cells, by initiating cellular scaffold repopulation and catalyzing a host specific regenerative response that may additionally mediate paracrine signaling pathways to modulate recruitment of resident stem and progenitor populations, and regulate inflammation, immune response, apoptosis and fibrosis (Caplan, 2009). The ability to engineer neo-organs *without* a requirement for the isolation, expansion and manipulation of defined stem cell populations has greatly facilitated the commercial viability of this platform technology and will continue to do so in the future (Basu and Ludlow, 2010).

2.8 Role of extracellular matrix (ECM) in regeneration

An analysis of the bladder submucosal matrix generated by bladder smooth muscle cells identified multiple key paracrine signaling factors including VEGF, BMP4, PDGF-ββ, KGF, TGF-β1, IGF, bFGF, EGF and TGF-α as resident components (Chun et al., 2007). This observation might suggest that the presence of living cells is entirely superfluous for triggering regeneration, which would greatly lower cost-of-goods for neo-organ production. A synthetic scaffold may be matured with an autologous, committed cell population to facilitate deposition of ECM, decellularized and implanted within the host to induce organ regeneration. Alternatively, defined ECM components or growth factors may be introduced into the scaffold to recapitulate key aspects of organogenesis (Sahoo et al, 2010). However, studies to date confirm that ECM alone in the context of, for example, an acellular matrix such as small intestinal submucosa or bladder mucosa, is insufficient to induce organogenesis *in vivo* (Doede et al., 2009; Dorin et al., 2008). It is likely that living cells are required to modulate a more sustained, physiologically relevant regenerative response.

2.9 Conclusions & clinical outlook

We may summarize the key factors affecting the commercial viability of any tubular organ tissue engineering or regenerative technology platform as follows:

- A biodegradable, synthetic scaffold based on well-characterized materials (e.g., PGA, PLGA) that is reproducibly manufactured with defined characteristics.
- A population of committed cells (typically smooth muscle cells), easily isolatable and expandable, is required for scaffold seeding to trigger regeneration *in vivo*.
- A purified population of defined stem cells is neither needed nor desirable. Cost of goods increases significantly with the requirements to monitor stem cell potential and to control directed differentiation.
- Engineering an entire organ *in vitro* or within the peritoneal cavity is neither needed nor desirable. Instead, an *in vitro* cell seeded scaffold is adequate to trigger the body's innate regenerative response, resulting in *de novo* organogenesis *in vivo*.

The outlook for tubular organ regeneration in the clinic is promising. We have seen how the foundational technology platform consisting of a biodegradable scaffold nucleated with a population of committed smooth muscle cells, first pioneered for the bladder, may be applied towards regeneration of multiple, functionally disparate tubular organs. Proof of concept studies in human patients have already been successfully completed with neo-bladder, trachea and vascular grafts. However, the neo-bladder continues to represent the pioneering regenerative neo-organ, with Phase 2 clinical trials currently underway [www.clinicaltrials.gov/ct2/show/NCT00512148?term=tengion&rank=4; www.clinicaltrials.gov/ct2/show/NCT00419120?term=tengion&rank=5]. Additionally, clinical trials are underway for the development of a neo-urinary conduit, a bladder derivative neo-organ designed to facilitate urinary diversion and developed through foundational principles closely resembling those used to engineer the neo-bladder replacement [www.clinicaltrials.gov/ct2/show/NCT01087697?term=Tengion&rank=3].

3. Tissue engineering of solid organs

3.1 Introduction

Though mammalian fetuses are capable of spontaneous regeneration of damaged organs and tissues until the third trimester, adult mammals typically respond to trauma by initiating reparative healing mechanisms. These are characterized by wound contraction, extensive fibrosis and scar tissue formation mediated by the migration and re-organization of myo-fibroblasts, as demonstrated during the response of the heart to ischaemic injury (Boudoulas and Hatzopoulos, 2009). True organ regeneration as commonly observed in certain species of fish and amphibian is associated with an absence of the fibrotic response and concomitant reconstitution of the three dimensional laminar or parenchymal organization of the regenerating organ, as exemplified by the regeneration of adult mammalian liver in response to lobectomy.

The objective of *de novo* organ regeneration through tissue engineering or regenerative medicine is to use defined combinations of cells and biomaterials to catalyze a principally regenerative response towards organ trauma while simultaneously ameliorating any tendency towards fibrotic repair. The *de novo* regeneration of a complete organ in mammals has been demonstrated for laminarly organized, tubular organs including bladder, trachea and vessels of the vasculature (Basu and Ludlow, 2010) as well as recently for the small intestine (Basu et al., 2011c). However, the regeneration of solid organs presents a unique challenge, requiring the organization of highly specialized cell types into complex three dimensional micro-architectures within a parenchymal matrix. Additionally, the regenerating solid neo-organ must synergize with developing elements of the vasculature, lymphatic and nervous systems throughout neo-organ morphogenesis.

In this section, we will highlight the latest progress towards regeneration of solid organs in mammals, with an emphasis on identifying commonalities of methodology that may drive the creation of an organ regeneration platform broadly applicable towards multiple solid organ systems (Figure 2). We will focus on identifying aspects of the regenerative methodology most conducive towards translation into commercial process development and manufacture.

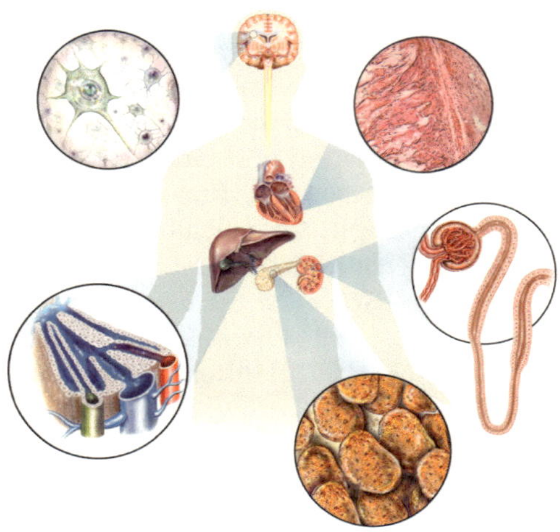

Fig. 2. Solid organ systems potentially amenable to commercial tissue engineering strategies. From top, clockwise: central nervous system, heart, kidney, pancreas, liver.

3.2 Kidney

Kidney provides an excellent illustration of the considerable technical difficulties associated with solid organ regeneration. Numerous specialized cell types including podocytes, mesangial cells, endothelial cells, fibroblasts, epithelial cells and numerous stem and progenitor cell populations are organized across the renal parenchyma into discrete, specialized functional units or nephrons, which serve to selectively filter electrolytes from the vasculature (Vize *et al.*, 2003; Basu *et al.*, 2010).

Efforts to trigger regeneration within animal models of ischaemic or chronic renal disease have typically centered around the isolation, expansion and re-introduction of defined populations of mesenchymal, embryonic or renal stem cells, potentially capable of site-specific engraftment and directed differentiation along multiple renal lineages as well as facilitating the creation of a regenerative micro-environment through paracrine signaling mechanisms (reviewed by Sagrinati *et al.*, 2008; Hopkins *et al.*, 2009).

Such cell therapy methodologies have had generally mixed results, with little if any evidence supporting site specific engraftment and directed differentiation as a mechanism of action by exogenously applied stem or progenitor populations. Introduction of stem cells within the kidney by direct injection into the renal vasculature or renal parenchyma leads to apoptosis or efflux of the majority of applied cells from the target organ within days of implantation (Togel *et al.*, 2008). Uncontrolled differentiation of stem cells that actually do engraft may also represent a significant technical challenge and potential regulatory obstacle against successful commercialization (Kunter *et al.*, 2007).

Reconstitution of kidney mass, a central prerequisite for true *de novo* organ regeneration, will likely require a cell/biomaterial scaffold composite facilitating the morphogenesis of glomeruli and tubules derived from seeded primary renal cell populations, as well as the migration and directed differentiation of host derived renal stem and progenitor cells within the scaffold framework, followed by progressive vascularization and innervation of the developing neo-organ, deposition of extracellular matrix and the reconstitution of a renal parenchyma.

To this end, it may be reasoned that biomaterials derived from the native kidney are best suited to retain renal specific elements of the extracellular matrix (ECM) potentially capable of modulating renal morphogenesis as well as the three dimensional parenchymal micro-architecture unique to the kidney. Application of sodium dodecyl-sulphate (SDS) detergent to whole rat kidneys in one recent report was shown to facilitate the removal of renal cells while maintaining the functional integrity of the ECM as well as the overall three dimensional parenchymal organization of the kidney. Murine pluripotent embryonic stem (ES) cells were applied to this renal scaffold via the renal artery or ureter, and the cell seeded biomaterial composite matured under static or pulsatile culture *without* the application of inductive differentiation agents.

This approach permitted an evaluation of the specific effect of native renal-derived ECM in modulating the directed differentiation of the ES cell population. After arterial seeding, it was observed that ES cells localized within the Bowman's capsule by 4 days post-seeding and within the associated vasculature and renal cortex by day 10. Niche specific localization of ES cells was accompanied by concomitant acquisition of the renal markers Pax-2 and Ksp-cadherin as evidenced by histochemical and RT-PCR approaches (Ross *et al.*, 2009). The effectiveness of SDS relative to other detergents in preserving components of the ECM as well as details of the renal micro-architecture was confirmed independently in comparative studies of the effect of decellularization within monkey kidneys with multiple detergent agents (Nakayama *et al.*, 2010).

Although it is not unreasonable to assume that the organ specific ECM and three dimensional histo-architecture associated with scaffolds procured by decellularization of native organs may be ideally suited to direct the potential regeneration of that specific organ type, this strategy may not be appropriate for a commercially viable solid organ regeneration platform. Apart from difficulties associated with the procurement of cadaveric organs for scaffold assembly, decellularization is a generally uncontrolled process not easily subject to scale-up, process development and quality assurance. Monitoring and verification of the extent of cell loss will substantially increase cost of goods associated with manufacture. Furthermore, native scaffolds retain the potential for immunogenicity despite the presumed absence of cells (Zhou *et al.* 2010). Reliable and reproducible manufacturability of solid neo-organs will likely require the application of synthetic biomaterials with defined physical and chemical properties. With these considerations granted, progress towards creation of biosynthetic scaffolds appropriate for the regeneration of renal architecture was provided by the demonstration that tubular and glomerular structures spontaneously self-assembled from primary rat renal cell populations within one week of growth in collagen I gels (Joraku *et al.*, 2009).

The ability to introduce specific biomimetic peptides and defined proteolytic cleavage sites within the context of gel based biomaterials raises the intriguing possibility of controlling the morphogenesis of tubules, glomeruli or other renal structural units to modulate defined functional outcomes. For example, polyethylene glycol (PEG)-based hydrogels engineered with protease degradation sites and controlled densities of RGD peptide or laminin bioligands have been found to regulate epithelial morphogenesis of cysts derived from MDCK cells, such that cysts grown within ligand functionalized gels demonstrated an increased frequency of lumen formation and unambiguous baso-lateral polarization compared to those grown in unmodified hydrogels (Chung *et al.*, 2008). Therefore, a true renal augment designed to trigger regeneration of glomeruli, tubules, EPO secreting fibroblasts (Paliege *et al.*, 2010) or other key renal cell populations may potentially be envisioned as an injectable hydrogel containing functionalized matrix optimized to catalyze this defined regenerative outcome. Such methodologies alleviate potential concerns regarding the requirement for major surgical intervention within the diseased organ.

Alternatively, a neo-kidney augment may be contemplated as a semi-permanent, implantable, cell/biomaterial composite that upon introduction within or adjacent to the renal parenchyma of a diseased organ, provides a foundational framework for regeneration of tubular or other renal superstructures as well as potentially establishing a regenerative microenvironment through paracrine mediated recruitment of native stem and progenitor populations as well as amelioration of inflammatory, fibrotic and apoptotic cascades (Caplan 2009). To this end, certain species of polyester fleeces have demonstrated the capacity to facilitate the growth of renal tubules within the context of a fibrous artificial interstitium.

In this system, a heterogenous primary renal cell population incorporating stem and progenitor cells was extracted from the sub-capsular space of embryonic rabbit kidneys. The isolated cell population was sandwiched between two layers of polyester fleece and maintained within a perfusion culture system in the presence of aldosterone, a hormone involved in the renin/angiotensin axis (Vize *et al.*, 2003). Spontaneous generation of tubular structures was observed with concomitant expression of key functional markers including cingulin and Na^+/K^+ ATPase. The regenerated tubules appeared to interact with the polyester fibers within the context of the artificial interstitium.

The authors speculate that these cell seeded polyester scaffolds may be multiplexed by horizontal "tiling" or "paving" as well as vertical "piling" to create renal superstructures supporting the continued morphogenesis of renal tubules *in vitro*. The authors suggest that these compounded renal units could potentially form the basis of a true neo-kidney augment upon implantation within the sub-capsular space between the renal capsule and the renal parenchyma (Roessger *et al.*, 2009; Minuth *et al.*, 2008), although our observations of the mechanical properties of the renal capsule associated with diseased, fibrotic human kidneys suggest that this approach may not be feasible.

Nevertheless, this methodology illustrates one key criterion for commercial viability; synthetic polyester fleeces are leveraged for regeneration of renal tubular superstructures, without the requirement for extracellular matrix components derived from decellularized kidney or from other naturally occurring sources. In addition, the application of a defined, serum/BPE (Bovine Pituitary Extract) free media for tubule growth and maintenance additionally serves to facilitate large scale process development. Conversely, it remains to be ascertained whether cells derived from human kidney tissue in general and from diseased human organs in particular, are capable of supporting the spontaneous assembly of tubular structures *de novo* within the context of polyester or other synthetic polymer based biomaterial.

At Tengion, development of the Neo-Kidney Augment (NKA™), a cell/biomaterial composite for renal tissue engineering, has focused on step-wise identification of bioactive cellular components and biomaterials amenable to implantation within renal parenchyma. To this end, we have leveraged principles discussed throughout this chapter, identifying lineage committed, primary renal cell populations as therapeutically bioactive agents capable of mediating functional rescue of aspects of disease phenotype within small animal clinical models of chronic kidney disease (Kelley et al., 2010; Presnell et al., 2010). Similarly, our evaluation of biomaterials compatible with renal parenchyma has led us towards application of gelatin hydrogels as the best tolerated biomaterial scaffold for renal tissue engineering (Basu et al., 2011b). Studies are currently underway to evaluate the ability of bioactive, primary renal cell/hydrogel biomaterial composite constructs (NKA™) to catalyze functional rescue of disease in small and large animal clinical models of chronic kidney disease.

3.3 Heart

Mammalian heart provides one of the clearest demonstrations of the inability of most mammalian solid organs to regenerate. Cardiac ischemia typically results in extensive fibrosis, scarification and loss of function (Boudoulas and Hatzopoulos, 2009). In contrast, zebrafish and other non-mammalian vertebrates are capable of complete regeneration and reconstitution of function upon removal of up to 20% of the ventricle, leading to speculation that an understanding of the mechanism of action underlying regeneration within model organisms such as zebrafish may identify analogous mechanisms that may be leveraged within mammals to trigger cardiac regeneration. To this end, dedifferentiation and proliferation of existing cardiomyocytes was shown to be the principal mechanism of regeneration following ventricular amputation within zebrafish (Jopling *et al.*, 2010).

This modality of action notwithstanding, tissue engineering approaches towards construction of functional mammalian heart have generally focused on decellularization of cadaveric organs to provide scaffold structures for reseeding and implantation. For example, neonatal rat cardiac or aortic endothelial cells were used to seed a decellularized cadaveric rat heart. Upon growth within customized bioreactors for 8 days, evidence of spontaneous contractility was observed. Pump functionality of up to 2% of adult was successfully reconstituted (Ott *et al.*, 2008).

However, although of interest as proof of concept, the requirement for cadaveric organs as a source of biomaterials may ultimately limit the usefulness of this methodology for commercial development and application within the clinic. As we have seen with the kidney, decellularization is a difficult procedure to monitor during quality assurance regimens, and there can be no guarantee that the resultant tissue engineered composite will lack immunogenicity upon implantation (Zhou *et al.* 2010). Furthermore, it remains to be demonstrated whether cardiac cells derived from adult human tissue are capable of repopulating a scaffold to regenerate a functional organ. Finally, the requirement for tissue maturation within pulsatile bioreactors will likely substantially increase cost of goods for tissue engineered cardiac neo-organs.

These criticisms aside, the engineering of synthetic scaffolds that recapitulate defined cardiac structures such as valves and chambers remains technically challenging (Fong *et al.*, 2008). We speculate that triggering the dedifferentiation and subsequent proliferation of existing cardiomyocyte populations by synthetic biomaterial composites containing defined biomimetic peptides and/or autologously derived lineage committed cell populations with smooth muscle cell-like properties may ultimately prove to be the more commercially feasible approach for regeneration of cardiac substructure and ultimately neo-organ regeneration.

Evidence supporting the idea that biomaterial composites seeded with lineage committed cell populations may represent a potential solid organ regeneration platform comes from studies of biodegradable scaffolds nucleated with human ES cell derived cardiomyocytes alone or cardiomyocytes, endothelial cells and embryonic fibroblasts. Upon implantation within rat heart, more extensive vascularization was observed from tri-culture seeded constructs when compared to scaffolds seeded with ES cell derived cardiomyocytes alone (Lesman *et al.*, 2010). In addition, acellular or mesenchymal stem cell (MSC) seeded SIS (small intestinal submuscosa) grafts have been implanted on the epicardial surface of a rabbit model of myocardial infarct. Resultant ventricular functionality and histopathology were significantly more improved in MSC seeded relative to acellular grafts.

Finally, the use of injectable gels derived from decellularized human or porcine pericardium to trigger migration and proliferation of cardiomyocytes and cardiac progenitors has also been explored (Seif-Naraghi *et al.*, 2010). Such synthetic hydrogels incorporating biomimetic peptides or containing committed smooth muscle, endothelial or other fully differentiated cell type may serve as a commercially feasible organ regeneration platform to trigger cardiac regeneration. These regenerative platforms most likely leverage paracrine and ECM-based signaling to recreate a regenerative micro-environment and thereby facilitate the regenerative response.

3.4 Liver

The regenerative potential of the liver is unparalleled among mammalian organs. Adult liver progenitors are thought to be defined by the population of "oval cells" capable of reconstituting both hepatocytes and biliary epithelium upon mobilization by an appropriate regenerative signal (Zaret & Grompe, 2008). However, as with the heart, liver regeneration does not appear to leverage discrete, organ specific pools of stem and progenitor cells, rather, operating through the increased proliferation of existing, mature hepatocytes. Unfortunately, this magnified proliferative capacity has not translated into the ready expandability of hepatocyte populations *in vitro*. Mammalian hepatocytes remain notoriously difficult to maintain and expand in culture (Underhill *et al.*, 2007). Regardless, regenerative medicine and tissue engineering approaches to reconstitute the liver have typically focused on the isolation and expansion of mature, adult hepatocytes as a cell source for biomaterials seeding or, alternatively, on the directed differentiation of ES, iPS, adipose or bone marrow derived MSC towards a hepatic lineage followed by 3D culturing.

Morphogenesis of the mammalian liver is triggered by induction of the embryonic endodermal epithelium by adjacent mesodermal populations (reviewed by Zaret & Grompe,

2008). Mimicking these early developmental signaling events through co-culture of hepatocytes with mesenchymal cell populations such as bone marrow-derived stem cells (BMSC) or even 3T3 fibroblasts results in significant enhancement of hepatic functionality as evidenced by prolonged maintenance of hepatocyte specific morphology and enhanced secretion of albumin (Nahmias *et al.*, 2006). This observation raises the possibility that biomaterials seeded with mesenchymal cell populations may function as a potential solid organ regeneration platform, acting to facilitate the proliferation and functionality of hepatocytes *in vivo*.

Evidence to this effect was provided by studies of Nagase analbuminea rats implanted with corraline hyaluronic acid (HA) ceramic disks seeded either with freshly isolated rat hepatocytes alone, or rat hepatocytes together with bone marrow derived mononuclear cells (Takeda *et al.*, 2005). Hepatocytes cultured in the presence of BMSC secreted significantly more albumin into the media during *in vitro* culture relative to hepatocytes in monoculture. These effects were recapitulated *in vivo* within 4 weeks in analbuminic rats, within which HA-based biomaterials seeded with both BMSC and hepatocytes were associated with significantly greater blood serum albumin levels relative to monoculture controls. Finally, implantation of co-cultured biomaterials within mice presenting chemically induced liver damage resulted in complementation of blood serum albumin as well as increased levels of serum IL-6.

This reconstitution of function notwithstanding, no evidence was provided regarding tissue regeneration *in vivo* within the context of the biomaterial, or whether the presence of hepatocytes is in fact essential. A clear demonstration of hepatic tissue regeneration and functional complementation using synthetic biomaterials seeded with BMSC or some other mesenchymal cell population would be of significant commercial interest. As with heart and kidney, we speculate that paracrine action by factors released from the seeded mesenchymal cells may be adequate to trigger angiogenesis and vascularization of the biomaterials as well as to stimulate the proliferation of the resident hepatocyte population. In keeping with the suggestion of paracrine mechanisms, it may additionally be possible to leverage the resident pluripotent progenitor populations normally resident in the liver by implantation of a biomaterial within the canals of Hering to provide additional space for the development of regenerated tissue (Zhang *et al.*, 2008).

Alternate cell sources of hepatocyte like cells may offer a potential substitute for mediating catalysis of liver regeneration. Hepatocyte like cells may be derived from bone marrow or adipose primary cells through a multi-step directed differentiation regimen attempting to phenocopy key signaling events during hepatic morphogenesis through the controlled application of recombinant bioactive factors and small molecules including hepatocyte growth factor (HGF), basic fibroblast growth factor (bFGF) and oncostatin-M (OMS) (Talens-Visconti *et al.*, 2006). These pseudo-hepatocytes display a characteristic polygonal morphology, demonstrate expression of key hepatocyte associated transcriptional and protein markers and secrete albumin into the media.

Similarly, cells derived from the stromal vascular fraction of adipose can be driven to acquire hepatocyte-like characteristics by application of analogous multi-stage differentiation protocols. Importantly, both adipose derived hepatocytes as well as undifferentiated adipose stromal vascular fraction derived cells are capable of engraftment within chemically damaged liver *in vivo,* where they appear to form cords of tissue within the hepatic parenchyma and acquire hepatocyte specific functionality, as demonstrated by the *in situ* expression of albumin (Ruiz *et al.*, 2010). In this study, engraftment by human adipose derived stromal cells within the liver of a SCID mouse recipient was examined, with functionality being monitored through the use of antibodies specifically recognizing human albumin. The differentiation of bone marrow derived primary cells into hepatocyte-like cell types appears to be facilitated by growth in three dimensions, as shown by the enhanced secretion of albumin, urea, transferrin, serum glutamic pyruvic transaminase and serum oxaloacetate aminotransferase from hepatocyte-like cells grown in three-dimensions on polycaprolactate (PCL) scaffolds relative to similar populations maintained in two-dimensional monoculture (Ruiz *et al.*, 2010).

The trans-differentiation of bone marrow cells towards a hepatic phenotype may also be accomplished by leveraging paracrine and ECM mediated signaling mechanisms between existing hepatic cells and bone marrow cells. For example, human bone marrow MSC may be incubated in plates containing HepG2 derived ECM and HepG2 derived conditioned media for up to 30 days to facilitate the acquisition of a hepatocyte-like phenotype (Tai *et al.*, 2009). These hepatocyte like cells were seeded onto an RGD (arginine-glycine-aspartic acid) modified chitosan alginate polyelectrolyte fibrous non-woven scaffold prior to implantation within a rat liver lobectomy model presenting with 70% removal of liver mass. Post-implantation analysis of the cell seeded biomaterials within 1-2 weeks confirmed the expression of key hepatic markers. Detection of human albumin within rat serum was also demonstrated. No histological evidence for regeneration was provided, although this outcome is unlikely over such a short time period. Taken together, these results provide evidence for functional complementation *without* clear demonstration of *de novo* hepatic tissue regeneration.

An alternative, methodology for regeneration of hepatic tissue is based on the engineering of contiguous monolayers of primary hepatocytes through culturing on temperature responsive surfaces (Ohashi *et al.*, 2007). The temperature sensitive polymer $PIPAA_m$, upon decrease in temperature to below 32°C, rapidly hydrates triggering the spontaneous detachment of cultured hepatocytes in the form of discrete, uniform sheets. Stacking of multiple hepatic tissue sheet monolayers within the subcutaneous space of mice resulted in the formation of a significant hepatic tissue mass, with histologically meaningful micro-architecture, vascularization and functionality as assayed by the presence of glycogen. Importantly, ectopic engineered hepatic tissue was capable of reacting to a regenerative stimulus (2/3 liver resection) by significantly increased levels of proliferative activity. Furthermore, hepatocytes organized as stacked monolayer sheets generated significantly greater functional neo-organ volume compared to the same number of hepatocytes introduced into the subcutaneous cavity within an injectable Matrigel matrix. Such injectable hepatocyte matrices have been proposed as alternate regenerative stimuli for liver (Fiegel *et al.*, 2009).

3.5 Pancreas

Strategies for regeneration of pancreas have focused almost exclusively on the *de novo* regeneration of pancreatic β-cell populations. Morphogenesis of the pancreas broadly resembles hepatogenesis, with induction of the endodermal epithelium of embryonic foregut triggered by adjacent mesenchymal cell populations (Zaret & Grompe, 2008). As a result, methodologies developed to mediate the directed differentiation of ES or adult derived stem and progenitor cells towards a pancreatic lineage leverage many of the same key developmental morphogens as those formulated to mediate acquisition of hepatic phenotype.

The forced expression of certain pancreatic transcription factors and/or treatment with defined growth factors is sufficient to trigger acquisition of insulin transcription within non-pancreatic cell populations as well as within pancreatic ductal and acinar cells, though typically not at levels comparable to true β-cells. Lineage transdifferentiation and dedifferentiation towards a progenitor phenotype have been proposed as mechanisms of action underlying the observed *de novo* presentation of β-cell specific phenotypes *in vitro*. Evidence is also accumulating that these mechanisms may operate *in vivo* to modulate true pancreatic regeneration (reviewed by Juhl *et al.*, 2010).

Strategies to specifically target pancreatic ductal and acinar cells for delivery of defined transcription factors selected to mediate dedifferentiation and reacquisition of a β-cell specific phenotype, while useful as proof of concept, may not represent a commercially viable solid organ regeneration platform. Similarly, methodologies oriented around the directed differentiation of ES or adult stem cells will considerably increase cost of goods, with process development and quality assurance issues focused on the evaluation of stem cell potential and the extent and completeness of the directed differentiation process (**see Table 1**). To this end, it may be worthwhile exploring whether tissue engineering approaches for regeneration of pancreas may be designed to leverage the innate regenerative potential inherent within the organ itself. Although not as dramatic as the liver, regeneration of mammalian pancreas has been observed under certain circumstances (Cano *et al.*, 2008).

Stem Cells	Committed Cells
Considerable initial variability in proliferative capacity and multi-lineage differentiation potential Expansion in culture leads to loss of differentiation potential. Monitoring multi-lineage differentiation potential is lengthy and expensive Directed differentiation is an uncontrolled, inefficient process. Only a proportion of cell population acquires lineage specific characteristics Requirement for inductive cytokines to direct lineage specific differentiation leads to significant increase in cost of goods Multiple molecular, proteomic and functional tests required to evaluate stem and differentiation potential. Substantial increasein time, cost, labor. Tests may be misleading and unreliable (Montzka et al., 2009) Long term effects of inductive cytokines on cells implanted *in vivo* not known. Potential for transformation, teratoma formation	Straightforward, readily isolatable cell sources, less subject to donor variability Readily expanded in culture without loss of lineage specific characteristics No requirement to monitor stem cell potential or directed differentiation No requirement for recombinant cytokines, substantially decreasing cost of goods No regulatory concerns regarding effects of recombinant cytokines on cell transformation Significant reduction in time required to expand committed cell population Application of committed smooth muscle cells (example) for organ regeneration demonstrated across multiple organ systems within *in vivo* models No possibility of abnormal in vivo differentiation or teratoma formation

Table 1. Potential of Stem and Committed Cell Populations for Application in Commercially Viable Solid Organ Regeneration Platforms

Autologously derived pancreatic islet cells or alternate cell sources presenting a β-cell like phenotype may be engineered within an appropriate gelatinous matrix or other biosynthetic scaffold. Considerable effort is currently being invested in developing pancreatic islet cell encapsulation technology, which may be manifested as a vascular device, microcapsule, tubular or planar membrane chamber or sheet architecture (reviewed by Sambanis, 2007). Encapsulation techniques facilitate the delivery of allogeneic or cadaveric islets cells by modulation of the host immune response. Recent developments in encapsulation methodologies include the application of bioactive hydrogels with functionalized moieties designed to improve β-cell survival and secretion of insulin (Lin and Anseth 2009) as well as to modulate inflammation (Su *et al.*, 2010). Implantation of PEG tubes containing rat islet cultures maintained on acellular pancreatic matrix was observed to lead to partial rescue of insulin secretory activity (De Carlo *et al.*, 2010).

Alternatively, pancreatic islet cells may be expanded through growth over polyglycolic acid (PGA) scaffolds. Scaffold seeded cells may then be further matured into tissue engineered islets within a thermo-responsive gel prior to harvesting and implantation beneath the kidney capsule of streptozotocin (STZ) induced diabetic mice, triggering a subsequent return to the normo-glycaemic condition (Kodama *et al.*, 2009). Similarly, transplantation of pancreatic islets grown over a biodegradable scaffold composed of a vicryl fleece with polydioxanone backing and implanted within a canine total pancreatectomy model resulted in normo-glycaemia without the requirement for exogenous insulin injection (n=4), in one case up to 5 months post-implantation. In contrast, dogs receiving an equivalent mass of islets without scaffold did not become normo-glycaemic at any time (Kin *et al.*, 2008).

The sheet architecture solid organ regeneration platform pioneered for application in liver regeneration (Ohashi *et al.*, 2007) has also been applied towards ectopic regeneration of pancreatic tissue (Shimizu *et al.*, 2009). In this study, isolated rat pancreatic islets were expanded over laminin 5 coated $PIPAA_m$ plates. Implantation of the tissue engineered pancreatic sheets within the subcutaneous space of rats led to reconstitution of pancreatic-like tissue structures within 7 days post-implantation. As with the liver, the authors speculate that stacking of pancreatic islet cell sheets may lead to regeneration of an increased mass of pancreatic tissue. It is noteworthy that none of these studies have examined the regeneration of native pancreas *in situ*. This may be a function of the additional challenges inherent in modulating regeneration *in situ*, including the requirement for vascularization and oxygenation specific to the internal volume of regenerating solid organs located deep within the peritoneal cavity. It will be of significant interest to evaluate whether islet cells implanted within biomaterials and ligated to lobectomized pancreas are capable of catalyzing partial *de novo* organ regeneration.

3.6 Spleen

Although the regeneration of spleen may not be clinically or commercially relevant, the principles developed for engineering these organs may have broader implications for solid organ regeneration. To this end, progress towards regeneration of spleen has been reported through the leveraging of platform technologies successfully applied towards engineering of small intestine and stomach, these representing examples of tubular, laminarly organized organs with fundamentally distinct micro- and macro-architecture compared to the spleen or other solid organs (Grikscheit *et al.*, 2008).

In this approach, organoid units generated by the incomplete digestion of rat splenic tissue were seeded within tubular PGA/PLA (poly-lactic acid) scaffolds, and the resultant composites implanted within the omentum of the peritoneal cavity of splenectomized rats. This ectopic tissue engineered spleen demonstrated splenic tissue organization as well as providing protection against *Pneumococcal* induced septicemia. Interestingly, spleen slices cultured ectopically within omentum also mediated formation of quasi-spleen like structures capable of providing protection against induced septicemia.

3.7 Central nervous system

Potentially the best defined examples of a solid organ regeneration platform demonstrating evidence of cellular regeneration *in situ* within a damaged organ *and* catalyzed by implantation of a cell/biomaterial composite at the injury site comes from brain. Tissue engineering of brain and spinal cord typically involves the introduction of gel-based biomaterials within the brain that may be nucleated with neuronal stem cells and may additionally be supplemented with paracrine signaling factors such as vascular endothelial growth factor (VEGF), brain derived neurotrophic factor (BDNF) or components of the extracellular matrix including laminin and fibronectin.

For example, a biomimetic hydrogel incorporating matrix metalloproteinase (MMP) degradation sites, a laminin derived peptide and the neurotrophic factor BDNF was shown to direct the *in vitro* differentiation of seeded mesenchymal stem cells towards a neuronal lineage. Subsequently, this biomimetic hydrogel was introduced into the intrathecal space within a rat model of spinal cord injury. Although no histological evidence for neuronal regeneration was presented, rats treated with the biomimetic hydrogel showed the greatest improvement in tests of locomotion compared to rats treated with non-biomimetic control hydrogels (Park *et al.*, 2009).

In another approach, defined brain defects were created within rat models and subsequently implanted with a gelatin based scaffold impregnated with VEGF. Histological evidence was provided as evidence for migration of endothelial, astroglial and microglial cells within the periphery of the scaffold at 30 days post-implantation (Zhang *et al.*, 2008). This observed regenerative effect was found to be dependant on the presence of VEGF. A demonstration of the application of a non-gel based scaffold for tissue engineering of brain is the use of poly-lactic-*co*-glycolic acid (PLGA) micro-particles in the 50-200 μm size range. These neuro-scaffolds were seeded with neural stem cells and introduced into the brain cavity of stroke induced rats by directed injection under magnetic resonance imaging (MRI) guidance. Within 7 days post-implantation, the neural stem cells had dispersed within the scaffold, and presented as a tightly packed mass at the center of the biomaterial, but with a broader distribution resembling a honey-comb like structure towards the periphery. The graft displayed a mixed population of neuronal, astrocytic and stem cell specific markers, together with evidence of inflammation, but little if any vascularization within the body of the biomaterial. No evidence was provided that the tissue engineered brain tissue had any functional significance in terms of impact to the rat stroke model (Bible *et al.*, 2009).

4. Conclusions

When examined together, platform technologies for regeneration of solid organs remain largely as academic proof of concept. Factors needing to be addressed for successful commercialization include the following:

- Synthetic biomaterials, readily manufactured with defined physical and chemical properties, approved by FDA for implantation within the human body.

- Avoidance of decellularized scaffolds that require cadaveric organ sources and may be potentially immunogenic.
- Use of committed cell types, not stem or progenitor cell populations that require extensive monitoring of stem potential and directed differentiation.
- Autologous cell sources where possible facilitate FDA acceptance of introduction of cellular material.
- Emphasis on a platform approach: cell/biomaterial composites that upon delivery, trigger a broad regenerative response within multiple solid organs.

This is in marked contrast to analogous platform systems developed for tubular organs such as bladder and bladder derivatives, which are currently undergoing Phase I/II clinical trials (Basu & Ludlow, 2010). As we have seen, there are few published reports documenting the *in situ* regeneration of a solid organ in response to damage as a function of the implantation of a cell/biomaterial composite. The majority of published reports have focused on tissue engineering approaches towards solid organ regeneration using the peritoneal cavity or the subcutaneous space as a living bioreactor to facilitate the vascularization of the regenerating composite that is a prerequisite for organogenesis. However, it is our position that such a methodology may not be conducive to large scale clinical application, nor does it represent a commercially viable developmental strategy.

In this regard, a comparison with organ regeneration platforms developed for application towards the regeneration of laminarly organized tubular organs may be helpful. The use of committed smooth muscle cells seeded onto a biodegradable scaffold of synthetic polymer such as PGA, PLGA or PCL provides an example of a commercially viable organ regeneration platform that has been successfully applied towards regeneration of bladder and bladder derivatives. The serial stacking and piling of polyester fleeces nucleated with renal primary cell populations (Roessger *et al.*, 2009; Minuth *et al.*, 2008) may represent a viable solid organ regeneration platform amenable to process development, large scale manufacture and industrial quality assurance regimens, provided the caveats discussed earlier are addressed. Similarly, serial stacking of tissue sheets engineered by monolayer formation over temperature responsive $PIPAA_m$ surfaces as has been demonstrated for liver and pancreas may be viable for commercial development beyond proof of concept.

Regenerative platforms that focus on committed cell populations such as hepatocytes, pancreatic islet cells and cells derived from the stromal vascular fraction of adipose *instead* of stem and progenitor cells have the potential to substantially reduce cost of goods by avoiding technical challenges associated with the isolation and expansion of stem cell populations, maintenance and monitoring of stem cell potential, monitoring and characterization of directed differentiation protocols and the costs associated with recombinant cytokines required to drive directed differentiation. Furthermore, introduction of mesenchymal stem cells within the renal parenchyma of rat models of glomerulonephritis has resulted in abnormal differentiation towards glomerular adipocytes, raising the potential for additional regulatory headaches. It remains unclear how the potential for mis-directed differentiation within the solid organ parenchyma may be definitively eliminated (Kunter *et al.*, 2007).

The observation that adipose derived stromal vascular fraction cells are capable of engraftment *in vivo* within liver and present acquisition of hepatic functionality raises the possibility that readily isolatable adipose cells implanted within the context of synthetic biomaterials or as serially stacked tissue sheets may be capable of stimulating the innate regenerative potential latent within liver, and potentially, additional solid organs. To this end, it has been reported that conditioned media from MSC enhance the survival and functionality of pancreatic islet cells following transplantation in diabetic rats (Park *et al.*, 2010).

There is no evidence to indicate that the stem cell potential of these cell types is directly connected to their secretomic profiles: it is likely that fibroblasts or smooth muscle cells derived from these "MSC"-like populations may function just as effectively for paracrine signaling during the regenerative process. Such a combination of readily isolatable, committed cell types complexed with a biomaterial that may be reliably manufactured, has defined physical and chemical properties, and is already acceptable to FDA and other regulatory agencies for implantation within the human body would represent the ideal platform technology for the commercially viable regeneration of solid organs.

To summarize, factors impacting successful commercialization of solid organ regenerative technologies include:

- Identification of a synthetic scaffold structure that is broadly relevant to solid organ parenchyma, regardless of organ type.
- Committed cell types that may be capable of triggering organ regeneration within the context of an implanted scaffold.
- Alternatively, common approaches for the isolation of organ specific cell populations capable of mediating regeneration within the context of an implanted scaffold. For example, cell populations may be isolated from multiple solid organs by density centrifugation under standard conditions prior to scaffold seeding.
- Standardized methodologies for combining organ derived or organ independent cell populations with biosynthetic scaffolds to create implantable composites, for example, stacking or piling of cell sheets.
- A focus on methodologies that avoid prolonged maturation periods within bioreactors, but rely instead on triggering the body's innate regenerative potential to stimulate neo-organogenesis.
- Identification of common mechanisms of action being leveraged across multiple organ types through common regeneration platforms.

5. References

Andrade C.F. *et al.* (2007). Cell based tissue engineering for lung regeneration. *Am J Physiol Lung Cell Mol Physiol* 292, L510-518

Atala A. *et al.* (2006). Tissue engineered bladders for patients needing cystoplasty, *Lancet* 367, 1241-1246

Araki M. *et al.* (2009). Development of a new tissue engineered sheet for reconstruction of the stomach. *Artificial Organs* 33, 818-826

Asnaghi M.A. *et al.* (2009). A double-chamber rotating bioreactor for the development of tissue engineered hollow organs: from concept to clinical trial. *Biomaterials* 30, 5260-5269

Basu J & Ludlow JW (2010). Platform technologies for tubular organ regeneration, *Trends in Biotechnology*, 28(10): 526-533

Basu J, Genheimer C, O'Reilly DD, *et al.*, (2010). Distribution and analysis of stem and progenitor cell populations in large mammal and human kidneys. *FASEB J* 24: lb35

Basu *et al.*, (2011a). Expansion of the human adipose-derived stromal vascular fraction yields a population of smooth muscle-like cells with markedly distinct phenotypic and functional properties relative to mesenchymal stem cells. *Tissue Eng Part C* 17(8): 843-860

Basu *et al.*, (2011b). Functional evaluation of primary renal cell/biomaterial Neo-Kidney Augment prototypes for renal tissue engineering. *Cell Transplantation* doi: 10.3727/096368911X566172

Basu *et al.*, (2011c). Regeneration of rodent small intestine tissue following implantation of scaffolds seeded with a novel source of smooth muscle cells. *Regen Med*, In Press

Bible E, Chau DY, Alexander MR, *et al.*, (2009). The support of neural stem cells transplanted into stroke induced brain cavities by PLGA particles. *Biomaterials* 30: 2985-94

Boudoulas KD and Hatzopoulos AK. (2009). Cardiac repair and regeneration: the Rubik's cube of cell therapy for heart disease. *Dis Model Mech* 2: 344-58

Cano DA, Rulifson IC, Heiser PW, *et al.*, (2008). Regulated beta-cell regeneration in the adult mouse pancreas. *Diabetes* 57: 958-66

Caplan AI. (2009). Why are MSCs therapeutic? New data, new insight. *J Pathol* 217: 318-324

Campbell G.R. *et al.* (2008). The peritoneal cavity as a bioreactor for tissue engineering visceral organs: bladder uterus and vas deferens. *J Tissue Eng Regen Med* 2, 50-60

Chen K.L. *et al.* (2010). Bioengineered corporal tissue for structural and functional restoration of the penis. *PNAS* 107, 3346-3350

Chun S.Y. *et al.* (2007). Identification and characterization of bioactive factors in bladder submucosal matrix. *Biomaterials* 28, 4251-4256

Chung IM, Enemchukwu NO, Khaja SD, *et al.*, (2008). Bioadhesive hydrogel microenvironments to modulate epithelial morphogenesis. *Biomaterials* 2008, 29: 2637-2645

Cortiella J *et al.*, (2010). Influence of acellular natural lung matrix on murine embryonic stem cell differentiation and tissue formation. *Tissue Eng Part A* 16, 2565-2580

De Carlo E, Baiguera S, Conconi MT, *et al.*, (2010). Pancreatic acellular matrix supports islet survival and function in a synthetic tubular device: in vitro and in vivo studies. *Int J Mol Med* 25: 195-202

De Filippo R.E. *et al.* (2008). Tissue engineering a complete vaginal replacement from a small biopsy of autologous tissue. *Transplantation* 86, 208-214

Doede T. *et al.* (2009). Unsuccessful alloplastic esophageal replacement with porcine small intestinal submucosa. *Artificial Organs* 33, 328-333

Dorin R.P. *et al.* (2008). Tubularized urethral replacement with unseeded matrices: what is the maximum distance for normal tissue regeneration? *World J Urol* 26, 323-326

Fiegel HC, Kneser U, Kluth D, *et al.*, (2009). Development of hepatic tissue engineering. *Pediatr Surg Int* 25: 667-673

Fong PM, Park J, Breuer CK. (2008). Heart Valves. *Principles of Tissue Engineering*, 3rd edition

Go T. *et al.* (2010). Both epithelial cells and mesenchymal stem cell derived chondrocytes contribute to the survival of tissue-engineered airway transplants in pigs. *J Thorac Cardiovasc Surg* 139, 437-443

Gong Z. and Niklason, L.E. (2008). Small diameter human vessel wall engineered from bone marrow derived mesenchymal stem cells (hMSCs). *FASEB J* 22, 1635-1648

Grikscheit TC, Sala FG, Ogilvie J, *et al.*, (2008). Tissue engineered spleen protects against overwhelming pneumococcal sepsis in a rodent model. *J Surg Res* 149: 214-8

Hashish M. *et al.* (2010). Surgical implantation of a bioengineered internal anal sphincter. *J Ped Surg* 45, 52-58

Hibino N. *et al.* (2010). Late term results of tissue engineered vascular grafts in humans. *J Thorac Cardiovasc Surg* 139, 431-436

Hopkins C, Li J, Rae F, *et al.*, (2009). Stem cell options for kidney disease. *J Pathol.* 217: 265-281

Jack G.S. *et al.* (2009). Urinary bladder smooth muscle engineered from adipose stem cells and a three dimensional synthetic composite. *Biomaterials* 30, 3259-32

Jayo M, et al. (2008) a. Early cellular and stromal responses in regeneration versus repair of a mammalian bladder using autologous cell and biodegradable scaffold technologies. *J Urol* 180, 392-397

Jayo M, et al., (2008) b. Long term durability, tissue regeneration and neo-organ growth during skeletal maturation with a neo-bladder augmentation construct. *Regen Med* 3, 671-682

Jopling C, Sleep E, Raya M, *et al.*,(2010). Zebrafish heart regeneration occurs by cardiomyocyte dedifferentiation and proliferation. *Nature* 464, 606-611

Joraku A, Stern KA, Atala A, *et al.*, (2009). In vitro generation of three dimensional renal structures. *Methods* 47: 129-33

Juhl K, Bonner-Weir S, Sharma A. (2010). Regenerating pancreatic β-cells: plasticity of adult pancreatic cells and the feasibility of in-vivo neogenesis. *Curr Opin Organ Transplant* 15: 79-85

Kasimir M.T. *et al.* (2006). Decellularization does not eliminate thrombogenicity and inflammatory stimulation in tissue engineered porcine heart valves. *J Heart Valve Dis* 15, 278-286

Kelley *et al.*, (2010). Tubular cell enriched subpopulation of primary renal cells improves survival and augments kidney function in a rodent model of chronic kidney disease. *Am J Physiol Renal Physiol* 299(5):F1026-39

Kin T, O'Neil JJ, Pawlick R, *et al.*, (2008). The use of an approved biodegradable polymer scaffold as a solid support system for improvement of islet engraftment. *Artif Organs* 32: 990-3

Kodama S, Kojima K, Furuta S. *et al.*, (2009). Engineering functional islets from cultured cells. *Tissue Eng Part A* 15: 3321-3329

Komura M *et al.* (2008). An animal model study for tissue engineered trachea fabricated from a biodegradable scaffold using chondrocytes to augment repair of tracheal stenosis. *J Pediatr Surg* 43, 2141-2146

Komura M *et al.* (2008). Human tracheal chondrocytes as a cell source for augmenting stenotic tracheal segments: the first feasibility study in an in vivo culture system. *Pediatr Surg Int* 24, 1117-1121

Kunter U, Rong S, Boor P, *et al.*, (2007). Mesenchymal stem cells prevent progressive experimental renal failure but maldifferentiate into glomerular adipocytes. *J Am Soc Nephrol* 18: 1754-64

Lesman A, Habib M, Caspi O, *et al.*, (2010). Transplantation of a tissue engineered human vascularized cardiac muscle. *Tissue Eng Part A* 16: 115-125

Lin CC and Anseth KS. (2009). Glucagon like peptide 1 functionalized PEG hydrogels promote survival and function of encapsulated β-cells. *Biomacromolecules* 10: 2460-7

Lin G. *et al* (2008). Defining stem and progenitor cells within adipose tissue. *Stem Cells Dev* 17, 1053-1063

Macchiarini P. *et al.* (2008). Clinical transplantation of a tissue-engineered airway. *Lancet* 372, 2023-2033

Metharom P. *et al* (2008). Myeloid lineage of high proliferative potential human smooth muscle outgrowth cells circulating in blood and vasculogenic smooth muscle like cells in vivo. *Atherosclerosis* 198, 29-38

Minuth WW, Denk L, Castrop H. (2008). Generation of tubular superstructures by piling of renal stem/progenitor cells. *Tissue Engineering Part C*, 14: 3-13

Mondrinos M *et al.* (2006). Engineering three dimensional pulmonary tissue constructs. *Tissue Eng* 12, 717-728

Mondrinos M.J. *et al.* (2007). A tissue engineered model of fetal distal lung tissue. *Am J Physiol Lung Cell Mol Physiol* 293, L639-650

Montzka K. *et al.* (2009). Neural differentiation potential of human bone marrow derived mesenchymal stromal cells: misleading marker expression. *BMC Neurosci* 10:16

Nahmias Y, Casali M, Barbe L, *et al.*, (2006). Liver endothelial cells promote LDL-R expression and the uptake of HCV-like particles in primary rat and human hepatocytes. *Hepatology* 43: 257-265

Nakase Y. *et al.* (2006). Tissue engineering of small intestinal tissue using collagen sponge scaffolds seeded with smooth muscle cells. *Tissue Eng* 12, 403-412

Nakase Y. *et al.* (2007). Endocrine cell and nerve regeneration in autologous in situ tissue engineered small intestine. *J Surg Res* 137, 61-68

Nakase Y. *et al.* (2008). Intrathoracic esophageal replacement by in situ tissue engineered esophagus. *J Thorac Cardiovasc Surg* 136, 850-859

Nakayama KH, Batchelder CA, Lee CI, *et al.*, (2010). Decellularized rhesus monkey kidney as a three dimensional scaffold for renal tissue engineering. Tissue Eng Part A

Ohashi K, Yokoyama T, Yamato M, *et al.*, (2007). Engineering functional two and three dimensional liver systems in vivo using hepatic tissue sheets. *Nat Med* 13: 880-885

Ott HC, Matthiesen TS, Goh SK *et al.*, (2008). Perfusion decellularized matrix: using nature's platform to engineer a bioartificial heart. *Nature Med* 14: 213-221

Ott HC. *et al.* (2010). Regeneration and orthotopic transplantation of a bioartificial lung. *Nat. Med.* 16, 927-933

Paliege A, Rosenberger C, Bondke A, *et al.*, (2010). Hypoxia-inducible factor-2 alpha expressing interstitial fibroblasts are the only renal cells that express erythropoietin under hypoxia-inducible factor stabilization. *Kidney Int* 77: 312-8

Park KS et al., (2010). Trophic molecules derived from human mesenchymal stem cells enhance survival, function and angiogenesis of isolated islets after transplantation". *Transplantation* 89: 509-517

Petersen T.H. *et al.* (2010). Tissue-engineered lungs for in vivo implantation. *Science* 329, 538-541

Presnell S *et al.*, (2011). Isolation, characterization and expansion methods for defined primary renal cell populations from rodent, canine and human normal and diseased kidneys. *Tissue Eng Part C* 17(3): 261-73

Roessger A, Denk L, Minuth WW. (2009). Potential of stem/progenitor cell cultures within polyester fleeces to regenerate renal tubules. *Biomaterials* 30: 3723-3732

Roh J. *et al.* (2010). Tissue engineered vascular grafts transform into mature blood vessels via an inflammation-mediated process of vascular remodeling. *PNAS* 107, 4669-4674

Ross EA, Williams MJ, Hamakazi T, *et al.*, (2009). Embryonic stem cells proliferate and differentiate when seeded into kidney scaffolds. *J Am Soc Nephrol* 20: 2338-2347

Ruiz JC, Ludlow JW, Sherwood S, *et al.*, (2010). Differentiated human adipose derived stem cells exhibit hepatogenic capability in vitro and in vivo. *J Cell Physiol*

Sakuma T. *et al.* (2009). Mature, adipocyte derived, dedifferentiated fat cells can differentiate into smooth muscle like cells and contribute to bladder tissue regeneration. *J Urol* 182, 355-365

Sagrinati C, Ronconi E, Lazzeri E, *et al.*, (2008). Stem-cell approaches for kidney repair: choosing the right cells. *Trends Mol Med.* 14: 277-285

Sala F.G. *et al.* (2009). Tissue engineered small intestine and stomach form from autologous tissue in a preclinical large animal model. *J Surg Res.* 156, 205-212

Sahoo S. *et al.* (2010). Growth factor delivery through electrospun nanofibers in scaffolds for tissue engineering applications. *J Biomed Mater Res A* PMID #20014288

Sambanis A. (2007). Bioartificial pancreas. *Principles of Tissue Engineering*, 3rd edition

Seif-Naraghi SB, Salvatore MA, Schup-Magoffin PJ, *et al.*, (2010). Design and characterization of an injectable pericardial matrix gel: a potentially autologous scaffold for cardiac tissue engineering. *Tissue Eng Part A* 16:2017-2027

Shigemura N. *et al.* (2006). Lung tissue engineering technique with adipose stromal cells improves surgical outcome for pulmonary emphysema. *Am J Respir Crit Care Med* 174, 1199-1205

Shimizu H, Ohashi K, Utoh R, *et al.*, (2009). Bioengineering of a functional sheet of islet cells for treatment of diabetes mellitus. *Biomaterials* 30: 5943-5949

Somara S. *et al* (2009). Bioengineered internal anal sphincter derived from isolated human internal and sphincter smooth muscle cells. *Gastroenterology* 137, 53-61

Su J, Hu BH, Lowe WL, *et al.*, (2010). Anti-inflammatory peptide-functionalized hydrogels for insulin-secreting cell encapsulation. *Biomaterials* 31: 308-314

Tai BC, Du C, Gao S, *et al.*, (2009). The use of a polyelectrolyte fibrous scaffold to deliver differentiated hMSCs to the liver. *Biomaterials* 31: 48-57

Takeda M, Yamamoto M, Iscda K, *et al.*, (2005). Availability of bone marrow stromal cells in three dimensional coculture with hepatocytes and transplantation into liver damaged mice. *J. Bioscience and Bioengineering* 100: 77-81

Talens-Visconti R, Bonara A, Jover R, *et al.*, (2006). Hepatogenic differentiation of human mesenchymal stem cells from adipose tissue in comparison with bone marrow mesenchymal stem cells. *World J Gastroenterol.* 12: 5834-5845

Tian H et al., (2010) a. Myogenic differentiation of human bone marrow mesenchymal stem cells on a 3D nano fibrous scaffold for bladder tissue engineering. *Biomaterials* 31, 870-872

Tian H et al., (2010) b. Differentiation of human bone marrow mesenchymal stem cells into bladder cells: potential for urologic tissue engineering. *Tissue Eng Part A* PMID 20020816

Togel F, Yang Y, Zhang P, *et al.*, (2008). Bioluminesence imaging to monitor the in vivo distribution of administered mesenchymal stem cells in acute kidney injury. *Am J Physiol Renal Physiol* 295: F315-321

Toyoda M. et al. (2009). Characterization and comparison of adipose tissue derived cells from human subcutaneous and omental adipose tissues. *Cell Biochem Funct* 27, 440-447

Underhill GH, Khetani SR, Chen AA, *et al.*, 2007. Liver. *Principles of Tissue Engineering*, 3rd edition

Urita Y. *et al.* (2007). Regeneration of the esophagus using gastric acellular matrix: an experimental study in a rat model. *Pediatr Surg Int* 23, 21-26

Vize PD, Woolf AS, Bard JBL. (2003). *The Kidney: from normal development to congenital disease.* Editors, Academic Press

Wang C. *et al.* (2010). A small diameter elastic blood vessel wall prepared under pulsatile conditions from polyglycolic acid mesh and smooth muscle cells differentiated from adipose derived stem cells. *Biomaterials* 31, 621-630

Zaret KS and Grompe M. (2008). Generation and regeneration of cells of the liver and pancreas. *Science* 322: 1490

Zhang L, Theise N, Chua M *et al.* (2008). The stem cell niche of human livers: Symmetry between development and regeneration. *Hepatology* 48: 1598-1607

Zhang H, Kamiya T, Hayashi T, *et al.*, (2008). Gelatin-siloxane hybrid scaffolds with vascular endothelial growth factor induces brain tissue regeneration. *Curr Neurovasc Res* 5: 112-117

Zhou J *et al.* (2010). Impact of heart valve decellularization on 3D ultrastructure, immunogenicity and thrombogenicity. *Biomaterials* 31, 2549-2554

Zhou J, Fritze O, Schleicher M, *et al.* (2010). Impact of heart valve decellularization on 3D ultrastructure, immunogenicity and thrombogenicity. *Biomaterials* 31, 2549-2554

12

Self-Organization as a Tool in Mammalian Tissue Engineering

Jamie A. Davies
University of Edinburgh
UK

1. Introduction

The end goal of most efforts in tissue engineering is the production of an artificial tissue or organ that is as similar as possible to the corresponding natural structure. So far, most approaches to this have involved combining cells with artificially-sculpted, spun or printed scaffolds. The approach works well for anatomically-simple, matrix-rich structures such as connective tissue, both in culture and in vivo. The visually-striking example of an engineered 'human ear' on the back of a mouse (Cao *et al.*, 1997) brought much public attention to the idea. Scaffold-based tissue engineering has since found valuable clinical use in the production of new cartilage (Andereya *et al.*, 2006), ligaments (Vunjak-Novakovic *et al.*, 2004), vessels (Lovett *et al.*, 2010), bladder wall (Atala, 2011) and nipple (Cao *et al.*, 1998).

Some of the most significant clinical requirements for effective tissue engineering concern not matrix-rich, simple tissues such as connective tissue, but very intricately-arranged complex organs that consist of many cell types, precisely located and in intimate contact with one another. Outstanding amongst these, in terms of clinical urgency, is the kidney, a fragile organ that regenerates itself very poorly, and which is damaged irreversibly by a large range of toxins, including some medicines. The demand for transplantable kidneys far exceeds their supply: in the UK alone, there are about 6,500 people on the waiting list, many leading fairly miserable lives in which they spend many hours per week hooked up to a dialysis machine.

Being able to engineer organs such as kidney and pancreas promises a very positive impact on the lives of many patients, particularly if the engineering could be done from the patient's own stem cells. There are, though, significant problems in extending scaffold-based techniques to organs such as these. The kidney, for example, consists of at least sixty-four distinct cell types (Little *et al.*, 2007) and these are arranged not haphazardly but in very precise order along intricately folded and branched tubules, vessels and stroma (Fig 1). Even if a scaffold could be laid down by some highly-developed three-dimensional printing process to pattern accurately the basement membranes of each of a hundred thousand nephrons, ten thousand collecting ducts and a corresponding number of vessels, it is difficult to see how cells would enter each tube in the appropriate order to populate each segment with the correct type of cell. Kidneys do not normally develop by cells moving into a pre-made scaffold, so there is no reason to suppose that their cells would have evolved the ability to do this even if a scaffold could be provided for them. In normal life, there may be some limited movement of cells along tubules as they are replenished from stem cell populations (Lindgren *et al.*, 2011) but here the movement is from a stem cell niche part way along each nephron (at the 'neck' of the Bowman's capsule), not from the open end of the ureter, artery and vein, which could be the only openings in an anatomically-realistic 3-D printed kidney.

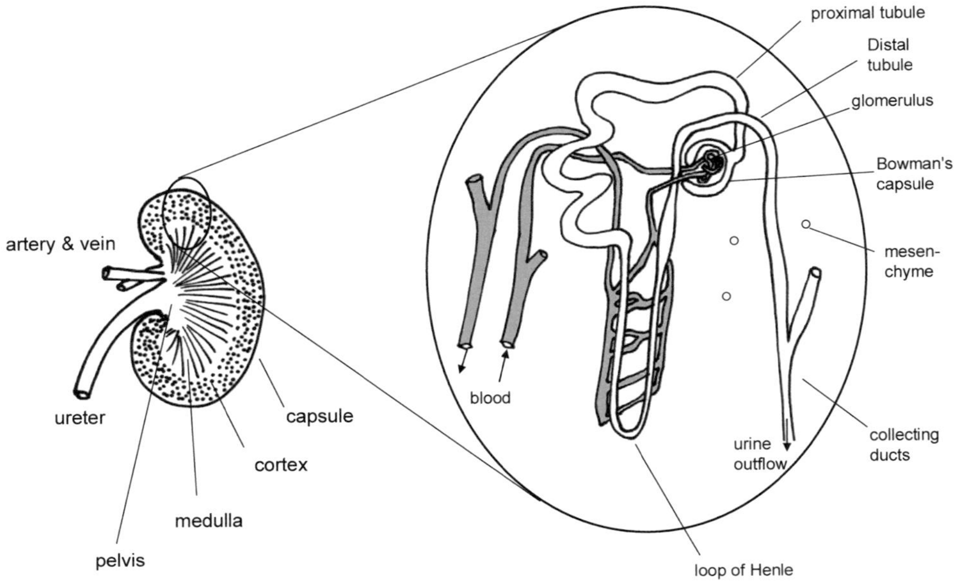

Fig. 1. The basic anatomy of the kidney (in this case mouse: human kidneys have multiple segments). The figure on the left shows the gross anatomy of the organ, while the detail shows one of the c. 2,000 nephrons of the organ (100,000 - 1,000,000 in human). Many of the tissue types present in the nephron include several different types of cells, for example the glomerulus contains at least five, and in each case these cell types are organized precisely.

It may turn out to be possible, one day, to print the cells and the matrix together at the required resolution to put everything in the correct location at the time of manufacture. There are formidable technical challenges to this endeavour, however, and it is sensible to explore alternatives.

The approach that is most radically different to developing ever more intricate 3-D micro-fabrication techniques is to minimize the requirement for engineering by making maximal use of cells' inherent abilities to organize themselves. The underlying idea is to work, as far as possible, with the flow of normal development; to engineer a system to turn a simple suspension of cells, such would be produced by stem cell culture, into a properly-arranged early foetal form of the organ in question, and then to engineer an environment that lets that "foetal organ" grow and mature in the usual manner. There are three main challenges in this; (a) production of the correct type of committed stem cells in the first place, (b) self-organization of a suspension of these cells into a "foetal organ" and (c) transplant of that into a recipient in a way that allows it to grow and mature. Most of this chapter will concentrate on step (b), engineering by self-organization, but steps (a) and (c) will be discussed briefly towards the end.

2. The self-organizing abilities of cells

The self-organizing abilities of cells arise from two broad classes of mechanism; biophysical and developmental (Davies, 2005). Self-organization through simple biophysics makes no demands on there being a 'developmental programme' or on cells responding to each other's signals; instead it works purely on the current properties of the cells. Different cell types bear different types and quantities of molecules that mediate adhesion between cells or that mediate adhesion between cells and matrix. As always, if the components of a system (cells, in this case) are free to change their relationships, they will tend towards an arrangement that minimizes free energy. By definition, unbound adhesion molecules are in a state of higher free energy than bound ones (if this were not true, they would not promote adhesion because they would not naturally bind), and the difference in free energy is greater for higher affinity interactions. Free energy is therefore minimized by a state in which the maximum high-affinity binding is able to take place.

When only one cell type is present, free energy is minimized either by cells sticking together as much as possible in an aggregate, as happens when cell-cell affinity is greater than cell-matrix affinity (Fig 2a), or by cells dispersing themselves so that they are completely surrounded by matrix, as happens when cell-matrix affinity is higher (the two-dimensional equivalent of this is the spreading of cells on a culture dish to which they are highly adhesive). Where cell-cell affinity is higher than cell-matrix affinity, and where two cell types are present, differences in the affinity of cells for their own kind and for the other kind can drive cell sorting (Steinberg, 1962b). Where each cell type has a higher affinity for its own kind than for another, mixtures of cells will spontaneously separate, the cell type with the highest mutual affinity being surrounded by their less adhesive neighbours (Fig 2b) (Steinberg, 1962a; Foty & Steinberg, 2005). Where each type of cell sticks better to the other type and one type is much less common than the other, the resulting arrangement will be a 'salt-and-pepper' mixture in which the less common type is surrounded by the more common. Where the numbers are equal, stripes would be predicted to result.

The situation becomes more complicated still where cells are polarized, as simple epithelia are, so that their apical surfaces are barely adhesive, their lateral surfaces are strongly adhesive to similar cells and their basal surfaces are most adhesive to matrix. Here, epithelial cells tend to produce cysts or tubes in which lateral cell-cell contacts are maximized, as are basal cell-matrix interactions. The direction of the apico-basal polarity of cells in the walls of these cysts depends on the availability of suitable matrix: if an adhesive matrix is available, they will polarize basal side outwards; if it is not, they will polarize basal side inwards (Wang *et al.*, 1990). Any mesenchymal cells present will be located on the outside of the cysts (Fig 2c).

It cannot be over-emphasized that the arrangements described above arise from simply biophysics and have nothing to do with any 'developmental programmes'. To emphasize this point, it is worth noting that the arrangements emerge in computer simulations that model the current state of the cells but include no information whatever about any changes in gene expression (Takano *et al.*, 2003; Krupinski *et al.*, 2011; Agarwal, 1995). Of course, a 'developmental programme' may underlie the reason that different cells express different adhesion molecules in the first place, but that is as far as it goes: once established, those differences are enough to drive cells into specific, predictable arrangements (as long as they are free to move).

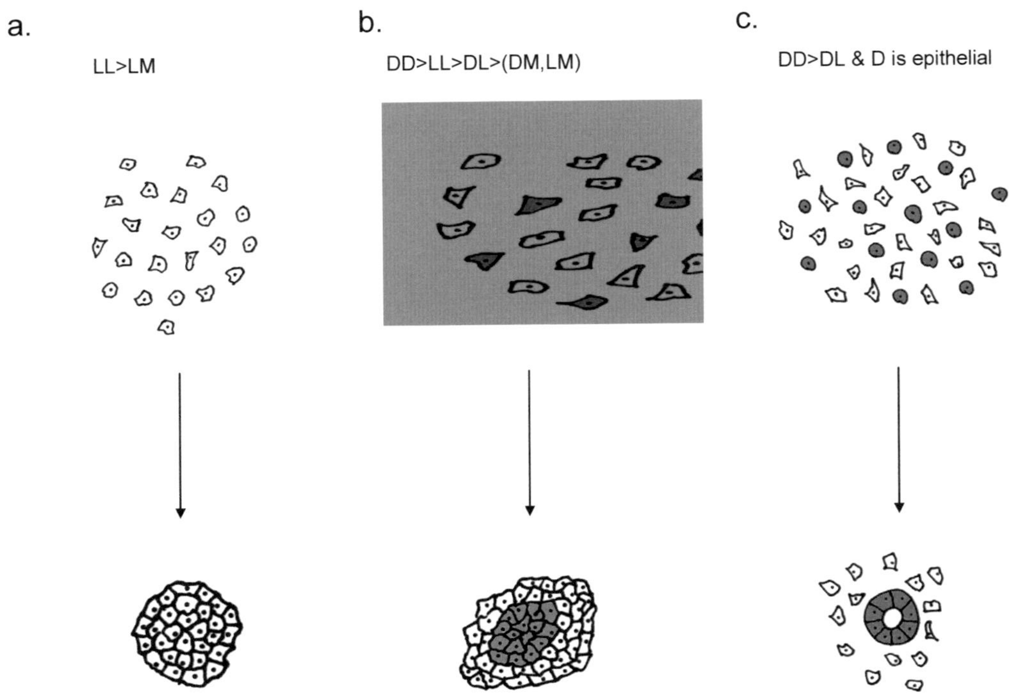

Fig. 2. Adhesion-mediated cell sorting. (a) depicts a group of identical cells, drawn in a light colour and called 'L' for this reason, whose mutual adhesion (LL) is greater then their adhesion to the matrix (LM): these cells aggregate together to maximize their areas of contact and minimize unbound adhesion molecules. (b) depicts a mixture of cells, dark (D) and light (L). In this case, the mutual adhesion of dark cells (DD) is stronger than the mutual adhesion of light cells (LL), but any cell-cell interactions are stronger than cell-matrix ones. Under these circumstances, the cells aggregate, but with the dark cells in the middle. (c) Where the dark cells are epithelial, and primed to achieve apico-basal polarization when in contact with one another, the result is an epithelial cyst surrounded by mesenchyme. References for these effects are given in the main text.

The second class of self-organizing mechanism is more classically developmental and involves cells changing their states over time. In vertebrates, at least, rather little developmental change is cell-autonomous. Rather, most is controlled by signals received from the environment, which is dominated by the secretions of other cells. The cells of a developing organ multiply, die, move and differentiate in response to the signals of other cells around them. The behaviours available to a cell are set by its current internal state (eg gene expression), but which of these potential behaviours are triggered, how much and when, are determined by neighbours. In the kidney, for example, the branching epithelium that will become the urine collecting duct system grows only when it receives signals, such as glial cell-derived neurotrophic factor (GDNF) (Sainio *et al.*, 1997), from the cap of mesenchymal cells that covers each branch tip. The cells of that mesenchymal cap proliferate in response to signals, such as Wingless homologue 9b (Wnt9b) (Karner *et al.*, 2011) produced by the developing collecting duct system. At first sight, this mutual encouragement has the look of a positive feedback system, and so it is when everything is in balance. If the sizes of the tissues happen to be out of balance, however, it can work as negative feedback that steers the system back where it should be. If a collecting duct tip has branched too early, for example, and there is not enough cap mesenchyme to properly service two new branches, there will not be enough of a signal from it to maintain proliferation in the new collecting duct tips, and they will therefore be forced to wait until proliferation of the cap mesenchyme has caught up. Similarly, if the cap mesenchyme has become too large, its further proliferation will have to wait until the branches have caught up again.

Negative feedback processes such as the one outlined above are very common in organ development: in the kidney alone, there are similar processes to balance production of excretory nephrons and stroma and to balance production of vessels and glomeruli (Davies & Fisher, 2002). The self-correcting nature of these systems is an important element of natural self-organization of tissues in foetal life, reducing the need for unrealistic accuracy and making the growth of very large animals, such as us, possible. Importantly for the context of this chapter, they mean that a tissue engineer may not have to produce an engineered "foetal organ" that is exactly, 100% identical to what would exist in a real foetus. Rather, there is reason to hope that even a rough approximation to an early foetal organ will be enough to kick-start the cells' own abilities to correct errors and converge automatically on the normal anatomy.

This, then, is the theoretical background to a method of tissue engineering that aims to allow cells to do almost everything for themselves, with the minimum of human intervention. The rest of this chapter will describe the progress we have made in applying these principles to the problem of engineering "foetal kidneys". Given that the approach will work with the normal processes of development, it is necessary describe briefly how kidneys normally develop in an embryo before considering using this information for engineering purposes.

3. Normal kidney development

In its earliest stages, the anatomy of a normal foetal kidney rudiment is very simple: it consists of an unbranched epithelial tube, the ureteric bud (itself a side-branch from the Wolffian duct, outside the kidney), surrounded by mesenchyme. Over the next few days of normal development, the ureteric bud will undergo rounds of branching and growth to produce a tree-like collecting duct system. The mesenchyme close to the bud tip will condense to form a 'cap' over the tip. As the tips divide, the cap will tear in two, so that each new tip carries away a small cap of its own. Some of the signals that ensure these two components keep pace with each other and maintain co-location have been described in the section 2 of this chapter. As well as maintaining itself, the cap also sheds groups of cells that will undergo a mesenchyme-to-epithelial transition: these will become excretory nephrons (Fig 3).

For much of renal development, roughly two nephrons form for each collecting duct branching event, although the first-formed nephrons later disappear. The nephrons form first as small cysts, and these then undergo a stereotyped series of morphogenetic events, progressing through the so-called 'comma-shaped' and 'S-shaped' stages. As they do, they become segmented into Bowman's capsule, proximal tubule, Loop of Henle and distal tubule. The distal tubule connects to the nearby collecting duct, and the Loop of Henle extends towards the inside, or medulla, of the organ. Within the Bowman's capsule, cells become specialized for urine filtration. As they do so, they secrete vascular endothelial growth factor (VEGF), which acts as a chemo-attractant for endothelial cells, bringing capillaries to what will become the glomerulus of each nephron (Tufro, 2000). Similar chemo-attraction, based on angiopoietin, is thought to attract blood vessels to the Loops of Henle (Kolatsi-Joannou *et al.*, 2001).

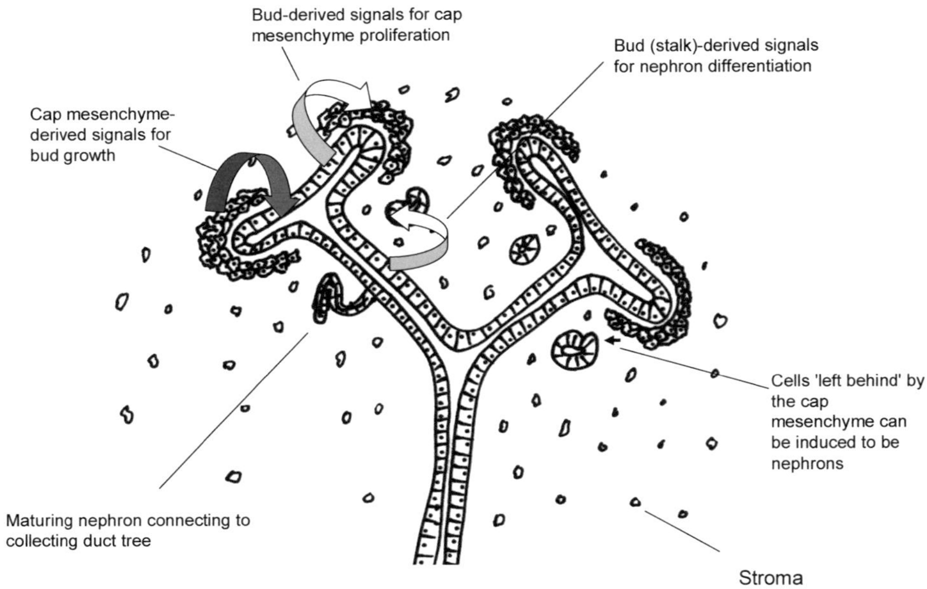

Fig. 3. A very simplified diagram of kidney development, showing how nephrons derive from cap mesenchyme, and showing how signalling between the branching ureteric bud/ collecting duct tree and the mesenchyme patterns the organ and ensure that the compartments proliferate in pace with each other

If a kidney rudiment is removed from a mouse and maintained in simple organ culture conditions, almost all of the above events happen normally, if a little slowly. The exception is the formation of a system of blood vessels, probably because they normally enter the kidney as a branch from the aorta (as the future renal artery) and from the vena cava (as the future renal vein): there is some evidence for an endogenous source of endothelia too (Loughna *et al.*, 1998). The fact that a kidney rudiment will mature basically normally in culture is very important to the topic of this chapter, because it implies that the few cell types involved 'know' everything they need to know, without any need for information to be provided by the rest of the embryo. All we may need to engineer is the very simple bud-and-mesenchyme combinations from which the kidney develops, and then let nature take it course.

4. Engineering a kidney from a suspension of renogenic cells

The ultimate goal of this work is the production of clinically-useful kidneys from appropriately committed stem cells. In this context, 'appropriately committed' means one population identical to the stem cells that maintain the mesenchyme (some of whose daughters give rise to nephrons etc) and another population identical to the stem cells in the ureteric bud (that give rise to new bud tips and the stalks they lay down behind them). There have been some encouraging developments towards directing mouse embryonic stem (ES) cells toward a renal fate (Kim & Dressler, 2005), but there is not yet a reliable method for producing the required cell states at high efficiency.

To experiment with ways of engineering "foetal kidneys" from suspensions of stem cells, we therefore began not with ES cells, but with cells isolated directly from early foetal kidney rudiments, which are therefore known to be in exactly the required state. The stage chosen was E10.5-E11.5, when the ureteric bud and mesenchyme are present but no nephrons have yet begun to form. Depending on definitions, a kidney rudiment of this stage can be considered to consist of either two cell types (metanephrogenic mesenchyme and ureteric bud), or of three (metanephrogenic mesenchyme, ureteric bud tip and ureteric bud stalk). Given that we have shown ureteric bud tip and stalk cells to be inter-convertible in both directions (Sweeney *et al.*, 2008), this chapter will consider there to be just two cell types. We begin by using enzymes to disperse cells of kidney rudiments into a simple suspension of isolated cells, to simulate a harvest from a culture dish or FACS machine that would be the output of a reliable method for programming stem cells to become renogenic, when such a method is finally developed.

The immediate goal is therefore to allow these cells to come together to re-create something sufficiently like a real kidney rudiment that it can go on and develop as if it really were one. The mesenchymal and epithelial cells are known to express very different sets of adhesion molecules (see www.gudmap.org). It is therefore a reasonable assumption that if they are simply brought together into a random lump, then biophysical mechanisms might lead to their sorting out into epithelial cysts and surrounding mesenchyme. This turns out to be true: if the suspension of cells is re-aggregated by simple centrifugation and then cultured on a polycarbonate filter (which is not adhesive enough for cells to spread out on it, so they remain as a lump), epithelial cells find one another and form one or more cysts, surrounded by mesenchyme (Unbekandt & Davies, 2009). There is, though, a very great loss of cells in this process, probably because cells separated form their normal cell-cell and cell-matrix contacts tend to undergo elective cell death in a process called anoikis (Frisch & Screaton, 2001).

Inhibition of the Rho-dependent kinase, ROCK, is known to protect some cell types from elective cell death (Watanabe *et al.*, 2007). In particular, ROCK inhibitors used to study the effect of the Rho-ROCK-myosin pathways on renal tubule morphogenesis also result in reduced apoptosis (Meyer *et al.*, 2006). Together, these observations suggest that pharmacological inhibitors of ROCK might be effective in preventing the massive death in kidney rudiment cell dispersal and re-aggregation experiments. So it proved: inhibition of ROCK using the drugs H1152 or Y27632 resulted in very much improved survival, resulting in re-aggregates with many epithelial cysts expressing ureteric bud markers, surrounded by mesenchyme (Unbekandt & Davies, 2009). This proved beyond reasonable doubt that the mixed cells from kidney rudiments are able to organize themselves back into their basic tissue types. Unfortunately, the system as described so far has two major defects from the point of view of making a more mature kidney: first, the mesenchyme cells fails to form nephrons and second, there are many ureteric 'cysts' rather than one branched ureteric bud.

Our studies on the effect of ROCK inhibition on the morphogenesis of renal epithelia had already suggested that nephron formation requires normal ROCK function at various critical stages, including the very earliest (Lindstrom and Davies, unpublished). Nephron formation does not begin in a normal foetal kidney until about one day later than the stages at which kidneys were harvested for the dispersal and reaggregation experiments described in the paragraph above. The danger of anoikis would be expected to be highest when cells are dispersed, before they have come together again to find suitable neighbours, while the danger to nephron formation would be relevant only after cells have found each other well enough for the signalling loops that induce nephron formation to be up and running. This reasoning suggested that it might be possible to use ROCK inhibition for temporary 'life support' during the first 24h of the reaggregation experiment, and then to remove it and rely on cells' new contacts to keep them alive and healthy. When this is done, the result is good survival and good formation of nephrons. The nephrons go through all of the normal morphological stages of development, such as the comma and S-shaped stages, and they connect to nearby ureteric bud/ collecting duct 'cysts'. Furthermore, they express markers for Bowman's capsule, proximal tubule, distal tubule etc at the expected times and in the correct places (Unbekandt & Davies, 2009).

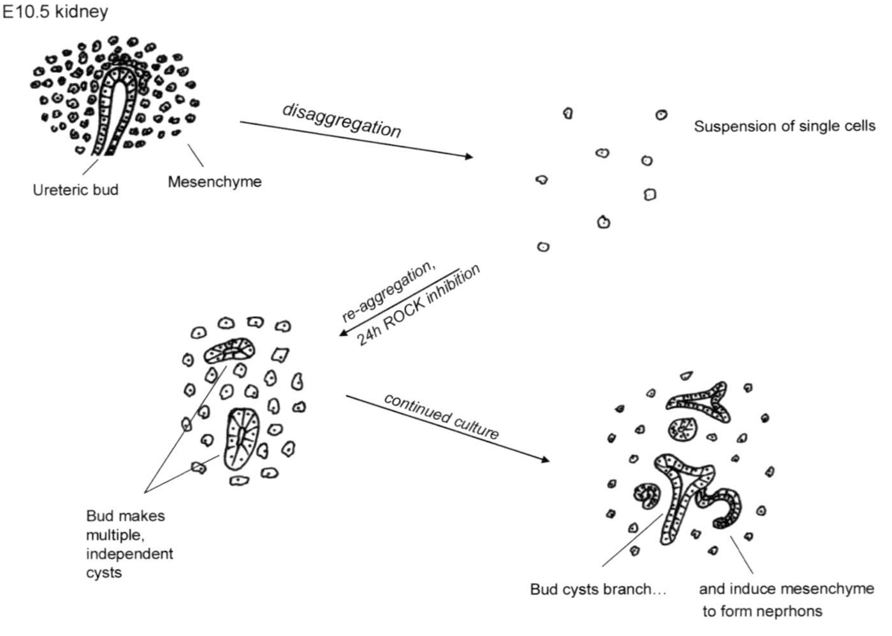

Fig. 4. The basic method of reconstructing renal tissue by re-aggregation of dispersed progenitor cells. The method illustrated here, based on Unbekandt and Davies 2010, has the disadvantage that multiple small ureteric bud 'tree-lets' are formed: an improved method is illustrated in Fig 6.

Under the conditions just described, many of the small ureteric bud 'cysts' form extended tubules that branch, each becoming a 'tree-let'. (Fig 4) This suggests that the basic cell biology of ureteric bud/collecting duct morphogenesis is running normally, but the presence of many 'tree-lets' rather than one tree is abnormal and would be functionally useless for two reasons. The first is that, in a normal kidney, nephrons drain their urine to the branches of a single collecting duct tree and these branches drain in turn to the original 'trunk' of the tree, the ureter. Having nephrons connecting to lots of small, isolated 'tree-lets' would provide no means of egress for urine. The second reason is more subtle, but still critical. The architecture of the growing collecting duct tree imposes a large-scale order on the kidney, in particular a cortex in which blood-filtering glomeruli are located and a medulla into which the Loops of Henle dip, and through which the collecting ducts pass down from the cortex toward the ureter. When the kidney is working, the cortical interstitium is of normal salinity but the medulla is very salty, and this saltiness is critical to the kidney's ability to recover water and to concentrate urine. If a kidney had its nephrons scattered everywhere, without the clear cortico-medullary distinction imposed by arrangement around a single collecting duct tree, concentration of urine could not take place.

The fundamental problem with the simple reaggregation system is that small epithelial cysts are stable. Once enough epithelial cells have come together to form a small cyst, there is no reason for them to leave in favour of moving again to find a larger mass (indeed, the biophysics of the situation would prevent them from leaving, which would entail breaking energetically-favourable mutual adhesions). This is an example of the well-known physical phenomenon of a local minimum: a system that has reached a moderately energetically-desirable state but cannot reach an even more desirable state, because all possible routes between these two states would involve being in a temporarily less desirable one (Fig 5). Multiple small aggregates therefore form, and the system is stuck in this state.

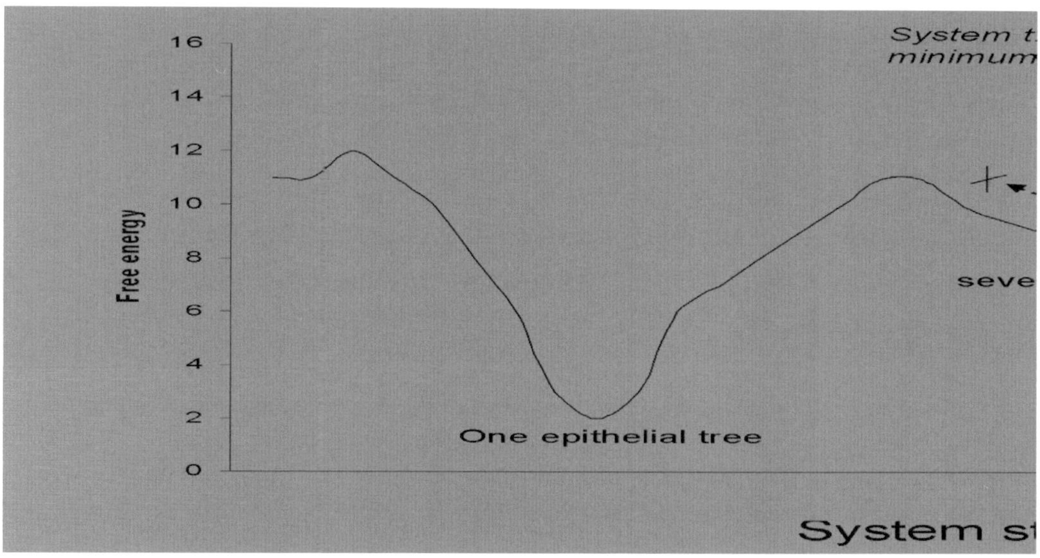

Fig. 5. The problem of the system being trapped in a local minimum. In re-organizing themselves from a random mix into one in which epithelial cells adhere to one another in separate small cysts surrounded by mesenchyme, the system succeeds in lowering its free energy. In order to reach the most favourable possible state (lowest on the diagram), though, the cells of the cysts would have to let go of one another and travel to join in with another cyst. This would mean a temporary move to a less favourable state than they initially found. The system therefore becomes trapped into a state that is moderately favourable, but not the best possible.

The idea of solving this problem by placing only a very few epithelial cells in the culture to begin with, although theoretically-possible, is unlikely to be useful because the total volume of the culture would have to be very small to allow the epithelial cells to find one another at all by random wandering (there is no evidence for long range chemotactic attraction between these cells, and no reason to suppose from normal development that such attraction should exist). Also, making an extremely small rudiment is not likely to be maximally useful from the point of tissue engineering. We have therefore devised a system of serial culture, in which multiple cysts are allowed to form in the first culture, and one of these is then combined with a fresh suspension of mesenchyme cells for a second culture (Ganeva *et al.*, 2011). When this is done, the cyst develops into a branching tubule. The overall result is an engineered "foetal kidney" that is arranged, as it should be, around a single branched ureteric bud/ collecting duct system (Fig 6) (Ganeva *et al.*, 2011).

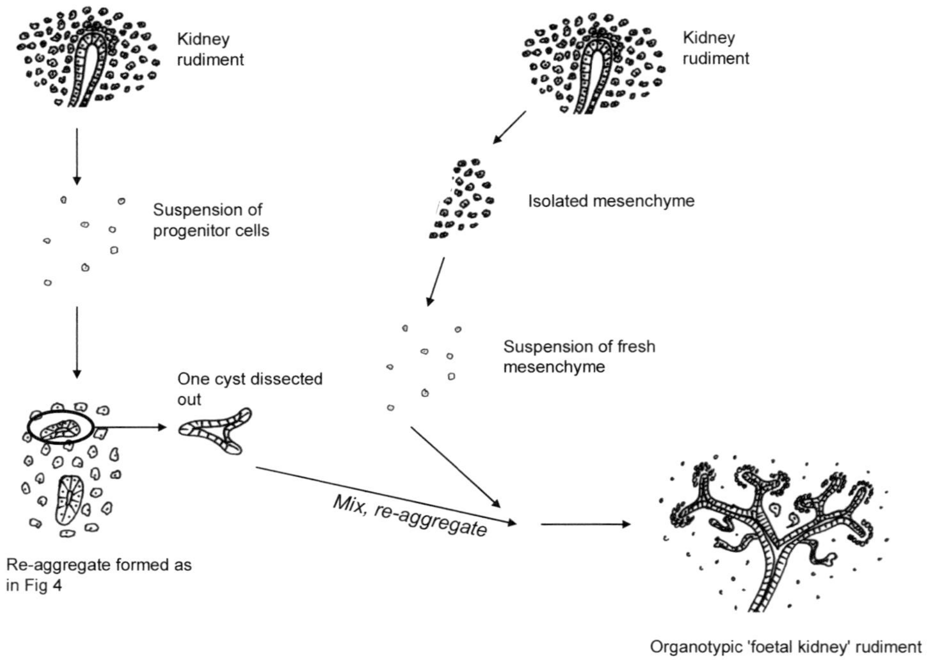

Fig. 6. An improved method for reconstructing kidney rudiments by re-aggregation from progenitors (Ganeva et al. 2011). A conventional re-aggregation is first performed, as in Fig 4, and then just one ureteric bud cyst is isolated from it and combined with a suspension of fresh renal mesenchyme cells free of ureteric bud. Under these circumstances, the result is a kidney organized normally around a single ureteric bud/ collecting duct tree.

5. Connecting to a blood supply

Like intact foetal kidney rudiments grown in organ culture, the engineered "foetal kidneys" produced by the method describe above lack a vascular system. This is a major omission for an organ, the main purpose of which is to filter blood and which receives about a fifth of the cardiac output. Normal developing kidneys attract endothelial ingrowth by secreting molecules such as VEGF, particularly from specialized cells that will go on to form the filtration barrier in the glomeruli (Tufro, 2000). If a mouse kidney rudiment is cultured on the chorioallantoic membrane of a bird's egg, this chemoattraction is sufficiently strong to induce ingrowth of blood vessels from this membrane: indeed, the ease with which quail cells can be distinguished from those of mouse allowed this system to be used to demonstrate that glomerular endothelia arise from ingrowing vessels rather than from the nephron itself (Sariola *et al.*, 1983).

Our preliminary data suggests that self-organized "kidney rudiments" produced by the re-aggregation method described here can attract blood vessels in a similar way. When they are cultured on a chick egg chorioallantoic membrane, the rudiments attract ingrowth of vessels that is clearly visible at a gross level and sections of the resulting hybrid tissue show the formation of what appear to be glomeruli.

6. Self-organized organ rudiments in a clinical context

The production of a tiny analogue of a normal foetal kidney is clearly a long way from the functional, adult organ that is actually required. Is effort in this direction therefore a waste of time compared with efforts to build an adult organ directly?

I would argue that it is anything but a waste of time, for several reasons. The first is that we *know* that it is possible for a kidney to build itself from its foetal form, at least in a foetal environment, because that is what happens in normal life: enthusiasts for direct engineering can only *hope* that their aim is possible in any environment at all. The second is that we already know that it is possible to transplant normal foetal kidney rudiments into an immature or adult recipient and have those rudiments grow and become functional. So far, function is partial and not enough to sustain life for a long period (Rogers & Hammerman, 2004; Rogers *et al.*, 1998) , but these are early days and the technique continues to improve. Given how well kidneys grow in culture, in isolation from the foetal environment, the problems are probably not those of missing signals from the rest of a foetal body. It is more probable that they are to do with what is a very immature organ having to stand blood pressures and levels of blood oxygenation far in excess of the foetal situation while it is growing, and then having too poor a blood supply to function in its ectopic location. These problems may be soluble by surgery, possibly moving the graft from place to place as it grows, so that it begins its life in a protected site with low blood flow and pressure but ends up grafted to the renal artery.

The third reason is that the time-course of many renal diseases is very slow; in these conditions, a child is typically diagnosed before the age of four with a condition that will require transplantation only in their teens or twenties. There is therefore plenty of time for a transplanted 'foetal' kidney to grow to mature function before the host kidney completely collapses. Indeed, with even partial function being shared, the host kidney may last a great deal longer.

Formidable challenges certainly remain. There is a pressing need to improve the functional maturation of transplanted foetal kidneys, and to test the self-organized rudiments the same way. There is also, of course, the need for a method for production of renogenic stem cells but even this is probably easier than finding methods for the production of the many different cell types that would be needed for direct engineering of an adult organ.

7. Potential for application to other organs

There is every reason to suppose that the techniques that we have applied to the kidney can be applied equally well to other visceral organs of the basically glandular type - that is, organs that consist of many repetitions of fine-grained anatomy. Lung, pancreas, prostate, mammary gland and salivary gland all develop in a very similar way and are organized around a branching epithelial tree, while liver and testis development are not wholly dissimilar although their tubes form by a hollowing-out process rather than invasive branching. In each case, the organ begins to develop from two cell types, epithelium and mesenchyme, that have self-renewing stem cell properties and that differ markedly in their expression of adhesion molecules. A suspension of such cells would be expected to sort into epithelial ducts surrounded by mesenchyme, and the signal exchange between such tissues would be expected to mimic that found in normal development and therefore to organize the growth of an engineered 'foetal organ'. We have indeed shown this to be the case with lung (Unbekandt and Davies, 2010).

Whether the engineering of other 'foetal organs', followed by their transplantation to mature in situ, is likely to be clinically useful is governed mainly by the likelihood of their being able to mature. Kidneys are unusual, in that many renal diseases progress slowly and a patient with very weak renal function can be kept alive for very long periods by dialysis, allowing time for a transplanted rudiment to mature before any demands are made of it. The same may be true of pancreas in the context of diabetes, where the patient's existing pancreas maintains exocrine function and injected insulin substitutes for endocrine function until the transplanted rudiment is ready to take over. Similarly, mammary gland function is obviously dispensible (being absent in the male half of the population anyway). It is, on the other hand, harder to see how the scheme would work for lungs, for which there is no adequate long-term substitute.

Engineering gonads by re-aggregation of dispersed cells offers intriguing possibilities in fields other than medicine. At the cell suspension stage it is possible to mix cells from different sources: the result is a fine-grained chimaeric organ (Unbekandt and Davies, 2010). It is also, in principle, possible to subtract cells from the suspension stage by FACS or magnetic sorting. Putting these ideas together, it might be possible to disaggregate the cells of an early gonad rudiment from the foetus of a common (domestic) animal, sort away the germ cells, replace them with germ cells from a rare, endangered species that does not breed well in captivity, and transplant the resulted engineered gonad rudiments back in to a domestic host. In this way, the rare genome of a highly-endangered species might be propagated in domestic animals which are happy to mate in captive conditions, to produce offspring of the endangered species. Placental immunology and other physiological considerations will almost certainly impose limits on how different the species involved might be before an embryo cannot be carried, but the approach may work between related species if there is nothing else to be done. Obviously, conservation of species by protection of habitat is a much better course of action, but realistically many habitats are being lost and will take a long time to be regenerated even if there is the will. In this interval, a 'Noah's Ark' based on propagation of genomes in surrogate domestic animals might provide an emergency stop-gap solution.

8. Conclusion

In summary, data obtained so far suggests that the idea of using cells' own abilities for self-organization to produce an immature organ progenitor, with a view to placing this in a host to mature, is a viable avenue of research. This is especially true for organs whose failure is slow enough to allow time for transplanted organs to grow. In making maximal use of cells' own abilities and in deliberately minimizing the requirement for detailed engineering, the approach contrasts with many others in this book. Which will win out in the end, or whether the techniques will co-exist for different problems, remains to be seen.

9. Acknowledgements

The author would like to acknowledge financial support for the work described herein, from the National Centre for 3Rs, BBSRC and the European Union. He is grateful to Peter Hohenstein, Mathieu Unbekandt, Nils Lindström and Veronika Ganeva for their helpful discussions.

10. References

Agarwal, P. (1995) Cellular segregation and engulfment simulations using the cell programming language. *J Theor.Biol*, 176, 79-89.

Andereya, S., Maus, U., Gavenis, K., Muller-Rath, R., Miltner, O., Mumme, T. & Schneider, U. (2006) [First clinical experiences with a novel 3D-collagen gel (CaReS) for the treatment of focal cartilage defects in the knee]. *Z.Orthop.Ihre Grenzgeb.*, 144, 272-280.

Atala, A. (2011) Tissue engineering of human bladder. *Br.Med Bull*, 97, 81-104.

Cao, Y., Vacanti, J.P., Paige, K.T., Upton, J. & Vacanti, C.A. (1997) Transplantation of chondrocytes utilizing a polymer-cell construct to produce tissue-engineered cartilage in the shape of a human ear. *Plast.Reconstr.Surg.*, 100, 297-302.

Cao, Y.L., Lach, E., Kim, T.H., Rodriguez, A., Arevalo, C.A. & Vacanti, C.A. (1998) Tissue-engineered nipple reconstruction. *Plast.Reconstr.Surg.*, 102, 2293-2298.

Davies, J.A. & Fisher, C.E. (2002) Genes and proteins in renal development. *Exp Nephrol*, 10, 102-13.

Davies, J. (2005) *Mechanisms of Morphogenesis*. Academic Press.

Foty, R.A. & Steinberg, M.S. (2005) The differential adhesion hypothesis: a direct evaluation. *Dev Biol*, 278, 255-263.

Frisch, S.M. & Screaton, R.A (2001) Anoikis mechanisms. *Curr.Opin.Cell Biol*, 13, 555-562.

Ganeva, V., Unbekandt, M. & Davies, J.A. (2011) An improved kidney dissociation and re-aggregation culture system results in nephrons arranged organotypically around a single collecting duct system. *Organogenesis*, 7.

Karner, C.M., Das, A., Ma, Z., Self, M., Chen, C., Lum, L., Oliver, G. & Carroll, T.J. (2011) Canonical Wnt9b signaling balances progenitor cell expansion and differentiation during kidney development. *Development*, 138, 1247-1257.

Kim, D. & Dressler, G.R. (2005) Nephrogenic factors promote differentiation of mouse embryonic stem cells into renal epithelia. *J Am Soc Nephrol*, 16, 3527-3534.

Kolatsi-Joannou, M., Li, X.Z., Suda, T., Yuan, H.T. & Woolf, A.S. (2001) Expression and potential role of angiopoietins and Tie-2 in early development of the mouse metanephros. *Dev Dyn.*, 222, 120-126.

Krupinski, P., Chickarmane, V. & Peterson, C. (2011) Simulating the Mammalian blastocyst - molecular and mechanical interactions pattern the embryo. *PLoS.Comput.Biol*, 7, e1001128.

Lindgren, D., Bostrom, A.K., Nilsson, K., Hansson, J., Sjolund, J., Moller, C., Jirstrom, K., Nilsson, E., Landberg, G., Axelson, H. & Johansson, M.E. (2011) Isolation and characterization of progenitor-like cells from human renal proximal tubules. *Am J Pathol.*, 178, 828-837.

Little, M.H., Brennan, J., Georgas, K., Davies, J.A., Davidson, D.R., Baldock, R.A., Beverdam, A., Bertram, J.F., Capel, B., Chiu, H.S., Clements, D., Cullen-McEwen, L., Fleming, J., Gilbert, T., Houghton, D., Kaufman, M.H., Kleymenova, E., Koopman, P.A., Lewis, A.G., McMahon, A.P., Mendelsohn, C.L., Mitchell, E.K., Rumballe, B.A., Sweeney, D.E., Valerius, M.T., Yamada, G., Yang, Y. & Yu, J. (2007) A high-resolution anatomical ontology of the developing murine genitourinary tract. *Gene Expr.Patterns.*

Loughna, S., Yuan, H.T. & Woolf, A.S. (1998) Effects of oxygen on vascular patterning in Tie1/LacZ metanephric kidneys in vitro. *Biochem.Biophys Res Commun.*, 247, 361-366.

Lovett, M., Eng, G., Kluge, J.A., Cannizzaro, C., Vunjak-Novakovic, G. & Kaplan, D.L. (2010) Tubular silk scaffolds for small diameter vascular grafts. *Organogenesis*, 6, 217-224.

Meyer, T.N., Schwesinger, C., Sampogna, R.V., Vaughn, D.A., Stuart, R.O., Steer, D.L., Bush, K.T. & Nigam, S.K. (2006) Rho kinase acts at separate steps in ureteric bud and metanephric mesenchyme morphogenesis during kidney development. *Differentiation*, 74, 638-647.

Rogers, S.A., Lowell, J.A., Hammerman, N.A. & Hammerman, M.R. (1998) Transplantation of developing metanephroi into adult rats. *Kidney Int.*, 54, 27-37.

Rogers, S. & Hammerman, M. (2004) Prolongation of Life in Anephric Rats following de novo Renal Organogenesis. *Organogenesis*, 1, 22-25.

Sainio, K., Suvanto, P., Davies, J., Wartiovaara, J., Wartiovaara, K., Saarma, M., Arumae, U., Meng, X., Lindahl, M., Pachnis, V. & Sariola, H. (1997) Glial-cell-line-derived neurotrophic factor is required for bud initiation from ureteric epithelium. *Development*, 124, 4077-4087.

Sariola, H., Ekblom, P., Lehtonen, E. & Saxen, L. (1983) Differentiation and vascularization of the metanephric kidney grafted on the chorioallantoic membrane. *Dev Biol*, 96, 427-435.

Steinberg, M.S. (1962a) Mechanism of tissue reconstruction by dissociated cells. II. Time-course of events. *Science*, 137, 762-763.

Steinberg, M.S. (1962b) On the mechanism of tissue reconstruction by dissociated cells. I. Population kinetics, differential adhesiveness and the absence of directed migration. *Proc Natl.Acad Sci U.S.A*, 48, 1577-1582.

Sweeney, D., Lindstrom, N. & Davies, J.A. (2008) Developmental plasticity and regenerative capacity in the renal ureteric bud/collecting duct system. *Development*.

Takano, R., Mochizuki, A. & Iwasa, Y. (2003) Possibility of tissue separation caused by cell adhesion. *J Theor.Biol*, 221, 459-474.

Tufro, A. (2000) VEGF spatially directs angiogenesis during metanephric development in vitro. *Dev Biol.*, 227, 558-566.

Unbekandt, M. & Davies, J.A. (2009) Dissociation of embryonic kidneys followed by reaggregation allows the formation of renal tissues. *Kidney Int.*.

Vunjak-Novakovic, G., Altman, G., Horan, R. & Kaplan, D.L. (2004) Tissue engineering of ligaments. *Annu.Rev Biomed.Eng*, 6, 131-156.

Wang, A.Z., Ojakian, G.K. & Nelson, W.J. (1990) Steps in the morphogenesis of a polarized epithelium. I. Uncoupling the roles of cell-cell and cell-substratum contact in establishing plasma membrane polarity in multicellular epithelial (MDCK) cysts. *J Cell Sci*, 95 (Pt 1), 137-151.

Watanabe, K., Ueno, M., Kamiya, D., Nishiyama, A., Matsumura, M., Wataya, T., Takahashi, J.B., Nishikawa, S., Nishikawa, S., Muguruma, K. & Sasai, Y. (2007) A ROCK inhibitor permits survival of dissociated human embryonic stem cells. *Nat Biotechnol.*, 25, 681-686.

13

Scaffolds for Tissue Engineering Via Thermally Induced Phase Separation

Carlos A. Martínez-Pérez, Imelda Olivas-Armendariz, Javier S. Castro-Carmona and Perla E. García-Casillas
Tissue Engineering and Regenerative Medicine Group, Institute of Engineering and Technology, Autonomous University of Juarez City, UACJ
Mexico

1. Introduction

Tissue engineering is a multidisciplinary field that uses engineering materials called scaffolds to support cell seeding and biochemical factors with the aim to regenerate biological function of a tissue or organ, in this process typically there are involve three main components. Biochemical Factors, such as growth factors, proteins that stimulate proliferation and differentiation cell. Cells which can perform the appropriate tissue functions regenerating the lost or damage tissue, and scaffolds that will act as artificial extracellular matrix that provides mechanical support to cells.

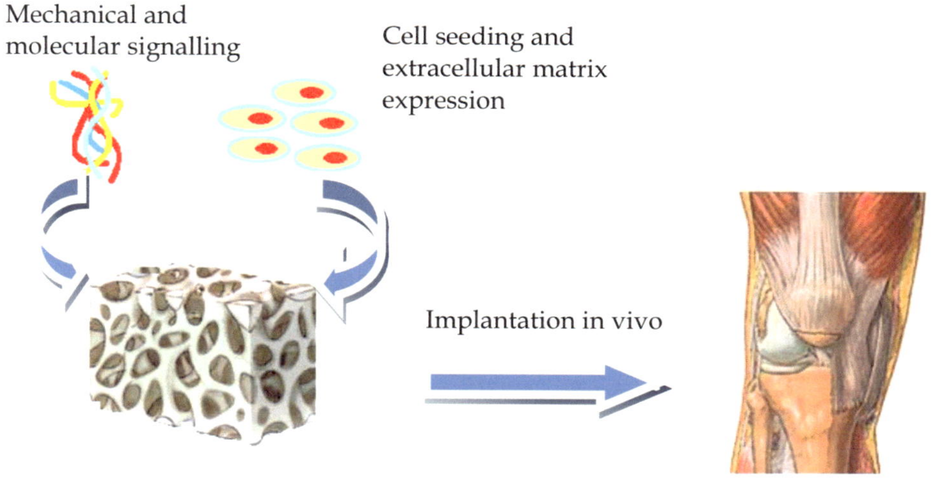

Fig. 1. Schematic representation of a scaffold loaded with cell, growth factor and other molecules for tissue regeneration.

Tissue engineering scaffolds are traditionally compose of porous polymer materials the sustained three-dimensional (3D) growth of cells that is of particular interest in tissue engineering because they can be potentially tailored to mimic the natural extracellular matrix (ECM) in terms of the structure, chemical composition, and mechanical properties. The scaffolds that play a key role in tissue engineering require certain characteristics as high interconnected porosity in order to allow the vascularization process, appropriate distribution and pore size according to the tissue to be regenerated, sufficient initial mechanical strength, biodegradability and biocompatibility. These materials have to provide a suitable environment for the adherence, proliferation and cell differentiation and guide tissue development. Although, there are many approaches to obtain scaffolds to fulfill all requirements, it still represents a big challenge, because it is not clear what defines an ideal scaffold/cell construct, even for a specific tissue type due to its the complexity that include material composition, porous architecture like type, size and interconnection of the pores, structural mechanics, surface properties, degradation properties, products of degradation, biocomopatibility, kinetic of scaffold degradation, etc. This complexity can be appreciated schematically in figure 2 (Hutmacher et al., 2008).

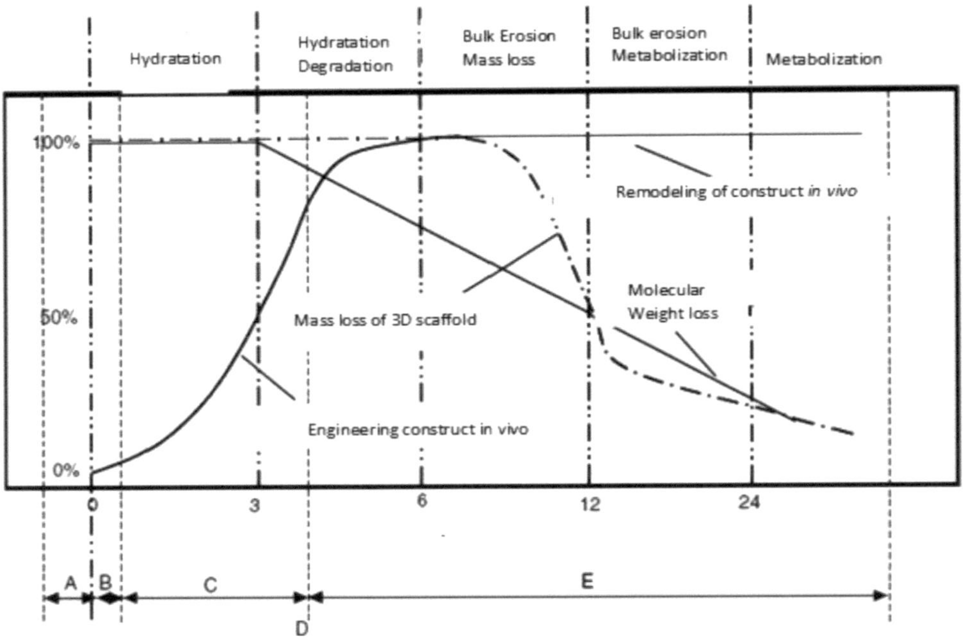

Fig. 2. Illustration of the complex interdependence of molecular weight loss and mass loss of 3D scaffold matrix plotted against time frame for tissue engineering transplant.

Thermally induced phase separation (TIPS) has been explored in order to produce a well interconnected porous structure as tissue engineering scaffold (Olivas et al., 2010; Martel et al., 2010; Ma P. 2008; Patist et al. 2004; Ahmad et al., 2010). This technique is based on changes in thermal energy to induce the de-mixing of a homogeneous polymer solution into a multi-phase system domain by a quench route. When the phase separation occurs, the homogenous solution separates in a polymer-rich phase and solvent-rich phase either by solid-liquid de-mixing or liquid-liquid phase separation mechanism. After the solvent is extracted and depending upon the system and phase separation conditions, different morphologies and characteristics of the materials can be obtained: closed or open-pore

material, spheres, powders, bead-like morphology, etc. Most of the works for tissue engineering seek scaffolds with open pore and well interconnected morphology.
One of the most attractive characteristics of TIPS over other techniques is the formation of not only an intrinsically interconnected polymer network, but also an interconnected porous space in one simple process that is scalable, fast and controllable. TIPS is thus a very convenient methodology for fabricating porous materials as scaffold architectures that can be obtained by means of the manipulation of processing parameters and system properties. A variety of polymers scaffolds as Polylactic Acid (Chen J et al., 2010; He L et al., 2009); Polyurethane (Guan J et al.,2005; Martinez et al., 2006; Fromstein et al. 2002), Polycaprolactone[15] and others have been prepared by TIPS technique, also blends of polymers[Martel et al., 2010; Maquet et al., 2001), and composite of polymers with nanohydroxyapatite (Liu et al 2009), Carbon nanotubes (Olivas et al., 2010; Jell et al., 2008) and other have been successfully fabricated.

a) b) c)

Fig. 3. Pictures of scaffolds prepared via thermally induced phase separation a) Chitosan, b) and c) Chitosan-Ag nanoparticles.

2. Scaffold fabrication

There are many methods to prepare porous scaffolds, such as electrospinning (Demir M et. al., 2002; Lee K et. al. 2003), porogen leaching (Fujimoto K et. al.,1993; Fromstein et. al., 2002;), phase inversion(Kowligi R et. al., 1998), laser excimer (Doy K et. al. 1996) and thermally induced phase separation (TIPS) to mention a few. Each processing technology has its advantages and disadvantages, for example electrospinning is a relatively inexpensive technique to produce fibers with diameters size of nanometer to few microns with enhanced mechanical properties, and also the high aspect ratio would improve the cell proliferation. However, it is difficult to make a large volume scaffold. The porogen leaching method has the advantage of controlling pore sizes by manipulating the size of the salt particulate, also is a relatively inexpensive technique. However, scaffold fabricated by porogen leaching can have limited interconnectivity and residual salt, which would affect cell proliferation. Phase inversion method can result in low interconnectivity and difficulty in controlling the pore size (Kowligi R et. al., 1998). The laser excimer method can make scaffolds with well define pores with a size and shape very homogenous but achieving connectivity remains a challenge (Doy K et. al. 1996). Thermally induced phase separation method offers the ability to control pore size by varying the preparation conditions and also provides means to control pore structure, additionally the scaffold can be molded into a range of shapes and sizes (Jianjun G et. al., 2005; Martinez Perez C et. al. 2006). It is beyond the scope of this chapter to cover all scaffold fabrication techniques available. Hence the aim of this chapter is to provide an overview of the TIPS technique which it has been employed to produce a range of scaffolds for tissue engineering.

2.1 Thermally induced phase separation

Thermally induced phase separation has gained significant attention from scientific and practical point of view that has led TIPS technique to be applied in the fabrication of microporous membranes or microcellular foams from medicine and the chemical industry, scaffolds for tissue engineering, and as a drug carrier for controlled release. Microporous membranes are used in a very wide range of applications. Examples include the use in hemodialysis process, kidney artificial (Ulbrich, 2006), the removal of bacteria and viruses (Qui & Matsuyama, 2010), the treatment of waste water (Yave et al., 2005), oil-water separation (Funk et al, 2008), the use in batteries (Vanegas et al. 2009; Cui et al, 2008), gas separation (Funk & Lloyd 2008), etc. Adjusting TIPS parameters, such as types of polymers, solvent or no solvent ratio, polymer concentration, thermal quenching could be obtained materials with distinctive morphologies according to their applications (Aristéia & Felisberti, 2009; Luo et al.,2008).

TIPS technique is based on changes in thermal energy to induce the de-mixing of a homogeneous polymer solution into a two or multi-phase system domain. When the phase separation occurs, the homogenous solution separates in a polymer-rich phase and polymer-poor phase usually by either exposure of the solution to another immiscible solvent or cooling the solution below a binodal solubility curve where a liquid–liquid phase separation or solid–liquid de-mixing mechanism can be presented. After that, the solvent is extracted by liofilization and depending upon the system and phase separation conditions, different morphologies and characteristics of the materials can be obtained. A typical temperature-composition phase diagram for a binary polymer–solvent system with an upper critical solution temperature is presented in Figure 4. As can see there are two curves, the binodal that represents the thermodynamic equilibrium of liquid–liquid de-mixing and the spinodal curve. When the temperature of a solution is above the binodal curve, the polymer solution is homogeneous. A polymer-rich phase and a solvent-reach phase coexist in a solution in the L–L demixing region (Vandeweerdt and Berghmans, 1991; Williams and Moore, 1987). The maximum point, at which both the binodal and the spinodal curves merge, is the critical point of the system. The area under the spinodal curve is the unstable region, and the area located in the zone between the binodal and spinodal curves is the metastable region. L–L demixing in the metastable region displays a poor connected stringy or beady morphology consequence of a nucleation and growth mechanism for polymer concentration lower or higher than the critical point concentration, respectively (Matsuyama et al., 2000; Sperling, 2003). On the other hand, if the system is quenched into the unstable region, the L-L phase separation takes place by a spinodal mechanism that results in a well-interconnected porous structure. In figure 5 a SEM picture of Polyurethane scaffold prepared in the different region of the phase diagram is shown where it can be appreciated the different types of morphologies than can be acquired.

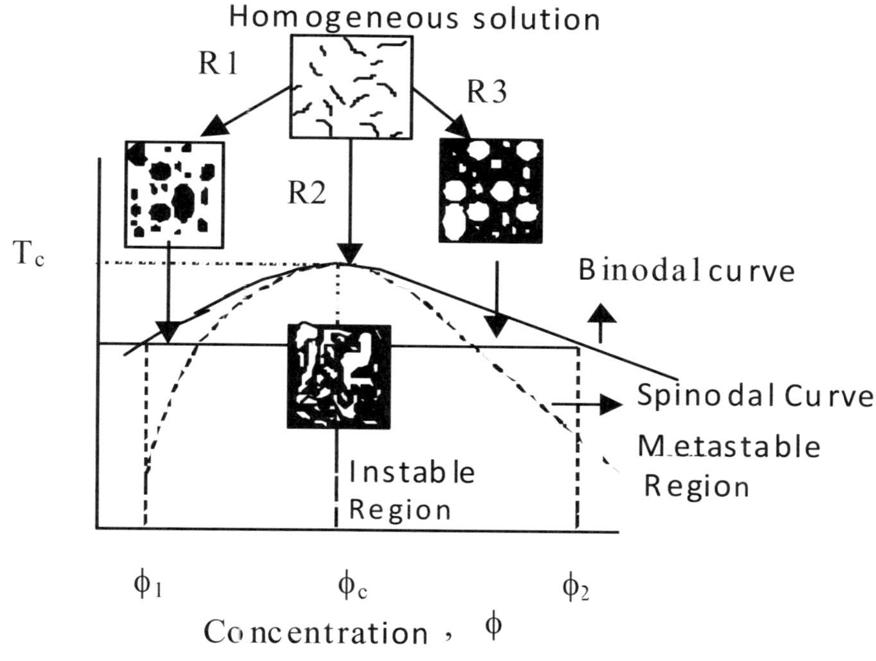

Fig. 4. A schematic representation of a binary phase diagram of a polymer solution.

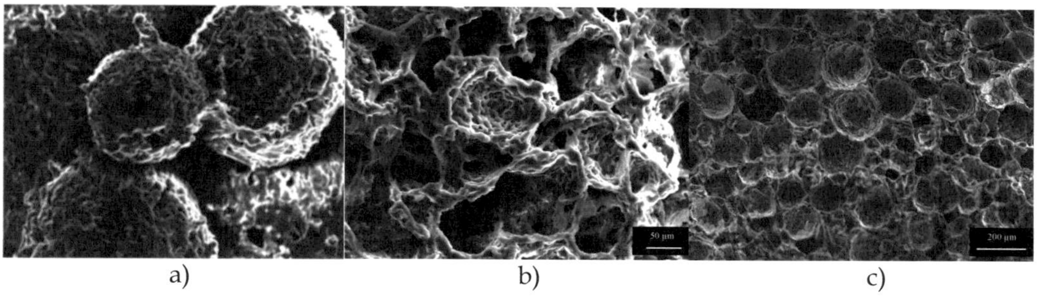

Fig. 5. SEM pictures of PU obtained a) at the metastable zone in the left side of the phase diagram, b) at the inestable zone, c) at the metastable zone in the right side of the phase diagram.

The spinodal curve is the line at which the Gibbs free energy of mixing second derivative is equal to zero, and it divides the two-phase region into unstable and metastable zones. For scaffolds for tissue engineering that require an open and well interconnected porosity it is preferable to take the freezing under the spinodal curve where the distribution, shape and size pore, interconnectivity can be determined by a delicate balance of several parameters such as molecular weight of the polymers, concentration, quenching route, type of solvent and additives as nonsolvent composition.

These parameters have a great influence at the early stage of the phase separation, but in the later stage, the coalescence of phase separated droplets continuously proceeds minimizing the interfacial free energy associated with the interfacial area, which is called the coarsening process. This effect is induced by a differential interfacial tension exerted between the two phase separated domains. It was demonstrated that the coarsening process results in pore size enlargement primarily via Ostwald ripening, coalescense, or a hydrodynamic flow mechanism (Mooney et al., 1996; Lo et al., 1996). Thus, the coarsening process should be carefully considered as a kinetic parameter to control the size and shape of the pore. In order to obtain macroporous, it is desirable to work with the coarsening effect; it will induce the pore enlargement. The coarsening process, however, concomitantly tends to generate more closed pores; thus, it is important to optimize various TIPS parameters to achieve an interconnected and open macroporous open structure. Large pore size and well interconnected structure are critical parameters for cell seeding and neovascularization when implanted in vivo.

Some researchers have taken advantage of the solid-liquid phase separation induced by the solidification of the solvent before the liquid-liquid separation can occur, they have prepared microtubular orientantion-structured scaffold like is shown in figure 6. By using this technique the orientation structure of the scaffold is guided by solvent crystallization under certain temperature gradient. The results of these studies showed that when the concentration of polymer solution increased the thickness of the wall of formed microtubules scaffold increased and the diameter of microtubules decreased (see table 1), that the diameter of the microtubules scaffolds were reduced as the temperature gradient increased, and also the polymer type affect the morphology of the scaffold. It was considered that the dependence of the diameter of microtubules on temperature gradient is in agreement with the crystallization theory. By using a lower temperature-cooling agent, a higher temperature gradient would perform to lead formation of much more crystal nucleus with a higher crystallization speed. As result of this, a great number of smaller sized crystals had been produced. In Accordance with others studies (Cao Y et al., 2006; Maquet et al., 2001; Guan et al., 2005) in which open porous and interconnected architecture was obtained, and, porosities of 80-96% and pore size of 14 to 250 μm that can be vary with the polymer concentration, quenching temperature and polymer type. Another study with a binary system used two different types of TIPS processes: uniform quench and non uniform quench in a model binary polymer solution (Lee et al., 2004). The results showed that the spatial concentration profiles and patterns confirm the formation of isotropic morphology

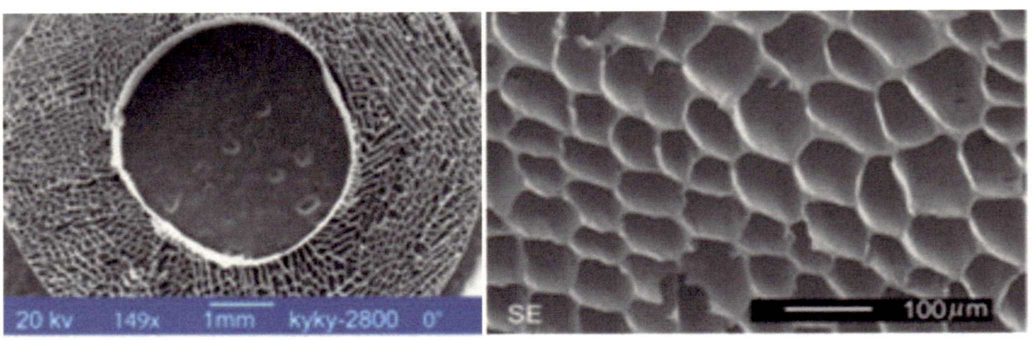

Fig. 6. View of a microtubular orientation-structured scaffold. (Yang et al., 2006; Hu et al., 2008).

from a uniform quench while anisotropic morphology forms from non-uniform quench and indicate the lower-temperature regions of the spatial temperature gradient containing higher droplet density, larger droplets size when the phase separation process was allowed to proceed for long period of time, and the morphological analysis of the shape factor shows that the formation of the droplet shape is independent on the spatial temperature gradient.

Polymer /super cooling-temperature	Concentration of polymer in solvent	Porosity (%)	Average density of scaffold (g/cm³)	Average diameter of microtubules (μm)	References
PLGA/-10°C	5	90.4	0.1316	120	(Hu et al.,2008)
	7	83.8	0.1637	100	
	9	78.4	0.2201	80	
PLLA /-20°C	3 (w/w%)	96	0.051	137	(Yang et al., 2006)
	5	94	0.078	122	
	7	92	0.099	117	
	9	90	0.128	114	
PLGA/-20°C	3	96	0.051	123	(Yang et al., 2006)
	5	93	0.094	118	
	7	92	0.105	113	
	9	90	0.128	109	

Table 1. Morphology of microtubules orientation-structured scaffolds on concentration polymer

3. Polymer scaffolds prepared via TIPS

As it was mention above that biodegradable polymers are suitable for scaffolds. Synthetic and natural biodegradable materials such as Poly (lactide) (PLA), Poly (L-lactic acid) (PLLA), Poly (D,L-lactic acid-co-glycolide acid) (PDLG), Poly (D,L-lactide) (PDLLA), Poly (lactide-co-glycolide) (PLGA), Poly (lactide-co-caprolactone) (PLCL), Poly (ε-caprolactone) (PCL), Polyurethane (PU), Poly (ether ester urethane) (PEEUU), collagen, gelatin, and chitosan have been used in the fabrication of scaffolds. These have been designed in order to be applied in spinal cord regeneration (Maquet et al., 2001), bone and cartilage regeneration (Barroca, et al. 2010; Olivas et al., 2009), blood vessel (Hu et al., 2008), incontinence problems (Ahmadi et al., 2011), peripheral nerves (Aijun et al., 2005), and perennial fistula repair (Heshaw et al., 2010), among others. These had been performance with appropriate structural and functional properties, with the goal to mimic the extracellular matrix (ECM) of the tissue to be repair. The ECM is a natural scaffold for tissue and organ morphogenesis, maintenance, and reconstruction following injury and it is a vital, dynamic and indispensable component of all tissue and organs. Therefore, it is a challenge for tissue engineering, the construction of artificial structures that lead the constructive remodeling of injured or missing tissues or organs, since it regulate cell behavior, such as attachment, migration, proliferation, and differentiation. Even more, it must have a controlled degradation, interconnected porosity with high porosity, appropriate pore size, and mechanical properties the most closely to the tissue to be repaired.

Schugens (1996) prepared biodegradable scaffolds of Polylactic Acid (PLA) by the TIPS method using dioxane as solvent. Instead of the L-L demixing mechanism, a solid-liquid mechanism take place due to the solidification of solvent and it gave a highly anisotropic tubular morphology with a ladder-like microporous structure. A porous structure obtained in this manner is highly anisotropic with relatively small pores (Yang et al., 2006). Further studies have indicated that by introducing water as a non-solvent, L–L de-mixing can be realized in the polyester–dioxane–water ternary system, and that macroporous polyester scaffolds can be obtained. The morphology of the resulting scaffold is strongly dependent on the phase separation behavior of the ternary system.

By increasing the polymer concentration except when there is no presence of non-solvente the temperature of the cloud point and the temperature where the L-L de-mixing take place increases considerably, also higher the non-solvent concentration, higher the L-L de-mixing temperature (Hua et al., 2002, 2003). The L–L demixing temperature of a crystalline PLLA ternary system is higher than that of amorphous PDLLA and PLGA ternary systems. However, the effect of the molecular weight of the polymer seems to be less prominent (Hua et al., 2003). The phase behavior of the system is also be affected by adding various compounds (Hua et al., 2001; Shin et al., 2005). It has been demonstrated that by adding a surface active material such as Pluronic F127, a tri-block polymeric surfactant, the interfacial energy between two phases can be reduced, effectively stabilizing the porous structure of the scaffolds (Nam & Park, 1999). The addition of NaCl can shift the binodal curve to a higher temperature and therefore create a larger operable domain for spinodal decomposition (Hua et al., 2001).

Additionally, researchers have introduced non-solvent as water to induce a liquid-liquid phase separation having a polymer-solvent-water ternary system. A liquid-liquid phase separation occurs when the temperature of polymer solution is decreased and depended of the thermal driving force. The cloud-point temperature was highly dependent on water content, polymer concentration and the freezing point of the system was nearly independent of polymer concentration and water content (Barroca et al., 2010; Chen et al., 2010; Jun et al., 2003). The cloud point temperature increases as the polymer concentration increases and when the molecular weight of the polymer increases because it reduces the polymer-solvent interaction. When a polymer blend is prepared and its polymers have similar solubility parameters the phase separation of the polymers behaves similar as they behave by themselves. When a semi-crystalline polymer is part of the solution, the phase separation is more complicated because the polymer potentially crystallizes involving the gelation temperature as another important parameter in the morphology of the scaffolds because semicrystalline polymers, generally, form a gel as a result of liquid-liquid phase separation.

The interlocking of small crystal agglomerates may play a key role for the formation of the gel when the solvent is removed a highly porous structured is obtained. The effects of the quench route on the phase in a study (PLLA-dioxane-water) by maintaining the demixing temperature in the unstable region for a period of time and then quenching to -196°C, researchers have found that the gelation induced by liquid-liquid demixing is essential to maintain scaffolds with uniform interconnected macroporous. The porosities of the scaffolds decreased with increases of polymer concentration and this occur by the change of the phase separation mechanism and by the extent of the coarsening process involved. The effect of molecular weight and the aging time on the scaffold morphology is important because the morphology with the higher molecular weight polymer was consistently better organized.

The less uniform pore structure (lower molecular weight) was likely to be caused by coarsening of two separated phases since the viscosity effect is one of the factors responsible for a less organized pore structure. On the other hand, with short aging time, a large pore size due to the large thermodynamic driving forces obtained. The microporous scaffolds have a highly interconnected macroporous range from 20 to 300µm and exhibited isotropic morphology (table 2).

Polymer	Polymer concentration w%	Weight ratios of solvent/water	Porosity	Pore size	Morphology	References
PLA	7.5	9010 85/15	80 84	< 50 50	Isotropic lacy structure, interconnected pores	(Chen et al., 2010)
PLGA	9	87/13		> 70	Regular and highly interconnected macroporous structure	(Jun et al.,2003)
PLLA	4.5	87/13		20-50		(Tanaka et al.,2008)
PCL/PLLA	8-15	90/10		20-300	Good connective structure	(Tanaka et al.,2008)
PCL	10	87/13		50-100	Good connective structure	(Tanaka et al.,2008)
PLGA/PLLA	9	87/13		50-200	Well interconnected macroporous structure	(Chul et al., 2005)

Table 2. Morphology of scaffolds, water content and polymer concentration.

Collagen is the main component of the ECM in many tissues and the nanofibrous architecture of collagen is important for cell adhesion, proliferation and differentiation. . In order to obtain various structures of nanofibrous scaffolds prepared via TIPS process (Fig. 7), researches have made several modifications to the process and combined it with others process such as porogen leaching (Liu et al., 2009), inner layer (Aijun et al., 2005), and electrospinning (He et al., 2009). This process involved the dissolution of polymer in solvent, phase separation, and polymer gelation at low temperature. However, immunogenicity and pathogen transmission associated with collagen is a concern. Therefore, several works have been focused in the fabrication of collagen-like nanofibrous scaffolds. In recent work nanofibrous gelatin scaffolds were produced by combining the thermally induced phase separation and porogen leaching by Liu and collaborators (Liu et al., 2009) . Gelatin is derived from collagen by acidic or basic hydrolysis and by its chemical composition is very similar to that of collagen. The gelation depends on the temperature, the solvent, and the polymer concentration in the solution. They obtained fiber diameter ranged from 50nm to

500nm, which is on the same scale as natural collagen fibers, the average fiber diameter did not change with different gelatin concentrations. The scaffolds prepared with 7.5% (w/v) gelatin solution had a low density, 97.51% of porosity and well defined macropores with pore size of 250-420μm. During the process water and ethanol solvent mixtures were used to dissolve gelatin. The addition of ethanol was a critical step in creating the nanofibrous structure because when gelatin was dissolved in water alone, it could only form smooth surface structure after phase separation. The addition of certain amount of ethanol in gelatin solution resulted in the formation of gelatin structure after phase separation. Furthermore, when increased the amount of ethanol in the solvent mixture resulted in poor solubility of gelatin. For this reason, the ethanol/water ratio in gelatin solution was very important during the fabrication of nanofibrous. The in vitro analysis indicated the gelatin fibrous scaffolds enhanced cell adhesion and proliferation. In other study the researchers fabricated fiber porous and tubular chitosan scaffolds for guided peripheral nerves and blood vessel tissue regeneration by combining inner layer and TIPS. The scaffolds had a biphasic wall structure, with fibrous inner layer and semipermeable outer layer. The inner diameter was 2.5 mm and outer diameter was 4.5mm. In vitro characterization shows that the scaffolds have a very good cytocompatibility.

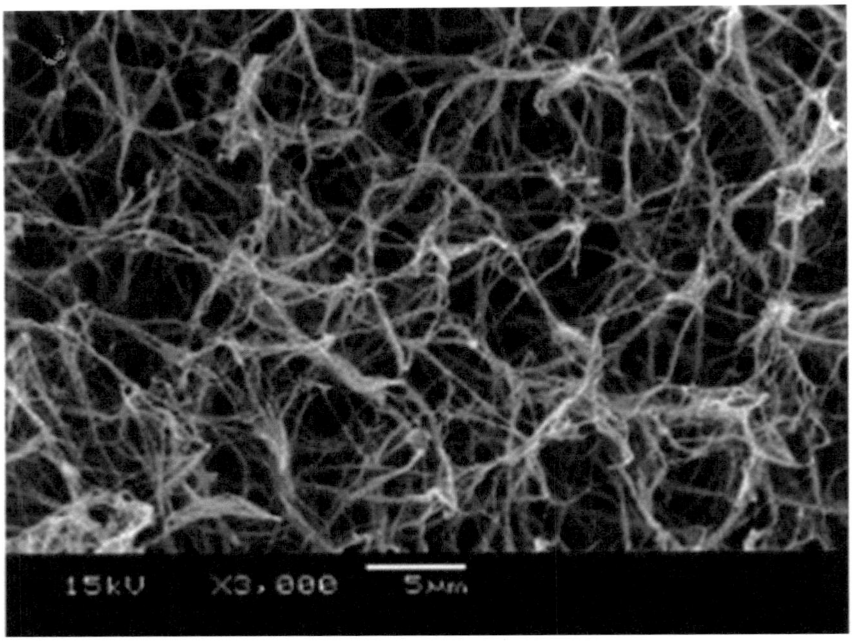

Fig. 7. SEM image of nanofibrous structure fabricated by TIPS (Beachley et al. 2010).

He Liu et. al., prepared PLLA nanofibrous scaffolds with 3D macro and microporous structures by liquid-liquid phase separation from PLLA-dioxane-water system showing that aging in the gel status was essential for the formation of a nanofibrous network. The coalescence of the solvent rich drops also occurred when aging was performed in order to reduce the surface tension. This coarsening effect resulted in pore size enlargement. The extent of coarsening affected the size and morphologies of macro or micro pores in the scaffold. High quenching temperature favors the pore enlargement, due to the formation of larger solvent rich droplets. High water content in the system increase the pore size but also can decrease the pore homogeneity. The gradual addition of water as non solvent increase the gelation point and generate macropores (up to 300 μm) and micropores with a high interconnection. The enlarged pore sizes with the increase of water content is likely due to the combined effects of weaker polymer diluents interaction, larger quenching depth, and lowered viscosity in the co-solvent system. The formation of nanofibrous structure was related to the liquid-liquid phase separation and the crystallization kinetics of polymer affected by the cooling rate.

4. Particulate filled composite scaffolds

Filled composite scaffolds produced by thermally induced phase separation are other alternative materials which exhibit pore anisotropy that it has shown to support migration, adhesion, spreading and viability of cells (Blaker et al., 2005; Barroca et. al. 2010;). Commonly, these scaffolds consist of biodegradable polymer with inorganic particles, such as bioactive glass and hydroxyapatite. The composite has the advantages of the two types of material and for example the bioactivity, control degradation kinetics and the mechanical properties can be improved. Two polymer-diluent zeolite was reported by Funk (Funk CV et. al., 2008) where they showed the that zeolite particle to have a significant effect on the droplet growth and final cell size of that microporous membrane which depend on the particle loading and processing condition. Nanohydroxyapatite-filled PLGA scaffolds were prepared by Huang and collaborators (Huang et al., 2008) by TIPS technique where the influence of Nanohydroxyapatite (NHA) content on the microstructure and properties of the composite was studied. It was found that the pore size of the PLGA/NHA scaffolds decreases with the increase of PLGA concentration and HA content. Also, the mechanical properties and water absorption ability of the composite scaffolds were considerably improved with the content of NHA. PLGA/NHA scaffolds exhibited significantly higher cell growth, alkaline phosphatase activity than PLGA scaffolds, having in their system the best results with 10 wt.% NHA. PLLA was prepared via TIPS with the incorporation of micro and nanohydroxyapatite, NAH had better results that of micrometric size and compare with PLLA scaffolds, NAH/PLLA scaffolds showed highest compressive strength (8.67 MPa) with 85.06% porosity which is comparable to the high end of compressive strength of cancellous bone (2–10 MPa) (Nejati et al., 2008). Also it was found that the incorporation of NHA, the scaffolds have more regular microarchitecture due to its more interfacial area, surface reactivity and ultra-fine structure. This suggests that the developed nHAP/PLLA scaffold via TIPS fulfill most of the requirements as a suitable bone substitute for bone tissue engineering applications.

Carbon Nanotubes (CNT) has also been incorporated in scaffolds for tissue engineering (Jell et al. 2008; Olivas et al. 2010). Jell and collaborators design a route via TIPS to manufacture three-dimensional, highly porous polyurethane containing CNT where the mechanical properties were improved significantly with the addition until 5% of CNT, also the composite scaffolds were not only found to be non-toxic but also induce phenotypic changes that may enhance wound healing and bone formation in vivo. Olivas and coworkers also found that the incorporation of small amount of MWCNT will significantly improve the mechanical properties of the scaffolds.

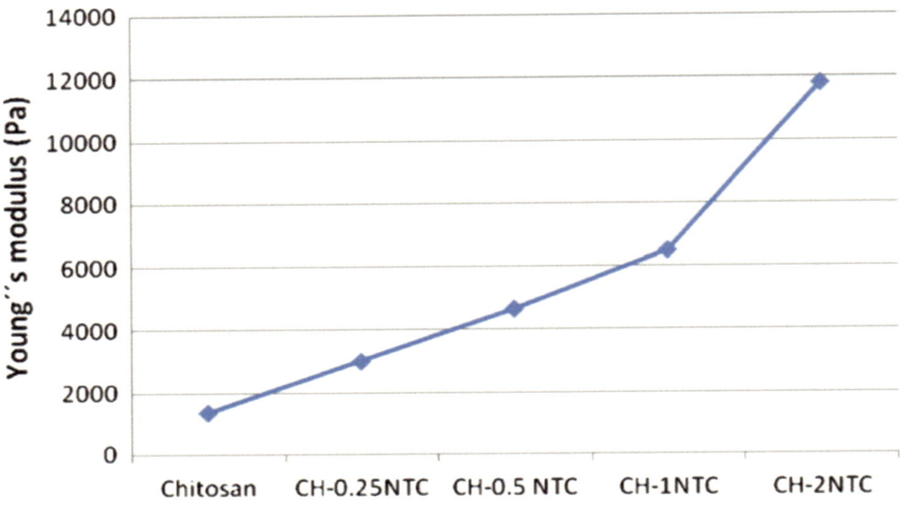

Fig. 8. Variation of the Young's modulus of the Chitosan and Chitosan–MWCNT Composites.

In Fig 9a-d shows FE–SEM pictures of Chitosan-MWCNT scaffolds prepared by TIPS containing 0, 0.5, 1 and 2 wt% of MWCNT, respectively; a highly porous morphology can be appreciated which reach until 92%; the pore size is between 100 and 300 μm. For scaffold utility, the pore size, porosity and interconnectivity become important parameters for analysis where macroporos between 50 and 300 μm are most useful for cell and tissue ingrowth, and smaller pores contribute to solute diffusion. Here, in accordance with Nejati (Nejati et al., 2008) where it was found that the incorporation of NHA improves the micro architecture of the scaffolds, as the MWCNT concentration is increase a better microarchitecture of the chitosan scaffolds can be appreciated. The pores have a better circle form than the scaffolds of pure Chitosan that present and more irregular morphology.

The porosity and size pore of the CH–MWCNT composites are in the appropriate range for application in tissue engineering. In Fig. 9 FE-SEM pictures of the composites at higher magnification are presented. CNTs were embedded at the polymer matrix and appeared uniformly distributed throughout the scaffold. Some MWCNT emerge from the pore surface, forming a rough, hairy texture. The alignment of the CNTs to certain directions into the matrix and in the surface perpendicular to it could be attributed to the phase separation during the freeze-drying processes. Even when the CNTs are emerging from the surface, they are coated by Chitosan showing the compatibility between them.

The Chitosan-MWCNT scaffolds were evaluated in vitro by seeding osteoblasts in its structure, also the composite scaffolds were it was found that these scaffolds promote osteoblasts cell proliferation and, also it was observed by SEM the formation of calcium phosphate crystals on the scaffolds surface after just a few days of culture. Biomineralization of the scaffolds is a good sign that it could be enhance bone formation in vivo.

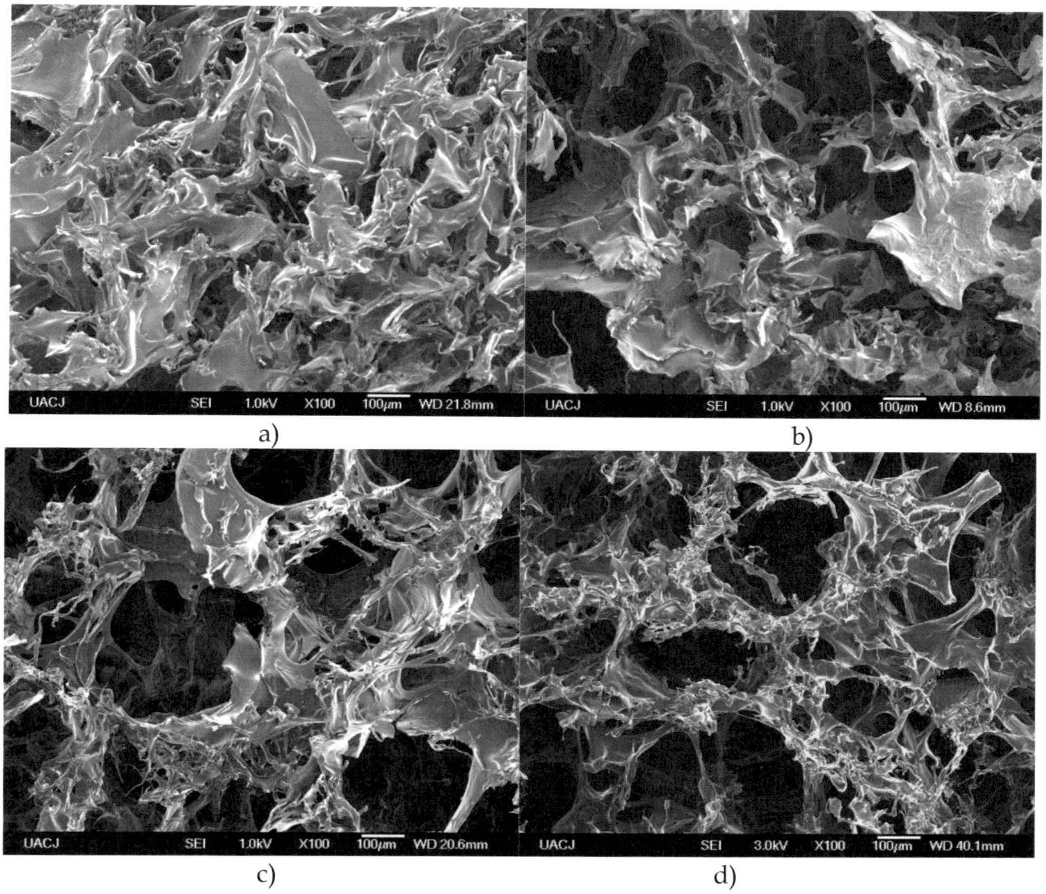

Fig. 9. SEM pictures of Chitosan with MWCNT (a) Chitosan alone b) 0.5 wt% of MWCNT; (c) 1% wt% and (d) 2 wt% of MWCNT.

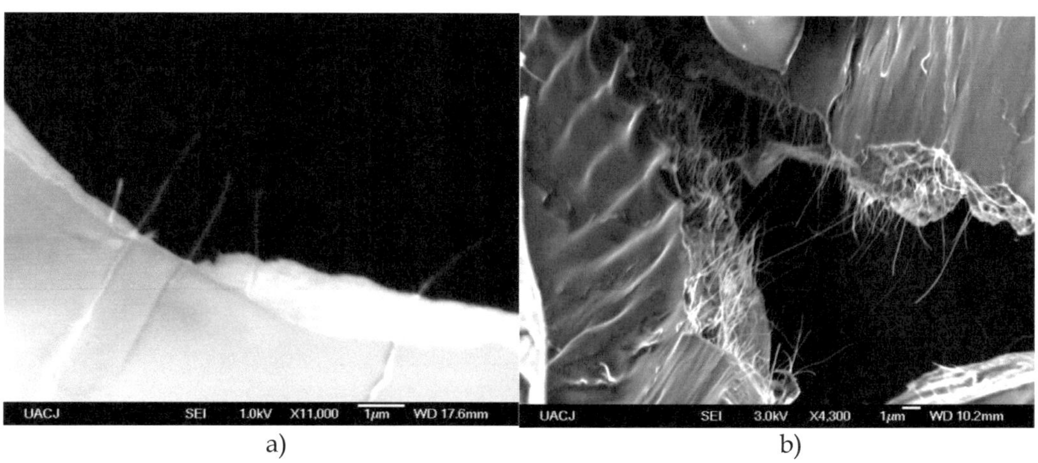

Fig. 10. a Chitosan-0.5% MWCNT composite. b) Chitosa-1% MWCNT composite.

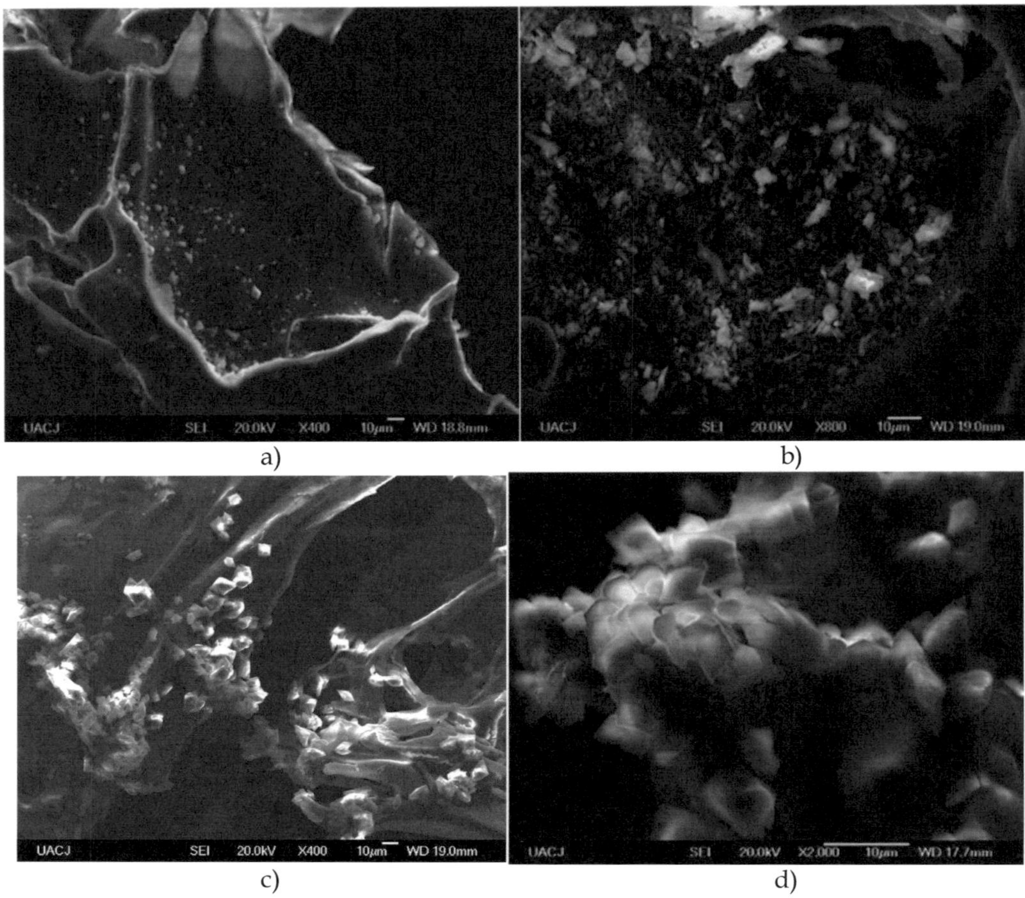

Fig. 11. Scaffolds biomineralization after it has been seeded with osteoblasts. a) Chitosan-1%MWCNT after 1 week of culture. b) Chitosan-1%MWCNT after 2 weeks of culture; c) Chitosan-2%MWCNT after 1 week of culture; d) Chitosan-2%MWCNT after 2 weeks of culture

5. Biomaterials for biomelecules delivery prepared by TIPS

Porous biodegradable microspheres and spheres for tissue engineering and drug delivery applications have been made by TIPS (Blacker et al., 2008; Keshaw et al., 2010). Microporous spheres are likely to be advantageous in the application to inaccessible wounds, such as fistulae. Highly interconnected porous structure with pore size ranged from 5 to 50 µm in diameter will allow efficient packing into the wound cavity whilst their ability to stimulate vascularization will promote tissue regeneration. PLGA microspheres produced by TIPS using dimethyl carbonate as a solvent and rapid quenching in -196°C have a characteristic morphology of foams formed by liquid-liquid phase separation. These Microspheres have a highly anisotropic channel-like morphology and internal ladder structure. The microspheres are monodisperse due to the consistent droplet formation. Reducing polymer concentration, reduce viscosity and decrease the size of droplets formed. Blacker and collaborators successfully loaded these microspheres to delivery proteins and antibodys (Blacker et. al. 2008).

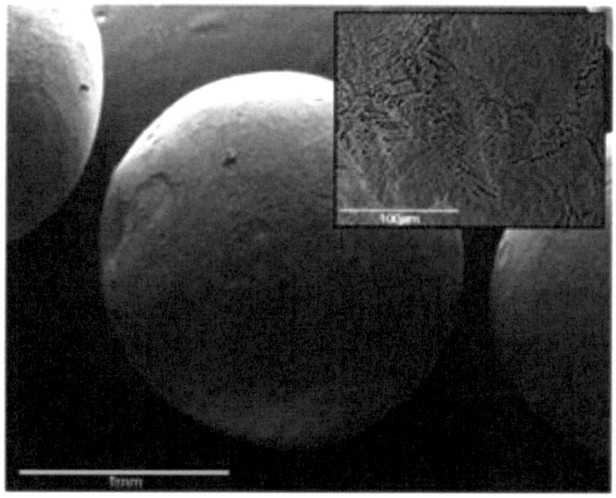

Fig. 12. SEM picture of well-formed PLGA microspheres by TIPS (Blacker J et. al. 2008).

Also, highly porous poly(D,L-lactic-co-glycolic acid) (PLGA) scaffolds were fabricated by a TIPS technique to deliver plasmid DNA in a controlled manner. Plasmid DNA was directly loaded into the inner pore region of the scaffold during the TIPS process. By optimizing the parameters, PLGA scaffolds releasing plasmid DNA over 21 days were successfully fabricated. Plasmid DNA released from the scaffolds fully maintained its structural integrity and showed comparable transfection efficiency to native plasmid DNA demonstrating that this kind of materials can be used for tissue engineering where a combined gene delivery strategy is required (Woo et. al. 2004). On the other hand, Niu and coworkers (Niu et. al., 2009) prepared a composite via TIPS, porous NHA/collagen/poly(L-lactic acid)/chitosan microspheres loaded with BMP-2 derived synthetic peptide with the aim that it can be delivered in controlled way. Dioxane was used as the solvent for PLLA. In order to not affect the morphology of the scaffolds the containing quantities of chitosan microsphere must be remain less than 30%. Also as the other scaffolds that are particulate-filled the compressive modulus of the composite scaffolds increased from 15.4 to 25.5 MPa, while the compressive strength increased from 1.42 to 1.63 MPa as the microspheres contents increased from 0% to 50%. Also, the results indicated that the degradation rate of the scaffolds was increased with the enhancement of CMs dosage and finally, it was shown that the synthetic peptide is release in a temporally controlled manner, depending on the degradation of both incorporated chitosan microspheres and PLLA matrix. In vitro bioactivity assay revealed that the encapsulated synthetic peptide was biologically active as evidenced by stimulation of rabbit marrow mesenchymal stem cells (MSCs) alkaline phosphatase (ALP) activity.

6. Conclusions

A key element in tissue engineering is the scaffold since it enables the fabrication of artificial tissue that mimic that to be restore or regenerate. Scaffolds give support for cells, growth factors, and other biomolecules that eventually will enhance the formation of new tissue in the human body, the scaffold should fulfill several requirements such as nocarcinogenic, nontoxic, nonantigenic, it also must be biocompatible, biodegradable, also is preferred to be bioactive. Besides the biomaterial issues, the structure and morphology of the scaffolds is very important. They should be highly porous, with defined size and shape pore, well interconnected porosity to allow vascularization. Thermally induced phase separation is a technique that can produced scaffolds with the desired characteristics. It can be fabricate in the desired shape and also can be scalable for fabrication of materials with enough quantity and quality. The process can be tuned just adjusting some parameters of the process and give materials with the appropriate morphology and structure for tissue engineering. In the process it can be incorporated another inorganic materials to enhance its properties, even can be loaded with growth factors and others biomolecules that make this technique one of the most adequate for scaffolds fabrication for tissue engineering.

7. References

Ahmad, R.,Mordan N., Forbes A and Day R.(2011) Enhanced attachment, growth and migration of smooth muscle cells on microcarriers produced using thermally induced phase separation. *Acta Biomaterialia,* 7: 1542-1549.

Aijun, A., Qiang A., Wenling C., Chang Z., Yandao G., Nanming Z., and Xiufang Z. (2005) Fiber-Based Chitosan tubular scaffolds for soft tissue engineering: Fabrication and in vitro evaluation. *Tsinghua Science and Technology,* 10: 449-453.

Aristéia, J., and Felisberti M. (2009) Pororus polymer structure obtained via the TIPS process from EVOH/PMMA/DMF solutions *Journal of Membrane Science,* 344: 237-243.

Barroca, N., Daniel-da-Silva S., Vilarinho P., and Fernandes M. (2010) Tailoring the morphology of high molecular weight PLLA scaffolds through bioglass addition. *Acta Biomaterialia,* 6: 3611-3620.

Beachley V., and Wen X. (2010) Polymer nanofibrous structures: Fabrication, biofunctionalization, and cell interactions. *Progress in Polymer Science,* 35: 868-892.

Blake, J., Knowles J., and Day R. (2008) Novel fabrication techniques to produce microspheres by thermally induced phase separation for tissue engineering and drug delivery. *Acta Biomaterialia,* 4: 264-272.

Blaker, J., Maquet V., Jerome R., Boccaccini A., and Nazhat S. (2005) Mechanical properties of highly porous PDLLA/Bioglass composite foams as scaffolds for bone tissue engineering. *Acta Biomaterialia,* 1: 643-652.

Blaker, J., Nazhat S., maquet V., and Boccaccini A. (2011) Long-term in vitro degradation of PDLLA/Bioglass bone scaffolds in acellular simulated body fluid. *Acta Biomaterialia,* 7 : 829-840.

Brandt, H. X. (2008). Bioactivation of knitted cellulose scaffolds by strontium. *Cellulose,* 15, 275-283.

Cao, Y., Mitchell G., Messina A., Price L., Thompson E., Penington A., Morrison W., O´Connor A., Stevens G., and Cooper-White J. (2006) The influence of architecture on degradation into three-dimensional poly(lactic-co-glycolic acid) scaffolds in vitro and in vivo. *Biomaterials,* 27: 2854-2864.

Chen, J., Ling S., and Tsay R. (2010) A morphological study of porous polylactide scaffolds prepared by termally induced phase separation. *Journal of the Taiwan Institute of Chemical Engineers,* 41: 229-238.

Chi Yena , Hongyan Hea, James Leea L. , Winston W.S., Ho (2009) Synthesis and characterization of nanoporous polycaprolactone membranes via thermally- and nonsolvent-induced phase separations for biomedical device application *Journal of Membrane Science,* 343 :180-188.

Chu, K., Sup B., Heung J., Gwan T., Do J., and Sung D. (2005) A facile preparation of highly interconnected macroporous PLGA scaffolds by liquid-liquid phase separation II. *Polymer,* 46: 3801-38C8.

Cui, Z., Xu Y., Zhu L., YuWang J., Xi Z., and Zhu B. (2008) Preparation of PVDF/PEO-PPO-PEO blend microporous memebranes for lithium ion batteries via thermally induced phase separation process. *Journal of Membrane Science,* 325: 957-963.

Demir , MM, Yilgor I, Yilgor E, Erman B. (2002) Electrospinning of polyurethane fibers. *Polymer,* 43: 3303-9.

Hutmacher D, Woodfiel T, Dalton P & Lewis J. (2008). Scaffold design and fabrication, In: *Tissue Engineering* Clemens Van Blitterswijk, pp. 403-454, Academic Press, ISBN:978-0-12-370869-4, Canada.

Doi , K, Nakayama Y, Matsuda T. (1996) Novel compliant and tissuepermeable microporous polyurethane vascular prosthesis fabricated using an excimer laser ablation technique. *Journal of Biomedical Materials Research,* 31:27-33.

Doi , K. (1996). Novel compliant and tissue permeable microporous polyurethane vascular prótesis fabricated using an excimer laser ablation technique. *Journal of Biomedical Materials Research,* 31, 27-33.

Fromstein , JD, Woodhouse KA. (2002);Elastomeric biodegradable polyurethane blends for soft tissue application. *Journal of Biomaterials Scie Polymer Edn,* 13:391-406.

Fujimoto , K, Minato M, Miyamoto S, Kaneko T, Kikuchi H, Sakai K, Okada M, Ikada Y (1993) Porous polyurethane tubes as vascular graft. *Journal of Applied Biomaterials,* 4:347-54.

Funk , C., and Lloyd D. (2008) Zeolite-filled micropororus mixed matrix (ZeoTIPS) memebranes prediction of gas separation performance.*Journal of Membrane Science,* 313: 224-231.

Funk , C., Beaver B., and Lloyd D. (2008) Effect of particulate filler on cell size in membranes formed via liquid-liquid thermally induced phase separation. *Journal of Membrane Science,* 325: 1-5.

Guan, J., Fujimoto K., Sacks M., and Wagner W. (2005) Preparation and characterization of highly pororus biodegradable polyurethane scaffolds for tissue applications. *Biomaterials,* 26: 3961-3971.

He L. , Zhang Y., Zeng X., Quan D., Liao S., Zeng Y., Lu J., and Ramakrishna S. (2009) Fabrication and characterization of poly (L-lactic acid) 3D nanofibrous scaffolds with controlled architecture by liquid-liquid phase separation from a ternary polymer-solvent system. *Polymer,* 50: 4128-4138.

Heijkants, R. G. J. C. , van Calck R. V. , de Groot J. H., Pennings A. J. , Schouten A. J. , . van Tienen T. G , Ramrattan N. , Buma P. and R. P. H. Veth (2004) Design, synthesis and properties of a degradable polyurethane scaffold for meniscus regeneration. *Journal of Materials Science: Materials in Medicine,* 15:423-427.

Hu X. , Shen H., Yang F., Bei J., and Wang S. (2008) Preparation and cell affinity of microtubular orientation-structured PLGA (70/30) blood vessel scaffold. *Biomaterials,* 29: 3128-3136.

Hua F. J., G. E. Kim, J. D. Lee, Y. K. Son, and D. S. Lee (2002) Macroporous Poly(L-lactide) Scaffold 1 Preparation of a Macroporous Scaffold by Liquid-Liquid Phase Separation of PLLA-Dioxane-Water System. Journal of Biomedical Materials Research Part B, 63: 161-168.

Hua F. J., T. G. Park, and D. S. Lee, (2003) A Facile Preparation of Highly Interconnected Macroporous Poly(DL-Lactic Acid-Co-Glycolic Acid) (PLGA) Scaffolds by Liquid-Liquid Phase Separation of a PLGA-Dioxane-Water Ternary System. *Polymer*, 44, 1911-1920.

Huang K, REN J, CHEN,REN T and ZHOU X (2008) Preparation and Properties of Poly(lactide-co-glycolide) (PLGA)/Nano-Hydroxyapatite (NHA) Scaffolds by Thermally Induced Phase Separation and Rabbit MSCs Culture on Scaffolds. *Journal of Biomaterials Applications*, 22:409-432.

Jell, G., Verdejo R, Safinia L., Shaffer S. P., Stevens M. M., Bismarck A. (2008) *Journal of Materials Chemistry*, 18: 1865-1872

Jianjun G, Kazuro L. Fujimoto, Michael S. Sacks and William R. Wagner. (2005) Preparation and characteriza-tion of highly porous, biodegradable polyurethane scaffolds for soft tissue applications*Biomaterials*, 26:3961-3971.

Keshaw , H., Thapar N., Burns A., Mordan N., Knowles J., and Forbes A. (2010) Microporous collagen spheres produced via thermally induced phase separation for tissue regeneration. *Acta Biomaterialia*, 6: 1158-1166.

Kowligi , RR, von Maltzahn WW, Eberhart RC. (1998) Fabrication and characterization of small-diameter vascular prostheses. *Journal of Biomedical Materials Research*, 22:245-56.

La Carrubba, V., Carfì Pavia F, Brucato V. and Piccarolo S. (2010) *International Journal of Materials Forming*, 1:567-570.

Lee K ., Chan P., and Feng X. (2004) Mophology development and characterization of the phase-separated structure resulting from the thermal-induced phase separation phenomenon in polymer solutions under a temperature gradient. *Chemical Engineering Science*, 59: 1491-1504.

Lee KH , Kim HY, RyuYJ, Kim KW, Choi SW. (2003) Mechanical behavior of electrospun fiber mats of poly(vinyl chloride)/ polyurethane polyblends. Journal of Polymer Science Polym Phys, 41:1256-62.

Liu X. , and Phase Ma P. (2009) Separation, pore structure, and properties of nanofibrous gelatin scaffolds. *Biomaterials*, 30:4094-4103.

Liu X., Smith L., Hu J., and Ma P. (2009) Biomimetic nanofibrous gelatin/apatite composite scaffolds for bone tissue engineering. *Biomaterials*, 30:2252-2258.

Lo H, Kadiyala S, Guggino E, Leong K (1996) Poly(L-lactic acid)foams with cell seeding and controlled-release capacity *Journal of Biomedical Materials Research*, 30:475-484.

Luo, B., Li Z., and Wang X. (2008) Formation of anisotropic micoporous isotactic polypropylene (iPP) memebrane via thermally induced phase separation. *Desalination*, 233:19-31.

Ma PX, Zhang RY (2001) Microtubular architecture of biodegradable polymer scaffolds. *Journal of Biomedical Research*, 56:469-477.

Ma *P* X (2008) Biomimetic materials for tissue engineeringAdvanced Drug Delivery Reviews, 60:184-198.

Martel -Estrada A, Martínez-Pérez C.A, Chacón J. G, García-Casillas P, Olivas I (2010) Synthesis and thermo-physical properties of chitosan/poly(dl-lactide-co-glycolide) composites prepared by thermally induced phase separation. *Carbohydrate Polymers*, 81:775-783.

Martel-Estrada A., Martínez-Pérez C., Chacón-Nava J., García-Casillas P., and Olivas-Armendáriz I. (2011) In vitro bioactivity of chitosan/poly(D,L-lactide-co-glycolide) composites. *Materials Letters*, 65: 137-141.

Martínez , Pérez C. A., Garcia-Casillas P. E., Martínez-Villafañe A., Romero-García (2006) Porous Biodegradable polyurethane scaffolds prepared by thermally induced phase separation. *Journal of Advanced Materials* (2006) Special ed. 1, 5-11.

Mooney D, Baldwin D, Such N, Vacanti J, Lauger R. (1996) Novel approach to fabricate porous sponges of poly(D,L-lactic-co-glycolic acid) without the use of organic solvents. *Biomaterials*, 17:1417-1422.

Nam, Y. S. and T. G. Park,(1999)Biodegradable Polymeric Microcellular Foams by Modified Thermally Induced Phase Separation Method. *Biomaterials*, 20: 1783-1790.

Nejati E, Mirzadeh H, Zandi M (2008) Synthesis and characterization of nano-hydroxyapatite rods/poly(L-lactide acid) composite scaffolds for bone tissue engineering. *Composites: Part A* , 39: 1589–1596.

Niu X, Feng Q, Wang M, Guo X and Zheng Q. (2009)Porous nano-HA/collagen/PLLA scaffold containing chitosan microspheres for controlled delivery of synthetic peptide derived from BMP-2 *Journal of Controlled Release*, 134: 111-117.

Olivas, Armendariz I, García-Casillas P. E, Martínez-Villafañe A., and C.A. Martínez-Pérez C. A. (2009)Synthesis and Characterization of porous polyurethane-chitosan blends *Cellular Polymers*, 28:159-172.

Olivas , Armendáriz I, García-Casillas P, Martinez-Sánchez R, Villafañe A.M and Martínez-Pérez C.A, (2010) Chitosan/MWCNT composites prepared by thermal induced phase separation. *Journal of Alloys and Compounds*, 495: 592-595..

Qiu Y., and Matsuyama H. (2010) Preparation and characterization of poly (vinyl butyral) hollow fiber membrane via thermally induced phase separation with diluents polyethylene glycol. *Desalination*, 257: 117-123.

Rowlands, A. S., Lim S. A., Martin D., Cooper J (2007) Polyurethane-poly(lactic-co-glycolic) acid composite scaffolds fabricated by thermally induced phase separation *Biomaterials*, 28: 2109-2121.

Shin, K. C., B. S. Kim, J. H. Kim, T. G. Park, J. D. Nam, and D. S. Lee, (2005)A Facile Preparation of highly interconnected macroporous PLGA scaffolds by liquid-liquid phase separation II, *Polymer*, 46, 3801-3808.

Sperling, L. H. (2003) Concentrated solutions and phase separation behavior", Introduction to Physical Polymer Science, John-Wiley and Sons, Singapore.

Tanaka T, Tsuchiya T, Takahashi H, Taniguchi M, Ohara M and Lloyd D (2009) *Journal of Chemical Engineering of Japan*, 39:144-153.

Tanaka , T., Eguchi S., Saitoh H., Taniguchi M., and Lloyd D. (2008) Microporous foams of polymer blends of poly(L-lactic acid) and poly(caprolactone). *Desalination*, 234: 175-183.

Ulbricht, M. (2006)Advanced fuctional polymer membranes. *Polymer*. 47: 2217-2262.

Vandeweerdt, P. and H. Berghmans, (1991) Temperature-Concentration Behavior of Solutions of Polydisperse, Atactic Poly(Methyl Methacrylate) and its Influence on the Formation of Amorphous, Microporous Membranes. *Macromolecules*, 24, 3547 (1991).

Vanegas, M., Quijada R., and Serafini D. (2009) Microporous memebranes prepared via thermally induced phase separation from metallocenic syndiotactic polypropylenes. *Polymer*, 50: 2081-1086.

Williams, J. M. and Moore J, (1987)Microcellular Foams: Phase Behavior of Poly(4-Methyl-1-Pentene) in Diisopropylbenzene. *Polymer,* 28:1950-1958.

Woo K, Chul K, Kim S, Jeong J and Park T (2004) *Journal of Biomaterials Science Polymer Edn*, 15:1341–1353.

Wu L., (2006) Room-temperature¨injection molding/particulate leaching approach for fabrication of biodegradable three-dimensional porous scaffolds. *Biomaterials*. 27: 185-191.

Xiaohua Liu, Peter X. Ma (2010) The nanofibrous architecture of poly(L-lactic acid) -based functional copolymers. *Biomaterials,* 31:259-269.

Xiaohua Liua, Youngjun Wona, Peter X. M (2006) Porogen-induced surface modification of nano-fibrous poly(L-lactic acid) scaffolds for tissue engineering *Biomaterials*, 27:3980-3987.

Xixue , Hu, Hong Shen, Fei Yang, Jianzhong Bei, Shenguo Wang (2008) Preparation and cell affinity of microtubular orientation-structured PLGA(70/30) blood vessel scaffold. *Biomaterials* 29: 3128-3136.

Xue , Y. (2009).Temperature and loading rate effects on tensile properties of kenaf bast fiber bundles and composite. *Composites. Part B* , 40: 189-196.

Yang, F., Qu X., Cui W., Bei J., Yu F., Lu S., and Wang S.(2006) Manufacturing and morphology structure of polylactide type microtubules orientation-structured scaffolds. *Biomaterials*, 27: 4923-4933.

Yave , W., Quijada R., Ulbricht M., and Benavente R. (2005) Syndiotactic polypropylene as potential material for the preparation of porous membranes via thermally induced phase separation (TIPS) process. *Polymer*, 46: 11582-11590.

14

Nano-Doped Matrices for Tissue Regeneration

Leonardo Ricotti, Gianni Ciofani, Virgilio Mattoli and Arianna Menciassi
The Biorobotics Institute, Scuola Superiore Sant'Anna, Pontedera (Pisa)
Center for Micro-Bio-Robotics @ SSSA, Istituto Italiano di Tecnologia, Pontedera (Pisa)
Italy

1. Introduction

Tissue engineering can be defined as an interdisciplinary field applying the principles of engineering and life sciences towards the development of biological substitutes that restore, maintain, or improve tissue function (Langer & Vacanti, 1993). To accomplish this aim, a balanced combination of cell culture growth with supporting biomaterials is required, as well as the introduction of bioactive agents able to enhance and direct cell and tissue aggregation. Regenerative medicine has been sometimes looked as an extension of tissue engineering, but it can be considered nowadays one of the major interdisciplinary scientific challenges, aiming at regenerating the soft and hard tissues, organs, and nerves damaged or responsible for main human disabilities. The definition given by Kaiser marks the difference of such field from others, and its specific link with pathologies care: "*a new branch of medicine will develop that attempts to change the course of chronic disease and in many instances will regenerate tired and failing organ systems*" (Kaiser, 1992). The central focus of regenerative medicine is represented by human cells. These may be somatic, adult or embryo-derived stem cells and, recently, induced-pluripotent stem cells.

In both tissue engineering and regenerative medicine, the role of scaffolds is predominant, constituting the framework for cell attachment, proliferation, and differentiation. Biodegradable natural and synthetic polymers, as well as some non-biodegradable polymers have been extensively studied and used for the 2D and 3D reconstruction, as well as for the healing of different tissue typologies (Jagur-Grodzinski, 2006). Several requirements have been identified as crucial for the production of tissue engineering scaffolds: (1) the scaffold should possess interconnecting pores of appropriate scale to favour tissue integration and vascularisation; (2) it should be made from material with controlled biodegradability or bioresorbability, allowing the new forming tissue to replace the scaffold; (3) it should have appropriate surface chemistry to favour cellularattachment, proliferation and differentiation; (4) it should own adequate mechanical properties to match the intended site of implantation and handling; (5) it should not induce any adverse response; (6) it should be easy to fabricate into a variety of sizes and shapes (Hutmacher, 2001). Furthermore, it is known that the principal objective of a scaffold is to recapitulate extracellular matrix (ECM) function in a temporally coordinated and spatially organized structure, and a key issue is to encode required biological signals within the scaffold, in order to control the main cellular processes (Causa *et al.*, 2007).

In order to mimick the nanometric organization of biological structures, many attempts have been made in the latest years in order to obtain synthetic or natural scaffolds having nanometric-sized cues, able to better direct cell behaviour. Nanotechnologic processes paved the way for such challenge, allowing for the preparation of matrices with pores, grooves, fibers and other structures in a submicrometric range of sizes. This allowed a broad range of new insights, like the investigation of the effect of submicron topography on cell adhesion, migration and differentiation, or the evaluation of the differences, in terms of protein adsorption and morphology, on materials showing a nanometric roughness in comparison with more flat or micrometric-roughness-characterized substrates. Furthermore, the introduction of nanocomposites made by bulk materials (*e.g.*, polymers) with nanoparticles embedded in their structure, allowed the fabrication of new structured nanoscale materials having promising properties for tissue engineering applications. In general, the ability to control the assembly of nanoparticles into discrete organized clusters inside a polymeric matrix is of broad interest in the field of nanotechnology, as magnetic, electronic and optical behaviours of complex, three-dimensional nanocomposites are highly dependent on the size of the nanoparticles, as well as on the distance between them (Sanyal *et al.*, 2004). Such nanoparticle-mediated properties become biologically interesting when integrated into a scaffold for tissue engineering or regenerative purposes; the incorporation of nano-sized objects into degradable or not-degradable polymeric networks, in fact, may provide a more favourable synthetic microenvironment to more closely mimic natural tissue physiology, with the addition of certain physical stimuli.

In the present chapter, some techniques and preparation methods to obtain nano-doped matrices are described, together with the chemical and physical properties of the fabricated nanoscale material. The encouraging results obtained by many research groups using polymeric scaffolds doped with various nanoparticle typologies are then reported, highlighting the specific advantages that the inclusion of such nanoparticles brings to this technology. The main aspects regarding protein adsorption, cell adhesion, proliferation and differentiation on the described scaffolds are discussed, also trying to envision future applications and challenges related to these materials. The conclusion aims at strengthening what emerges from recent studies, namely that nano-doped scaffolds can be considered a new and promising instrument for cell and tissue growth and regeneration, with the possibility to provide tuned and controlled physical stimuli to the biological culture, in order to enhance or direct its behaviour.

2. The importance of "nano" in scaffold design

Nanomaterials are materials with basic structural units, grains, particles, fibers or other constituent components smaller than 100 nm in at least one dimension (Siegel & Fougere, 1995), and they include nanoparticles, nanoclusters, nanocrystals, nanotubes, nanofibers, nanowires, nanorods, nanofilms, *etc.* The intrigue of nanotechnology relies on the ability to control material properties by assembling such materials at the nanoscale, allowing for the construction of new devices with peculiar properties. In fact, after decreasing material size into the nanoscale, there is a dramatic increase in surface area and surface roughness to volume ratios, thus leading towards superior physiochemical (*i.e.*, mechanical, electrical, optical, catalytic, magnetic, *etc.*) properties.

Several investigators recently came to the conclusion that in order to achieve a breakthrough in the fields of tissue engineering and regenerative medicine, it is necessary to mimic the natural biological processes (Tu & Tirrell, 2004; Stupp, 2005). From this perspective, the importance of nanomaterials becomes clear, since natural tissues or organs are nanometer in dimensions and cells directly interact with nanostructured ECM matrix. Bone, for example, is a nanocomposite consisting of a protein based soft hydrogel template (constituted by collagen or other non-collagenous proteins (laminin, fibronectin, vitronectin, *etc.*) and water) and hard inorganic components like hydroxyapatite (HA). The bone matrix is composed for the 70% of nanocrystalline HA, which is typically 20-80 nm long and 2-5 nm thick (Kaplan *et al.*, 1994). Other protein components in the bone ECM have also nanometric size, contributing to create a nanoscale environment surrounding and affecting mesenchymal stem cell, osteoblast, osteoclast, and fibroblast adhesion, proliferation and differentiation. Cartilage is a flexible connective tissue composed of a small cellular component (the chondrocytes) and a dense nanostructured ECM rich in collagen fibers, proteoglycans and elastin fibers. Cartilage is able to lubricate joints and withstand static and dynamic loads remarkably well; such properties can be well understood treating cartilage ECM as a composite medium, with a proteoglycans phase exerting swelling pressure and collagen phase resisting it (Basser & Horkay, 2005).

The use of nanomaterials for bone and cartilage regeneration allowed the achievement of exciting results, well reviewed in (Zhang & Webster, 2009). This success is due to the dimensional similarity of materials to bone/cartilage tissue, but also to their special surface properties (topography, surface wettability, and surface energy) due to their greater surface area and roughness compared to conventional or micron structured materials. An important mechanism underlying the increased cellular response on nanostructured materials relies in protein adsorption. As known, cell behaviour is strongly influenced by serum protein adsorption on the biomaterials surface; a nanostructuration allows an increased adsorption of specific proteins (fibronectin, vitronectin, laminin, *etc.*) before cells adhere on implants, regulating cell behaviour and dictating tissue regeneration. Furthermore, the surface topography determines not only the amount of protein adsorbed, but even its functional interconnectivity (Fig. 1).

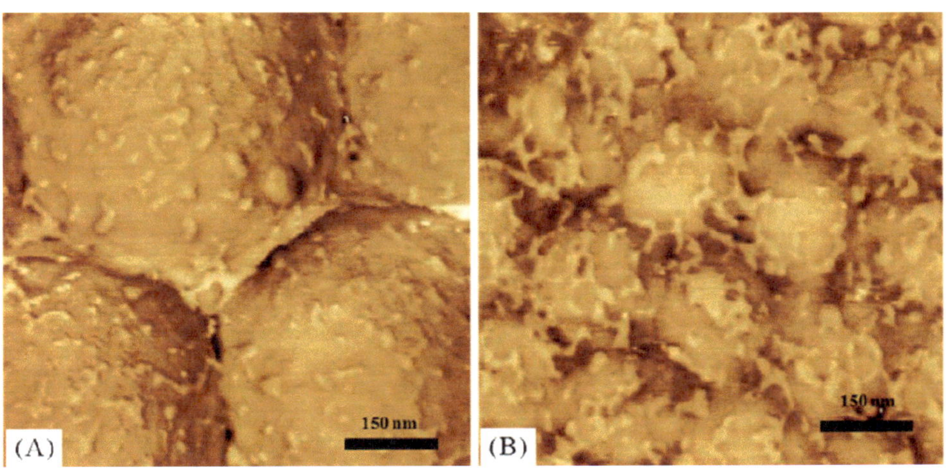

Fig. 1. Atomic force microscope of PLGA nanospheres with different dimensions coated with fibronectin (5 μg/ml). (A) Phase image of fibronectin adsorbed on PLGA with 500 nm surface features, revealing no interconnectivity between proteins. (B) Phase image of fibronectin adsorbed on PLGA with 200 nm surface features, revealing significant interconnectivity between proteins. Images from (Miller *et al.*, 2007). Reproduced with permission from John Wiley and Sons.

Collagen in bone and cartilage is a triple helix self-assembled into nanofibers 300 nm in length and 1.5 nm in diameter; many recent efforts have been therefore dedicated to exploring the influence of novel biomimetic nanofibrous or nanotubular scaffolds on regenerative medicine, by following a bottom-up self-assembly process. Carbon nanotubes/nanofibers (CNTs/CNFs), for example, due to their superior mechanical and electrical properties, are ideal scaffold candidates for bone tissue engineering applications (Tran *et al.*, 2009).

Vascular tissue is a layered structure characterized by several nanostructured feature, due to the presence of collagen and elastin in the vascular ECM. Even for this kind of tissue, nanomaterials have a strong influence on regeneration performances, like demonstrated by (Choudhary *et al.*, 2007), reporting that vascular cell adhesion and proliferation were greatly improved on nanostructured Ti compared to conventional Ti.

Nanomaterials have also demonstrated to be useful for damaged nerves healing. Central nervous system (CNS) and peripheral nervous system (PNS) show different repair procedures after injury: for the PNS, the damaged axons usually regenerate and recover by means of proliferating Schwann cells; for the CNS, it is much more difficult to re-extend and re-innervate axons, due to the absence of Schwann cells. Moreover, due to the influence of astrocytes, meningeal cells and oligodendrocytes, a thick glial scar tissue forms around neural biomaterials, preventing proximal axon growth and inhibiting neuron regeneration. To be effective in PNS and CNS regeneration, biomaterials should therefore have excellent cytocompatibility, mechanical and electrical properties. Nanotechnology allowed in the latest years the development of novel and improved neural tissue engineering materials, with the design of nanofiber/nanotube scaffolds (Fig. 2) with exceptional cytocompatibility and conductivity properties to boost neuron activities (Mattson *et al.*, 2000; Gheith *et al.*, 2005), as well as with the introduction of piezoelectric nanoparticles-mediated neuron stimulation, with the aim of enhancing neurite outgrowth in treated cells (Ciofani *et al.*, 2010).

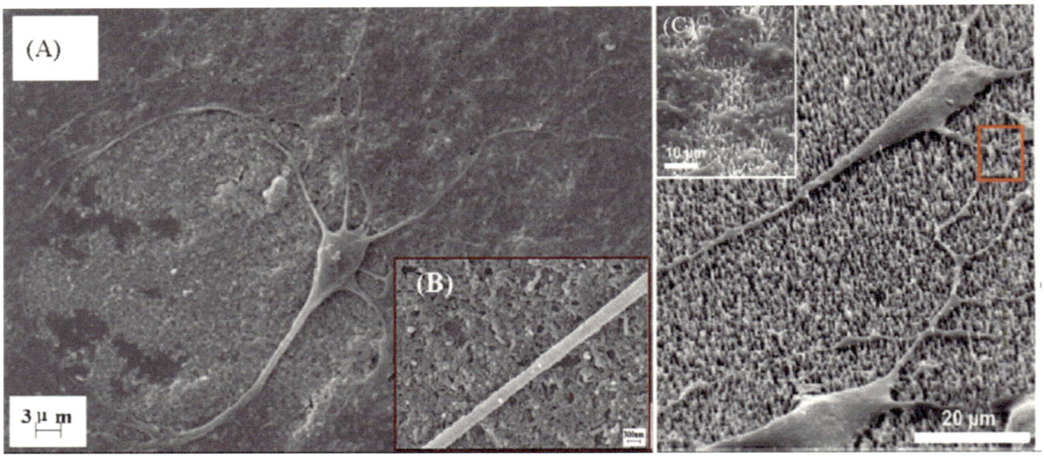

Fig. 2. a Neonatal hippocampal neurons adhering on purified multi-walled carbon nanotubes-coated glass substrates, showing extended neuritis after 8 days of culture; (B) single neurite in close contact to carbon nanotubes; (C) PC12 neural cells grown on vertically aligned carbon nanotubes coated with polypyrrole. Images from (Zhang & Webster, 2009). Reproduced with permission from Elsevier.

The regeneration of other tissues and organs, like bladder, muscle, skin, kidney, liver, pancreas and immune system has been addressed using nanotechnology approaches (Nukavarapu, *et al.*, 2008; Khademhosseini *et al.*, 2008). Nanomaterials more efficiently improve such tissues regeneration, for the same reasons mentioned above (biologically inspired roughness, increased surface energy, selective protein adsorption, *etc.*).

3. Preparation methods and properties of doped matrices

Composite materials made of nanoparticles in a polymeric matrix have been under investigation for at least one decade, not only for biological applications (only recently proposed), but also to develop smart devices, like nonlinear optical materials. A study of Pavel and MacKay (Pavel & Mackay, 2000) described the production of randomly dispersed (and non-aggregated) cadmium sulfide nanoparticles in a transparent polymer matrix of poly(methyl metacrylate) (PMMA). The process was based on a reverse micellar system, followed by polymerization to produce a solid inorganic/organic composite, and it was the first report in the literature of such technique to obtain a "one-system" synthesis of nanoparticles dispersed in a polymer matrix.

In general, the field of polymer matrix-based nanocomposites started to emerge with the recognition that exfoliated clays could yield significant mechanical properties advantages as a modification of polymeric systems. Clays are naturally occurring minerals, mainly aluminosilicates, having a sheet-like (layered) structure, and consisting of silica SiO_4 tetrahedra bonded to alumina AlO_6 octahedra in a variety of ways. Nanocomposites can be obtained by following different methods, such as in situ polymerization, solution, and latex methods. Melt processing has been also recognized an appreciable method, as it is considered more economical, more flexible for formulation, and it involves compounding and fabrication facilities commonly used in commercial practice (Paul & Robeson, 2008). During the fabrication process, a key aspect in the polymer-clay (or organoclay) interaction is the affinity that the polymer segments have for the silicate surface. Specific surfactants can be used, in order to induce a greater exfoliation; in all cases, the best exfoliation is achieved when the structure of the surfactant and the process parameters are optimized (Fig. 3(A)). As already mentioned, the most evident effect of clay-based reinforcement is an increase of the mechanical properties of the material, that has been demonstrated to be superior if compared with microcomposites or bulk materials (Fig. 3(B)).

Electrophoretic deposition (EDP) is a cost-effecting and efficient processing method to produce ceramic nanocomposites and laminates. It can be seen as two combined processes: first the migration of charged particles dispersed in a liquid medium (electrophoresis) under an applied electric field and secondly the coagulation process of the particles at the electrode. Good dispersion and stability of the suspension are essential factors to obtain a homogeneous deposition. Cho and co-workers described the co-deposition of carbon nanotubes (CNTs) and TiO_2 nanoparticles on a stainless steel substrate by EDP (Cho *et al.*, 2008). During this process, both CNTs and titania particles move towards the deposition electrode, once an electrical field is applied, creating a region close to the electrode surface, specifically called "infiltration trajectory". Here, the charge of the deposited CNT films influences the motion of the charged particles; under the effect of the repulsive forces due to the surrounding CNTs, the particles will follow the path with the fewest possible obstacles until reaching the next interstice between adjacent CNTs. The result is a homogeneous distribution of CNTs and TiO_2 throughout the whole coating thickness, with a high amount of porosity. The authors did not perform quantitative evaluation of physical properties, but they envisioned the interesting mechanical, electrical conductivity, and photocatalytic properties of these coatings, useful for biomedical applications.

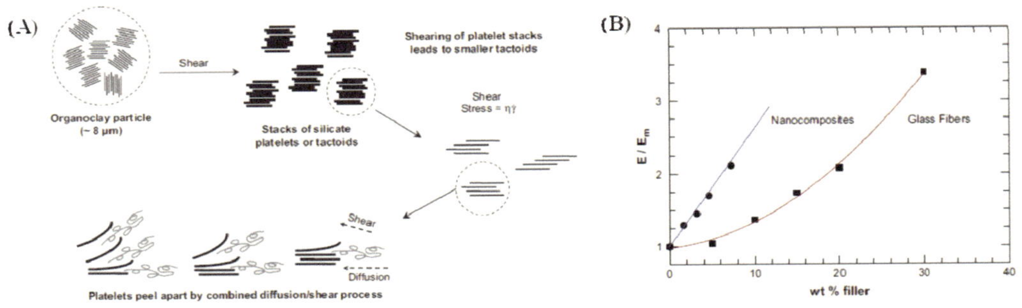

Fig. 3. a Mechanism of organoclay dispersion and exfoliation during melt processing; (B) comparison of modulus reinforcement (relative to matrix polymer) between nanocomposites based on nanoclays and glass fibers (aspect ratio ~20) for a nylon 6 matrix. Images from (Paul & Robeson, 2008). Reproduced with permission from Elsevier.

In order to replicate natural structures on the nano-scale level, electrospinning is a common and inexpensive method that allows the easy fabrication of random or ordered nanofibrous constructs. A variety of materials can be processed by electrospinning, and a certain level of control can be achieved over the desired nano-structures. In the development of bone grafts, nanofibrous scaffolds have been used as fillers, and to bridge critical size defects; the fabrication of nanofiber-calcium phosphate composites has been a step forward. Since minerals in the native bone are predominantly found on the surface, some researchers have deposited hydroxyapatite nanoparticles on the surface of electrospun fibers. The most commonly used surface mineralization techniques include soaking the scaffold in simulated body fluid (SBF) (Wei & Ma, 2006) and the alternating dipping method (Ngiam *et al.*, 2009). Of these two techniques, the alternating dipping method has the advantage of significantly faster mineral deposition, requiring hours to complete rather than days for the SBF soaking technique.

The mineralization of 3D block nanofibrous scaffolds is different. In fact, 2D membranes have most part of their surface exposed to the solution, while 3D block scaffolds require the solution to reach the inner core, requiring a modification of the alternating dipping technique. Teo and co-workers described the fabrication of 3D electrospun poly(L-lactic acid) scaffolds and their mineralization by means of two approaches: the static mineralization (soaking the scaffold for 1h in a mineral solution) and the flow mineralization (placing the scaffold in a perfusion chamber, and using a circulating peristaltic pump to periodically provide the scaffold with the mineral solution) (Teo *et al.*, 2011). The analysis of mechanical properties revealed that the presence of minerals significantly increased both compressive strength and compressive modulus of the scaffolds. The compressive strength of the static mineralized scaffolds was 2.4 times greater than that of the unmineralized scaffolds, while the modulus was 1.9 times greater. Such analysis also revealed that the mineralization technique has a significant influence on the mechanical properties of the scaffolds, being the compressive strength and modulus of the flow mineralized scaffolds 6.1 times and 2.8 times greater, respectively, than those of the static mineralized scaffolds. Interestingly, these properties are not related to the overall mineral content (found to be 43.3% in flow mineralized scaffolds, and 62.3% in static mineralized scaffolds), but to the distribution of minerals inside the 3D structure of the scaffold.

The use of nanoparticles (mainly nano-hydroxyapatite), nanofibers and nanotubes in polymer or bioceramic matrices in order to produce nanometric features on the surface of 3D scaffolds has been recently well reviewed (Meng *et al.*, 2010). Table 1 shows advantages and disadvantages of the main used techniques to produce nanocomposite scaffolds.

Nanostructured surfaces	Method of fabrication	Advantages	Disadvantages
Nanoparticles	Solvent-casting/salt-leaching technique	Controlled porosity, controlled pore size and simple operation	Limited interpore connectivity, using organic solvents
	Modified rapid prototyping technique	Ability to produce complex products quickly	Difficulty to design and fabricate scaffolds with fine microstructures, low porosity
	Freeze-drying and freezing -lyophilization	Highly porous structures, high pore interconnectivity	Limited to small pore sizes
	Self-dispersing technique	Good dispersion, immobilization of nanoparticles on the surface of 3D scaffolds by chemical bonding	Relatively long processing time, use of organic solvents
	LbL method	Multilayers incorporating proteins, drugs, and growth factors in mild conditions, strong adhesion between each layer, easy to control many processing variables during the preparation	Dissolving problems of nanoparticles in aqueous media, requiring organic solvents
	Paste extruding deposition Process	Complete pore interconnectivity, macroshape control, 3D interconnected pore structure	Limited to small pore sizes, use of organic solvents
	Micro stereolithography MSTL	Controlling the shape of composite structures with higher resolution, well developed 3D interconnected pore structure	Shrinkage problems

Nanostructured surfaces	Method of fabrication	Advantages	Disadvantages
	Layer manufacturing	Well-interconnected pores, improved surface roughness	Use of organic solvents, problems with residual solvent
	Nano-emulsion and selective laser sintering techniques	Nanosized particles are well encapsulated in microspheres, complete pore interconnectivity, macroshape control	High processing temperatures, limited to small pore sizes
	TIPS and subsequent solvent Sublimation	High porosity with pore anisotropy and high pore interconnectivity	Use of organic solvents and long time to sublime solvent
Nanofibers	Particle leaching and Electrospinning	Homogeneous nanofiber morphology can be generated spontaneously in a 3-D macroporous and nanofibrous structure	Relatively long processing time, problems with residual solvent
	Fiber bonding and Electrospinning	Combination of nano and microfibers in the same 3D scaffold architecture, mimicking the physical structure of ECM	Limit range of polymers, use of organic solvents
	Electrospinning and LbL Methods	Controlled fiber layer thickness, fiber diameter, and fiber orientation, ability to create complex hierarchical architectures	Use of high-voltage apparatus
	DPMD and electrospinning	Ability to fabricate highly functionalized 3D scaffolds with an open porous network, a controllable shape and a biocompatible nanofibrous inner architecture	Relatively high processing temperature for the workable range
	Electrospinning with 3D Collecting	Controlled patterned architectures and 3D configurations	Difficulties to control many parameters
	Phase separation and particle leaching techniques	Controlled macropore shape and size by particles, interpore opening size by assembly conditions, pore wall morphologies by phase separation parameters	Limited to a few polymers, longer processing time, unable to produce long and continuous fibers with control over fiber orientation

Nanostructured surfaces	Method of fabrication	Advantages	Disadvantages
	Self assembly	Close resemblance to biological processes	Complex process, limited to a few polymers, inability to control
CNT assemblies	EPD	Simple process, homogeneous nano and microstructures possible, controlled thickness of CNT layers in different substrates	Needs to use organic solvents in some cases
	Chemical modification	Controlled surface chemistry, good dispersion, direct covalent bonding with the matrix	Use of organic solvents
	Solvent casting with the inclusion of CNTs	Achieves mechanical integrity, electrical conductivity	Limited by the complexity in the design, construction of the mold

Table 1. Methods of fabrication of 3D nanocomposite scaffolds and corresponding advantages and disadvantages of each strategy. Table from (Meng *et al.*, 2010). Reproduced with permission from John Wiley and Sons.

The hybrid technique developed by Park and co-workers, based on a combination of direct polymer melt deposition (DPMD) and electrospinning (Park *et al.*, 2008) is particularly interesting, even if not including nanoparticles but nanofibers in a polymer matrix. The aim of the DPMD process was to determine the overall architecture of the 3D scaffold, by producing polycaprolactone (PCL) microfibers able to provide mechanical support and overall shape of the device; to this aim, a stainless steel syringe with a micronozzle, a syringe heating device, a compressed air dispenser and a three-dimensionally moving micropositioning system were used. The electrospinning process allowed to obtain PCL and PCL/collagen nanometric fibers applying voltage to a 22-gauge needle on a syringe pump at an infusion speed of 10 μm/min and using a grounded collector. By combining DPMS and electrospinning a hybrid process was proposed, consisting of two-step sequential process (Fig. 4(a)). As a woodpile-like structure was fabricated in a layer-by-layer approach, the nanofiber matrices were inserted between each layer of the microfibrous structure. Subsequent microfiber layers combined with nanofiber matrices were repeatedly laminated onto the previously combined layers, in order to fabricate a 3D complex structure. (Fig. 4(B) and 4(C)).

The beauty of this and other similar approaches is the ability to efficiently simulate the 3D fibrous extracellular environment, allowing the envisioning of challenging applications in tissue engineering and regeneration.

Another hybrid approach was proposed by Erisken and co-workers, that described a methodology integrating a twin screw extrusion process with electrospinning, and allowing the dispersion of nanoparticles into polymeric binders and the generation of nanoparticle-incorporated fibers and nanofibers. In particular, they dispersed β-tricalcium phosphate (β-

TCP) nanoparticles into PCL matrices, in order to generate biodegradable non-woven meshes for tissue engineering applications (Erisken *et al.*, 2008-a). The fiber size was in the range 200 - 2000 nm, and mechanical characterization of the scaffolds revealed that the ultimate tensile strength at break of the meshes increased from 0.47 to 0.79 upon the incorporation of the β-TCP nanoparticles.

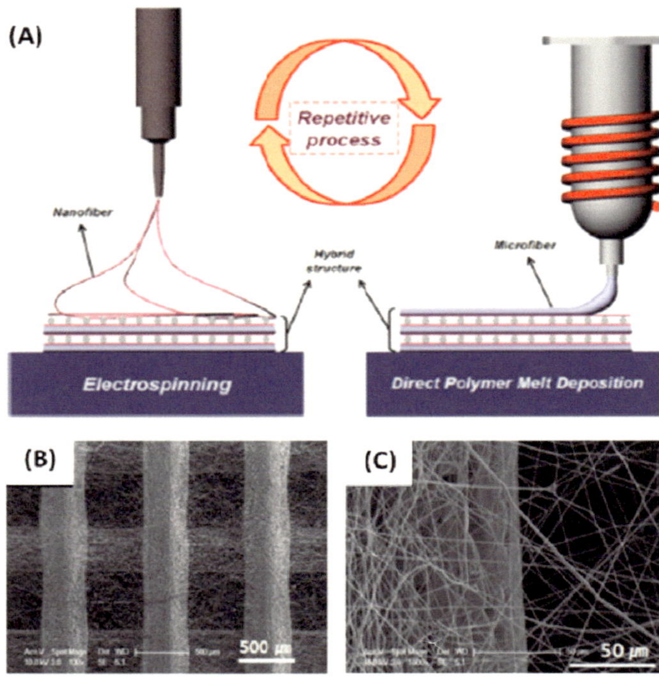

Fig. 4. (A) The DPMD/electrospinning hybrid process. A 3D structure can be built by alternating such processes; (B) and (C) photographs of hybrid layers composed by microfibers and electrospun nanofibers. Images from (Park *et al.*, 2008). Reproduced with permission from Elsevier.

More recently, Jack and co-workers reported the fabrication and characterization of nano-sized hydroxyapatite (HA)/poly(hydroxybutyrate-co-hydroxyvalerate) (PHBV) polymer composite scaffolds, developed by means of a modified thermally induced phase-separation technique (Jack *et al.*, 2009). The presence of the HA nanoparticles significantly affected the thermodynamic state of the polymer-solvent solutions, resulting in a completely different porous architecture upon solidification. EDS characterization confirmed that the clusters observed in the SEM images were indeed HA particles (Fig. 5(A), (B), and (C)).
In order to investigate the mineralization ability, the scaffolds were immersed in SBF for an overall period of 2 weeks, evaluating the mechanical performances before and after this treatment. The results showed that the immersion of the pure PHBV scaffolds in SBF for up to 2 weeks does not significantly change their mechanical performances, confirming that the polymer matrix does not significantly degrade in such an environment over the observed period. In addition, the incorporation of the HA nanoparticles leads to a marked increase in both the stiffness and strength of the scaffolds (Fig. 5(D) and (E)).

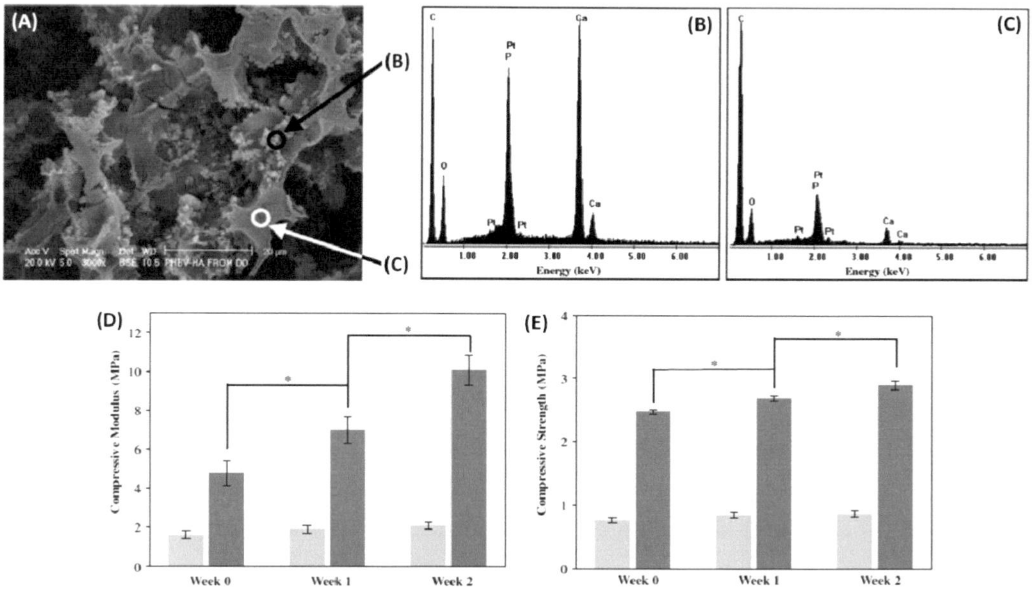

Fig. 5. (A) SEM image showing the PHBV scaffold doped with HA nanoparticles. The circles indicate the approximate regions in the scaffold that the EDS spectra (B) and (C) were collected from; compressive moduli (D) and compressive strength (E) of the PHBV (light grey) and HA/PHBV (dark grey) scaffolds following 0, 1 and 2 weeks of immersion in SBF. $*=p<0.05$. Images from (Jack *et al.*, 2009). Reproduced with permission from Elsevier.

As described, biocompatibility and structural stability are two important aspects for engineering functional tissues. An interesting grafted collagen scaffold with a collagen triple-helix shape was developed and reinforced with Al_2O_3-ZrO_2 nanoparticles, in order to strengthen the mechanical properties of the matrix (Cao *et al.*, 2006). Such modification also enhanced the thermal stability of the collagen matrix, bringing the authors to envision biomedical and bionic applications of this hybrid material. Regarding the influence of nanoparticles inclusion on the degradability of a matrix, Liu and co-workers showed that titania nanoparticles alter PLGA degradation, when dispersed in the polymeric structure (Liu *et al.*, 2006). To produce the scaffolds, PLGA pellets were dissolved in chloroform and nanophase titania powder was added to give a 30/70 ceramic/polymer weight ratio. The composite mixture was then sonicated using different output power settings (and thus obtaining different samples). After sonication, the suspension was cast into a Teflon petri dish, evaporate in air at room temperature for 24 h and dried in air vacuum chamber for 48 h. After incubation in PBS for several days, the weight loss for the doped samples was significantly lower than that of pure PLGA samples. Moreover, results showed that the nanophase titania dispersed in PLGA can improve the structural stability and buffer the harmful pH variations.

4. Current applications in tissue engineering and regenerative medicine

The development of innovative materials useful as scaffolds for regenerative medicine and tissue engineering purposes tries to address, as we described, the mimicking of natural ECM in terms of structure, chemical composition, and mechanical properties. In the latest decade, an intense research on polymeric matrices doped with several kinds of nanoparticles has been carried on, trying to focus on the possible *in vitro* and *in vivo* biomedical applications of these materials. For example, an interesting electrically conductive biodegradable composite material made of polypyrrole (PP) nanoparticles and poly(D,L-lactide) (PDLLA) was synthesized in 2004 and tested *in vitro* with human skin fibroblasts (Shi *et al.*, 2004). It is well known that electrical stimulation is capable of modifying cellular activities, like cell migration, cell adhesion, DNA synthesis, and protein secretion. An electrically conductive scaffold is therefore interesting for the regulation of these activities, but it also must satisfy other basic features, as biocompatibility and degradability. Such aim was achieved in this study, including electrically conductive PP nanoparticles in a biocompatible and degradable PDLLA matrix, easy to process and to test with cell cultures. Human skin fibroblasts were cultured on the doped scaffolds, applying electrical currents in the range 0-800 μA. Cell growth was affected by such stimulation, showing a maximum peak of growth in correspondence to a 10 μA current stimulation.

The development of doped matrices can be useful not only for tissue regeneration, but also for the enhancement of materials antimicrobial acitivity. It is the case of the cellulose acetate nanofibers containing silver nanoparticles produced by Son and co-workers. They demonstrated for the first time that polymer nanofibers containing Ag nanoparticles on their surface could be produced by UV irradiation of the elctrospun fibers with small amounts of silver nitrate ($AgNO_3$) (Son *et al.*, 2006). The number and size of Ag nanoparticle on the matrix surface was found to continuously increase with the irradiation time, and that the particles with an average size of 21 nm exhibited a strong antimicrobial activity.

Specific applications for bone tissue engineering were described by some works in 2006 and 2007, reporting the results obtained by fabricating doped matrices and testing their biological activity by immersing them in standard culture media and simulated body fluid (SBF), or evaluating their *in vitro* response with osteoblasts-like cells or with human mesenchymal stem cells (hMSCs) (Li *et al.*, 2006; Gerhardt *et al.*, 2007; Sui *et al.*, 2007; Torres *et al.*, 2007; Zhang *et al.*, 2007). The first example of Li *et al.* is constituted by electrospun silk fibroin scaffolds containing bone morphogenetic protein 2 (BMP-2) and nanoparticles of hydroxyapatite. TEM images confirmed that nHAP particles were successfully embedded in the silk fiber scaffold, being well oriented along the fiber axis in some regions, but more aggregated in other regions. Human bone marrow-derived MSCs were cultured for up to 31 days on nHAP- and BMP-2-containing silk scaffolds, on only BMP-2-containing silk scaffolds, and on bare silk scaffolds as control, using an osteogenic medium. The scaffolds containing BMP-2 supported higher calcium deposition and enhanced transcript levels of bone-specific markers than on the controls; furthermore, the coexistence of BMP-2 and nHPA in the scaffold resulted in the highest calcium deposition and up-regulation of the transcript levels of the bone-related markers. Gerhardt and co-workers developed PDLLA scaffolds filled with 0, 5 and 30 wt% TiO_2 nanoparticles by means of solvent casting. They showed that the titania nanoparticles were not directly exposed on the composite surface, but rather embedded in the PDLLA matrix, without forming agglomerates even at the highest concentration. The bioactivity of such scaffolds was evaluated by immersing them in supersaturated simulated body fluid for up to 3 weeks and evaluating the formation of hydroxyapatite on the material surface. HA nanocrystals with an average diameter of 40 nm were formed on the 30 wt% TiO_2 composite films after 2 weeks, while only some traces of HA crystals appeared on pure PDLLA and low titania content films after 3 weeks. The effect of TiO_2 nanoparticles on the metabolic activity of MG-63 osteoblast-like cells was also evaluated, finding that particle concentration up to 100 μg/ml had no significant effect on MG-63 cell viability. Sui *et al.* developed PLLA/HA hybrid membranes *via* electrospinning of a PLLA/HA dispersion. The authors found that the inclusion of HA nanoparticles implied a considerably increment in the roughness and in the mechanical properties of the matrix, as well as a decrease in its degradation rate in water. MG-63 osteoblast-like cells were cultured on PLLA and PLLA/HA membranes, finding a higher cell adhesion and growth in the latter substrates. A three-dimensional matrix was introduced by Zhang and co-workers, that described the fabrication of PLGA hollow fiber membranes using a wet phase-inversion approach and doped with HA nanoparticles. The aim was to obtain an aligned, bioactive and biodegradable scaffold mimicking the natural histological structure of human long bone, and they obtained an anisotropic membrane with a rough outer skin and a smooth inner skin (Fig. 6(A)). The structure showed a highly porous morphology with a unique asymmetric finger-like macroviod structure ideal to promote neovascularisation (Fig. 6(B)). pH values of the culture media at different time-points were evaluated for pure PLGA membranes and PLGA/HA membranes, finding that they were closer to neutral for the composite matrices; nanoHA powders in the composite structures may facilitate to neutralize the culture media. The bioactivity of such 3D scaffolds was also evaluated, by immersing them in SBF for several days, finding a significantly higher biomineralization rate for the nanodoped matrix in comparison with the pure PLGA scaffold. Finally, Torres *et al.* tested the mechanical properties and bioactivity of porous PLGA/TiO_2 nanoparticle-filled foams produced by thermally induced solid-liquid phase separation (TIPS) and subsequent solvent sublimation. By manipulating the parameters of the TIPS process, they were able to control the porous architecture, determining the presence of orientated tubular macropores (Fig. 6(C), (D), and (E)). The bioactivity of the scaffolds was also demonstrated by immersion in SBF for up to 28 days, and confirming the formation of hydroxyapatite crystals on the scaffold surface.

The aim of doped biomaterials is not always to promote an enhancement of cell adhesion and proliferation; in some cases, a specific application requires the inhibition of the activity of a certain cell population. In the central nervous system (CNS), astroglial cells proliferate and produce an ECM protein-rich glial scar to isolate any implant (for example a neural electrode) and, in general, regeneration after injury in the CNS is inhibited primarily due to astrocytic glial scar formation. It was therefore proposed a novel composite material made of polyurethane (PU) and doped with zinc oxide (ZnO) nanoparticles as nerve guidance channel material able to significantly reduce astroglial cell adhesion and proliferation (Seil & Webster, 2008). Importantly, ZnO nanoparticles possess piezoelectric properties, that make them promising for neural regeneration applications. Regarding astrocyte behaviour on such materials, the collected data showed a reduced ability of these cells to adhere and proliferate on ZnO nanoparticle PU composites with higher nanoparticle concentrations. The authors finally proposed that the piezoelectric properties of such composites may provide the stimulatory cues necessary to promote neural cell function.

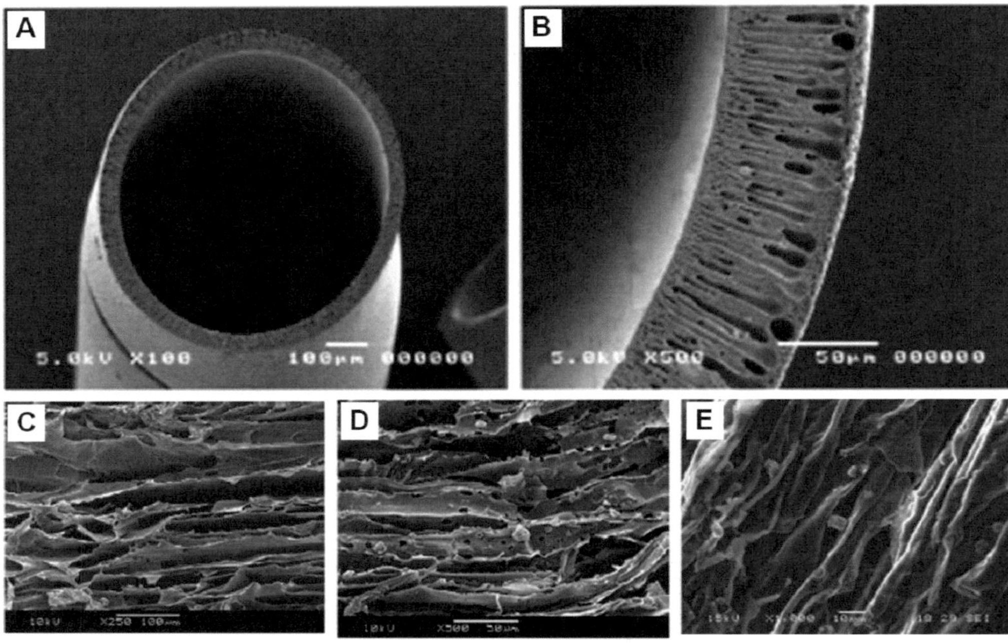

Fig. 6. (A), (B): PLGA hollow fiber membrane doped with 30% nanoHA; (C), (D), and (E): foam scaffolds made of plain PLGA(C), PLGA + 5 wt% TiO_2 (D), and PLGA + 20 wt% TiO_2 (E). Some TiO_2 agglomerates are visible in the 20 wt% foam. Images from (Zhang *et al.*, 2007) and (Torres *et al.*, 2007). Reproduced with permission from Elsevier

In the same year, several studies regarding bone tissue engineering were published (Erisken *et al.*, 2008-b; Guan *et al.*, 2008; Nejati *et al.*, 2008; Schneider *et al.*, 2008; Duan *et al.*, 2008; Wei *et al.*, 2008). An interesting PHB-based scaffold containing nanosized hydroxyapatite was developed by Guan and co-workers by gas-jet/electrospinning. Bone marrow stroma cells (BMSCs) were tested on the material, finding that both proliferation rate and alkaline phosphatase (ALP) activity increased were higher on the nHAP/PHB scaffolds surface than on standard tissue culture plates and pure PHB scaffolds surface. Rod-shaped nano-hydroxyapatite and nHAP/PLA composite scaffolds were developed by Nejati *et al.* The rod shaped nHAP particles (37-65 nm in width and 100-400 nm in length) were similar to natural bone apatite in terms of chemical composition and structural morphology and they were used to build PLA nanocomposites using thermally induced phase separation method (Fig. 7(H), and (I)). Such scaffolds were found to be comparable with cancellous bone in terms of microstructure and mechanical strength, making them suitable for bone tissue engineering applications. Rat mesenchymal stem cells (MSCs) were tested on the material, finding that nHAP/PLA scaffolds appeared biocompatible and non-cytotoxic to the cells. Furthermore, after seven days of culture, round shape cells were found on the surface of pure PLA scaffolds (Fig. 7(L)), while cells on the nanocomposite scaffolds exhibited spindle shaped morphology and migrated through the pores (Fig. 7(M)). A flexible, cotton wool-like PLGA/amorphous tricalcium phosphate (ATCP) nanocomposite was prepared by Schneider and co-workers by electrospinning. It was shown that ATCP nanoparticles clearly influenced the fiber morphology (Fig. 7(A), (B), (C), and (D)), also enhancing the biomineralization of the scaffold after immersion in SBF (Fig. 7(E), and (F)). Human mesenchymal stem cells (hMSCs) were cultured on the composite matrix (Fig. 7(G)), finding a good proliferation rate and an enhanced production of ALP and osteocalcin in comparison with the controls.

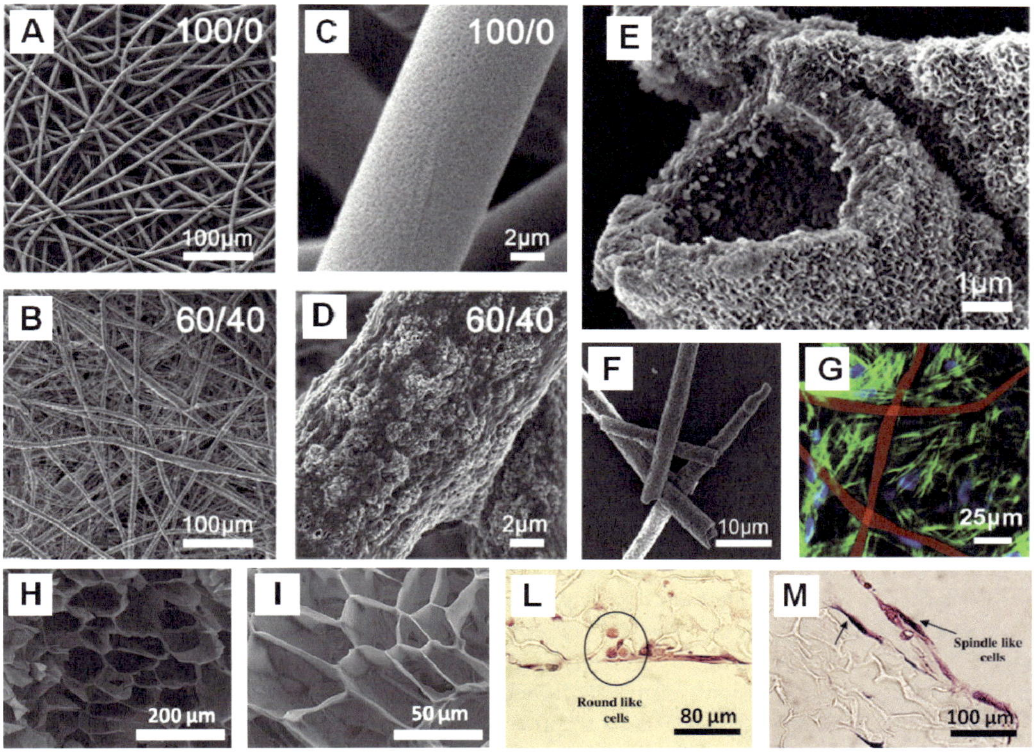

Fig. 7. (A), (C): SEM images of electrospun pure PLGA scaffold; (B), (D): SEM images of electrospun PLGA/ATCP (60/40) nanocomposite; (E), (F): SEM image of extracted PLGA/ATCP (60/40) tubes after 45h of immersion in SBF. The high-magnification image reveals a wall thickness of ~ 1 µm; (G): hMSCs morphology after 4 weeks of culture on the nanocomposite PLGA/ATCP scaffold. In blue: cell nuclei, in green: cytoskeletal actin, in red: scaffold architecture; (H): SEM image of pure PLA scaffold; (I): SEM image of nHAP/PLA composite scaffold; (L): optical microscope photographs of MSCs attached to the pure PLA scaffold; (M): optical microscope photographs of MSCs attached to the nHAP/PLA scaffold. Images from (Schneider *et al.*, 2008) and (Nejati *et al.*, 2008). Reproduced with permission from John Wiley and Sons and Elsevier.

A functionally graded non-woven mesh of polycaprolactone incorporated with tricalcium phosphate nanoparticles was fabricated by Erisken *et al.* using a new hybrid twin-screw extrusion/electrospinning process. With such method, the concentration of nanoparticles could be tailored in a targeted manner, changing the scaffold properties. MC3T3-E1 mouse preosteoblast cells were cultured on the matrix, showing a good proliferation and well differentiating into osteoblasts, converting the PCL-β-TCP non-woven composite meshes into a bone tissue-like constructs. Duan and collegues used the selective laser sintering technology to produce Ca-P nanoparticle filled poly(hydroxybutyrate-co-hydroxyvalerate) (PHBV) microspheres, as bioresorbable osteoconductive composite scaffolds. Ca-P particles had a plate-like morphology with about 10 μm in length and approximately several nanometers in thickness; the composite matrix was prepared by means of emulsion solvent evaporation method. Neither *in vitro* and *in vivo* experiments were performed, but the authors foresaw that, once inside the human body, the Ca-P particles superficially embedded on the microspheres could promote the formation of bone-like apatite and enhance bone formation, while the internal Ca-P particles could stiffen the composite and gradually release calcium and phosphate ions with the degradation on the PHBV matrix. Wei *et al.* used the solvent casting methods to fabricate PDLLA films with 0 and 20 wt% TiO_2 nanoparticles and with 20 wt% TiO_2 mixed with 5 wt% micrometer-sized Bioglass® particles. *In vitro* bioactivity studies were carried out using SBF and culturing osteoblast-like MG-63 cells. The PDLLS films containing different concentrations of TiO_2 and Bioglass® particulate inclusions showed no effect on cell viability after 7 days of incubation, and they showed promising properties of bioactivity, making them attractive for bone tissue engineering.

A novel injectable scaffold for tissue engineering was developed in 2009 by Huang and collegues (Huang *et al.*, 2009). The opportunity to use an injectable scaffold is attractive for *in vivo* perspectives, as it minimizes patient discomfort, risk of infection, scar formation and the cost of treatment. The developed substrate was made of nano-hydroxyapatite/collagen (nHAC) loaded on chitosan/β-glycerophosphate matrix. Bone-marrow-derived mesenchymal stem cells were tested, evaluating their proliferation capability. The composite material showed no toxicity and it did not interfere with the proliferation capability of MSCs, making it an interesting solution for bone tissue engineering, above all for its injectability (ability to flow out totally from a syringe).

Another matrix for bone regeneration was fabricated by Thein-Han and co-workers (Thein-Han *et al.*, 2009), using a silicone rubber doped with nanohydroxyapatite and a process involving uniform dispersion of nHA *via* shear mixing and ultrasonication, followed by compounding at sub-ambient temperature and high-pressure solidification when the final curing reaction occurred. MC3T3-E1 mouse pre-osteoblast cells were cultured on the scaffold, finding that the cell density on the composite matrix was significantly higher than that on the pure silicone rubber. Furthermore, immunofluorescence staining revealed that cells on the composite matrix showed larger vinculin focal contacts with higher expression level at the edges, in comparison with the non-doped material. Cytoskeletal actin organization was also more prominent on the composite scaffold.

Regarding liver tissue engineering, titania/chitosan composite scaffolds were prepared by Zhao *et al.* through a freeze-drying technique (Zhao *et al.*, 2009). Doped matrices with 0.3 of TiO_2/chitosan weight ratio showed the best mechanical performances, maintaining an interconnected pore structure. Hepatic immortal cell line HL-7702 was tested on these structures, detecting liver-specific functions, such as albumin secretion and urea synthesis. No differences were found between the doped scaffold and the pure chitosan matrix, in terms of albumin and urea production; FESEM micrographs of cells cultured on the different scaffolds revealed an better sustained attachment of hepatocytes on the doped matrix, probably thanks to the improved mechanical characteristics, implying the potential application of this material in liver tissue engineering.

Single-walled carbon nanotubes (SWNTs) are unique in structure and function, and they have recently received significant attention due to their potential to create nanostructured conductive materials. A composite SWNT/collagen material was recently proposed as

conductive peripheral nerve regeneration matrix (Tosun & McFetridge, 2010). Neuron-like PC12 cells were tested on collagen and SWNT/collagen scaffolds, finding an enhanced proliferation rate on the doped matrices, and a normal expression of p53 gene, indicating the absence of nanoparticle-induced DNA damage.
In another study, the influence of nanodoped matrices on cardiac and mesenchymal stem cells (CSCs and MSCs) was evaluated, using PLGA films loaded with TiO_2 and CeO_2 nanoparticles (Mandoli *et al.*, 2010). The hybrid nanoCeO_2/PLGA scaffolds showed a parallel surface pattern, with a spacing of about 250 μm between the ridges (Fig. 8(A)). Better CSC and MSC proliferate activity was observed for CeO_2 composites with respect to either TiO_2-added on unfilled PLGA films (Fig. 8(E)), together with a clear cell alignment in the direction of the ridges (Fig. 8(B), (C), and (D)).

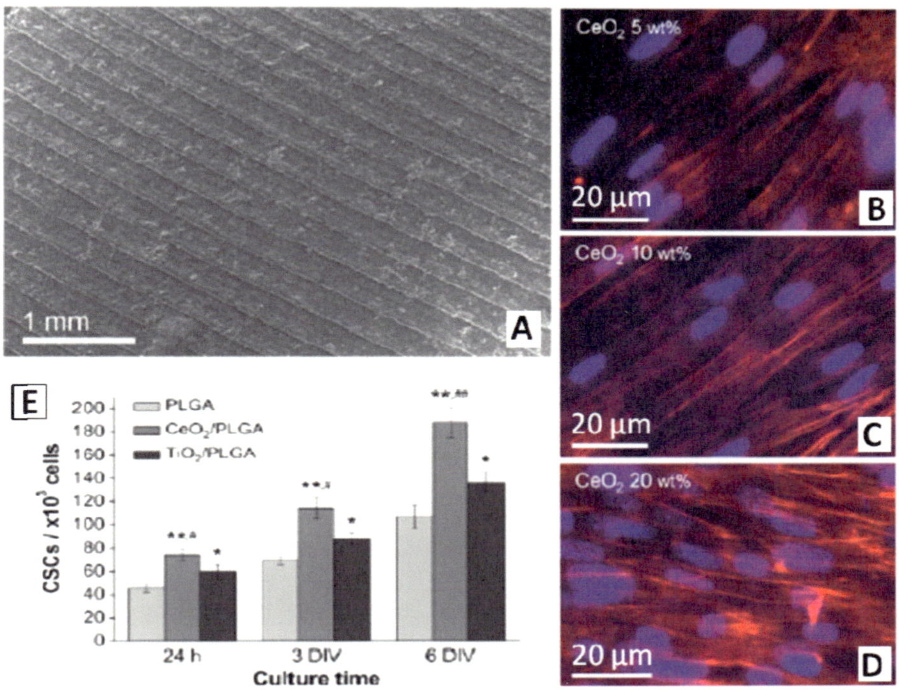

Fig. 8. (A): SEM image of parallel-organized 20% CeO_2/PLGA composite; (B), (C), and (D): CSC seeded on the composite substrates having different nanoparticle concentrations; (E): proliferation assay for the CSC culture on the different scaffolds (pure PLGA, CeO_2/PLGA and TiO_2/PLGA). Images from (Mandoli *et al.*, 2010). Reproduced with permission from John Wiley and Sons.

Skeletal muscle tissue engineering was also addressed using a nanodoped matrix-based approach (McKeon-Fischer & Freeman, 2010). PLLA and gold (Au) nanoparticles were electrospun to create three composite scaffolds: 7% Au-PLLA, 13% Au-PLLA, and 21% Au-PLLA. Rat primary skeletal muscle cells were cultured on the three scaffolds and on the control (pure PLLA matrix), finding that they showed low proliferation capability on the doped scaffolds. However, as demonstrated by the authors, this was not due to the nanoparticle toxicity, but to an enhanced cell fusion into myotubes and a higher level of cell differentiation. The electrical conductivity of the scaffolds (increasing with the increase of nanoparticle concentration) and the good cell behavior on them are promising for skeletal muscle regeneration.

More recently, H9c2 rat cardiomyocytes were cultured on PLGA scaffolds loaded with different concentrations of barium titanate nanoparticles (BTNPs) (Ciofani *et al.*, 2011). BNTPs have previously proven to be non-toxic even at high concentrations, and they have intriguing properties, related to their piezoelectric nature that could influence some cell activities. The PLGA/BTNPs were found to have increased mechanical properties in comparison with bare PLGA scaffolds, with a Young's modulus linearly increasing with the BNTPs concentration and the surface roughness was also increasing due to the nanoparticle presence. H9c2 cells showed an increased proliferation on the doped matrices, probably related to the different protein adsorption due to the different surface roughness. The authors hypothesize that BNTPs piezoelectricity could also have a role in such cell behavior, also envisioning future nanoparticle-mediated scaffold stimulation.

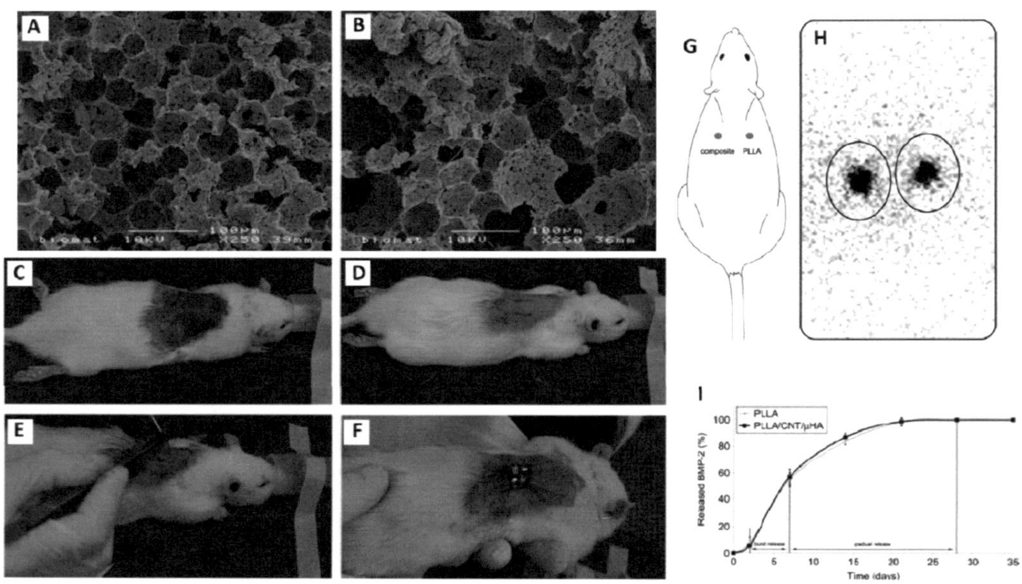

Fig. 9. (A) and (B): SEM images of PLLA and PLLA-CNT-μHA, respectively; (C), (D), (E), and (F): Overview of the subcutaneous implantation of the scaffolds on the back of the rat; (G): schematic overview of the scaffold placement; (H): scintigraphic image of the *in vivo* measurement of released ^{131}I-labelled BMP-2 in rat 14 days after implantation; (I): *in vivo* release profile of ^{131}I-labelled BMP-2 from the scaffolds. Images from (van der Zande *et al.*, 2011). Reproduced with permission from John Wiley and Sons.

An interesting nanocomposite for wound dressing applications was fabricated using β-chitin and silver nanoparticles (Kumar *et al.*, 2010). Such scaffold showed high porosity, which increased for the composite matrix. This is due to the inclusion of the nanosilver solution with the original β-chitin hydrogel; when the sample was freeze-dried, the water in the silver colloid, which got incorporated into the hydrogel, evaporated and the vacant space was left as pores in the scaffold, ultimately resulting in a highly porous composite. Antibacterial studies revealed higher susceptibility of Gram-negative bacteria to nanosilver, with a bacteria inhibition zone larger for higher nanosilver concentrations. The scaffolds were also demonstrated to be cell-friendly, without significant toxicity induced on cultured cells. The β-chitin/nanosilver composites were therefore proposed as novel and promising wound dressing material.

A promising *in vivo* study was carried out by van der Zande and co-workers (van der Zande *et al.*, 2011), that fabricated pure PLLA scaffolds (Fig. 9(A)) provided with bone morphogenetic protein-2 (BMP-2), to improve bone response. The problem to employ such material *in vivo* is that BMP-2 is clearly almost immediately released from the site of implantation by diffusion; to prolonge the retention of BMP-2 into the scaffold, the authors fabricated PLLA matrices doped with carbon nanotubes (CNTs) and microhydroxyapatite (μHA) (Fig. 9(B)).

Radiolabelled BMP-2 was loaded onto plain PLLA and composite PLLA-CNT-μ HA matrices, and the scaffolds were implanted subcutaneously for 5 weeks in rats (Fig. 9(C), (D), (D), and (E)). In contrast with authors' hypothesis, the *in vivo* release showed no differences between the composite and plain PLLA scaffolds. Nevertheless, although incorporated CNTs and μ HA in a PLLA scaffold did not alter the release pattern of BMP-2, they could still provide composite scaffolds with several other beneficial characteristics, as higher mechanical performances and enhanced bone formation.

5. Conclusion

The development of innovative materials useful as scaffolds for the sustained growth of cells is of particular interest in regenerative medicine and tissue engineering, because they can be potentially tailored to mimic the natural extracellular matrix in terms of structure, chemical composition, and mechanical properties.

In the past decades, an urgent necessity of innovative, "smart" materials for tissue engineering raised, in order to obtain constructs able to sustain and promote cell proliferation and tissue regeneration, and particular attention was dedicated to polymer/ceramics composites, in order to take advantage by the physico-chemical properties of these two classes of materials.

Nanostructured materials have been extensively explored in many biological applications because of their intriguing physical and chemical properties. In particular, the intrinsic optical, magnetic, and electrical properties owned by nanomaterials can offer remarkable opportunities of interaction with complex biological processes for several biomedical applications.

The advent of nanotechnology is set to accelerate development of improved and sophisticated smart material technologies. Researchers are now considering the possibilities of designing, altering, and controlling material structure at nanoscale levels in order to enhance material performance and process efficacy. The advancements in nanomaterials are expected to increase product quality and performance, and they are finding acceptance in diverse applications such as sensors and electronic devices. Nanosensor particles assist in creating tools for analyzing living cells and serve as reporters in industrial process monitoring. In the future, smart materials are likely to derive their success from nanotechnology that is likely to be instrumental in creating more varied, complex, and intelligent systems.

The advances and improvements in smart materials allow them to cater to a diverse set of applications, especially in the defense, aerospace, healthcare, electronics, and semiconductor industries. Although very few of these applications are at present commercially viable, their potential for future acceptance is enormous. Smart materials are particularly useful for cellular production: with the addition of cellular fluid and by regulating the cell's shape and mechanical conditions, smart materials can mimic these cells' interactions and exhibit effective results.

In our opinion, the combination of smart and nanomaterials, and the exploitation of their impressive chemical and physical properties, will constitute a fundamental step towards realistic clinical applications of tissue engineering and regenerative medicine.

6. References

Basser, P.J. & Horkay, F. (2005). Toward a constitutive law of cartilage : a polymer physics perspective. *Macromolecular Symposia*, Vol.227, Is.1, pp. 53-64.

Cao, Y., Zhou, Y., Shan, Y., Ju, H. & Xue, X. (2006). Triple-helix scaffolds of grafted collagen reinforced by Al_2O_3-ZrO_2 nanoparticles. *Advanced Materials*, Vol.18, Is.14, pp. 1838-1841.

Causa, F., Netti, P.A. & Ambrosio, L. (2007). A multi-functional scaffold for tissue regeneration : the need to engineer a tissue analogue. *Biomaterials*, Vol.28, No.34, pp. 5093-5099.

Cho, J., Schaab, S., Roether, J.A. & Boccaccini, A.R. (2008). Nanostructured carbon nanotube/TiO_2 composite coatings using electrophoretic deposition (EDP). *Journal of Nanoparticle Research*, Vol.10, pp. 99-105.

Choudhary, S., Haberstroh, K.M. & Webster, T.J. (2007). Enhanced functions of vascular cells on nanostructured Ti for improved stent applications. *Tissue Engineering: Part A*, Vol.13, Is.7, pp. 1421-1430.

Ciofani, G., Danti, S., D'Alessandro, D., Ricotti, L., Moscato, S., Bertoni, G., Falqui, A., Berrettini, S., Petrini, M., Mattoli, V. & Menciassi, A. (2010). Enhancement of neurite outgrowth in neuronal-like cells following boron nitride nanotube-mediated stimulation. *ACS Nano*, Vol.4, No.10, pp. 6267-6277.

Ciofani, G., Ricotti, L., Menciassi, A. & Mattoli, V. (2011). Preparation, characterization and *in vitro* testing of poly(lactic-co-glycolic) acid/barium titanate nanoparticle composites for enhanced cellular proliferation. *Biomedical Microdevices*, Vol.13, pp. 255-266.

Duan, B., Wang, M., Zhou, W.Y. & Cheung, W.L. (2008). Synthesis of Ca-P nanoparticles and fabrication of Ca-P/PHBV nanocomposite microspheres for bone tissue engineering applications. *Applied Surface Science*, Vol.255, pp. 529-533.

Erisken, C., Kalyon, D.M. & Wang, H. (2008). A hybrid twin screw extrusion/ electrospinning method to process nanoparticle-incorporated electrospun nanofibers. *Nanotechnology*, Vol.19, No.16, pp. 165302-165310.

Erisken, C., Kalyon, D.M. & Wang, H. (2008). Functionally graded electrospun polycaprolactone and β-tricalcium phosphate nanocomposites for tissue engineering applications. *Biomaterials*, Vol.29, pp. 4065-4073.

Gerhardt, L.C., Jell, G.M.R. & Boccaccini, A.R. (2007). Titanium dioxide (TiO_2) nanoparticles filled poly(D,L lactic acid) (PDLLA) matrix composites for bone tissue engineering. *Journal of Material Science : Materials in Medicine*, Vol.18, No.7, pp. 1287-1298.

Gheith, M.K., Sinani, V.A., Wicksted, J.P., Matts, R.L. & Kotov, N.A. (2005). Single-walled carbon nanotube polyelectrolyte multilayers and freestanding films as a biocompatible platform for neuroprosthetic implants. *Advanced Materials*, Vol.17, Is.22, pp. 2663-2670.

Guan, D., Chen, Z., Huang, C. & Lin Y. (2008). Attachment, proliferation and differentiation of BMSCs on gas-jet/electrospun nHAP/PHB fibrous scaffolds. *Applied Surface Science*, Vol.255, pp. 324-327.

Huang, Z., Tian, J., Yu, B., Xu, Y. & Feng, Q. (2009). A bone-like nano-hydroxyapatite/collagen loaded injectable scaffold. *Biomedical Materials*, Vol.4, pp. 55005-55012.

Hutmacher, D.W. (2001). Scaffol design and fabrication technologies for engineering tissues-state of the art and future perspectives. *Journal of Biomaterials Science, Polymer Edition*, Vol.12, pp. 107-124.

Jack, K.S., Velayudhan, S., Luckman, P., Trau, M., Grøndahl, L. & Cooper-White, J. (2009). The fabrication and characterization of biodegradable HA/PHBV nanoparticle-polymer composite scaffolds. *Acta Biomaterialia*, Vol.5, Is.7, pp. 2657-2667.

Jagur-Grodzinski, J. (2006). Polymers for tissue engineering, medical devices, and regenerative medicine. Concise general review of recent studies. *Polymers for Advanced Technologies*, Vol.17, pp. 395-418.

Kaiser, L.R. (1992). The future of multihospital systems. *Topics in Health Care Financing*, Vol.18, No.4, pp. 32-45.

Kaplan, F.S., Hayes, W.C., Keaveny, T.M., Boskey, A.Einhorn, T.A. & Iannotti, J.P. (1994). In : *Orthopedic basic science*, Simon, S.R. (Ed.), pp. 127-185, Rosemont.

Khademhosseini, A., Borenstein, J., Toner, M. & Takayama, S. (2008). *Micro and nanoengineering of the cell microenvironment : technologies and applications*, Artech House, Norwood.

Kumar, P.T.S., Abhilash, S., Manzoor, K., Nair, S.V., Tamura, H. & Jayakumar, R. (2010). Preparation and characterization of novel β-chitin/nanosilver composite scaffolds for wound dressing applications. *Carbohydrate Polymers*, Vol.80, pp. 761-767.

Langer, R. & Vacanti, J.P. (1993). Tissue Engineering. *Science*, Vol.260, No.5110, pp. 920-926.

Lee, J.W., Ahn, G.S., Kim, D.S. & Cho, D.W. (2009). Development of nano- and microscale composite 3D scaffolds using PPF/DEF-HA and micro-stereolithography. *Microelectronic Engineering*, Vol.86, Is.4-6, pp. 1465-1467.

Li, C., Vepari, C., Jin, H.J., Kim, H.J. & Kaplan, D.L. (2006). Electrospun silk-BMP-2 scaffolds for bone tissue engineering. *Biomaterials*, Vol.27, Is.16, pp. 3115-3124.

Liu, H., Slamovich, E.B. & Webster, T.J. (2006). Less harmful acidic degradation of poly(lacic-co-glycolic acid) bone tissue engineering scaffolds through titania nanoparticle addition. *International Journal of Nanomedicine*, Vol.1, No.4, pp. 541-545.

Mandoli, C., Pagliari, F., Pagliari, S., Forte, G., Di Nardo, P., Licoccia, S. & Traversa, E. (2010). Stem cells aligned growth induced by CeO_2 nanoparticles in PLGA scaffolds with improved bioactivity for regenerative medicine. *Advanced Functional Materials*, Vol.20, pp. 1617-1624.

Mattson, M.P., Haddon, R.C. & Rao, A.M. (2000). Molecular functionalization of carbon nanotubes and use as substrates for neuronal growth. *Journal of Molecular Neuroscience*, Vol.14, No.3, pp. 175-182.

McKeon-Fischer, K.D. & Freeman, J.W. (2010). Characterization of electrospun poly(L-lactide) and gold nanoparticle composite scaffolds for skeletal muscle tissue engineering. *Journal of Tissue Engineering and Regenerative Medicine*, In press - DOI : 10.1002/term.348.

Meng, D., Erol, M. & Boccaccini, A.R. (2010). Processing technologies for 3D nanostructured tissue engineering scaffolds. *Advanced Biomaterials*, Vol.12, No.9, pp. B467-B487.

Miller, D.C., Haberstroh, K.M. & Webster, T.J. (2007). PLGA nanometer surface features manipulate fibronectin interactions for improved vascular cell adhesion. *Journal of Biomedical Materials Research*, Vol.81A, Is.3, pp. 678-684.

Nejati, E., Mirzadeh, H. & Zandi, M. (2008). Synthesis and characterization of nano-hydroxyapatite rods/poly(L-lactide acid) composite scaffolds for bone tissue engineering. *Composites : Part A*, Vol.39, pp. 1589-1596.

Ngiam, M., Liao, S., Patil, A., Cheng, Z., Yang, F., Gubler, M., Ramakrishna, S. & Chan, C.K. (2009). Fabrication of mineralized polymeric nanofibrous composites for bone graft materials. *Tissue Engineering : Part A*, Vol.15, Is.3, pp. 535-546.

Nukavarapu, S.P., Kumbar, S.G., Nair, L.S. & Laurencin, C.T., In : Gonsalves, K.E., Halberstadt, C.R., Laurencin, C.T. & Nair, L.S. (Eds.) (2008). *Biomedical nanostructures*. John Wiley & Sons, Inc., New Jersey, pp. 377-407.

Park, S.H, Kim, T.G., Kim, H.C., Yang, D.Y. & Park, T.G. (2008). Development of dual scale scaffolds via direct polymer melt deposition and electrospinning for applications in tissue regeneration. *Acta Biomaterialia*, Vol.4, Is.5, pp. 1198-1207.

Paul, D.R. & Robeson, L.M. (2008). Polymer nanotechnology : nanocomposites. *Polymer*, Vol.49, pp. 3187-3204.

Pavel, F.M. & Mackay, R.A. (2000). Reverse micellar synthesis of a nanoparticle/polymer composite. *Langmuir*, Vol.16, pp. 8568-8574.

Sanyal, A., Norsten, T.B. & Rotello, V.M. (2004). *Nanoparticle polymer ensembles*. In : *Nanoparticles, building blocks for nanotechnology*, Springer, ISBN 0-306-48287-8, New York.

Schneider, O.D., Loher, S., Brunner, T.J., Uebersax, L., Simonet, M., Grass, R.N., Merkle, H.P. & Stark, W.J. (2008). Cotton wool-like nanocomposite biomaterials prepared by electrospinning : *in vitro* bioactivity and osteogenic differentiation of human mesenchymal stem cells. *Journal of Biomedical Materials Research, Part B : Applied Biomaterials*, Vol.84, No.2, pp. 350-362.

Seil, J.T. & Webster, T.J. (2008). Decreased astroglial cell adhesion and proliferation on zinc oxide nanoparticle polyurethane composites. *International Journal of Nanomedicine*, Vol.3, Is.4, pp. 523-531.

Shi, G., Rouabhia, M., Wang, Z., Dao, L.H. & Zhang, Z. (2004). A novel electrically conductive and biodegradable composite made of polypyrrole nanoparticles and polylactide. *Biomaterials*, Vol.25, Is.13, pp. 2477-2488.

Siegel, R.W. & Fougere, G.E. (1995). Mechanical properties of nanophase metals. *Nanostructured Materials*, Vol.6, No.1-4, pp. 205-216.

Son, W.K., Youk, J.H. & Park, W.H. (2006). Antimicrobial cellulose acetate nanofibers containing silver nanoparticles. *Carbohydrate Polymers*, Vol.65, Is.4, pp. 430-434.

Stupp, S.I. (2005). Biomaterials for regenerative medicine. *MRS bullettin*, Vol.30, No.7, pp. 546-553.

Sui, G., Yang, X., Mei, F., Hu, X., Chen, G., Deng, X. & Ryu, S. (2007). Poly-L-lactic acid/hydroxyapatite hybrid membrane for bone tissue regeneration. *Journal of Biomedical Materials Research Part A*, Vol.82, Is.2, pp. 445-454.

Teo, W.E., Liao, S., Chan, C. & Ramakrishna, S. (2011). Fabrication and characterization of hierarchically organized nanoparticle-reinforced nanofibrous composite scaffolds. *Acta Biomateriallia*, Vol.7, pp. 193-202.

Thein-Han, W.W. & Misra. R.D.K. (2009). Biomimetic chitosan-nanohydroxyapatite composite scaffolds for bone tissue engineering. *Acta Biomaterialia*, Vol.5, Is.4, pp. 1182-1197.

Thein-Han, W.W., Shah, J. & Misra, R.D.K. (2009). Superior *in vitro* biological response and mechanical properties of an implantable nanostructured biomaterial : nanohydroxyapatite-silicone rubber composite. *Acta Biomaterialia*, Vol.5, pp. 2668-2679.

Torres, F.G., Nazhat, S.N., Sheikh, S.H., Fadzullah, M., Maquet, V. & Boccaccini, A.R. (2007). Mechanical properties and bioactivity of porous PLGA/TiO_2 nanoparticle-filled composites for tissue engineering scaffolds. *Composite Science and Technology*, Vol.67, Is.6, pp. 1139-1147.

Tosun, Z. & McFetridge, P.S. (2010). A composite SWNT-collagen matrix : characterization and preliminary assessment as a conductive peripheral nerve regeneration matrix. *Journal of Neural Engineering*, Vol.7, pp. 66002-66012.

Tran, P.A., Zhang, L. & Webster, T.J. (2009). Carbon nanofibers and carbon nanotubes in regenerative medicine. *Advanced Drug Delivery Reviews*, Vol.61, Is.12, pp. 1097-1114.

Tu, R.S. & Tirrell, M. (2004). Bottom-up design of biomimetic assemblies. *Advanced Drug Delivery Reviews*, Vol.56, Is.11, pp. 1537-1563.

Van der Zande, M., Walboomers, X.F., Olalde, B., Jurado, M.J., Alava, J.I., Boerman, O.C. & Jansen, J.A. (2011). Effect of nanotubes and apatite on growth factor release from PLLA scaffolds. *Journal of Tissue Engineering and Regenerative Medicine*, Vol.5, pp. 476-482.

Wei, G. & Ma, P. (2006). Macroporous and nanofibrous polymer scaffolds and polymer/bone-like apatite composite scaffolds generated by sugar spheres. *Journal of Biomedical Materials Research*, Vol.78A, pp. 306-315.

Wei, J., Chen, Q.Z., Stevens, M.M., Roether, J.A. & Boccaccini, A.R. (2008). Biocompatibility and bioactivity of PDLLA/TiO_2 and PDLLA/TiO_2/Bioglass® nanocomposites. *Materials Science and Engineering C*, Vol.28, pp. 1-10.

Zhang, N., Nichols, H.L., Tylor, S. & Wen, X. (2007). Fabrication of nanocrystalline hydroxyapatite doped degradable composite hollow fiber for guided and biomimetic bone tissue engineering. *Materials Science and Engineering*, Vol.27, No.3, pp. 599-606.

Zhang, L. & Webster, T.J. (2009). Nanotechnology and nanomaterials: promises for improved tissue regeneration. *Nano Today*, Vol.4, pp. 66-80.

Zhao, L., Chang, J. & Zhai, W. (2009). Preparation and HL-7702 cell functionality of titania/chitosan composite scaffolds. *Journal of Materials Science: Materials in Medicine*, Vol.20, pp. 949-957.

15

The Role of Platelet Gel in Regenerative Medicine

Primož Rožman[1], Danijela Semenič[2] and Dragica Maja Smrke[2]
[1]*Blood Transfusion Centre*
[2]*Department of Surgical Infections, University Clinical Centre Ljubljana*
Slovenia

1. Introduction

For decades, platelet concentrates have been a standard transfusion component for the treatment of clotting disorders. They are prepared from the whole blood within 24 hours from collection from voluntary blood donors by means of differential centrifugation of platelet rich plasma (PRP) or by the apheresis and stored at room temperature with a maximum shelf life of 7 days. They are transfused to the patients who suffer from thrombocytopenia or some other platelet disorder in order to sustain the haemostatic function. Freezing platelets is feasible but results in poor recovery, although the platelets retain their clotting functions after thawing. The use of platelets is permanently increasing, due to advanced surgical and other therapies and they are considered vital products in blood banking today (Brecher, 2005).

Apart from their clotting functions, platelets have been shown to possess some other evolutionary highly conserved biological functions, such as immune defence, tissue forming and regeneration. These properties are now clinically used by the virtue of platelet gel preparations. Technically, the clinical precursor of platelet gel has been fibrin glue that was composed of two separate solutions - fibrinogen and thrombin. When mixed together, these agents were similar to the last stages of the clotting cascade in forming a fibrin clot. Fibrinogen has been obtained from pooled allogeneic single-donor plasma units and from the autologous blood of the patients and has been usually isolated by the process of cryoprecipitation. The additional thrombin component has been generally derived from commercial bovine sources. Some investigators have even added calcium chloride and/or antifibrinolytics (i.e., aminocaproic acid, aprotinin) to their preparations in order to enhance the clot formation. Although fibrin glue has been used in a variety of surgical procedures for the prevention of bleeding, it has been especially useful in heparinized patients undergoing cardiovascular procedures requiring extracorporeal circulation, as it does not need an intact hemostatic system to be effective. Fibrin glue has also been evaluated in the presealing of woven or knitted Dacron vascular grafts. The major drawback to its use has been attributed to the potential risk of transmitted serological disease from pooled and single-donor blood donors, so the patient's own blood was considered to be the safest preparation to prepare fibrin glue. Overall, fibrin glue is a useful adjunct to other methods to control bleeding in selected surgical patients (Thompson et al., 1988; Lee & Kang, 2011).

Autologous platelet rich plasma, which is by the activation of platelets converted to platelet gel (PG) was developed in the early 1970s as a by-product of multicomponent apheresis. The first published paper relating to the use of platelet-fibrinogen-thrombin mixture as a ″corneal adhesive″ is dated from 1975 (Rosenthal et al., 1975). In 1979, a report was published about platelet gel ″gel foam″ used to obtain sutureless nerve anastomosis (Fischer, 1979). Another publication in 1980, described the role of fibrin to drive fibroblast migration and collagen deposition and leading to granulation tissue formation (Brandstedt et al., 1980).

Articles published in the 1990s provided numerous insights into the basic mechanisms of the physiologic process of healing after injury had occurred. Some of the results provided by these papers were that fibroblasts are recruited at the injury site by soluble factors; that the platelet releasate contains a high level of chemotactic factors that recruit fibroblasts in a strictly dose-dependent manner, that the fibroblasts migrate along a fibrin-rich matrix; and that fibronectin is necessary for such migration to occur since it provides a conduit for transmigration. In their migration, fibroblasts adhere to the fibronectin fibres through alpha5beta1 and alpha5beta3 receptors. Proteases are necessary for fibroblast transmigration through the fibrin-rich matrix (the role of metalloproteases was depicted some years later). Interestingly, the fibrin/ fibronectin fibres retain growth factors released from the activated platelets within the injured site (Greiling & Clark, 1997).

In 1996, the physiological relationship between mesenchymal cells, fibrin and growth factors was shown to result in the establishment of granulation tissue. It was demonstrated that the formation of granulation tissue is strongly enhanced by the platelet releasate. The platelet releasate provides continuous stimulation for mesenchyme-derived cells to interact with fibrin, thus accelerating the development of granulation tissue (McClain et al., 1996). It is now confirmed that the human platelet lysate can induce the expansion of human mesenchymal stem cells *in vitro*, which will by omitting the animal derived albumin, change the cell cultivation approaches for regenerative medicine.

One of the first disciplines that readily accepted the new regenerative capacity of the PRP/PG was the maxillofacial surgery (Schallmoser et al., 2007). PRP was introduced to the maxillofacial community by Whitman et al. as an alternative to the fibrin glue in 1997 (Whitman et al., 1997). Marx et al. showed that by combining PRP with autologous bone in mandibular continuity defects resulted in significantly faster radiographic maturation and a histomorphometrically denser bone regeneration (Marx et al., 1998).

In 2000 - 2010 authors observed various healing effects of PG, for instance that chronic lower extremity ulcers (diabetic, post-traumatic, vascular) reepithelialized faster with the course of twice-daily wound treatment with platelet releasate as compared with similar wounds treated with a placebo. The supernatant or releasate of the platelet gel was able to promote dose-dependent proliferation and changes in gene expression as well as in metabolic activities related to protein synthesis (Junger et al., 2000;Valencia et al., 2001; Fang & Galiano, 2008; Bernuzzi et al., 2010; Ficarelli et al., 2008; Chen et al., 2010). The platelet concentrate was used for healing skin, soft tissue wounds, fractures, surgical injuries, diabetic ulcers, dental, oral and maxillofacial settings (Altman et al., 2001; Margolis et al., 2001; Anderson &Baker, 2003; Castro, 2004; Belli et al., 2005; Burkus, 2005; Carreon et al., 2005; Sclafani et al., 2005; Trowbridge et al., 2005; Brown et al., 2006; Driver et al., 2006; Savarino et al., 2006; Rozman & Bolta, 2007; Borzini & Mazzucco 2007; Smrke et al., 2007; Cieslik-Bielecka et al., 2008;).

Novel strategies of platelet gel application in cardiology were under investigation in the year 2004. Studies evaluated the use of platelet gel in the case of myocardial injury, in order to promote remodelling through the regeneration of myocytes, the induction of angiogenesis and the restoration of a normal extracellular matrix composition (Mogan & Larson, 2004; Balbo et al., 2010;).

PG/PRP was further intensively studied. Human umbilical vein endothelial cells (HUVECs) were isolated from umbilical cord veins and grown in appropriate conditions *in vitro*. The platelet gel-released supernatant was able to induce proliferation and to stimulate motility and the invasiveness of HUVECs. Higher concentrations of PG induced a reversion of the stimulatory processes. It became clear that different concentrations of the platelet growth factors do not have the same efficacy in inducing these processes: excessively high concentrations can have an inhibitory effect on the processes. Further studies are necessary to reach a better understanding of these complex biological processes (Rughetti et al., 2008).

PG treatment in 2008 revealed a unique capacity of articulating a pro-inflammatory and pro-angiogenic cytokine profile in human peripheral blood mononuclear cells, co-cultured with PG, which may partially explain the clinical success of PG application in a wide range of diseases (Naldini et al., 2008).

A 1-year retrospective review of a cardiac surgery patient in 2008 showed that the treatment group with PG significantly shortened the duration of intensive care and decreased the total length of stays as well as decreased postoperative blood loss, compared with the control group. In addition, the treatment group had no reported incisional wound infections (Englert et al., 2008; Gunaydin et al., 2008).

2. Platelet structure and physiology

Platelets are not "real" cells but rather non-nuclear cellular fragments derived from the megakaryocytes in the bone marrow through controlled cellular fragmentation, followed by their release into circulation (Jurk & Kehrel, 2005). A healthy adult aged between 30 and 70, weighing 70 kg, has in total about 1,300-1,500 grams of platelets (thrombocytes); 40% of them in the spinal bone marrow, 25% in the ribs and sternum, 15% in the pelvis, 10% in the skull, and 10% in other bones. Progenitor cells of the thrombocytic lineage develop in the same way as other immature cells of bloodlines in the bone marrow, originating from the common hematopoietic stem cell. When immature cells differentiate into directed stem cells they are morphologically recognizable as the progenitor cells of thrombocytic lineage. The progenitor cell is the megakaryoblast, with 21-50 mm diameter, which develops from the common myeloid progenitor named CFU-GEMM. The megakaryoblast gives rise to the promegakaryocyte and then the megakaryocyte - the largest cell in the bone marrow with 70-100 mm in diameter. Several growth factors are involved in the platelet differentiation, among which the thrombopoietin plays the most decisive role, inducing thrombopoiesis in a negative loop fashion by binding to the thrombocytes when they are present in sufficient numbers, thus decreasing their production from the HSCs in the bone marrow (See Figure 1).

There are 1/3 immature and 2/3 mature elements of megakaryocyte lineage in the bone marrow. Thrombocytes are formed by the breaking off of small fragments of the megakaryocyte cytoplasm. Each megakaryocyte gives rise to 2,000 to 4,000 thrombocytes. The development of the megakaryocyte from the megakaryoblast takes 4-5 days, and the life-span of a thrombocyte is 7-10 days. The number of thrombocytes in the peripheral blood is 150-350 x 10E9/L. If functionally classified, platelets consist of various molecules: membrane proteins 3%, signalling proteins 24%, cytoskeletal 15%, vesicular 6%, extracellular 4%, mitochondrial 7%, protein processing 22%, metabolic enzymes 6%, unknown 4%, miscellaneous 5.5% and nuclear proteins 3.5% (Garcia et al., 2004).

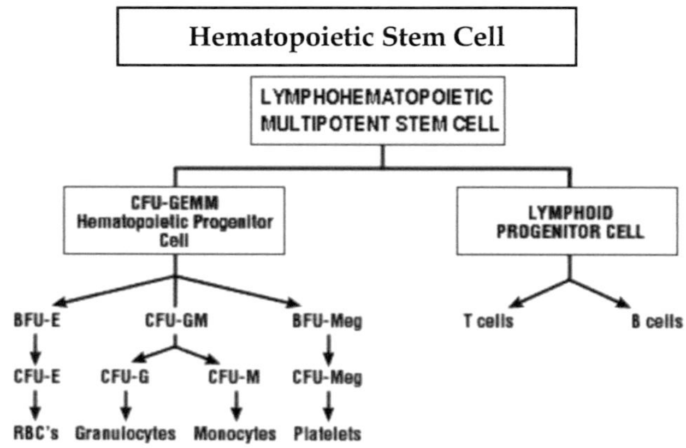

Fig. 1. Thrombopoiesis in the bone marrow

As mentioned previously, platelets are primarily specialized secretory cells that release the contents of their intracellular granules in response to activation. Beside their well known function in haemostasis where they prevent blood loss at the sites of vascular injury, they also release substances that promote tissue repair and influence the reactivity of vascular and blood cells in angiogenesis and inflammation (Jurk & Kehrel, 2005). At the site of the injury, platelets release an arsenal of potent inflammatory and mitogenic substances which are involved in all aspects of the wound healing process. Based on these facts, platelet releasate in the form of activated platelet gel has been extensively used for the topical *in loco* therapy of various clinical conditions (Borzini & Mazzucco, 2007). The purpose of this chapter is to review the current laboratory and clinical findings related to the therapy of these diseases with the use of platelet gel derived growth factors.

3. Platelet activation

Platelets are normally circulating in the blood in their resting form (Gresele at al., 2002). Non-activated platelets contain a complete array of presynthesized protein molecules important for platelet function, among which the high presence of cytoskeletal proteins, signalling proteins, membrane proteins, protein processing proteins and cytoskeleton regulatory proteins is noted (Garcia et al., 2004; Anitua et al., 2004). These molecules are synthesized already in the megakaryocytes and packaged into the granules. Three major storage compartments in platelets are alpha granules, dense granules and lyzosomes. The majority of the substances are contained in alpha granules (Rendu & Brohard-Bohn, 2001). (See Tab. 1).

Category	Term	Biological activities
Adhesive proteins	VWf + propeptide, Fg, Fn,Vn,TSP-1, laminin-8	Cell contact interactions, clotting, extracellular matrix composition
Clotting factors and associated proteins	Factor V/Va, factor XI, multimerin, gas6, protein S, high-molecular weight chininogen, antithrombin, tissue factor pathway inhibitor (TFPI)	Thrombin production and its regulation, angiogenesis
Fibrinolytic factors and associated proteins	Plasminogen, PAI-I, u-PA, osteonectin, α2-antiplasmin, histidine rich glycoprotein, TAFI, α2-macroglobulin	Plasmin production and vascular remodelling
Proteases and anti-proteases	Tissue inhibitor of metalloprotease-4 (TIMP-4), matrix metalloproteinases MMP-1,-2,-3, -4, -9, platelet inhibitor of FIX, protease nexin-2, C1 inhibitor, α1-antitripsin	Potentiation of aggregation, angiogenesis, vascular modelling, regulation of coagulation, regulation of cellular behaviour
Growth factors, cytokines and chemokines	PDGF, TGFβ1 and 2, EGF, IGF-1, VEGF (A and C), bFGF and FGF-2, hepatocyte GF, RANTES, IL-8, MIP-1α, growth regulated oncogene-α, ENA-78, MCP-3, angiopoietin-1, Il-1β, IGF BP-3, neutrophile chemo active protein	Chemotaxis, cell proliferation and differentiation, angiogenesis
Basic proteins and others	PF4, β-thromboglobulin, platelet basic protein, connective tissue activating peptide III, neutrophile activating peptide-2, endostatins	Regulation of angiogenesis, vascular modelling, cellular interactions. Endostatin is an endogenous inhibitor of angiogenesis and the growth of both primary tumors and metastasis.
Anti-microbial proteins	Thrombocidins	Bactericidal and fungicidal properties
Others	-chondroitin 4-sulfate, albumin, immunoglobulins	Diverse
Membrane glycoproteins	αIIbβ3, α β3, GPIb, PECAM-1, most plasma membrane constituents, receptors for primary agonists, CD40L, tissue factor, P-selectin	Platelet aggregation and adhesion, endocytosis of proteins, inflammation, thrombin generation, platelet-leukocyte interactions

Table 1. The contents of the platelet alpha-granules

When adhered to exposed endothelium or activated by agonists, platelet change their shape and secrete the contents of the granules (including ADP, fibrinogen and serotonin), which is followed by the platelet aggregation. The initiation of the signalling event within the platelet leads to the reorganization of the platelet cytoskeleton , which is seen as an extremely rapid shape change (See Figure 2).

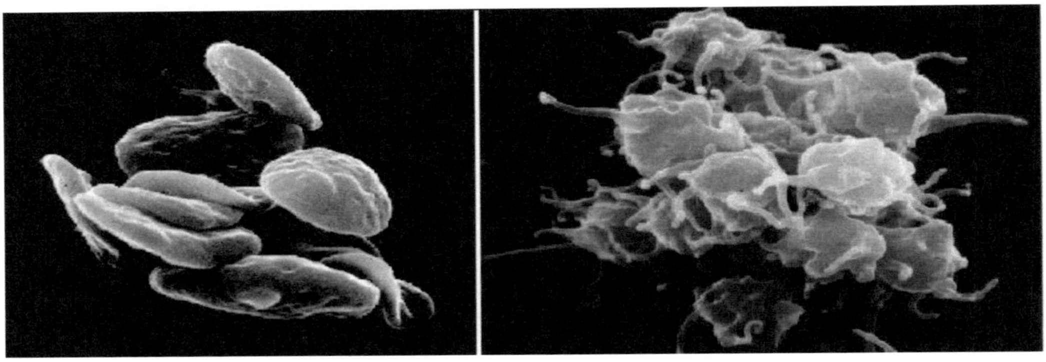

Legend: Resting platelets are smooth and disc shaped (left). Activated platelets have an irregular shape with many protruding pseudopodia (right).

Fig. 2. Platelets in their resting form and at the beginning of the activation.

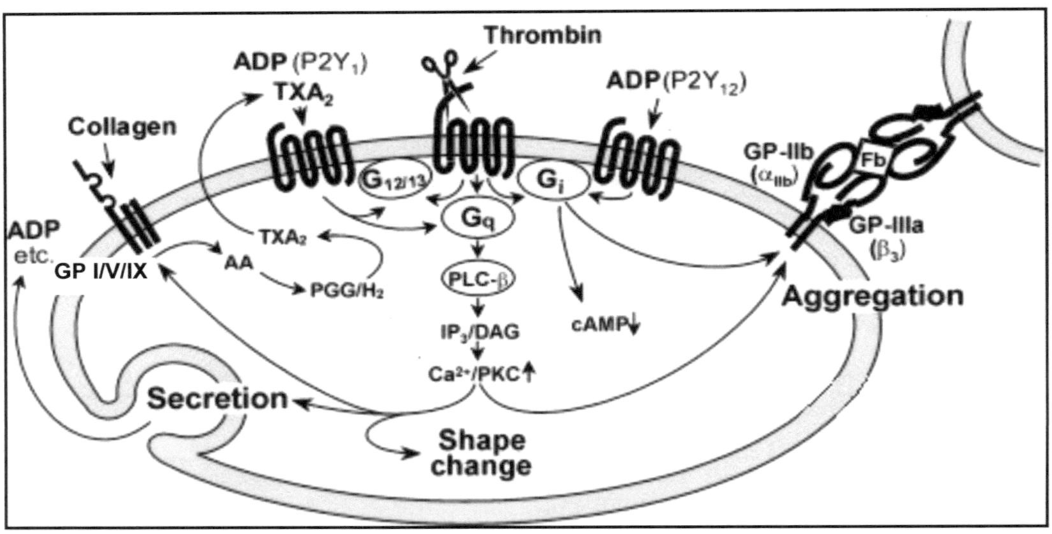

Fig. 3. Schematic representation of intracellular events during platelet activation - "outside in" and "inside out" signalling

Initiation of the signalling event within the platelet leads to the reorganization of the platelet cytoskeleton , which is seen as an shape change. Aggregation of platelets is mediated by molecules of fibrinogen or vWf, which connect platelets by bridging complexes of glycoprotein IIb/IIIa (integrin $\alpha_{IIb}\beta_3$) on adjacent platelets, forming a platelet aggregate. In order to bind the fibrinogen and vWf, GP-IIb/IIIa initially has to be converted from a low affinity/avidity state to a high affinity/avidity state by a process described as "inside-out signalling" that is initiated during and after platelet activation. Some of the substances released by these cells - in particular adenosine diphosphate, serotonin, thromboxane and others - can, in an autocrine and paracrine fashion, further enhance platelet activation and aggregation (See Figure 3.). Under these conditions there is also activation of the coagulation cascade; therefore thrombin is formed, which also markedly stimulates the platelet activation in a positive feedback loop fashion (Gresele et al., 2002).

4. Platelet receptors and antigens

It has been proposed that the original function of the mammalian platelets resembled the function of primitive haemocytes in certain lower organisms such as caterpillars, in which they mostly act as a defence against foreign organisms by adhering and aggregating to foreign bodies. This is probably the reason why human platelets also possess many features of classical inflammatory cells. As such, they can undergo chemotaxis, release potent inflammatory mediators such as platelet factor PF4 and histamine, they can phagocytose foreign particles, interact with neutrophils, stimulate basophiles, excrete a number of mediators, cytokines, growth factors and other proteins and even interact with bacteria and parasites. Platelet membranes also bear some immune molecules, such as MHC Class I molecules, Fc receptors and complement binding molecules. The platelet membrane also incorporates AB0 and some other polysaccharide blood group Antigens, which are very important in blood transfusion (Rožman, 2002).

The membrane of the resting platelet can modify considerably after the activation following adhesion. This stimulus causes numerous intracellular molecules that are a part of the lyzosomes, granules and other cellular compartments, to appear on the platelet surface. These molecules are therefore considered as markers of platelet activation. A short description of more than 45 distinctive platelet membrane structures found on resting platelets is given in Table 2. They are usually divided into five groups: adhesion molecules, immune molecules, receptors, blood group antigens and some other molecules.

An extensive review of all molecules found on platelets and their precursors is given in references (Kishimoto et al., 1997). The CD designations of all integral molecules of the platelet membrane, as agreed upon at the 9th HLDA (Human Leukocyte Differentiation Antigens) workshop in 2010., can also be found on the internet addresses: *http://hcdm.org*.

Many antigenic systems that are typical for erythrocytes and leukocytes, are also present on platelet membranes, including ABO and HLA antigens. Some others are exclusively found on platelets and their precursors. They are called human platelet alloantigens (HPA) and located on the platelet glycoproteins GPIa/IIa, GPIIb/IIIa and GPI/V/IX as a polymorphic part of the aminoacid chain. These polymorphisms are usually caused by a single nucleotide and the consequent aminoacid substitutions. The polymorphisms that cause alloimmunisation following transfusion and pregnancy were classified into the immunogenetic system of HPA (human platelet alloantigens) (Santoso & Kiefel, 1998). (see Table 3. and EMBL EBI immune polymorphysm database at the *http://www.ebi.ac.uk).*

For transfusion and transplantation, AB0, HPA and HLA antigens are important due to their capability of inducing an immune response in antigen- negative recipients. This could play a role in the allogeneic platelet gel settings, but surprisingly we could not find any immune consequences of the allogeneic platelet gel use in observed patients, treated topically with allogeneic platelets, as will be discussed later.

Group	SUBGROUP	Members	Characteristics and function
1. ADHESION MOLECULES	Integrins	CD41 (Integrin αIIb subunit, GPIIb)	Associates with GPIIIa into GPIIb/IIIa complex. The complex binds fibrinogen, vWf, fibronectin, vitronectin, thrombospondin
		CD61 (Integrin β3 subunit)	GPIIIa; associates with CD41 (into the GPIIb) or CD51 (into the vitronectin receptor)
		CD49a (Integrin α1 subunit)	Associates with CD29 (β1 subunit) into VLA-1 (GPIIa), binds to collagen and laminin
		CD49b (integrin α2 subunit)	GPIa or VLA-2α, associates with CD29 (GPIIa), binds collagen, laminin,
		CD49e (integrin α5 subunit)	α subunit of VLA-5 (vitronectin receptor), GPIc' subunit, of GPIc'/IIa, binds fibronectin
		CD49f (integrin α6 subunit)	VLA-6α subunit, GPIc, associates with CD29 (β1) into VLA-6 (CD49f/CD29), binds laminin
		CD29 (integrin β1 subunit)	Associates with CD49a in VLA-1 integrin (GPIIa) on platelets
		CD51 (integrin αV subunit)	Associates with CD61 into the vitronectin receptor. A receptor for RGD-containing proteins (fibrinogen, vWf, fibronectin, laminin, thrombospondin).
	Selectins	CD62 (P-selectin)	Binds sialyl Lewis^x, mediates the interaction of platelets with neutrophils, monocytes and a rolling interaction of neutrophils on endothelium
	Leucine-rich proteins (GPIb-IX-V complex)	CD42bα (GPIbα)	Binds to the von Willebrand factor, receptor for thrombin
		CD42bβ (CD42c) (GPIbβ)	Subunits of the GPIb-IX-V complex, bind to the von Willebrand factor
		CD42d (GPV)	
		CD42a (GPIX)	
	Ig – superfamily	CD54 (ICAM-1)	Interactions with integrins, receptor for rhinovirus
		CD31 (PECAM-1)	Interactions with integrins, inhibiting neutrophile migration
	CD44 variants	CD44 isoforms	Adhesion to the extracellular matrix
	TM4 superfamily	CD9	Tetra spanning membrane protein, regulation of integrin function
		CD165 (AD2, gp37, A108)	Adhesion molecule, thymocytes/ thymic epithelium
2. IMMUNE MOLECULES	MHC class I.	HLA-A, B, C	Immune recognition
	Fc receptors	CD32 (FcγRII)	Low affinity Fc receptor for aggregated Ig/immune complexes; activation
	C receptors	C1q-, C2-, C4-receptors	Complement receptors
	Regulation of complement activation	CD55 - DAF (decay acc. factor)	Blocking and deactivating components of complement and inhibition of lysis
		CD59 - MIRL	
		C8bp (C8-binding protein)	
		CD46 - MCP (membr. cof. prot.)	

3. RECEPTORS	Receptors for primary agonists	Medium affinity thrombin rec.	Mediate activation by agonistic ligands through different mechanisms
		High affinity thrombin rec.	
		Epinephrine receptor	
		TX A2 - thromboxane receptor	
		PAF receptor	
		ADP receptor	
		PAF receptor	
		Serotonin receptor	
		Vasopressin receptor	
	Inhibitory receptors	β2 adrenoreceptor	Inhibit platelet activation
		Adenosine receptor	
		Prostaglandin D2 (PGD2) rec.	
		Prostaglandin I2 (PGI2) rec.	
	Multipurpose receptors	CD36 (GPIV)	Adhesion receptor for collagen and thrombospondin, receptor for Plasmodium falciparum, macrophage scavenger receptor
	Other receptors	CD151 (PETA-3)	Activates platelets upon stimulation with MAbs
		CD110 (TPO R, MPL)	Thrombopoietin receptor, megakaryocyte proliferation and differentiation
		CD23 (FcεRII)	C-type lectin, low affinity receptor for IgE
4. BLOOD GROUP ANTIGENS	AB0	A,B,H	Oligosaccharides
	Le	Le^a, Le^b	
	Ii	I,i	
	P	P	
5. OTHER MOLECULES	Constitutive glycosphyngolipid	CDw17 (lactosyl ceramide)	Activation, release of granules
	Oligosaccharides	CD60	Oligosaccharide, present on gangliosides
	α-granules proteins	GMP-33	Translocated to the cell surface after activation
	Lyzosomes proteins	CD107a,b (Lyzosomes-associated membrane proteins LAMP-1&2)	Liposomal membrane protein translocated to the cell surface after activation
	TM4 superfamily – other members	CD63 (Granulophysin, LIMP)	Tetra spanning membrane protein, liposomal membrane protein translocated to the cell surface after activation
		CD82	Tetra spanning membrane protein, suppresses metastasis, transduce signals in B, T cells and monocytes
	GPI-linked proteins	CD109	Platelet activation factor, GR56, $Gov^{a/b}$ alloantigen on platelets

Table 2. Molecules on the non-activated platelet membrane

System	Antigen	Phenotype frequency*	Glycoprotein	Nucleotide Change*	Aminoacid change	CD
HPA-1	HPA-1a HPA-1b	97.9% 28.8%	GPIIIa	T^{176} C^{176}	Leucine^{33} Proline^{33}	CD61
HPA-2	HPA-2a HPA-2b	>99.9% 13.2%	GPIbα	C^{482} T^{482}	Threonine^{145} Methionine^{145}	CD42b
HPA-3	HPA-3a HPA-3b	80.95% 69.8%	GPIIb	T^{2621} G^{2621}	Isoleucine^{843} Serine^{843}	CD41
HPA-4	HPA-4a HPA-4b	>99.9% <0.1%	GPIIIa	G^{506} A^{506}	Arginine^{143} Glutamine^{143}	CD61
HPA-5	HPA-5a HPA-5b	99.0% 19.7%	GPIa	G^{1600} A^{1600}	Glutamic acid^{505} Lysine^{505}	CD49b
	HPA-6bw	0.7%	GPIIIa	G^{1544} A^{1544}	Arginine^{489} Glutamine^{489}	CD61
	HPA-7bw	0.2%	GPIIIa	C^{1297} G^{1297}	Proline^{407} Alanine^{407}	CD61
	HPA-8bw	<0.01%	GPIIIa	C^{1984} T^{1984}	Arginine^{636} Cysteine^{636}	CD61
	HPA-9bw	0.6%	GPIIb	G^{2602} A^{2602}	Valine^{837} Methionine^{837}	CD41
	HPA10bw	<1.6%	GPIIIa	G^{263} A^{263}	Arginine^{62} Glutamine^{62}	CD61
	HPA11bw	<0.25%	GPIIIa	G^{1976} A^{1976}	Arginine^{633} Histidine^{633}	CD61
	HPA-12bw	0.4%	GPIbβ	G^{119} A^{119}	Glycine^{15} Glutamic acid^{15}	CD42c
	HPA-13bw	0.25%	GPIa	C^{2483} T^{2483}	Threonine^{799} Methionine^{799}	CD49b
	HPA-14bw	<0.17%	GPIIIa	$AAG^{1909-1911}$ deletion	Lysine^{611} deletion	CD61
HPA-15	HPA-15a HPA-15b	81% 74%	CD109	C^{2108} A^{2108}	Serine^{703} Tyrosine^{703}	CD109
	HPA-16bw	<0.1%	GPIIIa	C^{497} T^{497}	Threonine^{140} Isoleucine^{140}	CD61
	HPA-17bw	<0.1%	GPIIb/IIIa	C^{662} T^{662}	Threonine^{195} Methionine^{195}	CD61
	HPA-18bw	<0.1%	GPIa	G^{2235} T^{2235}	Glutamine^{716} Histidine^{716}	CD49b
	HPA-19bw	<0.1%	GPIIIa	A^{487} C^{487}	Lysine^{137} Glutamine^{137}	CD61
	HPA-20bw	<0.1%	GPIIb	C^{1949} T^{1949}	Threonine^{619} Methionine^{619}	CD41
	HPA-21bw	<0.1%	GPIIIa	G^{1960} A^{1960}	Glutamic acid^{628} Lysine^{628}	CD61

**Nucleotide change - numbers are given in relation to the reference sequence in the NCBI database*

Table 3. Human platelet alloantigens (HPA) *(Modified from Metcalfe et al. 2003 and IPD HPA database http://www.ebi.ac.uk/ipd/hpa/freqs_1.html*

5. Platelet growth factors

As already stated, intact platelets contain alpha-granules and other granular bodies in which important mediator molecules are stored, and are released from the platelet following activation. Among these are the growth factors (GFs), which when released from activated platelets, exhibit an extensive tissue forming ability, such as the initiation and modulation of wound healing in both soft and hard tissues (Anitua et al., 2004; Harrison & Cramer, 1993; Lacoste et al., 2003; Weibrich et al., 2002). (See Table 1. and Table 4.)

IL-1ß, Interleukin 1beta	• stimulates lymphocyte proliferation • influences collagenase activity
IL-8, Interleukin 8	• chemotactic for all known types of migratory immune cells • specifically activates neutrophile granulocytes • mitogen for epidermal cells
EGF–ß, Epidermal growth factor (beta-Urogastron)	• stimulates re-epithelization, angiogenesis, and collagenase activity
bFGF/ FGF-2, Fibroblast growth factor (basic)	• stimulates angiogenesis, endothelial cell proliferation , collagen synthesis, wound contraction, matrix synthesis, epithelization, keratinocyte growth factor production
aFGF/ FGF-1, Fibroblast growth factor (acidic)	• most potent GF identified thus far for skin keratinocytes playing role in tissue repair following skin injuries.
KGF/ FGF-7 Keratinocyte growth factor	• growth and development of progenitors of granulocytes and macrophages, stimulates myeloblasts and monoblasts and triggers their irreversible differentiation, synergises with EPO in the proliferation of erythroid and megakaryocytic progenitor cells, in combination with another colony stimulating factor, M-CSF, synergically suppresses the generation of macrophage-containing cell colonies • for some types of blast cells acute myeloid leukemia acts as an autocrine mediator of growth. • a strong chemoattractant for neutrophils, enhances microbicidal activity, oxidative metabolism, and phagocytic activity of neutrophils and macrophages, improves their cytotoxicity
GM-SCF/ CSFα, granulocyte/macrophage colony-stimulating factor alpha	• stimulates both the proliferation and the differentiated function in osteoblasts.
IGF-1 Insulin-like growth factor	• a growth factor for normal fibroblasts, promotes the synthesis of collagenase and prostaglandin E2 in fibroblasts, may function also as an autocrine growth modulator for human chronic lymphocytic leukemia cells in vivo, an autocrine growth modulator for neuroblastoma cells, the autocrine growth-promoting activity is inhibited by IL4.
TNF- α Tumor necrosis factor alpha	• Besides its immune functions, TNF is a potent promoter of angiogenesis in vivo. TNF-alpha is a growth factor for human fibroblasts, where it promotes the synthesis of collagenase and prostaglandin E2

	• promotes the proliferation of astroglial cells and microglial cells and therefore may be involved in pathological processes such as astrogliosis and demyelinisation.
PDGF A Platelet derived growth factor A	• PDGF isoforms are potent mitogens for connective tissue cells, including dermal fibroblasts, arterial smooth muscle cells, chondrocytes and some epithelial and endothelial cells • chemotactic and mitogenic for fibroblasts and smooth muscle cells, neutrophils and mononuclear cells • activates TGF-β, stimulates neutrophils and macrophages, stimulates chemotaxis, mitogenesis of fibroblasts and smooth muscle cells, collagen synthesis and collagenase activity, angiogenesis • a major mitogen for connective tissue cells and certain other cell types • stimulates cell growth, but also changes cell shape and motility; induces the reorganization of the actin filament system and stimulates chemotaxis, i.e., a directed cell movement toward a gradient of PDGF
PDGF B Platelet derived growth factor B	• regulates the proliferation and the differentiation of multiple cell types. TGF found in platelets is subdivided into TGF α and β, which are the more generic connective tissue growth factors involved with matrix formation influencing osteoblasts to lay down bone matrix through the process of osteogenesis. • TGFβ1 and TGFβ2 activate fibroblasts, endothelial and osteoprogenitor cells, chondroprogenitor cells and mesenchimal stem cells. A chondroprogenitor cell will further differentiate and produce the matrix for cartilage. A mesenchymal stem cell stimulated to mitose provides wound healing cells
TGF-α, Transforming growth factor alpha	• stimulates mesenchymal, epithelial, and endothelial cell growth, endothelial chemotaxis
TGF-ß1, Transforming growth factor ß1	• stimulates monocytes to secrete FGF, PDGF, TNF- α , Interleukin-1 • stimulates fibroblast chemotaxis and proliferation • stimulates collagen synthesis • decrease dermal scarring
VEGF/ VEP Vascular endothelial growth factor	• highly specific mitogen for vascular endothelial cells • influences vascular permeability and is a strong angiogenic protein and probably also plays a role in neovascularisation under physiological conditions • in endothelial cells, VEGF induces the synthesis of von Willebrand factor. • a potent chemoattractant for monocytes and thus has procoagulatory activities. • in microvascular endothelial cells induces the synthesis of plasminogen activator and plasminogen activator inhibitor type 1. • induces the synthesis of the metalloproteinase, interstitial collagenase, which degrades interstitial collagen type 1, collagen type 2, and collagen type 3 under normal physiological conditions.

Table 4. Platelet growth factors (*Modified from http://www.copewithcytokines.de/cope.cgi)*

Growth factors modulate the functional activities of individual cells and tissues, mediate interactions between cells directly and regulate processes taking place in the extracellular environment in autocrine, paracrine, iuxtacrine or retrocrine fashion. GFs work by binding to specific cell surface receptors and can target cells in a number of recognized ways or modes. Release of these substances into the blood stream allows them to get to distant targets (endocrine mode), diffuse over short distances to affect other cells (juxtacrine mode), and to influence neighbouring cells (paracrine mode). Growth factors can even act on the cell in which they are produced (autocrine mode). These different modes are all likely to be operative during tissue repair.
The most important GFs released by the platelets are PDGF (the platelet derived growth factor), TGF-α & β (transforming growth factor alpha & β), EGF (epidermal growth factor), FGF (fibroblast growth factor), IGF (insulin growth factor), PDEGF (platelet derived epidermal growth factor), PDAF (platelet derived angiogenesis factor), IL-8 (interleukin-8), TNF-α (tumour necrosis factor alpha), CTGR (connective tissue growth factor), GM-CSF (granulocyte macrophage colony stimulating factor), KGF (keratinocyte growth factor), and Ang-2 (angiopoetin) (Frechette et al., 2005; Borzini & Mazzucco, 2005; Westerhuis et al., 2005; Everts et al., 2006) (see Table 4.). Beside these GFs, there are at least 60 different biologically active substances in the platelets that are involved in tissue repair mechanisms such as chemotaxis, cell proliferation and differentiation, angiogenesis, extracellular matrix deposition, immune modulation, antimicrobial activity, and remodelling (Borzini & Mazzucco, 2007). All these functions have been demonstrated in numerous *in vitro* (Phillips et al., 1994; Dankert, 1995; Brill et al., 2004; Cenni et al., 2005; Kark et al., 2006) and *in vivo* experiments (Debus et al., 2001; Kevy et al., 2004).

6. Therapeutic platelet products

Even the topically applied fibrin glue itself improves the haemostasis, reduces blood loss, and tightens the sealing of sutures due to fibrin formation, providing a stabilized matrix for ingrowing fibroblasts and therefore promoting the support of the granulation tissue, revascularization and re-epithelization. Further inclusion of growth factor containing platelets into this concept of tissue regeneration provided and increased tissue repair and regeneration. Various types of platelet products, such as platelet rich plasma, platelet-leukocyte gel and platelet gel, derived from autologous or allogeneic peripheral blood, have been tested for tissue repair.

6.1 Platelet-rich plasma

Platelet-rich plasma (PRP) is defined as a portion of the plasma fraction of autologous blood having a platelet concentration above baseline. Usual product of blood banking is the allogeneic platelet concentrate or platelet-rich plasma, obtained from random blood donors. Autologous platelet rich plasma preparations involve the procurement of 50-100ml of blood drawn from the patient at the point of care. The blood is processed using PRP devices yielding 5-10ml of concentrated PRP (3 to 5X baseline levels). PRP serves as a growth factor vehicle and has both mitogenic and chemotactic properties. It contains a high level of platelets and a full complement of clotting and growth factors (Lacci & Dardik, 2010). The PRP is then activated and applied to the wound site or mixed with bone material for implantation. This procedure is usually performed in under an hour by a well trained and qualified technician.

6.2 Platelet gel

Autologous platelet-gel (PG) is the result of harvesting one's own cells (platelets) that are concentrated by means of centrifugation and exposed to an agonist, which induces activation resulting in the release of intrinsic substances that are applied to the target area where they accelerate wound healing. Autologous PG is attractive because it concentrates a large number of biologically active substances, which are primarily proteins that participate in a complex series of mechanisms involved in inflammation and wound healing (Everts et al., 2006; Stammers et al., 2009; Tschon et al., 2011; Mehta & Watson, 2008). Lately, the possible use of allogeneic PG with similar results was demonstrated in our study (Smrke et al. 2007).

6.3 Platelet-leukocyte gel

Platelet-leukocyte gel contains high concentrations of platelets and leukocytes. As leukocytes play an important role in the innate host-defence, PLG might have antimicrobial properties. A unit of whole blood is centrifuged to obtain platelet-leukocyte-rich plasma (P-LRP). The PLG is then formed by mixing P-LRP with a thrombin-calcium chloride preparation. Similar to PG, the treatment with PLG provides a source of concentrated platelets, with granules that contain growth factors. In addition the high content of non-activated leukocytes, present in the PLG, promotes anti-microbial activity at the wound site through the destruction of bacteria and foreign materials and the removal of damaged tissue (Everts et al, 2006b; Everts et al. 2008).
Platelet gel can be nowadays prepared using a range of commercially available instruments from a number of manufactures. There are no uniform manufacturing standards for platelet gel in existence, which has disturbed the classical blood banking. The existing literature adds controversies to the use of PLT concentrates. When talking about platelets and their products, a great number of variables have to be considered. These variables are mainly related to the PRP preparation methods, the type of activators used, intra- and inter-species variability, types of pathology to be treated, the ways and times of administration and the association of PRP or PG with other treatments. The improved knowledge on the variables affecting therapeutic efficacy will surely help in addressing the best combination of factors implied in the different steps of PLT concentrate preparation and use (Mazzucco et al., 2009).

7. Activation of PRP and the formation of the gel

The typical source of platelet GFs is platelet-rich plasma (PRP) – Figure 4A. Platelet gel (PG) is formed after the PRP is exposed to a strong coagulation stimuli, which is usually the addition of thrombin and calcium (to counteract the anticoagulant citrate if plasma is citrated) (Mazzucco et al., 2004; De Somer et al., 2006) or some other fibrinogen-digestion based activator, such as batroxobine from the viper *Bothrops atrox* snake venom (Carr et al., 2003; Mazzucco et al., 2008).
Consequently, platelet activation and a coagulation cascade follow in a positive loop fashion, which finally results in the formation of a thrombus-like gelatinous substance –

platelet gel. (See Figures 4B, 4C). In the gel, activated platelets are trapped in a fibrin network and they release their active substances, which slowly diffuse into the surroundings. The final result - platelet gel - can be shaped according to the needs, put on different vehicles such as medical gauze or even advanced types of biocompatible carriers and scaffolds, such as fibrin or hyaluronic acid, and used topically. It can be used as a vehicle for the stem cells and ground cancellous bone transplant, allowing the formation and shaping of grafts with determined shapes (See Fig. 4D).

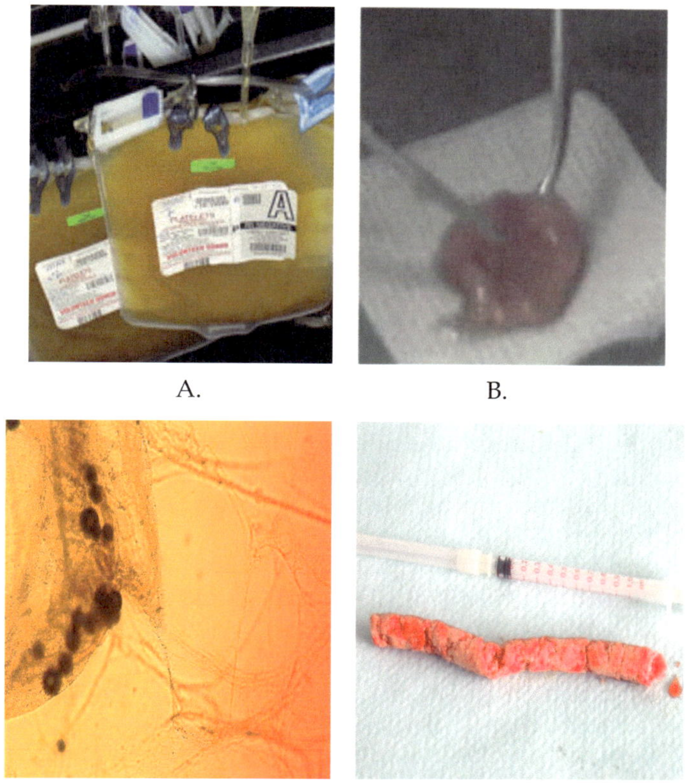

Legend: A. Platelet rich plasma; B.Platelet gel; C: Microscopical view of the platelet gel. Inverted microscope Nikon Eclipsete 300, 10x enlarged.; D: Graft with the shape used for mandibular deffects, containing PG, stem cells and cancellous bone

Fig. 4. Platelet gel

At the beginning autologous PRP/platelet gel was developed as a by-product of multicomponent apheresis (Rosenthal et al, 1975). Later several systems became available to prepare PRP and from this, the platelet gel. These systems produce two- to six-fold platelet - enriched concentrations. Very few data have been reported on the kinetics of growth factor release from PRP-gels. The bicavailability of growth factors in tissue healing depends on the amount of growth factors stored in the platelets but a portion of these is lost during platelet manipulation. In one report, three commercially available devices (Fibrinet, RegenPRP-Kit, Plateltex) and one manual procedure (home-made) were evaluated with reference to the resulting platelet concentration, growth factor content and the kinetics of growth factor release from the gel. Similar methods for platelet gel preparation have revealed different performances concerning the growth factor recovery and the kinetics of its release from the gel. It is unclear whether these noticeable differences are important for clinical management (Mazzucco et al., 2009).

Some research groups have successfully engineered kits for platelet gel formation, such as REGENLAB® and PLATELTEX®, which include sodium carboxymethylated cellulose fibres (Hydrofibre™) as a biocomposite-scaffold, entrapping human platelets in PRP. The PLATELTEX® activation kit, popular in Europe, ensures very rapid gelation and the production of platelet and fibrin gel for acute and chronic ulcer treatment. This reduces spillage of the gel during topical application. The new generation of PRP activation, provides for excellent platelet recovery, collection efficiency and PDGF-AB availability. Gelation of the PRP is induced by the enzyme batroxobin, which does not exert pharmacological activity after topical gel administration. In Europe, clinical trials have been authorised using PRP-gel preparation including activation with batroxobin. PRP-gel has not been considered a somatic cell therapy product.

Platelets treated with thrombin become activated and they release their growth factors quickly. Furthermore, thrombin-platelet interaction is a physiological mechanism that hastens the clot-retraction rate. On the contrary, platelets treated with batroxobin do not become activated; they are passively entrapped within the fibrin network, and their growth factor release occurs slowly. In these conditions, the clot retraction takes longer to occur. According to these differences between thrombin and batroxobin, it is expected that batroxobin-induced PRP activation will tailor the slow release of the platelet content, thus providing longer in loco availability of trophic factors. In selected clinical conditions, this durable anabolic factor availability might be preferable to a quick thrombin-induced growth factor release (Mazzucco et al, 2008).

An alternative means of PRP activation was documented in 2007, when PRP was pulsed with one 300ns pulse with an electric field of 30kV/cm, platelets aggregated and a platelet gel was produced (Zhang et al., 2008).

The lack of effective delivery and the efficient targeting specificity limits clinical applications of platelet gel. In 2008, heparin possessing PDGF binding domain was crosslinked to the collagen-based demineralized bone matrix (DBM) for the delivery of human PDGF. In *in vitro* experiments, heparin improves the binding of PDGF to collagen. The in *vitro* activity assay indicates that the collagen-heparin-PDGF (CH-PDGF) complex promotes human fibroblasts to proliferate on collagen gel. In addition, CH-PDGF stimulates cells to migrate into DBM scaffolds after implantation. The histological analysis shows that CH-PDGF promotes the vascularization of the implants. In summary, heparin-DBM/PDGF could prevent the diffusion of PDGF, prolong its activity, and promote the cellularization and vascularization of the scaffold (Sun et al., 2009).

Recent findings on multiple biological properties of human umbilical cord blood (UCB) and its high level of viral safety prompted some authors to investigate the characteristics of its platelets and the possibility of producing PG from cord blood. UCB-derived PG releases high levels of vascular endothelial growth factor (VEGF) and platelet-derived growth factor-BB (PDGF-BB), substantial amounts of fibroblast growth factor (FGF), hepatocyte growth factor (HGF) and transforming growth factor-beta 1 (TGFbeta1), and minimal amounts of PDGF-AB. These findings suggest that UCB-derived PG can be a preferable tool for tissue engineering applications where high levels of VEGF and PDGF may be desirable (Parazzi et al., 2010).

8. The results of recent clinical research

The results of clinical research indicate that platelet-derived growth factors act in synergy with plasma-derived factors to activate a complex network of autocrine functions that modulate and enhance healing. Therefore, autologous platelet rich plasma (PRP) in the form of activated platelet gel and recombinant morphogenic proteins have been used for healing all kinds of tissues with various results.

8.1 Chronic wounds

When the fibroblasts were sealed in the wound with fibrin to which platelet releasate was added, a cellular infiltrate that had characteristics of granulation tissue invaded the wound space. Platelet releasate appeared to be an additional requirement for the early induction of granulation tissue and for enhanced granulation tissue accumulation. It is now widely accepted that the correct strategy to promote the wound-healing cascade is to prepare an autologous PRP/platelet gel that contains growth factors, and administer it directly to sites of surgical interventions or injuries.

Patients suffering from diabetes mellitus, are seventeen times more likely to develop gangrene and five out of six major limb amputations occur in diabetic patients. In the United States, the diabetic foot problem accounts for 20% of all diabetic hospital admissions and 50% of all non-traumatic amputations. Ulcers occur in 15% of diabetics, and 6-20% of all hospitalized diabetic patients have foot ulcers. The topical use of platelet rich plasma may play an important role in the initiation of the repair process of chronic wounds. Many patients can be treated at home without difficulty, requiring only periodic outpatient examination (Steed et al., 1992; Millington and Norris, 2000; Borzini, 2006). Although some of the earlier studies resulted in poor outcomes (Reutter et al.,1999), newer studies tend to produce much better results.

8.2 Bone

Platelet gel is very effective in formation of new bone when applied together with cancellous bone containing stem cells for treatment of non healing of long bone fractures (Smrke et al., 2007). Platelet gel enhanced the bone formation contrary to the using of plasma alone as measured by quantity of incorporated bone both *in vitro* and *in vivo* (Geuze et al., 2009). In an article presented in 2010, 115 patients with finger amputations or wounds were treated with platelet gel. Loss of bone tissue represented an obstacle to the total tissue recovery, but the aesthetic results were satisfactory in nearly all cases, the recovery of soft tissue in all patients ranged from 80 to 100% (Balbo et al., 2010). However, platelet gel was ineffective for the spinal fusions where it failed to enhance fusion rate when added to autograft in patients undergoing instrumented posterolateral spinal fusion so several authors do not recommend its use to supplement autologous bone grafting during instrumented posterolateral spinal fusion (Castro, 2004; Burkus, 2005; De Somer et al., 2006; Savarino et al., 2006).

8.3 Tendons and ligaments

Tendons and ligaments naturally have poor blood supply. Damage most often occurs in the knee, ankle, shoulder, elbow, wrist, biceps, calf, hamstrings and Achilles tendons. The use of platelet gel has been generally well tolerated by the patients and it resulted in significant increase of healing parameters. Platelet injection therapy has changed the way orthopedic specialists treat sports injury patients. Usually PRP is injected into the damaged area using minimally invasive protocols that involve ultrasonically guided platelet rich plasma (PRP) injections. (Everts et al. 2007; Everts et al.,2008; Gardner et al., 2006).

8.4 Aesthetic surgery

There is especially extensive marketing for the use of PRP in aesthetic surgery and lifting (Bhanot & Alex, 2002; Powell et al., 2001). Private organisations tend to offer this therapy throughout the world. They offer PRP delivery by various instruments into the skin tissues, rendering good aesthetic effects (see Figure 5). The indications include facelifts, browlifts, the creation of flaps, blepharoplasty, mammaplasties, and abdominoplasties. PRP enhanced grafts generally had a higher graft acceptance and better graft retention than traditional fat grafting techniques, along with decreased swelling and bruising in the donor sites (Sclafani et al., 2005;Brown et al, 2006).

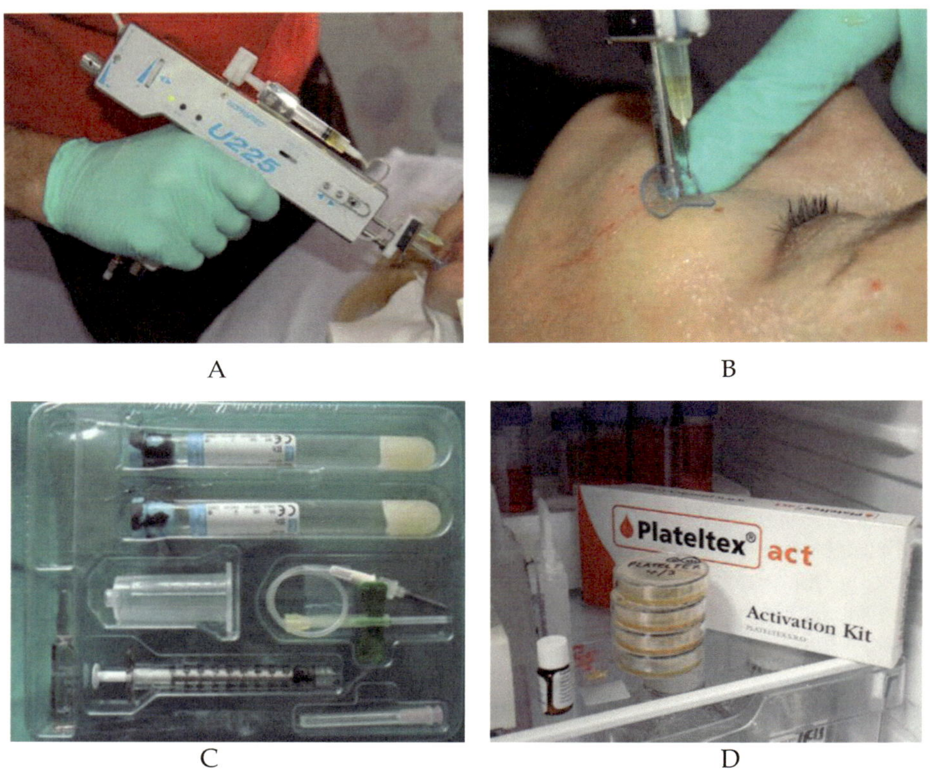

Legend: A: The instrument Mesogun for delivering PRP (U225)as marketed by company Boland; B: The needle depth can be accurately set for precise delivery; C: An example of kit for PRP preparation and application from Omnimed (Pty)., Ltd., JHB, South Africa; D: An example of platelet gel activation kit PLATELTEX® (http://www.bolandcell.co.za/platelet.htm)

Fig. 5. Use of autologous platelet-rich plasma (PRP).

8.5 Cardiovascular

Wound infection is a devastating complication after cardiovascular surgery. This can lead to surgical reoperation, prolonged intensive care unit stay, and increased mortality promoting the raising the cost of cardiac surgery while decreasing the patient quality of life. The antimicrobial action of white blood cells and platelets, both found in the platelet gel, may reduce the potential for infection to develop during cardiac surgery. Various protocols have documented such an improved healing of surgical wound in cardiac surgery (Mogan and Larson, 2004; Trowbridge et al,2005; Englert et al., 2008; Gunyaqdin et al., 2008)

8.6 Oral and craniofacial surgery

As mentioned, the oral surgeons were the leading community in the use of platelet gel for transplantation of artificial and natural grafts. Platelet gel reached a huge popularity in maxillofacial settings where it is considered standard treatment of bone defects, artificial endosseal implants, oral reconstructions, maxillary sinus augmentation, etc. (Marx et al., 1998; Belli et al., 2005; Pomerantz and Dutton, 2005; Mendez et al., 2006).

8.7 Other applications

Platelet gel has been successfully used for ocular diseases, such as for enhanced healing of corneal surface and ocular surgery (Koffler, 2006; Liu et al.,2006; Lee & Kang, 2011).

Lately, the studies of PG use addressed some unexpected fields of medicine. For instance, PG was used as a safe and effective tool in the management of mucosal skin lesions related to the graft- versus-host disease GVHD. Six patients with multiple lesions involving dermis, subcutaneous or oral mucosa and related to GVHD underwent PLT gel as local therapy. After the second PLT gel application, the pain disappeared in all cases and the granulation tissue was observed in the four patients, five of six patients showed a complete response, while one patient with a partial response died early from multiorgan failure (Picardi et al., 2010).

Platelet gel mixed with centrifuged fat tissue has been successfully applied for the patients affected by Parry - Romberg Syndrome (progressive hemifacial atrophy and volumetric deficit), which is an uncommon degenerative and poorly understood condition. It is characterized by a slow and progressive atrophy affecting one side of the face. The atrophy progresses slowly for several years and becomes stable. After stabilization of the disease, plastic surgery of autologous fat grafts can be performed. Study suggested a therapeutic plan comprised of two sequential treatments: acquisition of platelet gel from a small volume of blood (9 ml) followed by the Coleman technique for reconstructing the three-dimensional projection of the face contour, restoring the superficial density of the facial tissues. The results obtained prove the efficacy of these two treatments combined, and the satisfaction of the patient confirms the quality of the results. (Cervelli & Gentile, 2009).

Recently it is surprisingly becoming clear that platelets have a definite role in the cancer, which is probably due to their pro-angiogenesis stimulation and regulation of new blood vessel growth through numerous stimulators and inhibitors of angiogenesis, such as endostatin and VEGF, by several pathways. Cancer cells probably preferentially stimulate platelets to secrete their pro-angiogenic payload, which is counteracted by the anti-platelet agent aspirin that inhibits the platelet-mediated angiogenesis (Elisabeth et al., 2011).

9. Pros and cons of platelet gel therapies

The common advantages of the PG driven healing can be summarized as follows: there seems to be a complete absence of keloid formation, as far as pathologic hyperplasia, metaplasia or dyschromia of the healed lesion (Pomerantz et al., 2005, Lundblad & White, 2005). Other advantages are comfort and pain diminution (Pomerantz et al., 2005, Mendez et al., 2006), less oedema and ecchymosis (Powell et al., 2001; Mendez et al., 2006; Brown et al., 2006). Particularly valuable advantages are reported through literature: reduced mortality after cardiac surgery (Trowbridge et al., 2005), reduced infection rate after cardiac and orthopaedic surgery (Gilsanz et al., 2001; Salvadè et al., 2010), reduced transfusion supply and hospital stay after orthopaedic surgery (Gardner et al., 2006), reduced amputation rate in diabetic foot (Knighton et al., 1990). PRP therapy became very popular in the USA, which was due to the fact that it attracted enormous attention after the Super Bowl finals in 2009, when the athletes promoted it on TV (Donald, 2011).

Drawbacks of PG therapies are far less pronounced. It is interesting that besides the considerable variability of the clinical studies, PG therapy had practically no or only minute ill effects when compared to standard therapies. Owing to the high heterogeneity of the products, a need for standardization is a commonly shared sentiment. Heterogeneous production generates heterogeneous products which, in turn, have distinct biological properties and healing capacities (Zimmermann et al., 2003; Eppley et al., 2006; Fried et al., 2006).

Considering that heterogeneity involves products, clinical conditions, tissues to be cared for, treatment protocols and probably many other minor variables, authors strongly support the project for the establishment of a clinical database in order to collect enough data for multivariate analysis of the clinical results (Kantor & Margolis, 2000).

10. The dilemma of allogeneic vs. autologous platelet gel

In the majority of clinical applications autologous platelets were used for the platelet gel formation. As mentioned previously, autologous PG products are prone to enormous variability, resulting in scientific scepticism and disbelief regarding the actual value of this method. In our study in 2007, we used instead allogeneic PRP. We reported a case of a 50-year-old type II diabetic male with a comminuted fracture of the tibia and delayed union after insufficient initial osteosynthesis with a resulting pseudarthrosis, who was treated operatively by using a graft composed of allogeneic platelet gel mixed with autologous cancellous bone. Due to a history of diabetes, allogeneic instead of autologous platelets that were ABO and RhD matched, leukocyte depleted, irradiated and activated by human thrombin were used. No side effects were observed and no immune reactions such as platelet- or HLA-class I antibodies were detected (Smrke et al., 2007).

We explored several advantages of allogeneic platelets, as compared to the classical autologous setting. Allogeneic platelet units from the blood bank are available in larger quantities, they are safe and affordable, they are highly standardized in terms of platelet, residual leukocyte and red blood cell content, the centrifugal forces used for their isolation, the temperature of the centrifugation, techniques of separation and processing and the composition of the preservative solution, as mandated by international standards in USA and Europe. In addition, they are constantly resuspended on agitators in a standard manner (Brecher, 2005; Council of Europe, 2011). On the other hand, autologous platelet preparations are subject to enormous variability, which hinders serious clinical studies.

The extensive use of allogeneic platelets could, beside the regenerative medicine, influence the blood transfusion policies in future. The platelet units have a short shelf life of 3 - 7 days and are therefore prone to outdating, regardless of the best practices performed. In 2001 in USA, at least 1,234.000 out of a total of 12,898.000 produced platelet units (9.5%) were outdated and discarded (Sullivan et al, 2007). In another study, expiry rates of platelet units were different in various hospitals, ranging between 2% and 13.8% (Novis et al, 2002). It is clear from these figures that a considerable amount of platelet units are routinely discarded, which could be reversed by the utilization of frozen units in regenerative purposes, such as platelet gel preparation.

Besides, the use of platelet gel could have an important impact on economy of medicine. Let us focus only at the therapy of chronic wound with the platelet gel. Ulcer secondary to venous hypertension and venous insufficiency is the most common chronic wound with 57-80% incidence of all chronic wounds. Average incidence is 15-30 in 100.000 persons and they account for nearly 70% of all leg ulcers (Margolis et al., 2001). All these conditions are more prevalent in elderly population, which represents 20% of the total, a figure that is likely to increase to 25% by 2020. In addition, the increased occurrence and longevity of these ulcers are further compounded by the detrimental effects ageing has on the skin and the wound healing process. Chronic wounds significantly impair the quality of life of more than 9 million of people worldwide and account for a disproportionate share of healthcare expenditures: it is estimated that yearly treatment cost for health care systems averages 9 – 12 k€ per each patient. Even with increasing knowledge and the development of more sophisticated interventions, many clinicians encounter wounds that are hard-to-heal where, despite best efforts, wound healing is prolonged or never achieved (Moffatt, 2008). Indeed, chronic and hard-to-heal wounds are a challenge to the health care professionals. Thus, in the frame of progressively ageing population, the efficient therapy for chronic wounds and tissue reparation promises improving life quality and combating rising healthcare costs. Currently available and practiced therapies such as wound dressings, pentoxifylline, laser and intermittent pneumatic compression treatments applied in concomitance with compressive bandages have shown rather disappointing overall outcomes except some encouraging results obtained by combination of compression and bilayer of artificial skin (Palfreyman et al., 2006).

11. Conclusion

Platelets are intriguing cellular particles that have, in the course of human and animal evolution, beside their specialisation into the haemostasis, kept some other abilities of their predecessors - universal blood cells – the abilities of immune defence and tissue forming. For the purpose of tissue forming, platelets can also be used therapeutically. Autologous or allogeneic blood can be used for this purpose and centrifuged to produce a platelet concentrate, which is then further *in vitro* activated to form a thrombus-like gelatinous substance – the platelet gel. Many *in-vitro* studies have established that platelet- derived growth factors accelerate proliferation of an array of cells involved in soft and bony tissue regeneration. These effects have also been evaluated and confirmed by numerous *in vivo,* both in animal and human clinical studies. Consequently, we can say that the platelet-derived products currently represent a valuable therapeutic modality, offering opportunities for various applications in regenerative medicine and tissue engineering.

12. Acknowledgment

The authors acknowledge the kind assistance of Petra Pavlovčič in prepairing the manuscript.

13. References

Altman, D.G., Schulz, K.F., Moher, D., Egger, M., Davidoff, F., Elbourne, D., Gotzsche, P.C. & Lang, T. (2001). The revised CONSORT statement for reporting randomized trials: explanation and elaboration. *Ann Intern Med*, Vol.134, No.8, (April 2001), pp. 663-94, ISSN 0003-4819

Anderson, K. & Baker, S. (2003). Advances in facial rejuvenation surgery. *Curr Opin Otolaryngol Head Neck Surg*, Vol.11, No.4, (August 2003), pp. 256-60, ISSN 1068-9508

Anitua, E., Andia, I., Ardanza, B., Nurden, P. & Nurden, A.T. (2004). Autologous platelets as a source of proteins for healing and tissue regeneration. *Thromb Haemost*, Vol.91, No.1, (January 2004), pp. 4-15, ISSN 0340-6245

Balbo, R., Avonto, I., Marenchino, D., Maddalena, L., Menardi, G. & Peano, G. (2010). Platelet gel for the treatment of traumatic loss of finger substance. *Blood Transfus*, Vol.8, No.4, (October 2010), pp. 255-9, ISSN 1723-2007

Battinelli, E.M., Markens, B.A. & Italiano, J.E., Jr. (2011). Release of angiogenesis regulatory proteins from platelet alpha granules: modulation of physiological and pathological angiogenesis. *Blood*, Epub ahead of print June 16 (June 2011), doi: 10.1182/blood-2011-02-334524, ISSN 1528-0020

Belli, E., Longo, B. & Balestra, F.M. (2005). Autogenous platelet-rich plasma in combination with bovine-derived hydroxyapatite xenograft for treatment of a cystic lesion of the jaw. *J Craniofac Surg*, Vol.16, No.6, (November 2005), pp. 978-80, ISSN 1049-2275

Bernuzzi, G., Tardito, S., Bussolati, O., Adorni, D., Cantarelli, S., Fagnoni, F., Rossetti, A., Azzarone, M., Ficarelli, E.,Caleffi, E., Gazzola, G. & Franchini, M. (2010). Platelet gel in the treatment of cutaneous ulcers: the experience of the Immunohaematology and Transfusion Centre of Parma. *Blood Transfus*, Vol.8, No.4, (October 2010), pp. 237-47, ISSN 1049-2275

Bhanot, S. & Alex, J.C. (2002). Current applications of platelet gels in facial plastic surgery. *Facial Plast Surg* vol.18, No.1, (February 2002), pp.27-33, ISSN 0736-6825

Borzini, P. & Mazzucco, L. (2005). Tissue regeneration and in loco administration of platelet derivatives: clinical outcome, heterogeneous products, and heterogeneity of the effector mechanisms. *Transfusion*, Vol.45, No.11, (November 2005), pp. 1759-67, ISSN 0041-1132

Borzini, P., Mazzucco, L., Giampaolo, A. & Hassan, H.J. (2006). Platelet gel - the Italian way: a call for procedure standardization and quality control. *Transfus Med*, Vol.16, No.4, (August 2006), pp. 303-4, ISSN 0958-7578

Borzini, P. & Mazzucco, L. (2007). Platelet-rich plasma (PRP) and platelet derivatives for topical therapy. What is true from the biological view point? *ISBT Science Series*, Vol.2, No.1, (July 2007), pp. 272-81, ISSN 1751-2816

Brandstedt, S., Rank, F.& Olson, P.S. (1980). Wound-healing and formation of granulation-tissue in normal and defibrinogenated rabbits - an experimental-model and histological study. *Eur Surg Res*, Vol.12, No.1, (1980), pp. 12-21, ISSN 0014-312X

Brecher, M.E., (Ed.). (2005). *Technical Manual, 15th Edition*, American Association of Blood Banks, ISBN 1-56395-196-7, Bethesda

Brill, A., Elinav, H. & Varon, D. (2004). Differential role of platelet granular mediators in angiogenesis. *Cardiovasc Res*, Vol.63, No.2, (August 2004), pp. 226-35, ISSN 0008-6363

Brown, S.A., Appelt, E.A., Lipschitz, A., Sorokin, E.S. & Rohrich, R.J. (2006). Platelet gel sealant use in rhytidectomy. *Plast Reconstr Surg*, Vol.118, No.4, (September 2006), pp. 1019-25, ISSN 0032-1052

Burkus, J.K. (2005). Surgical treatment of the painful motion segment: matching technology with indications. *Spine*, Vol.30, No.16 Suppl, (August 2005), pp. 7-15, ISSN 0362-2436

Carreon, L.Y., Glassman, S.D., Anekstein, Y. & Puno, R.M. (2005). Platelet gel (AGF) fails to increase fusion rates in instrumented posterolateral fusions. *Spine*, Vol.30, No.9, (May 2005), pp. E243-46, ISSN 0362-2436

Castro, F.P., Jr. (2004). Role of activated growth factors in lumbar spinal fusions. *J Spinal Disord Tech*, Vol.17, No.5, (October 2004), pp. 380-4, ISSN 1536-0652

Carr, M.E., Jr., Carr, S.L., Tildon, T., Fisher, L.M. & Martin, E.J. (2003). Batroxobin-induced clots exhibit delayed and reduced platelet contractile force in some patients with clotting factor deficiencies. *J Thromb Haemost*, Vol.1, No.2, (February 2003), pp. 243-9, ISSN 1538-7933

Cenni, E., Ciapetti, G., Pagani, S., Perut, F., Giunti, A. & Baldini, N. (2005). Effects of activated platelet concentrates on human primary cultures of fibroblasts and osteoblasts. *J Periodontol*, Vol.76, No.3, (March 2005), pp. 323-8, ISSN 0022-3492

Cervelli, V. & Gentile, P. (2009). Use of platelet gel in Romberg syndrome. *Plast Reconstr Surg*, Vol.123, No.1, (January 2009), pp. 22e-33e, ISSN 0032-1052

Chen, T.M., Tsai, J.C. & Burnouf, T. (2010). A novel technique combining platelet gel, skin graft, and fibrin glue for healing recalcitrant lower extremity ulcers. *Dermatol Surg*, Vol.36, No.4, (April 2010), pp. 456-60, ISSN 1076-0512

Cieslik-Bielecka, A., Bielecki, T., Gazdzik, T.S., Cieslik, T., Szczepanski, T. (2008). Improved treatment of mandibular odontogenic cysts with platelet-rich gel. *Oral Surg Oral Med Oral Pathol Oral Radiol Endod*, Vol.105, No.4, (April 2008), pp. 423-9, ISSN 1079-2104

Council of Europe. (2011). *Guide to the Preparation, Use and Quality Assurance of Blood Components* (16th Edition), Council of Europe, ISBN 978-92-871-7022-4, Strasbourg

Crovetti, G., Martinelli, G., Issi, M., Barone, M., Guizzardi, M., Campanati, B., Moroni, M. & Carabelli, A. (2004). Platelet gel for healing cutaneous chronic wounds. *Transfus Apher Sci*, Vol.30, No.2. (April 2004), pp. 145–151, ISSN: 1473-0502

Dankert, J., van der W.J., Zaat, S.A., Joldersma, W., Klein, D. & Hess, J. (1995). Involvement of bactericidal factors from thrombin-stimulated platelets in clearance of adherent viridans streptococci in experimental infective endocarditis. *Infect Immun*, Vol.63, No.2, (February 1995), pp.663-71, ISSN 0019-9567

Debus, E.S., Schmidt, K., Geiger, D., Dietz, U.A. & Thiede A. (2001). Growth factors for preventing amputation in delayed wound healing. *Kongressbd Dtsch Ges Chir Kongr*, Vol.118, (2001), pp. 829-33, ISSN 1868-1050

Donald, T., (2011). Platelets take on new forms as sports medicine and wound-healing therapies. *AABB News*, Vol.13, No.3, (March 2011), pp. 18-23, ISSN 1523-939X

De Somer, F., De Brauwer, V., Vandekerckhove. M., Ducatelle, R., Uyttendaele, D. & Van Nooten, G. (2006). Can autologous thrombin with a rest fraction of ethanol be used safely for activation of concentrated autologous platelets applied on nerves? *Eur Spine J*, Vol.15, No.4, (April 2006), pp. 501-5, ISSN 0940-6719

Driver, V.R., Hanft, J., Fylling, C.P., Beriou, J.M. & the Autologel Diabetic Foot Ulcer Study Group (2006). A prospective, randomized, controlled trial of autologous platelet-rich plasma gel for the treatment of diabetic foot ulcers. *Ostomy Wound Manage*, Vol.52, No.6, (June 2006), pp. 68-70, 72, 74 passim, ISSN 0889-5899

Englert, S.J., Estep, T.H. & Ellis-Stoll, C.C. (2008). Postoperative surgical chest and leg incision sites using platelet gel: a retrospective study. *J Extra Corpor Technol*, Vol.40, No.4, (December 2008), pp. 225-8, ISSN 0022-1058

Eppley, B.L., Pietrzak, W.S. & Blanton, M. (2006). Platelet-rich plasma: a review of biology and applications in plastic surgery. *Plast Reconstr Surg*, Vol.118, No.6, (November 2006), pp. 147e-159e, ISSN 0032-1052

European Wound Management Association (EWMA). (2008). *Position document: Hard-to-heal wounds: a holistic approach.* MEP Ltd, London

Everts, P.A., Brown Mahoney, C., Hoffmann, J.J., Schonberger, J.P., Box, H.A., van Zundert, A. & Knape, J.T. (2006a). Platelet-rich plasma preparation using three devices: implications for platelet activation and platelet growth factor release. *Growth Factors*, Vol.24, No.3, (September 2006), pp. 165-71, ISSN 0897-7194

Everts, P.A., Devilee, R.J., Brown Mahoney, C., Eeftinck-Schattenkerk, M., Box, H.A., Knape, J.T. & van Zundert, A. (2006b). Platelet gel and fibrin sealant reduce allogeneic blood transfusions in total knee arthroplasty. *Acta Anaesthesiol Scand*, Vol.50, No.5, (May 2006), pp.593-9, ISSN 0001-5172

Everts, P.A., Hoffmann, J., Weibrich, G., Brown Mahoney, C., Schonberger, J.P., van Zundert, A. & Knape, J.T. (2006c). Differences in platelet growth factor release and leucocyte kinetics during autologous platelet gel formation. *Transfus Med*, Vol.16, No.5, (October 2006), pp. 363-8, ISSN 0958-7578

Everts, P.A, Jakimowicz, J.J., van Beek, M., Schönberger, J.P., Devilee, R.J., Overdevest, E.P., Knape, J.T.A. & van Zundert, A. (2007). Reviewing the structural features of autologous platelet-leukocyte gel and suggestions for use in surgery. *Eur Surg Res*, Vol.39, No.4, (April 2007), pp.199–207, ISSN 0014-312X

Everts, P.A., Devilee, R.J., Brown Mahoney, C., van Erp, A., Oosterbos, C.J., Stellenboom, M., Knape, J.T. & van Zundert, A. (2008). Exogenous application of platelet-leukocyte gel during open subacromial decompression contributes to improved patient outcome. A prospective randomized double-blind study. *Eur Surg Res*, Vol.40, No.2, (November 2008), pp. 203-10, ISSN 0014-312X

Ficarelli, E., Bernuzzi, G., Tognetti, E., Bussolati, O., Zucchi, A., Adorni, D. & De Panfilis, G. (2008).Treatment of chronic venous leg ulcers by platelet gel. *Dermatol Ther*, Vol.21, Issue Suppl. s1, (July/August 2008), pp. S13-7, ISSN 1396-0296

Fischer, H. (1979). A method of suture-free anastomosis of nerve transplantation is being reported, using facial nerve as the example. *Laryngol Rhinol Otol*, Vol.58, No.2 (February 1979), pp. 154-6, ISSN 0340-1588

Fried, D.W., Leo, J.J., Weber, F.P., Husain, M. & Cullen, J. (2006). Quantitative and qualitative analysis of platelet-rich plasma collection using the Haemonetics Cell Saver 5 in open heart surgery. *J Extra Corpor Technol,* Vol.38, No.3, (September 2006), pp. 235-40, ISSN 0022-1058

Fang, R.C. & Galiano, R.D. (2008). A review of becaplermin gel in the treatment of diabetic neuropathic foot ulcers. *Biologics*. Vol.2, No.1, (March 2008), pp. 1-12,. ISSN 1177-5475

Frechette, J.P., Martineau, I. & Gagnon, G. (2005). Platelet-rich plasmas: growth factor content and roles in wound healing. *J Dent Res*, Vol.84, No.5, (May 2005), pp. 434-9, ISSN 0022-0345

Garcia, A., Zitzmann, N. & Watson, S.P. (2004). Analyzing the platelet proteome. *Semin Thromb Hemost*, Vol.30, No.4 (August 2004), pp.485-9, ISSN 0094-6176

Gardner, M.J., Demetrakopoulos, D., Klepchick, P.R. & Mooar, P.A. (2007). The efficacy of autologous platelet gel in pain control and blood loss in total knee arthroplasty: An analysis of the haemoglobin, narcotic requirement and range of motion. *Int Orthop,* Vol.31, No.3, (June 2007), pp. 309-13, ISSN 0341-2695

Geuze, R.E., Wegman, F., Oner, F.C., Dhert, W.J. & Alblas J. (2009). Influence of endothelial progenitor cells and platelet gel on tissue-engineered bone ectopically in goats. *Tissue Eng Part A*, Vol.15, No.11, (November 2009), pp. 3669-77, ISSN 1937-3341

Gilsanz, F., Escalante, F., Auray, C. & Olbes, A.G. (2001). Treatment of leg ulcers in beta-thalassaemia intermedia: use of platelet-derived wound healing factors from the patient's own platelets. *Br J Haematol*, Vol.115, No.3, (December 2001), p.710, ISSN 0007-1048

Greiling, D. & Clark, R.A. (1997). Fibronectin provides a conduit for fibroblast transmigration from collagenous stroma into fibrin clot provisional matrix. *J Cell Sci*, Vol.110, (April 1997), pp. 861-70, ISSN 0021-9533

Gresele, P., Page, C., Fuster, V. & Vermylen, J. (Eds.). (2002). *Platelets in Thrombotic and Non-Thrombotic Disorders: Patophysiology, Pharmacology and Therapeutics*, Cambridge University Press, ISBN-13 978-0521802611, Cambridge

Gunaydin, S., McCusker, K., Sari, T., Onur, M., Gurpinar, A., Sevim ,H., Atasoy, P., Yorgancioglu, C. & Zorlutuna, Y. (2008). Clinical impact and biomaterial evaluation of autologous platelet gel in cardiac surgery. *Perfusion*, Vol.23, No.3, (May 2008), pp. 179-86, ISSN 0267-6591

Harrison, P. & Cramer, E.M. (1993). Platelet alpha-granules. *Blood Rev*, Vol.7, No.1, (March 1993), pp. 52-62, ISSN 0268-960X

Jünger, M., Steins, A., Hahn, M. & Hafner, H.M. (2000). Microcirculatory dysfunction in chronic venous insufficiency (CVI). *Microcirculation*, Vol.7, No.S1, (December 2000), pp. S3–S12, ISSN 1073-9688

Jurk, K. & Kehrel, B.E. (2005). Platelets: physiology and biochemistry. *Semin Thromb Hemost*, Vol.31, No.4, (2005), pp. 381-92, ISSN 0094-6176

Kantor, J. & Margolis, D.J. (2000). The accuracy of using a wound care specialty clinic database to study diabetic neuropathic foot ulcers. *Wound Repair Regen,* Vol.8, No.3, (May-June 2000), pp. 169-73, ISSN 1067-1927

Kark, L.R., Karp, J.M. & Davies, J.E. (2006). Platelet releasate increases the proliferation and migration of bone marrow-derived cells cultured under osteogenic conditions. *Clin Oral Implants Res*, Vol.17, No.3, (June 2006), pp. 321-7, ISSN 0905-7161

Kevy, S.V. & Jacobson, M.S. (2004). Comparison of methods for point of care preparation of autologous platelet gel. *J Extra Corpor Technol*, Vol.36, No.1, (March 2004), pp.28-35, ISSN 0022-1058

Knighton, D.R., Ciresi, K., Fiegel, V.D., Schumerth, S., Butler, E. & Cerra F. (1990). Stimulation of repair in chronic, nonhealing, cutaneous ulcers using platelet-derived wound healing formula. *Surg Gynecol Obstet*, Vol.170, No.1, (January 1990), pp. 56-60, ISSN 0039-6087

Kishimoto, T., Kikutani, H., von dem Borne, A.E.G.Kr., Goyert, S.M., Mason, D.Y., Miyasaka, M., Moretta, L., Okumura, K., Shaw, S., Springer, T.A., Sugamura, K. & Zola, H. (Eds.). (1997). *Leucocyte Typing VI: White Cell Differentiation Antigens*, Garland Publishing, ISBN 0815327455, New York, NY

Koffler, B.H. (2006). Autologous serum therapy of the ocular surface with novel delivery by platelet concentrate gel. *Ocul Surf*, Vol.4, No.4, (October 2006), pp.188-95, ISSN 1542-0124

Kunicki, T.J. & Nugent, D.J. (2006). Human platelet antigens, In: *Blood banking and transfusion medicine*, C.D. Hillyer, L.E. Silberstein, P.M. Ness & K.N. Anderson, (Eds.), pp. 63-80, W.B. Saunders Company, Philadelphia, PA ISBN 9780443065422,

Lacci, K.M. & Dardik, A. (2010). Platelet-rich plasma: support for its use in wound healing. *Yale J Biol Med.*, Vol.83, No.1, (March 2010), pp. 1-9, ISSN 0044-0086

Lacoste, E., Martineau, I. & Gagnon, G. (2003). Platelet concentrates: effects of calcium and thrombin on endothelial cell proliferation and growth factor release. *J Periodontol*, Vol.74, No.10, (October 2003), pp.1498-507, ISSN 0022-3492

Lee, J.H. & Kang, N.Y. (2011). Comparison of fibrin glue and sutures for conjunctival wound closure in strabismus surgery. *Korean J Ophthalmol*, Vol.25, No.3, (June 2011), pp.178-84, ISSN 1011-8942

Liu, L., Hartwig, D., Harloff, S., Herminghaus, P., Wedel, T., Kasper, K. & Geerling, G. (2006). Corneal epitheliotrophic capacity of three different blood-derived preparations. *Invest Ophthalmol Vis Sci*, Vol.47, No.6, (June 2006), pp.2438-44, ISSN 0146-0404

Lundblad, R.L. & White, G.C. (2005). The interaction of thrombin with blood platelets. *Platelets*, Vol.16, No.7, (November 2005), pp. 373-85, ISSN 0953-7104

Margolis, D.J., Kantor, J., Santanna, J., Strom, B.L. & Berlin, J.A. (2001). Effectiveness of platelet releasate for the treatment of diabetic neuropathic foot ulcers. *Diabetes Care*, Vol.24, No.3, (March 2001), pp. 483-8, ISSN 0149-5992

Marx, R.E., Carlson, E.R., Eichstaedt, R.M., Schimmele, S.R., Strauss, J.E. & Georgeff, K.R. (1998). Platelet-rich plasma: Growth factor enhancement for bone grafts. *Oral Surg Oral Med Oral Pathol Oral Radiol Endod*, Vol.85, No. 6, (June 1998), pp. 638-46, ISSN 1079-2104

Mazzucco, L., Cattana, E., Orecchia, S. & Borzini, P. (2004). Enhancing of fibroblast proliferation by platelet extract. *Vox Sang*, Vol.87, Issue Suppl. s3 (2004), pp. S2-S16, ISSN 0042-9007

Mazzucco, L., Balbo, V., Cattana, E. & Borzini, P. (2008). Platelet-rich plasma and platelet gel preparation using Plateltex. *Vox Sang,* Vol.94, No.3, (April 2008), pp. 202-8, ISSN 0042-9007

Mazzucco, L., Balbo, V., Cattana, E., Guaschino, R. & Borzini, P. (2009). Not every PRP-gel is born equal. Evaluation of growth factor availability for tissues through four PRP-gel preparations: Fibrinet, RegenPRP-Kit, Plateltex and one manual procedure. *Vox Sang,* Vol.97, No.2, (August 2009), pp. 110-8, ISSN 0042-9007

McClain, S.A., Simon, M., Jones, E., Nandi, A., Gailit, J.O., Tonnesen, M.G., Newman, D. & Clark, R.A., (1996). Mesenchymal cell activation is the rate-limiting step of granulation tissue induction. *Am J Pathol,* Vol.149, No.4, (October 1996), pp. 1257-70, ISSN 0002-9440

Mehta, S. & Watson, J.T. (2008). Platelet rich concentrate: basic science and current clinical applications. *J Orthop Trauma,* Vol.22, No.6, (July 2008), pp. 432–8, ISSN 0890-5339

Mendez, R., Lopez-Cedrun, J.L., Patino, B., Vazquez, I., Martin-Sastre R., Tellado, M.G. & Vela, D. (2006). Platelet-rich plasma (platelet gel) in secondary alveoloplasty in cleft patients. *Cir Pediatr,* Vol.19, No.1, (January 2006), pp. 23-6, ISSN 0214-1221

Metcalfe, P., Watkins, N.A., Ouwehand, W.H., Kaplan, C., Newman, P., Kekomaki, R., De Haas, M., Aster, R., Shibata, Y., Smith, J., Kiefel, V. & Santoso S. (2003). Nomenclature of human platelet antigens. *Vox Sang,* Vol.85, No.3, (October 2003), pp. 240–5, ISSN 0042-9007

Millington, J.T. & Norris, T.W. (2000). Effective treatment strategies for diabetic foot wounds. *J Fam Pract,* Vol.49, Suppl.11 (November 2000), pp. S40–S48, ISSN 0094-3509

Mogan, C. & Larson, D.F. (2004). Rationale of platelets gel to augment adaptive remodelling of the injured heart. *J Extra Corpor Technol,* Vol.36, No.2, (June 2004), pp.191-6, ISSN 0022-1058

Naldini, A., Morena, E., Fimiani, M., Campoccia, G., Fossombroni, V. & Carraro, F. (2008). The effects of autologous platelet gel on inflammatory cytokine response in human peripheral blood mononuclear cells. *Platelets,* Vol.19, No.4, (June 2008), pp. 268-74, ISSN 0953-7104

Novis, D.A., Renner, S., Friedberg, R.C., Walsh, M.K. & Saladino, A.J. (2002). Quality indicators of fresh frozen plasma and platelet utilization. *Arch Pathol Lab Med,* Vol.126, No.5, (May 2002), pp. 527-32, ISSN 0003-9985

Palfreyman, S.S.J., Nelson, E.A., Lochiel, R. & Michaels, J.A. (2006). Dressings for healing venous leg ulcers. *Cochrane Database Syst Rev,* Issue 3, (July 2006), Art. No. CD001103

Parazzi, V., Lazzari, L. & Rebulla P. (2010). Platelet gel from cord blood: a novel tool for tissue engineering. *Platelets,* Vol.21, No.7, (2010), pp. 549-54, ISSN 0953-7104

Phillips, G.D., Stone, A.M., Whitehead, R.A. & Knighton, D.R. (1994). Platelet derived wound healing factors (PDWHF) accelerate and augment wound healing angiogenesis in the rat. *In Vivo,* Vol.8, No.2, (March-April 1994), pp. 167-71, ISSN 0258-851X

Picardi, A., Lanti, A., Cudillo, L., Cerretti, R., Dentamaro, T., De Angelis, G., Ferraro, A., Di Veroli, A., Adorno, G. & Arcese, W. (2010). Platelet gel for treatment of mucocutaneous lesions related to graft-versus-host disease after allogeneic hematopoietic stem cell transplant. *Transfusion,* Vol. 50, No.2, (February 2010), pp. 501-6, ISSN 0041-1132

Pomerantz, J. & Dutton, J.M. (2005). Platelet gel for endoscopic sinus surgery. *Ann Otol Rhinol Laryngol* , 114, (2005), (699-704)

Powell, D.M., Chang, E. & Farrior, E.H. (2001). Recovery from deep-plane rhytidectomy following unilateral wound treatment with autologous platelet gel: a pilot study. *Arch Facial Plast Surg,* Vol.3, No.4, (October-December 2001), pp. 245-50, ISSN 1521-2491

Rendu, F. & Brohard-Bohn, B. (2001). The platelet release reaction: granules' constituents, secretion and functions. *Platelets,* Vol.21, No.5, (August 2001), pp. 261-73, ISSN 0953-7104

Reutter, H., Bort, S., Jung, M.F., Klyscz, T., Schippert, W., Zuder, D. & Junger, M. (1999). Questionable effectiveness of autologous platelet growth factors (PDWHF) in treatment of venous ulcers of the leg. *Hautarzt,* Vol.50 ,No.12, (December 1999), pp. 859-65, ISSN 0017-8470

Rosenthal, A.R., Harbury, C., Egbert, P.R. & Rubenstein, E. (1975). Use of a platelet-fibrinogen-thrombin mixture as a corneal adhesive: experiments with sutureless lamellar keratoplasty in the rabbit. *Invest Ophthalmol,* Vol.14, No.11, (November 1975), pp. 872-5, ISSN 0020-9988

Rožman, P. (2002). Platelet antigens. The role of human platelet alloantigens (HPA) in blood transfusion and transplantation. *Transpl Immunol,* Vol.10, No.2-3, (August 2002), pp.165-181, ISSN 0966-3274

Rožman, P. & Bolta, Z. (2007). Use of platelet growth factors in treating wounds and soft-tissue injuries. *Acta Dermatovenerol Alp Panonica Adriat,* Vol.16, No.4, (December 2007), pp. 156-65, ISSN 1318-4458

Rughetti, A., Giusti, I., D'Ascenzo, S., Leocata, P., Carta, G., Pavan, A., Dell'Orso, L. & Dolo, V. (2008). Platelet gel-released supernatant modulates the angiogenic capability of human endothelial cells. *Blood Transfus,* Vol.6, No.1, (January 2008), pp. 12-7, ISSN 1723-2007

Salvadè, A., Della Mina, P., Gaddi, D., Gatto, F., Villa, A., Bigoni, M., Perseghin, P., Serafini, M., Zatti, G., Biondi, A. & Biagi, E. (2010). Characterization of platelet lysate cultured mesenchymal stromal cells and their potential use in tissue-engineered osteogenic devices for the treatment of bone defects. *Tissue Eng Part C Methods,* Vol.16, No.2, (April 2010), pp.201-14, ISSN 1937-3384

Santoso, S. & Kiefel, V. (1998). Human platelet-specific alloantigens: update. *Vox Sang,* Vol.74, Suppl.2, (1998), pp. 249-253, ISSN 0042-9007

Savarino, L., Cenni, E., Tarabusi, C., Dallari, D., Stagni, C., Cenacchi, A., Fornasari, P.M,. Giunti, A. & Baldini, N. (2006). Evaluation of bone healing enhancement by lyophilized bone grafts supplemented with platelet gel: a standardized methodology in patients with tibial osteotomy for genu varus. *J Biomed Mater Res B Appl Biomater,* Vol.76, No.2, (February 2006), pp. 364-72, ISSN 1552-4973

Sclafani, A.P., Romo, T., Ukrainsky, G., McCormick, S.A., Litner, J., Kevy, S.V. & Jacobson, M.S. (2005). Modulation of wound response and soft tissue ingrowth in synthetic and allogeneic implants with platelet concentrate. *Arch Facial Plast Surg*, Vol.7, No.3, (May-June 2005), pp. 163-9, ISSN 1521-2491

Schallmoser, K., Bartmarm, C., Rohde, E., Reinisch, A., Kashofer, K., Emberger, W., Lanzer, G., Linkesch, W. & Strunk, D. (2007). A novel system for highly efficient clinical scale propagation of human mesenchymal stem cells with human platelet lysate. *Haematologica-the Hematology Journal*, Vol. 92, Suppl.2, (2007), pp. 317-8, ISSN 0390-6078

Smrke, D., Gubina, B., Domanovic, D. & Rozman, P. (2007). Allogeneic Platelet Gel with Autologous Cancellous Bone Graft for the Treatment of a Large Bone Defect. *Eur Surg Res*, Vol. 39, No.3, (March 2007), pp.170-4, ISSN 0014-312X

Stammers, A.H., Trowbridge, C.C., Marko, M., Woods, E.L., Brindisi, N., Pezzuto, J., Klayman, M., Fleming, S. & Petzold, J. (2009). Autologous platelet gel: fad or savoir? Do we really know? *J Extra Corpor Technol*, Vol.41, No.4, (December 2009), pp. P25-30, ISSN 0022-1058

Steed, D.L., Goslen, J.B., Holloway, G.A., Malone, J.M., Bunt, T.J. & Webster, M.W. (1992). Randomized prospective double-blind trial in healing chronic diabetic foot ulcers. CT-102 activated platelet supernatant, topical versus placebo. *Diabetes Care*, Vol.15, No.11, (November 1992), pp.1598–604, ISSN 0149-5992

Sun, B., Chen, B., Zhao, Y., Sun, W., Chen, K., Zhang, J., Wei, Z., Xiao, Z. & Dai, J. (2009). Crosslinking heparin to collagen scaffolds for the delivery of human platelet-derived growth factor. *J Biomed Mater Res B Appl Biomater.* Vol.91, No.1, (October 2009), pp. 366-72, ISSN 1552-4973

Sullivan, M.T., Cotten, R., Read, E.J. & Wallace, E.L. (2007). Blood collection and transfusion in the United States in 2001. *Transfusion*, Vol.47, No.3, (March 2007), pp. 385-94, ISSN 0041-1132

Thompson, D.F., Letassy, N.A. & Thompson, G.D. (1988). Fibrin glue: a review of its preparation, efficacy, and adverse effects as a topical haemostat. *Drug intell clin pharm*, Vol.22, No.12, (December 1988), pp. 946-52, ISSN 0012-6578

Trowbridge, C.C., Stammers, A.H., Wood, G.C., Murdock, J.D., Klayman, M., Yen, B.R., Woods, E. & Gilbert, C. (2005). Improved outcomes during cardiac surgery: a multifactorial enhancement of cardiopulmonary bypass techniques. *J Extra Corpor Technol*, Vol.37, No.2, (June 2005), pp. 165-72, ISSN 0022-1058

Tschon, M., Fini, M., Giardino, R., Filardo, G., Dallari, D., Torricelli, P., Martini, L., Giavaresi, G., Kon, E., Maltarello, M.C., Nicolini, A. & Carpi, A. (2011). Lights and shadows concerning platelet products for musculoskeletal regeneration. *Front Biosci (Elite Ed)*, Vol.3, (January 2011), pp. 96-107, ISSN 1945-0494

Valencia, I.C., Falabella, A., Kirsner, R.S. & Eaglstein, W.H. (2001). Chronic venous insufficiency and venous leg ulceration. *J Am Acad Dermatol*, Vol.44, No.3, (March 2001), pp. 401–21, ISSN 0190-9622

Weibrich, G., Kleis, W.K., Hafner, G. & Hitzler, W.E. (2002). Growth factor levels in platelet-rich plasma and correlations with donor age, sex, and platelet count. *J Craniomaxillofac Surg*, Vol.30, No.2, (April 2002), pp. 97-102, ISSN 1010-5182

Westerhuis, R.J., van Bezooijen, R.L. & Kloen, P. (2005). Use of bone morphogenetic proteins in traumatology. *Injury*, Vol.36, No.12, (December 2005), pp. 1405-12, ISSN 0020-1383

Whitman, D.H., Berry, R.L. & Green, D.M. (1997). Platelet gel: an autologous alternative to fibrin glue with applications in oral and maxillofacial surgery. *J Oral Maxillofac Surg*, Vol.55, No.11, (November 1997), pp. 1294-9, ISSN 0278-2391

Zhang, J.,Blackmore, P.F., Hargrave, B.Y., Xiao, S., Beebe, S.J. & Schoenbach, K.H. (2008). Nanosecond pulse electric field (nanopulse): a novel non-ligand agonist for platelet activation. *Arch Biochem Biophys*, Vol.471, No.2, (March 2008), pp. 240-8, ISSN 0003-9861

Zimmermann, R., Arnold, D., Strasser, E., Ringwald, J., Schlegel, A., Wiltfang, J. & Eckstein, R. (2003). Sample preparation technique and white cell content influence the detectable levels of growth factors in platelet concentrates. *Vox Sang*, Vol.85, No.4 (2003), pp. 283-9, ISSN 0042-9007

16

Regenerative Orthopedics

Christopher J. Centeno and Stephen J. Faulkner
The Centeno-Schultz Clinic
United States of America

1. Introduction

Mesenchymal stem cells (MSCs) are multipotent, adult stem cells that show clinical potential as therapeutic agents in regenerative medicine(Bruder et al., 1994; Barry, 2003; Alhadlaq & Mao, 2004; Gangji et al., 2005; Cha & Falanga, 2007). They are also known as marrow stromal cells and are derived from mesodermal tissues. The clonal nature of a sub-population of marrow nucleated cells was discovered in the 1960's (Becker et al., 1963). These cells were later assayed and renamed "colony forming fibroblasts"(Friedenstein et al., 1974). Experiments through the 1980's and 1990's demonstrated that environmental clues assisted MSCs in differentiating into different cell types. For example, culturing with ascorbic acid, inorganic phosphate, or dexamethasone could differentiate cells to osteoblasts, while culturing in the presence of TGF-beta caused cells to differentiate into chondrocytes (Alhadlaq & Mao, 2004). More recently it has been shown that bone marrow MSCs are actually a heterogeneous population of similar cells rather than one distinct cell type (Zhou et al., 2005). As a result, there is still no uniformly accepted definition of an MSC. Despite this issue, some groups have attempted to provide a definition which consists of adherence to plastic, MSC specific cell surface markers consistent with MSCs, as well as multi-lineage mesodermal tissue differentiation (Schauwer et al., 2010).

2. Cell source matters

MSCs can be easily isolated from many anatomic locations, including whole marrow aspirate, marrow mobilized whole blood, muscle biopsy, adipose liposuction aspirate, and other tissues (Alhadlaq & Mao, 2004). For orthopedic uses, these sources have been compared by many authors for their ability to heal bone and cartilage with differences being uncovered. As a rule, the closer the source tissue is to the target tissue being treated, the more effective the MSCs appear to be at differentiation to the target tissue type. For example, Vidal compared equine MSCs derived from bone marrow (bm-MSCs) vs. adipose tissue (a-MSCs) for chondrogenic potential and found that bm-MSCs produced a more hyaline like matrix and had improved glycosaminoglycan production (Vidal et al., 2008). Additional animal studies demonstrated that bm-MSCs produced better repair of a tibial osteochondral defect when compared to a-MSCs (Niemeyer et al., 2010). Keeping with this trend, Yoshimura determined that MSCs derived from the synovial tissue of the knee (closest to the target tissue of chondral cartilage) had better chondrogenesis than bm-MSCs (Yoshimura et al., 2007).

Significant controversy exists over whether adipose or bone marrow are better sources for orthopedic tissue repair (Frisbie et al., 2009). While adipose MSCs are more prevalent and are capable of orthopedic tissue differentiation, obtaining orthopedic tissues from these type of cells requires the use of considerably more growth factors. In addition, as stated above, the native chondrogenic potential of adipose derived MSCs doesn't appear to be as robust as bone marrow derived MSCs.

3. MSC culture expansion

A limited amount of cells can be obtained from any tissue. In many instances, the number that can be harvested from the source tissue is less than the quantity of cells needed for tissue repair. One method of obtaining more cells is culturing to obtain larger numbers. However, a delicate balance exists between length of time in culture (which produces more cells) and adverse consequences to the cells (such as genetic transformation).

MSCs are usually culture expanded via monolayer culture, which is a process that involves seeding a certain density of cells onto a specialized flask and allowing the MSCs to attach to a plastic surface and begin to form colonies. In this way, MSCs are selected from the marrow nucleated cell population through adherence. The MSC's that are adherent are then fed via a nutrient broth that is maintained above the growing cells. Oftentimes flasks are used with multiple surfaces to enhance cell production versus available incubator space. Because MSCs are contact inhibited, they will grow on this surface until they become confluent and then abruptly stop propagating. It is commonly believed that this "contact inhibition" property of MSCs is a key feature of their enhanced safety profile over other stem cell types that will propagate indefinitely without exhibiting contact inhibition. To keep MSCs proliferating in culture, when the colonies are near confluence, the non-adherent cells in the media are discarded and an enzyme is used to detach the MSCs from the plastic surface. The MSCs are then re-plated in a similar flask and fresh media added with this process being known as a "passage". Most MSCs in culture are grown to the 2nd to 5th passage, as some studies have shown decreased differentiation if MSCs are grown for prolonged periods in culture with a higher chance of genetic mutation (see Fig. 1)(Banfi et al., 2000; Crisostomo et al., 2006; Izadpanah et al., 2008). Thus MSC culture is a balance between being able to grow additional cells for therapy without culturing for prolonged periods. Most studies consider that a "pure" MSC population is obtained after approximately the second passage.

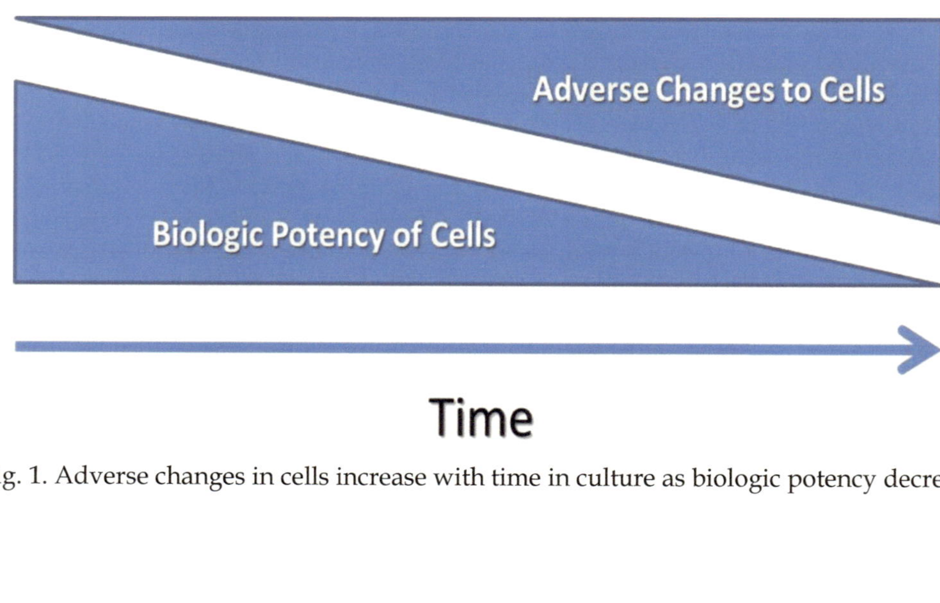

Fig. 1. Adverse changes in cells increase with time in culture as biologic potency decreases.

4. How MSC's effect tissue repair

Animal studies have demonstrated the multipotency of MSCs, and how they can differentiate into muscle, bone, cartilage, tendon, and various cells of internal organs. However, these cells also act via paracrine mechanisms to assist in tissue repair. For cell therapy, paracrine is defined as the production of certain growth factors and cytokines by the MSCs which can assist in tissue repair (Ladage et al., 2007). These growth factors include TGF-beta, VEGF, FGF, and other signaling factors that can help recruit other cells to the local area. In this way, MSCs can act as a coordinator of the repair response without having to act directly through differentiation to repair tissue. In fact, some have questioned whether most of the positive repair effects observed in experimental MSC therapies are due more to this paracrine signaling than differentiation of cells. In the end, it may not matter if MSCs act through differentiation or orchestrate the repair response, as long as the robustness of the repair meets the clinical need.

5. Donor vs. autologous MSC sourcing

Autologous stem cells obviously do not have the same communicable disease transmission risk as donor allogeneic cells. However, there may be practical reasons why donor cells are attractive. For example, some studies have shown a decreased differentiation potential for MSCs obtained from older patients (Zhou et al., 2008). In addition, somatic genetic variants (i.e. trisomy V and VII) have been demonstrated in the MSCs and osteoprogenitors of some patients with osteoarthritis (Broberg et al., 1998). In addition, allogeneic cells could be able to be mass produced in bioreactors, providing a ready supply of cells for therapy. On the other hand, some concerns have been raised about the use of allogeneic stem cells. As an example, Ueda recently discovered that stem cells transplanted from the bone marrow of elderly mice bred to have osteoporosis were able to induce osteoporosis in young healthy mice, indicating that the stem cells themselves may be a genetic disease vector (Ueda et al., 2007). In addition, many have argued that allogeneic MSCs are immune-privileged as they lack major histocompatibility complexes; however, Prigozhina has found that allogeneic MSCs lose their immunosuppressive potential in a mismatched setting (Prigozhina et al., 2008). In addition, Huang recently noted that MSCs transplanted for cardiac repair lost their immune-privileged status and transitioned to an immunogenic state after differentiation, limiting their usefulness in cardiac repair (Huang et al., 2010). In summary, while allogeneic MSCs may solve some issues such as cell availability on demand and mass distribution, they also present many hurdles that have to be overcome to allow lab to bedside translation.

6. MSC use in orthopedic specific conditions

6.1 Fibrocartilage-meniscus repair

The challenge in repairing the meniscus is due to the poor blood supply of the inner 2/3'rds (where many tears occur) versus the good blood supply of the outer 1/3 (red zone-where fewer non-healing tears are observed) (Hennerbichler et al., 2007). As a result of these differences in meniscus blood and nutrient supply, surgical repair of a meniscus tear is more likely to be successful in the outer 1/3 than the inner 2/3rd's. To overcome this limited healing ability of the inner 1/3rd of the meniscus, many surgeons often perform a partial mensectomy for tears of this area, which can lead to significant biomechanical abnormalities in the knee (Sturnieks et al., 2010). The promise of stem cells if used to enhance surgery may thus be in allowing the surgeon to repair the inner meniscus. For example, Izuta et al demonstrated that cultured MSCs may be able to overcome this problem of poor repair in the avascular zone. His group was able to demonstrate meniscus repair in the white zone when MSCs were transplanted into this area using a fibrin matrix (Izuta et al., 2005). Of note, Agung et al (2006) reported a murine model of intra-articular injection after acute injury of multiple knee structures, including the meniscus. This model demonstrated that for blind intra-articular injection (rather than the local adherent model proposed by Koga), the number of cells injected was related to their ability to be found in the meniscus. For example, at a dose of 1×10^6 MSCs, none were found in the injured meniscus but at a dose of 1×10^7 cells, MSCs were generally found in this area. This may fit well with Koga's hypothesis that MSCs act primarily where they physically attach, as a higher number of cells injected into the joint would make it more likely that cells would be able to attach at the site in need of repair (Koga et al., 2008). Horie et al (2009) reported that synovial derived MSCs that were injected into massive rat meniscus tears were able to differentiate and repair meniscal tissue. Of note, the authors also demonstrated that these cells did not migrate out of the knee to distant organs, further enhancing their theoretical safety profile for clinical use. Finally, Yamasaki et al (2005) validated that MSCs were capable of repopulating the meniscus. This investigator cryogenically treated the meniscus to kill the living cells and then reseeded it with cultured MSCs. The resulting tissue had appropriate biomechanical properties approximating the normal meniscus.

6.2 Hyaline cartilage: chondral defects

Some of the earliest models of cartilage repair used autologous, cultured chondrocytes (Brittberg et al., 1994). However, the complications of using chondrocytes for cartilage repair included hypertrophy, graft failure, long culture times, and the invasiveness of the implant procedure (Nejadnik et al., 2010). Because MSC's are multi-potent and have shown to have innate cartilage repair properties through both differentiation and paracrine signaling, animal models of cartilage repair using MSCs started to appear in the literature in the early 1990's (Caplan, 1991). In these studies an osteochondral defect (OCD) was created experimentally and the MSCs were implanted into the lesion, often in a hydrogel or other carrier or at times through local adherence (Minas & Nehrer, 1997; Buckwalter & Mankin, 1998; Angele et al., 1999; Johnstone & Yoo, 1999). Partial to robust healing of the OCD takes place over weeks to months (Alhadlaq & Mao, 2004). The cartilage produced by these cells was very much like native hyaline cartilage, but subtle differences have been observed (Katakai et al., 2009).

6.3 Biologic scaffolding

MSCs can be delivered to the lesion area in many different biologic scaffolds including hydrogels, fibrin, in native extra-cellular matrix, collagen, or in a suspension. A scaffold is a matrix with properties that support cell migration, attachment, three dimensional position, and engraftment. Based on unpublished data, we have noted that stiffer biologic scaffolds (like dense fibrin glue) tend to reduce MSC viability as they limit MSC movement through the material. On the other end of the spectrum is implanting MSCs without scaffolding. In this method, cells are delivered in a liquid suspension and slowly dripped on the lesion, using gravity dependent adherence to allow site attachment. For example, Fig 2 shows histology adapted from Koga et al (2008) demonstrating minimal cartilage repair with a control saline injection, minimal repair with MSC's injected intra-articular, and robust repair when MSCs were allowed to attach to the lesion via gravity.

Based on the published data showing the importance of MSC attachment to a lesion, we would hypothesize that exact placement of MSCs in a joint is very important. For example, blind intra-articular injection of MSCs or the placement of cells into the intra-articular cavity after surgery is likely to be less effective for facilitating repair than placement of cells at or near a lesion in need of repair. This exact placement of cells has been already used in daily orthopedics practice, for example in the Autologous Chondrocyte Implantation procedure (ACI) as well as in knee micro fracture. The knee ACI procedure depends on exact placement of cultured chondrocytes into an osteochondral defect through injection under a surgically prepared periosteal flap. The micro fracture procedure also relies on the surgeon creating holes in the osteochondral plate of an OCD (Osteochondral Defect) which allows whole marrow and a limited number of MSCs to clot in the defect.

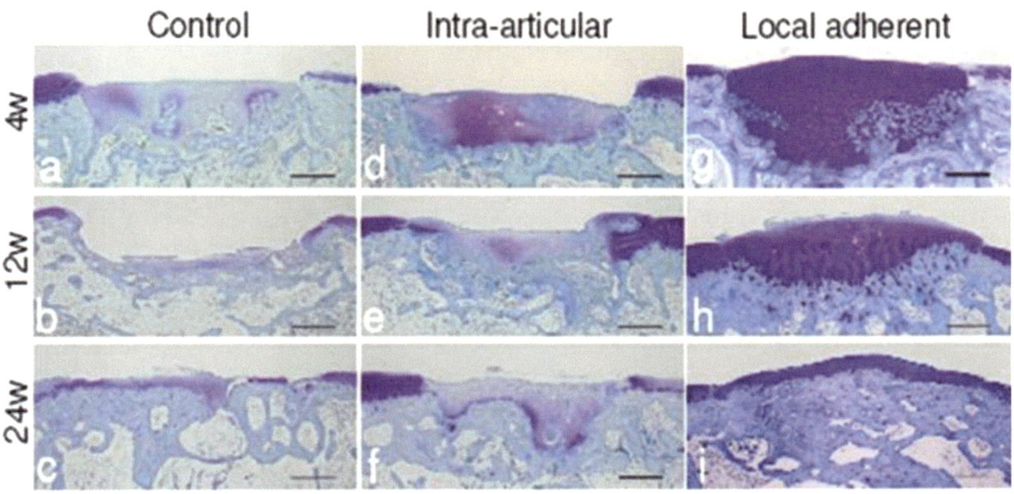

Fig. 2. Histology of chondral cartilage repair after staining, showing little repair in a control injection, some repair in a blind intra-articular injection of stem cells, and robust repair when cells were dripped on the lesion.

While existing cell based procedures for the knee have some reported success, both of these techniques seem to be more effective in younger, athletic patients. In addition, in both procedures, the cartilage produced by either chondrocytes or the low concentration of MSCs in a micro fracture clot tends more toward non-native fibrous cartilage versus the more hyaline like cartilage produced by higher concentrations of cultured MSCs (Mobasheri et al., 2009). Recently, McIlwraith et al have improved upon the standard micro fracture concept in an equine model by adding MSC's to the joint after surgery, demonstrating better cartilage repair with a combination of micro fracture plus MSCs than microfracture alone (McIlwraith, 2010). In this study, the repaired tissue was significantly firmer and had higher levels of aggrecan, a molecule that provides compressive stiffness to cartilage.

6.4 Tendon repair

Tendon tears are often difficult to treat without a high level of morbidity or re-rupture. This may be because of the high tensile strength tendons are required to bear as well as the fact that sutures in the tendon itself may reduce local blood flow, further impeding the repair process. Awad has published a rabbit model demonstrating that cultured MSCs were able to assist healing and produced better tendon appearance than non-MSC treated tendons. The MSC treated tendons had better maximum stress, modulus, and strain energy density as well as minor improvements in the histological appearance, including an increased number of tenocytes and larger and more mature-looking collagen fiber bundles (Awad et al., 1999). The same author later published that the seeding density of MSCs in the biologic scaffold and control over the rate of contraction of the healing tendon were important for successful MSC mediated tendon repair (Awad et al., 2000). Chong also demonstrated improved modulus in resected rabbit Achilles tendons treated with MSCs and morphometric changes, concluding that MSCs can improve the histological and biomechanical parameters in the early stages of tendon-healing (Chong et al., 2007). On the other hand, Gulotta was unable to demonstrate efficacy for an animal model of surgical rotator cuff tendon healing which showed no differences between MSC treated and untreated groups (Gulotta et al., 2009).

6.5 Intervertebral disc repair

Traditional spinal surgery treatments for degenerated intervertebral discs continue to show disappointing results (Deyo et al., 1993; Elias et al., 2000; Fritzell et al., 2003). While animal models of disc repair using MSCs are abundant, human data isn't readily available. For example, Sakai et al (2003, 2006) have published several animal models whereby MSCs are usually combined with atellocollagen and inserted into an experimentally created degenerative disc. This group of authors has observed encouraging improvements in MRI disc hydration, height, and morphology. Richardson et al (2004) and Risbud et al (2004) have investigated the co-culturing of MSCs with cells from the nucleus pulposis (NP) showing that this technique can produce partially differentiated cells that are capable of repopulating the NP in an animal model. Risbud et al (2004) has also used different methods for MSC differentiation toward the NP phonotype including using MSC exposure to hypoxia and TGF-beta in culture. Zhang et al (2005) has shown that MSCs injected into discs without pre-conditioning or co-culture can help to increase proteoglycan production in the NP. Finally, Miyamota et al (2010) recently demonstrated that intra-discal transplantation of synovial derived MSCs prevented disc degeneration through suppression of catabolic genes and perhaps proteoglycan production. In summary, while the results from animal models are impressive, questions remain as to whether a quadruped disc with its very different load characteristics can serve as an adequate model for bipedal disc repair. In addition, in all of the animal models studied to date, an artificially created degenerated disc (acute disc stab model) is used as a surrogate for the chronic degenerated discs normally encountered in patients (Yoshikawa et al., 2010).

7. Cell delivery in orthopedics

Delivery of cells into a joint to treat orthopedic injuries could take two common routes used daily in clinical practice: percutaneous injections and arthroscopic placement. Injecting cells into a confined space such as infiltrating into soft-tissues will likely result in the MSC's that stay local to the injection site. However, as discussed above, injecting in a large joint presents some concerns, as multiple animal models have shown that cells may or may not find their way into the damaged areas(Agung et al., 2006). Because MSC's function through local attachment to the damaged site, MSCs dripped on a lesion surgically or through a needle are another possible implant route. Other injection based methods may involve using MSCs tagged with ferrous nanoparticles and magnetic fields to encourage attachment to the damaged site (Kobayashi et al., 2008). Finally, since MSCs are capable of chemotaxis, placing certain growth factors on the injured tissue may result in more MSCs accumulating at the lesion (Fiedler et al., 2002).

Another challenge in MSC delivery is that most arthroscopic surgery is performed in a water environment. For cells in suspension, this presents a challenge, as the MSCs would easily be whisked away by the action of arthroscopy pumps meant clear debris from the operative site. To overcome this problem, Nejadnik et al (2010) have used a surgical approach similar to autologous chondrocyte implantation, where MSCs are placed in a dense hydrogel and sutured under a protective membrane. While this technique has promise, it's also more surgically invasive with likely higher morbidity than injection. Another alternative method is to adapt the MSC properties to be better suited for the water arthroscopy environment. For example, MSCs can be cultured to form a tissue engineered construct (TEC) by allowing the cells to produce their own extracellular matrix. The result is a pea sized implant with a paste like consistency that can be placed under water arthroscopy into a lesion or defect (Ando et al., 2007).

8. From the bench to the bedside in orthopedics

The clinical translation of MSCs from the lab to the bedside is clearly imminent. For example, our use of MSCs in orthopedics began in 2005 and we have published case studies in which positive MRI changes in cartilage and meniscus were observed in knees treated with culture expanded MSCs and hips treated with nucleated cell concentrates, corresponding with symptomatic improvement (Centeno et al., 2006; Centeno et al., 2008; Centeno et al., 2008). We have also noted that the complication rate of expanded MSC injection procedures is no greater than other needle-based interventional techniques directed at peripheral joints (Centeno et al., 2010). In a submitted and recently accepted publication on 339 patients, this safety profile was continued at up to three years post MSC reimplantation with over 200 MRI's of the reimplant sites showing no evidence of ectopic tissue or tumor formation (see Fig 3). Other authors have described similar safety profiles using more invasive surgical implant techniques. Wakatani described effective treatment of cartilage defects in 9 knees with culture expanded MSCs (Wakitani et al., 2007). These authors also published an 11-year prospective study of 45 knees (in 41 patients) treated with autologous cultured, bone marrow MSCs, with results indicating both safety and efficacy (Wakitani et al., 2010). Nejadnik recently described a comparison between surgically implanted chondrocytes versus cultured MSCs placed by needle in 72 knees of older patients.(Nejadnik et al., 2010) The MSC treated knees demonstrated good safety, less donor site morbidity, and better efficacy when compared with an autologous chondrocyte implantation procedure. Finally, Haleem has noted that autologous, cultured bm-MSCs re-implanted into articular cartilage defects in platelet rich fibrin demonstrated evidence of healed cartilage in most patients at 12 months post-operative.(Haleem et al., 2010)

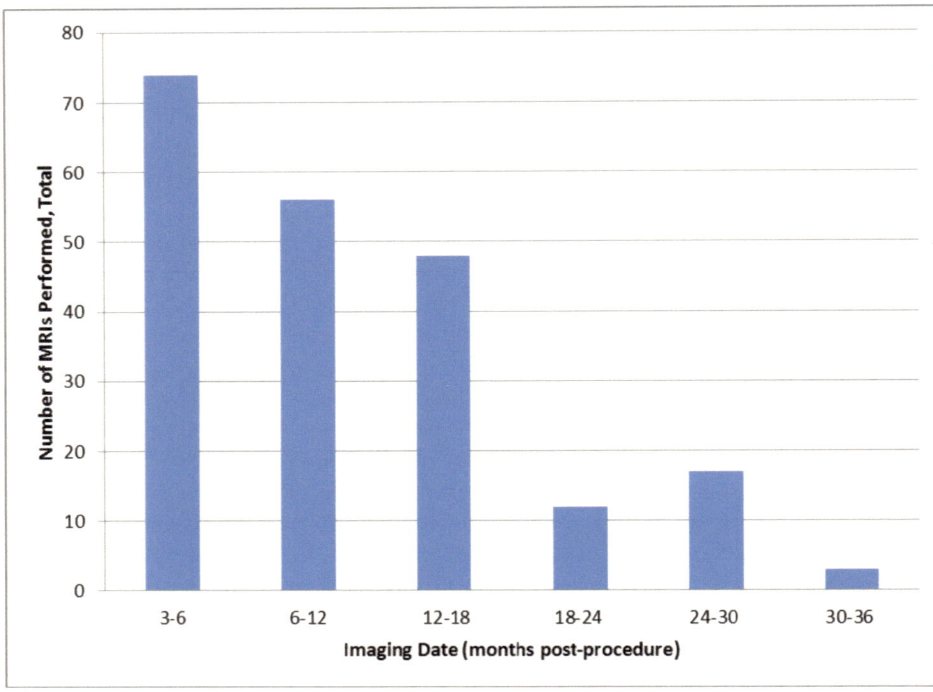

Fig. 3. Number of negative MRI follow-ups at re-implant sites for each end-point for 50 patients tracked after orthopedic MSC reimplantation.

While very little has been published on intervertebral disc repair in humans, some clinical data is available. Yoshikawa recently published on two patients who were treated with surgically implanted MSCs that were cultured using a serum free technique (Yoshikawa et al., 2010). After two years, no complications were noted and both patients showed modest improvements in vacuum phenomena on follow-up MRI. The only other human data of which we are aware is that produced by our group from 2005-2010 under IRB supervision and now being readied for publication (unpublished data). Our experience demonstrated that placing a bone marrow nucleated cell fraction (an enriched MSC population with other cells) into the disc via percutaneous means produced no measureable clinical or MRI results in patients with degenerative disc disease. Replicating the Sakai study (Sakai et al., 2003), where cultured MSCs were placed into the disc in a similar patient population also produced little measureable results. Finally, a third case series was performed where changes were made in culture and injection technique as well as the diagnosis being treated (changed from DDD to chronic disc bulge causing lumbar radiculopathy). This last model showed encouraging clinical and imaging results.

9. Implications in real world clinical applications

To consider the real world implications of viable cell based alternatives to more invasive orthopedic surgeries, total knee arthroplasty (TKA) is an apt model. Knee replacement surgery, also called knee arthroplasty, has been employed increasingly over the past 10 years as a means of treating symptomatic degenerative changes of the knee. It is estimated from discharge data from the Nationwide Inpatient Sample (NIS) of the Healthcare Cost and Utilization Project (HCUP), that the number of partial and total knee replacement procedures among U.S. patients 65 years and older increased dramatically from 178,653 in 2000 to 357,472 in 2008 (HCUP, 2008). Knee arthroplasty is associated with substantial mortality and morbidity; applying rates derived from 2000 data resulted in an estimated 4,964 TKA related deaths, 2,788 pulmonary emboli, 2,908 myocardial infarcts, and 4,670 cases of pneumonia (Mahomed NN, 2005). In contrast, we have recently submitted for publication a large case series of 202 knee and hip osteoarthritis patients treated with percutaneous injection of MSC's. As an example, while 2/3'rds of the knee patients were TKA candidates, only 6 knee patients reported TKA within an average total surveillance period of >2 years despite the injection. In addition, statistically significant differences in reported outcome between the treatment and an untreated control were observed. Finally, complications rates were minimal compared to TKA.

10. Regulatory processes

The regulatory environment in the United States and Europe for stem cells that are more than minimally manipulated considers these cells in the same regulatory category as mass produced drugs. This "drug" category includes cultured cells, using any cell for a non-homologous use (for example an adipose MSC for an orthopedic indication), and cells that have significant processing beyond a simple centrifugation. This regulatory posture has led to a prolonged approval process to bring these technologies from the bench to the bedside. While this approach may certainly make sense for mass produced cells being distributed en masse in vials, this same approach is also applied to autologous cells where the lot size is one patient. As a result, real world clinical knowledge on the clinical use of stem cells in patients is largely accumulating outside the U.S. and Europe. Many stem cell clinics have proliferated worldwide in countries with less stringent regulatory structures.

11. References

Agung, M., Ochi, M., Yanada, S., Adachi, N., Izuta, Y., Yamasaki, T. & Toda, K. (2006). Mobilization of bone marrow-derived mesenchymal stem cells into the injured tissues after intraarticular injection and their contribution to tissue regeneration. *Knee Surg Sports Traumatol Arthrosc*, Vol. 14, No. 12, pp. 1307-1314, ISSN 0942-2056.

Alhadlaq, A. & Mao, J.J. (2004). Mesenchymal stem cells: isolation and therapeutics. *Stem Cells Dev*, Vol. 13, No. 4, pp. 436-448.

Ando, W., Tateishi, K., Hart, D.A., Katakai, D., Tanaka, Y., Nakata, K., Hashimoto, J., Fujie, H., Shino, K., Yoshikawa, H. & Nakamura, N. (2007). Cartilage repair using an in vitro generated scaffold-free tissue-engineered construct derived from porcine synovial mesenchymal stem cells. *Biomaterials*, Vol. 28, No. 36, pp. 5462-5470, ISSN 0142-9612.

Angele, P., Kujat, R., Nerlich, M., Yoo, J., Goldberg, V. & Johnstone, B. (1999). Engineering of osteochondral tissue with bone marrow mesenchymal progenitor cells in a derivatized hyaluronan-gelatin composite sponge. *Tissue Eng*, Vol. 5, No. 6, pp. 545-554, ISSN 1076-3279.

Awad, H.A., Butler, D.L., Boivin, G.P., Smith, F.N., Malaviya, P., Huibregtse, B. & Caplan, A.I. (1999). Autologous mesenchymal stem cell-mediated repair of tendon. *Tissue Eng,* Vol. 5, No. 3, pp. 267-277.

Awad, H.A., Butler, D.L., Harris, M.T., Ibrahim, R.E., Wu, Y., Young, R.G., Kadiyala, S. & Boivin, G.P. (2000). In vitro characterization of mesenchymal stem cell-seeded collagen scaffolds for tendon repair: effects of initial seeding density on contraction kinetics. *J Biomed Mater Res,* Vol. 51, No. 2, pp. 233-240, ISSN 1097-4636.

Banfi, A., Muraglia, A., Dozin, B., Mastrogiacomo, M., Cancedda, R. & Quarto, R. (2000). Proliferation kinetics and differentiation potential of ex vivo expanded human bone marrow stromal cells: Implications for their use in cell therapy. *Exp Hematol,* Vol. 28, No. 6, pp. 707-715, ISSN 0301-472X.

Barry, F.P. (2003). Mesenchymal stem cell therapy in joint disease. *Novartis Found Symp,* Vol. 249, No. 86-96; discussion 96-102, pp. 170-104, 239-141.

Becker, A.J., Mc, C.E. & Till, J.E. (1963). Cytological demonstration of the clonal nature of spleen colonies derived from transplanted mouse marrow cells. *Nature,* Vol. 197, pp. 452-454, ISSN 0028-0836.

Brittberg, M., Lindahl, A., Nilsson, A., Ohlsson, C., Isaksson, O. & Peterson, L. (1994). Treatment of deep cartilage defects in the knee with autologous chondrocyte transplantation. *N Engl J Med,* Vol. 331, No. 14, pp. 889-895, ISSN 0028-4793.

Broberg, K., Hoglund, M., Lindstrand, A., Toksvig-Larsen, S., Mandahl, N. & Mertens, F. (1998). Polyclonal expansion of cells with trisomy 7 in synovia from patients with osteoarthritis. *Cytogenet Cell Genet,* Vol. 83, No. 1-2, pp. 30-34, ISSN 0301-0171.

Bruder, S.P., Fink, D.J. & Caplan, A.I. (1994). Mesenchymal stem cells in bone development, bone repair, and skeletal regeneration therapy. *J Cell Biochem,* Vol. 56, No. 3, pp. 283-294.

Buckwalter, J.A. & Mankin, H.J. (1998). Articular cartilage: degeneration and osteoarthritis, repair, regeneration, and transplantation. *Instr Course Lect,* Vol. 47, pp. 487-504.

Caplan, A.I. (1991). Mesenchymal stem cells. *J Orthop Res,* Vol. 9, No. 5, pp. 641-650, ISSN 1554-527X.

Centeno, C.J., Busse, D., Kisiday, J., Keohan, C., Freeman, M. & Karli, D. (2008). Increased knee cartilage volume in degenerative joint disease using percutaneously implanted, autologous mesenchymal stem cells. *Pain Physician,* Vol. 11, No. 3, pp. 343-353, ISSN 1533-3159.

Centeno, C.J., Busse, D., Kisiday, J., Keohan, C., Freeman, M. & Karli, D. (2008). Regeneration of meniscus cartilage in a knee treated with percutaneously implanted autologous mesenchymal stem cells. *Med Hypotheses,* Vol. 71, No. 6, pp. 900-908, ISSN 0306-9877.

Centeno, C.J., Kisiday, J., Freeman, M. & Schultz, J.R. (2006). Partial regeneration of the human hip via autologous bone marrow nucleated cell transfer: A case study. *Pain Physician,* Vol. 9, No. 3, pp. 253-256, ISSN 1533-3159.

Centeno, C.J., Schultz, J.R., Cheever, M., Robinson, B., Freeman, M. & Marasco, W. (2010). Safety and complications reporting on the re-implantation of culture-expanded mesenchymal stem cells using autologous platelet lysate technique. *Curr Stem Cell Res Ther,* Vol. No. 5, pp. 81-93, ISSN 1574-888X.

Cha, J. & Falanga, V. (2007). Stem cells in cutaneous wound healing. *Clin Dermatol,* Vol. 25, No. 1, pp. 73-78, ISSN 0738-081X.

Chong, A.K., Ang, A.D., Goh, J.C., Hui, J.H., Lim, A.Y., Lee, E.H. & Lim, B.H. (2007). Bone marrow-derived mesenchymal stem cells influence early tendon-healing in a rabbit achilles tendon model. *J Bone Joint Surg Am,* Vol. 89, No. 1, pp. 74-81, ISSN 0021-9355.

Crisostomo, P.R., Wang, M., Wairiuko, G.M., Morrell, E.D., Terrell, A.M., Seshadri, P., Nam, U.H. & Meldrum, D.R. (2006). High passage number of stem cells adversely affects stem cell activation and myocardial protection. *Shock,* Vol. 26, No. 6, pp. 575-580.

Deyo, R.A., Ciol, M.A., Cherkin, D.C., Loeser, J.D. & Bigos, S.J. (1993). Lumbar spinal fusion. A cohort study of complications, reoperations, and resource use in the Medicare population. *Spine,* Vol. 18, No. 11, pp. 1463-1470, ISSN 0362-2436.

Elias, W.J., Simmons, N.E., Kaptain, G.J., Chadduck, J.B. & Whitehill, R. (2000). Complications of posterior lumbar interbody fusion when using a titanium threaded cage device. *J Neurosurg,* Vol. 93, No. 1, pp. 45-52.

Fiedler, J., Roderer, G., Gunther, K.P. & Brenner, R.E. (2002). BMP-2, BMP-4, and PDGF-bb stimulate chemotactic migration of primary human mesenchymal progenitor cells. *J Cell Biochem,* Vol. 87, No. 3, pp. 305-312, ISSN 1097-4644.

Friedenstein, A.J., Deriglasova, U.F., Kulagina, N.N., Panasuk, A.F., Rudakowa, S.F., Luria, E.A. & Ruadkow, I.A. (1974). Precursors for fibroblasts in different populations of hematopoietic cells as detected by the in vitro colony assay method. *Exp Hematol,* Vol. 2, No. 2, pp. 83-92, ISSN 0301-472X.

Frisbie, D.D., Kisiday, J.D., Kawcak, C.E., Werpy, N.M. & McIlwraith, C.W. (2009). Evaluation of adipose-derived stromal vascular fraction or bone marrow-derived mesenchymal stem cells for treatment of osteoarthritis. *J Orthop Res,* Vol. 27, No. 12., pp. 1675-1680, ISSN 1554-527X.

Fritzell, P., Hagg, O. & Nordwall, A. (2003). Complications in lumbar fusion surgery for chronic low back pain: comparison of three surgical techniques used in a prospective randomized study. A report from the Swedish Lumbar Spine Study Group. *Eur Spine J,* Vol. 12, No. 2, pp. 178-189, ISSN 0940-6719.

Gangji, V., Toungouz, M. & Hauzeur, J.P. (2005). Stem cell therapy for osteonecrosis of the femoral head. *Expert Opin Biol Ther,* Vol. 5, No. 4, pp. 437-442.

Gulotta, L.V., Kovacevic, D., Ehteshami, J.R., Dagher, E., Packer, J.D. & Rodeo, S.A. (2009). Application of bone marrow-derived mesenchymal stem cells in a rotator cuff repair model. *Am J Sports Med,* Vol. 37, No. 11, pp. 2126-2133, ISSN 1552-3365.

Haleem, A.M., Singergy, A.A., Sabry, D., Atta, H.M., Rashed, L.A., Chu, C.R., El Shewy, M.T., Azzam, A. & Abdel Aziz, M.T. (2010). The Clinical Use of Human Culture-Expanded Autologous Bone Marrow Mesenchymal Stem Cells Transplanted on Platelet-Rich Fibrin Glue in the Treatment of Articular Cartilage Defects: A Pilot Study and Preliminary Results. *Cartilage,* Vol. 1, No. 4, pp. 253-261, ISSN 1947-6043.

HCUP. A. f. H. R. a. Quality. & 2008). Nationwide Inpatient Sample Issue.

Hennerbichler, A., Moutos, F.T., Hennerbichler, D., Weinberg, J.B. & Guilak, F. (2007). Repair response of the inner and outer regions of the porcine meniscus in vitro. *Am J Sports Med,* Vol. 35, No. 5, pp. 754- 762, ISSN 0363-5465.

Huang, X.P., Sun, Z., Miyagi, Y., McDonald Kinkaid, H., Zhang, L., Weisel, R.D. & Li, R.K. (2010). Differentiation of allogeneic mesenchymal stem cells induces immunogenicity and limits their long-term benefits for myocardial repair. *Circulation,* Vol. 122, No. 23, pp. 2419-2429, ISSN 1524-4539.

Izadpanah, R., Kaushal, D., Kriedt, C., Tsien, F., Patel, B., Dufour, J. & Bunnell, B.A. (2008). Long-term in vitro expansion alters the biology of adult mesenchymal stem cells. *Cancer Res,* Vol. 68, No. 11, pp. 4229-4238, ISSN 1538-7445.

Izuta, Y., Ochi, M., Adachi, N., Deie, M., Yamasaki, T. & Shinomiya, R. (2005). Meniscal repair using bone marrow-derived mesenchymal stem cells: experimental study using green fluorescent protein transgenic rats. *Knee,* Vol. 12, No. 3, pp. 217-223, ISSN 0968-0160.

Johnstone, B. & Yoo, J.U. (1999). Autologous mesenchymal progenitor cells in articular cartilage repair. *Clin Orthop Relat Res,* Vol. 367, S156-162.

Katakai, D., Imura, M., Ando, W., Tateishi, K., Yoshikawa, H., Nakamura, N. & Fujie, H. (2009). Compressive properties of cartilage-like tissues repaired in vivo with scaffold-free, tissue engineered constructs. *Clin Biomech (Bristol, Avon),* Vol. 24, No. 1, pp. 110-116, ISSN 1879-1271.

Kobayashi, T., Ochi, M., Yanada, S., Ishikawa, M., Adachi, N., Deie, M. & Arihiro, K. (2008). A novel cell delivery system using magnetically labeled mesenchymal stem cells and an external magnetic device for clinical cartilage repair. *Arthroscopy,* Vol. 24, No. 1, pp. 69-76, ISSN 1526-3231.

Koga, H., Shimaya, M., Muneta, T., Nimura, A., Morito, T., Hayashi, M., Suzuki, S., Ju, Y.J., Mochizuki, T. & Sekiya, I. (2008). Local adherent technique for transplanting mesenchymal stem cells as a potential treatment of cartilage defect. *Arthritis Res Ther,* Vol. 10, No. 4, pp. R84.

Ladage, D., Brixius, K., Steingen, C., Mehlhorn, U., Schwinger, R.H., Bloch, W. & Schmidt, A. (2007). Mesenchymal stem cells induce endothelial activation via paracine mechanisms. *Endothelium,* Vol. 14, No. 2, pp. 53-63.

Mahomed NN, B.J., Katz JN, Baron JA, Wright J, Losina E (2005). Epidemiology of total knee replacement in the United States Medicare population. *J Bone Joint Surg Am,* Vol. 87, No. 6, ISSN 1222-1228.

McIlwraith, C.W., Frisbie, D.D., Kisiday, J.D., Kawcak, C.E., Werpy, N.M., Rodkey, W.G., & Steadman, R.S. (2010) Use of bone marrow-derived culture-expanded mesenchymal stem cells to augment healing of chondral defects treated with microfracture In: *Proceedings of the 56th Annual Convention of the American Association of Equine Practitioners,* Moyer, W.A. pp. 27-28. Baltimore, Marlyand, USA.

Minas, T. & Nehrer, S. (1997). Current concepts in the treatment of articular cartilage defects. *Orthopedics,* Vol. 20, No. 6, pp. 525-538, ISSN 0147-7447.

Mobasheri, A., Csaki, C., Clutterbuck, A.L., Rahmanzadeh, M. & Shakibaei, M. (2009). Mesenchymal stem cells in connective tissue engineering and regenerative medicine: applications in cartilage repair and osteoarthritis therapy. *Histol Histopathol,* Vol. 24, No. 3, pp. 347-366, ISSN 1699-5848.

Nejadnik, H., Hui, J.H., Choong, E.P., Tai, B.C. & Lee, E.H. (2010). Autologous Bone Marrow-Derived Mesenchymal Stem Cells Versus Autologous Chondrocyte Implantation: An Observational Cohort Study. *Am J Sports Med,* Vol. 38, No. 6, pp. 1110-1116, ISSN 1552-3365.

Niemeyer, P., Fechner, K., Milz, S., Richter, W., Suedkamp, N.P., Mehlhorn, A.T., Pearce, S. & Kasten, P. (2010). Comparison of mesenchymal stem cells from bone marrow and adipose tissue for bone regeneration in a critical size defect of the sheep tibia and the influence of platelet-rich plasma. *Biomaterials,* Vol. 31, No. 13, pp. 3572-3579, ISSN 1878-5905.

Prigozhina, T.B., Khitrin, S., Elkin, G., Eizik, O., Morecki, S. & Slavin, S. (2008). Mesenchymal stromal cells lose their immunosuppressive potential after allotransplantation. *Exp Hematol,* Vol 36, No. 10, pp. 1370-1376, ISSN 0301-472X.

Sakai, D., Mochida, J., Yamamoto, Y., Nomura, T., Okuma, M., Nishimura, K., Nakai, T., Ando, K. & Hotta, T. (2003). Transplantation of mesenchymal stem cells embedded in Atelocollagen gel to the intervertebral disc: a potential therapeutic model for disc degeneration. *Biomaterials,* Vol. 24, No. 20, pp. 3531-3541, ISSN 0142-9612.

Schauwer, C.D., Meyer, E., Walle, G.R. & Soom, A.V. (2010). Markers of stemness in equine mesenchymal stem cells: a plea for uniformity. *Theriogenology,* Vol. 75, No. 8, pp. 1431-1443, ISSN 0093-691X.

Sturnieks, D.L., Besier, T.F. & Lloyd, D.G. (2010). Muscle activations to stabilize the knee following arthroscopic partial meniscectomy. *Clin Biomech (Bristol, Avon),* Vol. 26, No. 3. pp. 292-297, ISSN 1879-1271.

Ueda, Y., Inaba, M., Takada, K., Fukui, J., Sakaguchi, Y., Tsuda, M., Omae, M., Kushida, T., Iida, H. & Ikehara, S. (2007). Induction of senile osteoporosis in normal mice by intra-bone marrow-bone marrow transplantation from osteoporosis-prone mice. *Stem Cells,* Vol. 25, No. 6, pp. 1356-1363, ISSN 1549-4918.

Vidal, M.A., Robinson, S.O., Lopez, M.J., Paulsen, D.B., Borkhsenious, O., Johnson, J.R., Moore, R.M. & Gimble, J.M. (2008). Comparison of chondrogenic potential in equine mesenchymal stromal cells derived from adipose tissue and bone marrow. *Vet Surg,* Vol. 37, No. 8, pp. 713-724, ISSN 1532-950X.

Wakitani, S., Nawata, M., Tensho, K., Okabe, T., Machida, H. & Ohgushi, H. (2007). Repair of articular cartilage defects in the patello-femoral joint with autologous bone marrow mesenchymal cell transplantation: three case reports involving nine defects in five knees. *J Tissue Eng Regen Med,* Vol. 1, No. 1, pp. 74-79, ISSN 1932-7005.

Wakitani, S., Okabe, T., Horibe, S., Mitsuoka, T., Saito, M., Koyama, T., Nawata, M., Tensho, K., Kato, H., Uematsu, K., Kuroda, R., Kurosaka, M., Yoshiya, S., Hattori, K. & Ohgushi, H. (2010). Safety of autologous bone marrow-derived mesenchymal stem cell transplantation for cartilage repair in 41 patients with 45 joints followed for up to 11 years and 5 months. *J Tissue Eng Regen Med,* Vol. 5, No. 2, pp. 146-150, ISSN 1932-7005.

Yoshikawa, T., Ueda, Y., Miyazaki, K., Koizumi, M. & Takakura, Y. (2010). Disc regeneration therapy using marrow mesenchymal cell transplantation: a report of two case studies. *Spine (Phila Pa 1976),* Vol. 35, No. 11, pp. E475-480, ISSN 1528-1159.

Yoshimura, H., Muneta, T., Nimura, A., Yokoyama, A., Koga, H. & Sekiya, I. (2007). Comparison of rat mesenchymal stem cells derived from bone marrow, synovium, periosteum, adipose tissue, and muscle. *Cell Tissue Res,* Vol. 327, No. 3, pp. 449-462, ISSN 0302-766X.

Zhou, S., Greenberger, J.S., Epperly, M.W., Goff, J.P., Adler, C., Leboff, M.S. & Glowacki, J. (2008). Age-related intrinsic changes in human bone-marrow-derived mesenchymal stem cells and their differentiation to osteoblasts. *Aging Cell,* Vol. 7, No. 3, pp. 335-343, ISSN 1474-9726.

Zhou, Z., Jiang, E.L., Wang, M., Liu, Q.G., Zhai, W.J., Huang, Y., Wang, H.H. & Han, M.Z. (2005). [Comparative study on various subpopulations in mesenchymal stem cells of adult bone marrow]. *Zhongguo Shi Yan Xue Ye Xue Za Zhi,* Vol. 13, No. 1, pp. 54-58, ISSN 1009-2137.

17

Cell-Biomaterial Interactions Reproducing a Niche

Silvia Scaglione[1], Paolo Giannoni[2] and Rodolfo Quarto[3]
[1]CNR - National Research Council of Italy, IEIIT Institute, Genoa
[2]Advanced Biotechnology Center, Genoa
[3]DIMES, University of Genoa
Italy

1. Introduction

During the last years, several major progresses and improvements have been introduced in orthopaedic surgery as innovative and attractive approaches for potentially solving many of the limitations of the current therapies. Natural processes of bone repair are sufficient to restore the skeletal integrity for most lesions. However, this auto-regenerative potential has dimensional limits which require manipulation of natural healing mechanisms to be overcome.

In bone tissue engineering, biomaterials have been therefore proposed as scaffolds to direct and guide bone regeneration and to deliver stem cells in anatomical sites where the regenerative process is defective. Although material science technology has resulted in clear improvements, no ideal bone substitute has been developed yet and hence large bone defects still represent a major challenge for orthopaedic and reconstructive surgeons.

In the past years, scaffolds design has evolved from the obsolete, first-generation "spare-part" concept, to second-generation bio-inert "cell carriers", where biomaterials had to provide mechanical strength, durability, and possibly operate as cell delivery vehicles to achieve the regeneration of the target tissue- up to a third-generation of bio-functional materials that seek to incorporate instructive signals into scaffold nanostructures to modulate cellular functions, direct cell fate, and finally govern tissue regeneration in vivo. They should finally be resorbed in vivo, as soon the neo-formed tissue is able to fully substitute the graft.

The most intriguing concept in modern biomaterials is thus obtaining materials able to mimic a specific pre-existing microenvironment and, therefore, inducing cells to differentiate in a predetermined manner and to regenerate by themselves the desired tissue (i.e. bone tissue) according to physiological pathways.

In order to reach this ambitious task, intelligent biomaterials should be properly designed and an informative microenvironment, mimicking a physiological niche, provided. Any material has to be considered informative in the sense that its intrinsic nature (i.e. chemical composition) and structure (i.e. macro- micro- nano-architecture) will anyway transmit a signal that will be read and decoded by colonizing cells. We still know very little of how to create local microenvironments, or artificial niches, that will govern stem cells behaviour and their terminal fate. However, it has been highlighted in the last years that stem and progenitor cells are able to modify their behaviour and fate when loaded onto specific substrates (Hench and Polak 2002; Dalby, Gadegaard et al. 2007; Hunt 2008).

Among basic requirements for the design and generation of bone substitute materials, there is the development of biomimetic scaffolds with (i) an internal architecture able to favour cell migration and in vivo vascularisation, and (ii) a chemical composition permissive to cell attachment, selective differentiation and maintenance of cellular functions.

In this context, the bioengineering challenge become ambitious, since the complex cell-biomaterial interaction moves on multiple spatial and temporal scales. The micro-environmental cues, such as chemical environmental variables, are able to stimulate specific cellular responses at the molecular level already at early time points (Goshima, Goldberg et al. 1991; Ohgushi, Dohi et al. 1993; Fabbri, Celotti et al. 1995; Kon, Muraglia et al. 2000; Erbe, Marx et al. 2001; Endres, Hutmacher et al. 2003; Kasten, Luginbuhl et al. 2003; Livingston, Gordon et al. 2003; Niemeyer, Krause et al. 2003; Arinzeh, Tran et al. 2005; Kotobuki, Ioku et al. 2005; Kondo, Ogose et al. 2006; Fan, Ikoma et al. 2007; Mygind, Stiehler et al. 2007; Gigante, Manzotti et al. 2008; Ng, Tan et al. 2008; Bernhardt, Lode et al. 2009; Saldana, Sanchez-Salcedo et al. 2009). For example, ceramic scaffolds (i.e. hydroxyapatite) are able to induce a faster and more efficient cell adhesion. However, the cell-signalling pathways involved in the variation of gene expression are yet to be fully elucidated. Recently, it has been reported that cells loaded onto biomaterials are also able to decode the topographic cues of the scaffold, and respond to the shape of the micro- environment priming a specific cell differentiation pathway (Lenza, Vasconcelos et al. 2002; Dalby, Gadegaard et al. 2007; Huang, Lin et al. 2007; Wei and Ma 2008).

Tissue development and regeneration implies a spatio-temporal assembly of differentiating cells organized to create functional structures. This process is finely tuned, progressing gradually through cell-matrix and cell-cell interactions. Biomechanical forces generated by the contact among the differentiating cells within the tissue or with the ECM, have a profound impact on tissue growth, development, maintenance, and repair by providing the required metabolic support, strength, and endurance.

In this chapter, we will discuss about those biomaterials that are being designed and manufactured to gain the informative status necessary to drive proper molecular cross-talk and cell differentiation. In particular, we will explore: (i) how to develop intelligent informative scaffolds, (ii) how stem/progenitor cells decode biomaterials, (iii) promising bone substitutes in a tissue-engineering scenario.

2. How to develop informative scaffolds

Progress in biomaterials design and engineering are converging to enable a new generation of instructive materials to highlight as candidates for regenerative medicine. An emerging philosophy aims to overpass the traditional approach of recreating the complexity of living tissues ex vivo; in this context, the most ambitious strategy attempts to develop synthetic materials that establish key interactions with cells in ways that unlock the body's innate powers of organization and self-repair. The complex cell-biomaterial interaction moves on multiple spatial and temporal scales. Therefore, in order to influence effectively the cell behaviour, scaffolding materials must bear complex information, coded in their *physical* and *chemical* structures. In particular, bio-scaffolds must be properly designed to mimic the spatial organization of stem cells and their *niche* physiological structure.

Stem cell niche is defined as a dynamic microenvironment that balance stem cells activity to maintain tissue homeostasis and repair throughout the lifetime of the organism (Voog and Jones 2010). In principle, stem cells in their niche make decisions to either remain in a quiescent state, undergo self-renewal, or to exit the niche upon exposure to local or systemic stimuli. These signals are actively coordinated and presented in a temporally and spatially regulated manner. Proper microenvironmental cues given by the biomaterial may become "informative" for cells, stimulating specific cellular responses (Fig.1).

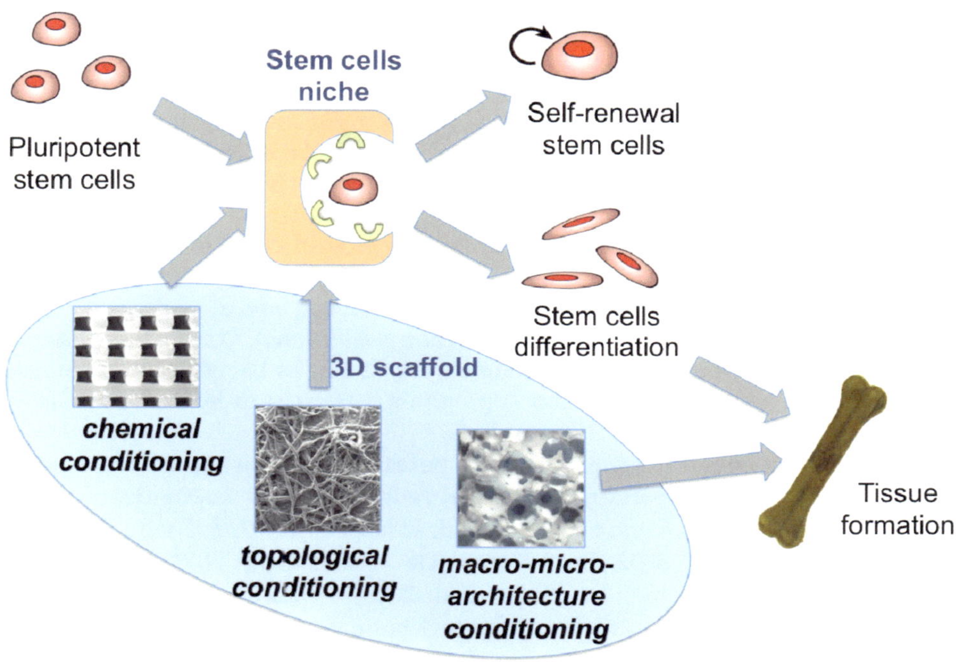

Fig. 1. Cell-biomaterial interaction.

The influence of chemical environmental variables on cell activity (i.e. *chemical conditioning*) was already probed (Goshima, Goldberg et al. 1991; Ohgushi, Dohi et al. 1993; Yuan, Yang et al. 1998; Yuan, Kurashina et al. 1999; Boo, Yamada et al. 2002; Kasten, Luginbuhl et al. 2003; Niemeyer, Krause et al. 2003; Arinzeh, Tran et al. 2005; Fan, Ikoma et al. 2007; Nakamura 2007; Guarino, Causa et al. 2008; Cheng, Ye et al. 2009; Saldana, Sanchez-Salcedo et al. 2009). For example, ceramic scaffolds (i.e. hydroxyapatite) allow a faster and more efficient cell adhesion (Goshima, Goldberg et al. 1991; Ohgushi, Dohi et al. 1993; Yuan, Yang et al. 1998; Yuan, Kurashina et al. 1999; Kasten, Luginbuhl et al. 2003; Arinzeh, Tran et al. 2005; Fan, Ikoma et al. 2007; Cheng, Ye et al. 2009; Saldana, Sanchez-Salcedo et al. 2009).
The synthesis of complex inorganic materials mimicking natural structures offers exciting avenues for the chemical construction of macrostructures and a new generation of biologically and structurally inspired scaffolds for tissue engineering.
Besides chemical conditioning, progenitor cells fate is also affecting by topographic cues of the scaffolds (i.e. *topological conditioning*). Recently, it has been reported that cells loaded

onto biomaterials are able to decode the topographic signals of the scaffold, and respond to the shape of the microenvironment priming a specific cell differentiation commitment (Dalby, Gadegaard et al. 2007). Thus, nanostructured biomaterials such as nanoparticles, nanofibers, nanosurfaces, and nanocomposites have gained increasing interest in regenerative medicine, since they offer a temporary ECM for regenerative cells (Hollister, Maddox et al. 2002; Balasundaram and Webster 2007; Wei and Ma 2008; Zhang and Webster 2009).

Recent studies have also shown that in the absence of adhesion peptides, cells interact with scaffolds by means of adsorbed protein, and in this regard topography and hydrophilicity are key considerations. For example, fibrous meshes with nanoscale fibre diameters have shown selective take-up of proteins relevant for cell attachment, such as fibronectin and vitronectin (Place, Evans et al. 2009). And whereas hydrophobic scaffolds tend to adsorb protein in sub-optimal configurations (with hydrophobic residues displaced towards the scaffold surface), hydrophilic polymers adsorb protein in a hydrated interfacial phase wherein the proteins are more likely to retain their native conformation (Place, Evans et al. 2009).

If the chemical and topographical cues imprint the progenitor/stem cell fate at early time points, by reproducing a proper stem cell niche, the 3D architecture design (i.e. pore size, total porosity, surface area) of the scaffolds plays a pivotal role at prolonged time of cell-biomaterial interaction (i.e. *macro-micro architecture conditioning*). During this phase, as soon cells are addressed towards their differentiative fate, it may be beneficial for a biomaterial to provide adequate space (porosity) and appropriate surface to foster and direct new tissue formation.

In the bone tissue engineering, for instance, pores are necessary to allow migration and proliferation of osteoblasts and mesenchymal cells, as well as vascularization (Gauthier, Bouler et al. 1998; Boyde, Corsi et al. 1999; Lu, Flautre et al. 1999; Chang, Lee et al. 2000; Hollister, Maddox et al. 2002; De Oliveira, De Aguiar et al. 2003; Karageorgiou and Kaplan 2005; Mastrogiacomo, Scaglione et al. 2006). In addition, macroporosity has a strong impact on the amount of newly formed bone tissue; moreover, a porous surface may improve mechanical interlocking between the implant biomaterial and the surrounding natural bone.

During the newly tissue formation, the internal structure of the scaffolds at a micro-macro scale may still influence the pattern of newly tissue formed: whenever a graded and geometrically ordered scaffold is offered to the cells as template for bone tissue regeneration, lamellar bone tissue is newly formed. Conversely, if neither biomechanical cues nor geometrical rules are applicable a bone tissue lacking in structural organization will be deposited within the implanted scaffolds (Scaglione et al. 2011).

2.1 Design of the chemical structure

A range of responses, such as cell adhesion, viability and differentiation, can be differentially affected by particular natural/synthetic substrates. Ideally, the chemical structure design of scaffolds for tissue engineering should meet the following criteria: (1) the surface should permit cell adhesion, promote cell growth, and allow the retention of differentiated cell functions; (2) the scaffolds should be biocompatible, neither the material nor its degradation by-products should provoke inflammation or toxicity in vivo; (3) the scaffold should be biodegradable and eventually eliminated.

In the last years, the generation of biologically and structurally inspired scaffolds mimicking the chemical construction of natural structures has been proposed and carried out. In many cases, biomimetic strategies do not set out to copy directly the structures of biological materials but aim to abstract key concepts from the biological systems that can be adapted within a synthetic context. The simplest biomimetic approach involves the design of single component systems that mimic the chemistry of the targeted biological material.

In the field of bone tissue engineering, a wide range of biomaterials, whose composition is such that they mimic natural bone, has been tested to stimulate ossification and to improve the osteogenic potential of osteo-progenitor cells. Calcium and phosphate ions are important components during the mineralization phase of the ossification process. Materials composed of calcium phosphate such as hydroxyapatite (HA) and tricalcium phosphate (TCP) are attractive candidates for bone substitutes (Bruder, Kraus et al. 1998; Gauthier, Bouler et al. 1998; Boyde, Corsi et al. 1999; Flautre, Anselme et al. 1999; Marcacci, Kon et al. 1999; Kon, Muraglia et al. 2000; Dong, Kojima et al. 2001; Dong, Uemura et al. 2002; Livingston, Gordon et al. 2003; Gauthier, Muller et al. 2005). They are also particularly advantageous for bone tissue engineering applications as they induce neither immune nor inflammatory responses in recipient organisms (Erbe, Marx et al. 2001; Livingston, Ducheyne et al. 2002; El-Ghannam 2005).

HA is a natural component of bone tissue and therefore has been considered the ideal material to build bone substitutes. The ceramic performs as a mechanical support, an osteomimetic surface and as a template for the newly formed bone tissue. On the other hand, cells recognize the ceramic surface as pre-existing bone (osteo-mimesis) and differentiate into osteoblasts depositing bone extracellular matrix.

HA coatings have been also proposed to improve the outcome of prosthetic implants, improving the interaction between natural bone and implanted device. Porous HA ceramics support bone formation by marrow mesenchymal stem cells in vitro and in vivo. However, its brittleness and poor resorbability limits its application in the regeneration and repair of bone defects.

To avoid these limitations, polymer materials have received increasing attention and have been widely used for tissue engineering applications; in addition to biodegradability they also offer an eased processability. There are two kinds of polymer materials: synthetic polymer, and naturally derived polymers. The main biodegradable synthetic polymers include polyesters, polyanhydride, polyorthoester, polycaprolactone, polycarbonate, and polyfumarate. The polyesters such as polyglycolic acid (PGA), polylactic acid (PLA), and their copolymer of polylactic-co -glycolic acid (PLGA) are most commonly used for tissue engineering. The naturally derived polymers include proteins of natural extracellular matrices such as collagen and glycosaminoglycan, alginic acid, chitosan, and polypeptides.

Biocompatible polymers have also been regarded as candidates for bone substitutes (Ren, Ren et al. 2005; Williams, Adewunmi et al. 2005; Jiang, Abdel-Fattah et al. 2006; Wu, Shaw et al. 2006; Bonzani, Adhikari et al. 2007). However, a number of practical problems still persist, such as the difficulty in controlling the in vivo degradation of bio-resorbable polymers, low efficiency of cell seeding, cytotoxicity of the breakdown products produced during scaffold degradation, in addition to poor mechanical properties, incomparable with natural hard tissues.

To overcome these limitations, ceramic/polymer composite materials have been explored (El-Amin, Botchwey et al. 2006; Kim, Park et al. 2006; Leung, Chan et al. 2006; Kretlow and Mikos 2007; Ren, Zhao et al. 2007). When used in blends with other polymers, HA particles exposed on the surface of scaffolds favour focal contact formation of osteoblasts. A bone-like mineral film, consisting mainly of calcium apatite, when layered onto the surface of polymeric-based substrates, does not achieve the same effect as when HA is incorporated into the bulk material.

Interestingly, heterogeneous composite scaffolds consisting of two distinct, but integrated layers, have been proposed to induce cells towards different lineages, (i.e. cartilage and bone) and possibly generate heterogeneous tissues, such as osteochondral grafts (Sherwood, Riley et al. 2002; Martin, Miot et al. 2007; Grayson, Chao et al. 2008; Tampieri, Sandri et al. 2008; Harley, Lynn et al. 2010; Kon, Delcogliano et al. 2010; Lynn, Best et al. 2010). Within these informative biomaterials, cells may recognize the differently designed surfaces of the graft as pre-existing bone/cartilage tissue (biomimesis) and deposit bone/cartilage extracellular matrix accordingly to their specific localization in the scaffold. Experimental evidences confirmed that the proper design of layered scaffolds containing distinct compositional and structural features that reflect the functional environment of the native tissues is able to address progenitor cells to alternate differentiation pathways, thus inducing a simultaneous regeneration of multiple tissues (Sherwood, Riley et al. 2002; Tampieri, Sandri et al. 2008; Harley, Lynn et al. 2010; Kon, Delcogliano et al. 2010; Lynn, Best et al. 2010).

2.2 Design of the architectural structure

Porous three-dimensional (3D) scaffolds fabricated from synthetic and naturally derived materials have been widely used in different tissue engineering applications, such as cartilage, bone, skin, and ligament. Depending on the specific targeted tissue, the internal architecture has been intelligently designed and the density, pore shape, pore size and pore interconnection pathway of the material predetermined. A key structural parameter, which is also common for a wide number of tissue substitutes, is the total porosity, which is mandatory for a massive cellular induced tissue formation within the implanted scaffold.

In the last decade, several methods have been developed to prepare these kinds of porous 3D scaffolds, including gas foaming (Mooney, Mazzoni et al. 1996; Harris, Kim et al. 1998), three-dimensional printing (Hutmacher 2000), phase separation (Schugens, Maquet et al. 1996; Nam and Park 1999) and porogen leaching (Mikos, Sarakinos et al. 1993).

The gas-foaming technique uses high-pressure CO_2 gas processing and dissolved gas molecules create the macropores noted post processing. The porosity and pore structure is dependent on the amount of gas dissolved in the polymer/ceramic structure, the rate and type of gas nucleation and the diffusion rate of gas molecules through the material to the pore nuclei. The advantages of this method are a large surface area for cell attachment and a rapid diffusion of nutrients in favour of cell survival and growth. The drawback of this method might be a lack of structural stability of the final scaffold.

To improve the pore structure, a combination of different techniques may be carried out, such as gas foaming and particulate leaching. After expansion, the salt particulates are leached out to yield macropores within the scaffold. The overall porosity and level of pore connectivity can be regulated by the ratio of material/salt particulates and the size of the salt particulates.

3D printing is a solid free-form fabrication process, which produces components by inkjet printing a binder into sequential powder layers. The part is built sequentially in layers. The binder is delivered to the powder bed producing the first layer, the bed is then lowered to a fixed distance, powder is deposited and spread evenly across the bed, and a second layer is built. This is repeated until the entire part, e.g. a porous scaffold, is fabricated.

The phase-separation technique is based on thermodynamic demixing of a homogeneous polymer-solvent solution into a polymer-rich phase and a polymer-poor phase, usually by either exposure of the solution to another immiscible solvent or cooling the solution below a bimodal solubility curve. Solvent is removed by freeze-drying, leaving behind the polymer as foam. Morphology is controlled by any phase transition that occurs during the cooling step, i. e. liquid-liquid or solid-liquid.

The porogen leaching method involves the casting of a mixture of polymer solution and porogen in a mold, drying the mixture, followed by a leaching out of the porogen with water to generate the pores. Usually, water-soluble particulates such as salts and carbohydrates are used as the porogen materials. The pore structures can easily be manipulated by controlling the property and fraction of the porogen, and the process is reproducible. This technique provides easy control of the pore structure and has been well established.

Such hierarchical porous architectures not only define the mechanical properties of the scaffold, but also the initial void space that is available for regenerating cells to form new tissues (including new blood vessels) as well as the pathways for mass transport via diffusion and/or convection (Gauthier, Bouler et al. 1998; Boyde, Corsi et al. 1999; Lu, Flautre et al. 1999; Chang, Lee et al. 2000; Hollister, Maddox et al. 2002; De Oliveira, De Aguiar et al. 2003; Karageorgiou and Kaplan 2005; Mastrogiacomo, Scaglione et al. 2006). While interconnected macroporosity of a biomaterial is important to provide sufficient space for cellular activity and tissue deposition, interactions between cells and biomaterials occur at the interface, i.e., the entire internal pore walls of a 3D scaffold. Microporosity is thus another key parameter of the architectural structure design of the scaffolds. Moreover, an incomplete pore interconnection or a limiting calibre of the interconnections could represent an important constraint to the overall biological system by limiting blood vessels invasion (Mastrogiacomo, Scaglione et al. 2006).

Besides macro-micro porosity, the design of a proper surface morphology/topography may directly and significantly affect cell-scaffold interactions and ultimately tissue formation and function (Woo, Chen et al. 2003; Smith and Ma 2004; Woo, Jun et al. 2007; Smith, Liu et al. 2008). Cells in vivo are exposed to adhesive contacts in all three-dimensions, thus bio-scaffolds must be organized to mimic the spatial organization of the stem cells niche. Extensive efforts have been therefore made to identify scaffolds that resemble the natural extracellular matrix (ECM). As well as requiring information from each other, cells derive a vast wealth of information from their environments, including the material that surrounds and separates them within tissues, the ECM. An informative material scaffold must take on this instructive role to some degree in order to maintain cell viability and control cell behaviour.

Advanced manufacturing techniques can be used to control the spatial sub-micrometric internal architecture in engineered tissues, manipulating the scaffold topography on the length scale of the stem cell niche and smaller. It has been demonstrated that nanofibrous polymeric scaffolds offer to the cells biomimetic configurations that resembles ECM collagen fibers in

their ability to support the differentiation of progenitor/stem cells along adipogenic, chondrogenic, and osteogenic lineages (Yang, Murugan et al. 2005; Badami, Kreke et al. 2006; Erisken, Kalyon et al. 2008; Bashur, Shaffer et al. 2009; Wise, Yarin et al. 2009).
Nanotechnology, or the use of nanomaterials, may help to realize materials mimicking surface properties (including topography, energy, etc) of natural tissues. For these reasons, different approaches toward the formation of nano-fibrous materials have emerged in the last years: self-assembly, electrospinning and phase separation (Jayaraman, Kotaki et al. 2004; Vasita and Katti 2006; Barnes, Sell et al. 2007; Smith, Liu et al. 2008; James, Toti et al. 2011). Each of these approaches is very different but has a unique set of characteristics which lends to its development as a scaffolding system with a potential to accommodate cells and guide their growth and subsequent tissue regeneration.
For instance, self-assembly can generate small diameter nano-fibers in the lowest end of the range of natural extracellular matrix collagen, while electrospinning has only generated large diameter nano-fibers on the upper end of the range of natural extracellular matrix collagen. Moreover, electro-spinning can be used to generate polymeric scaffolds with aligned nano-scale fibers that direct spatial adhesion and orientation of cells upon differentiation. Phase separation, on the other hand, has generated nano-fibers in the same range as natural extracellular matrix collagen and allows for the design of macropore structures.
In addition to the dimensional similarity to tissue compartments, nanomaterials also exhibit unique surface properties due to their significantly increased surface area and roughness compared to conventional or micron structured materials. Material surface properties mediate specific proteins adsorption and bioactivity, further regulating cell behaviour and tissue regeneration (Sato and Webster 2004; Balasundaram and Webster 2006; Liu and Webster 2006).

3. How stem/progenitor cells decode biomaterials

Although the identification of mesenchymal stem cells (MSC) is currently a matter of discussion, these cells have become attractive targets for clinical applications and a large number of studies on use in regenerative medicine have been produced (Tonti and Mannello 2008). The classic paradigm for tissue engineering considers seeding an appropriate cell source, like MSC, on or within a scaffold that facilitates cells growth, organization and differentiation into a specific and functional tissue.
Regardless of the topography and of the chemistry of the scaffolds, the constructs must also provide some level of physical support from the moment of implantation, to assist cell attachment and provide room for the deposition of new matrix, if needed for tissue reconstitution. This clearly implies close contact between matrix proteins, either of endogenous (cellular) or exogenous (secreted) origin, and the scaffold, a "play-of-three" that is extremely relevant for the cell.
In living tissues the main extracellular matrix constituents are comprised within a few macromolecule classes, such as collagens, elastin, proteoglycans, hyaluronic acid -and its derivatives- and adhesion glycoproteins (among which fibrinogen and fibronectin, tenascins and thrombospondins). Alternative splicing and secretion of different proportions of the ECM components allow the generation of a wide range of matrices, ranging from basal lamina to bone. Often the prototypical scaffolds for the cell-based repair of mesenchymal tissues (mainly cartilage or bone, or both), whether composed of ceramic or biodegradable

polymers, can be tailored to support cell adhesion and to degrade at rates coincident with new tissue development. However, on both scaffold types, the mechanism of cell adhesion is indirect and relies onto the deposition of extracellular matrix proteins by the seeded cells (Murphy, Hsiong et al. 2005; Chastain, Kundu et al. 2006). As a result, ceramic or biosynthetic scaffolds may lack at first the specificity of the original tissues, i.e. the exogenous proteic signals that ease cell adhesion, undermining the recognition of the local microenvironment by the cells and the consequent repair processes. For engineered tissue repair, then, it is critical to understand the mechanisms by which cells first recognize the surfaces and structures on which they are seeded and how these, in turn, may govern cell functions and influence cell-mediated remodelling events at the interfaces between the cell-seeded constructs and the host tissues.

3.1 The role of cell adhesion molecules as active mechanosensors

Cells normally sense the microenvironment elasticity as they anchor and pull on their surroundings. These processes relay in part on specific adhesion proteins- myosin, integrins, cadherins- able to transmit forces to the substrates. Considerable attention has been posed in understanding the cell responsiveness to external forces, ranging from fluid flow to stretching and twisting (Alenghat and Ingber 2002). However the cells respond also to the sensed resistance, whether it comes from normal tissue matrix or from synthetic substrata, with cytoskeletal alterations. The most recent literature points out to the existence of a feedback loop, in which cell-exerted forces are coupled to microenvironmental elasticity able to induce subsequent and additional changes to cellular responses (Discher, Janmey et al. 2005). Typically four protein families are involved in the adhesion processes: IgCAM, selectins, cadherins and integrins. The first three interact with complementary proteins/ligands on the partner cells surfaces, whereas integrins bind prevalently to extracellular matrix proteins. The expression of restricted isoforms of each of these classes of proteins allows specific interactions among cells and extracellular matrix, an essential requisite for embryonic development, tissue regeneration and force transmission. However, in the light of scaffold surface recognition, integrins and cadherins are of paramount importance for sensing the microenvironment external to the cell. Integrins are heterodimeric receptors made of two transmembrane chains, α and β, both contributing to binding specificity. A combinatorial strategy allows the vertebrate cells to express several sub-sets of integrins by selective combinations of 18 different α chains and additional 8 different β chains. At least 24 different dimers are so far known. The cytoplasmic tails of the intergrins, extracellularly linked to ECM components, bind to actin filaments of the cytoskeleton to generate focal contacts, through the concerted action of talin and vinculin. Another adaptor protein, paxillin, binds the integrins to recruit Src and FAK (focal adhesion) tyrosin-kinases, (Critchley 2000; Turner 2000). This chain of events promotes, within minutes, a rise of the intracellular Ca^{2+} concentration and the development of mature focal contacts, named focal adhesion sites, anchoring the actin stress fibers to the cell membrane. The tension generated by the organized stress fibers on the adhesion sites is maintained during cell movement and migration. Cell adhesion onto a specific substrate depends, then, on the integrin density on the cell surface, on the ligand concentration on the substratum surface and on their reciprocal affinity. The rapid association/dissociation of the integrin/ligand complexes allows the cell to redefine the interactions with the ECM during anchoring, movement and migration (Bercoff, Chaffai et al. 2003).

The importance of sensing the mechanical properties of the ECM within the cell surroundings has been clearly established in studies with tumor cells and fibrobroblasts (Discher, Janmey et al. 2005; Paszek, Zahir et al. 2005). Indeed tumors (often detected as a rigid mass within softer tissues) display a peculiar rigidity, in part due to the interstitial pressure caused by a perturbed vascular structure, in part due to fibrosis, but in part also due to an increase of the elastic module of transformed cells as a consequence of an altered cyto-architecture (Beil, Micoulet et al. 2003). A current understanding is that cells take advantage of actinomyosin contractility for dual interactions with the matrix. Cell contraction at integrin-based adhesion sites is essentially resisted by the matrix, and is followed by the accumulation of additional molecules at the sites involved. This process comes to a balance when tension forces are equilibrated at the cell-matrix interface. Regardless of the non-malignant or malignant nature of the cells, the mechanotransducing functions of integrins represent a major focus of several researches (Bershadsky, Balaban et al. 2003), Integrins are known to regulate Rho- and growth factors-ERK (Extracellular signal-regulated kinase) dependent growth (Lee and Juliano 2004). At the same time ERK influence ROCK (Rho-associated protein kinase) and myosin activity (Huang, Kamm et al. 2004). Interestingly matrix stiffness modulates growth factor signalling and Rho GTPase activity (Wang, Weaver et al. 1998); moreover, Rho activity is elevated in stiff tumors and is linked to cell invasivness, although single members of the Rho family are differently involved in branching and lamellipodia broadening (Vega, Fruhwirth et al. 2011); instead ROCK activation contributes to cell contractility by inhibiting depolimerization of actin filaments (Paszek, Zahir et al. 2005). In epithelial morphogenesis, for example, matrix stiffness clusters integrins and these, in turn, enhance ERK activation and ROCK-generated contractility and focal adhesion. Indeed non-malignant mammary epithelial cells can be induced to form normal polarized and growth-arrested acinar structures in basal membrane, laminin-containing collagen I gels that match stiffness of normal mammary gland stroma. However even a small increase in matrix stiffness (by changing the gel cross-linker ratios) significantly compromises tissue organization, inhibits lumen formation and disrupts adherens junctions, as testified by a diffused β -catenin and non co-localized E-cadherin and β -catenin. Integrin α3 β1 and talin are involved in the normal adhesion machinery, both on soft and stiff substrata; however epithelial cells interacting with a soft matrix assemble focal complexes and express high amounts of total and active Src family kinase, whereas on a stiff matrix, cells spread and assemble stress fibers, activate more ERK in response to growth factors, and form focal adhesion sites with FAK^{pY397} and vinculin (Paszek, Zahir et al. 2005). These events are in compliance with altered levels of integrin expression in stiff tumors (Guo, Ma et al. 2009) as well as in rigid 2D substrata with respect to 3D matrices (Yeung, Georges et al. 2005). Interestingly a similar behaviour was also observed in fibroblasts, where a phosphorylated FAK^{pY397} and vinculin were recruited to α5 β1 adhesion sites on stiff gels, although in this case force-dependent integrin aggregation precedes the appearance of FAK^{pY397} (Nicolas, Geiger et al. 2004). Force-dependent integrin anisotropy, as a result of matrix stiffness sensing, was demonstrated to be of relevance using integrin mutants (V737N) that promoted self-association through enhanced hydrogen bonding in the transmembrane domain of the protein. Although integrin clustering was not induced on stiff substrata, on soft gels V737N integrin-expressing cells spread significantly more, formed larger adhesion sites, expressed FAK^{pY397} and activated more ERK in response to growth factors (Paszek, Zahir et al. 2005). A mechanoregulatory circuit, then, integrates physical cues from the extracellular matrix with focal adhesion sites, through ERK- and

Rho-dependent cytoskeletal contractility, and regulates cells and tissue phenotype. Unbalance in integrin expression was also detected when placent-derived mesenchymal stromal cells were induced toward angiogenesis; VEGF-mediated adhesion and migration of placental-derived MSC onto fibronectin correlated with enhanced expression of α5 β1; at the same time anti- α5 or anti- β 1 antibodies inhibited angiogenesis when cells were cultured on chick corioallantoic membranes (Lee, Huang et al. 2009).

In fibroblasts fluorescence imaging has shown that F-actin fibers and stress fibers become increasingly organized if cells are cultured on increasingly stiffer substrates (Discher, Janmey et al. 2005). Contractile myotubes, instead, display strong focal adhesion and stress fibers when cultured on stiff gels or on glass micropatterns; they will however display actomyosin striation if cultured on top of a first layer of muscle cells (Discher, Janmey et al. 2005), Similarly, heart cells pulling on equally stiff heart cells can generate a positive feedback on their cytoskeletal organization that may not occur when the substratum is a scaffold or a different cell type. However variations and differences between cell types imply active and regulated responses, rather than a universal need of cells to exert traction forces; differences may depend in part on the expression and engagement of adhesion molecules, like it happens in the generation of shell-to-core cell aggregates obtained when randomly mixing two different cell types. Such an event was detected when mixing cardiomyocytes and retinal cells and is currently explained by the generation of surface tension at the interfaces of cell layers originated from low and high N-cadherin expressing cells (Discher, Janmey et al. 2005).

Cadherin-mediated interactions are prevalently homophylic, Ca2+-dependent and are responsible of driving cells into close contact through the organization of adherent junctions and desmosomes. Their common structural characteristic is the CAD domain, a folded structure organized in 7 β-sheets. Calcium ions bind between two CAD adjacent domains rendering the structure stiff, but the domains are free to rotate if the if the ions are not bound. Several cadherins display more than 5 extracellular CAD domains, arranged for *trans* or *cis* binding with other cadherin partners on the opposite cell (He, Cowin et al. 2003).

Cadherins contribute to growth contact inhibition; their cytoplasmatic tails bind adaptor proteins of the catenin family, linking cell-to-cell recognition to signal transduction pathways (Hamidouche, Hay et al. 2008). Cadherin-mediated adhesion was in fact linked to GTPase-cytoskeleton signalling (Delanoe-Ayari, Al Kurdi et al. 2004). Interestingly this process presents similarities with the self-organizing condensing mesenchymal cells that drive the growth of the pre-cartilagineous anlage in limb bud development through growth factor diffusion-processes and aptotaxis, although contribution from the latter seems relative (Christley, Alber et al. 2007).

In vitro chondrogenesis of MSC offers additional cues to dissect the pathways involved in cell shape-mediated commitment to differentiation. Chondrogenesis of MSC is induced when cells are cultured in high-density micromass pellets and stimulated with transforming growth factor β (Mackay, Beck et al. 1998). Rac1, another small GTPase, displayed a much higher activity in MSC undergoing smooth-muscle cell differentiation if compared to the same cells induced to chondrogenesis. Rac1 further regulated N-cadherin expression, a known requirement for smooth-muscle cell differentiation (Gao, McBeath et al. 2010). However it should be remembered that a dominant negative Rac1 was not sufficient to inhibit N-cadherin upregulation in the presence of TGF β 3, thus suggesting the possible presence of additional contributor proteins. The cytoplasmic domain of N-cadherin is indirectly linked to the cytoskeleton via the α-, β - and p120-catenin complex; deletion mutants of N-cadherin, lacking the β -catenin binding site, failed to support smooth-muscle cell differentiation of MSC, highlighting the role of these protein complexes in cell fate. Indeed cadherins are known to be implicated in mesenchymal condensation, although their expression is normally downregulated during chondrocytic differentiation (Oberlender and Tuan 1994). Possibly a concerted action of N-cadherin and Rac1 is necessary for myogenic differentiation of MSC, ensuring cell-to-cell contact; this may be a transient requirement that becomes unnecessary once cells get separated and encased in extracellular matrix during chondrogenic maturation.

3.2 Progenitor cells' lineage commitment: A matter of feeling

Transient signals, then, may be responsible of limited (time-wise and intensity-wise) cellular responses: other factors, such as soluble inducers in the growth medium, may also couple to matrix anchorage, as it was demonstrated for fibroblasts (Nakagawa, Pawelek et al. 1989), also derived from mesenchymal progenitors.

It is well known that differentiated cells of mesenchymal origin adhere and contract not only within soft tissues, but also on a variety of substrates in vitro, such as on collagen-coated acrylamide gels and glass (Engler, Sen et al. 2006). Such a wide range of possible adhesion substrates parallels a wide variation in matrix stiffness sensing, which in turns influences focal adhesion structures and the cell cytoskeleton (Cukierman, Pankov et al. 2001; Discher, Janmey et al. 2005). However, for tissue engineering purposes, pluripotent stem cells, rather than terminally differentiated ones, represent the gold standard for current and potential clinical applications (Peters, Schell et al. 2010; Ding, Shyu et al. 2011). In the last years it has been shown that stem cells or progenitor cells can be isolated from almost every tissue of the body (Bianco and Robey 2001), including menstrual blood (Ding, Shyu et al. 2011). Under the correct conditions, these cells can be stimulated to form new tissue, by using a simple biomaterials-based approach (Bianco and Robey 2001; Boo, Yamada et al. 2002; Cancedda, Bianchi et al. 2003; Barrilleaux, Phinney et al. 2006; Hutmacher, Schantz et al. 2007; Gigante, Manzotti et al. 2008; Hunt 2008)

In principle, stem cells in their niche undergo self-renewal, or exit the niche upon exposure to local or systemic stimuli. During tissue development and repair these signals are actively coordinated and are presented in a temporally and spatially regulated manner (Connelly, Garcia et al. 2008; Santiago, Pogemiller et al. 2009); for example the ECM surrounding osteogenically differentiating MSC is dynamically remodelled: biglycan is first detected in bone marrow surrounding MSC but not in unmineralized or in mineralized bone matrices; fibronectin and versican are observed in the regions of early mesenchymal condensation but they disappear in mature bone; decorin is present in unmineralized matrix but absent in mineralized bone (Hoshiba, Kawazoe et al. 2009). Indeed, bone marrow-derived mesenchymal stem cells (BMSC) represent a widely used class of progenitors, due to their ability to differentiate into several lineages (for ex, osteogenic, chondrogenic, myogenic or neurogenic), each of which is characterized by different matrix microenvironments and anchorage-dependent requirements (Goessler, Bieback et al. 2006; Djouad, Delorme et al. 2007; Boskey, Doty et al. 2008). At the tissue level, in fact, matrix stiffness accounts for distinctive ranges (Engler, Sen et al. 2006). The resistance that a cell feels when it deforms the ECM can be measured by the elastic constant of the matrix microenvironment, E, with values that range from 0.1-1.0 kPa (soft tissues, for example brain), to 1.0-20.0 kPa (muscle) up to >25.0 kPa for bone. By controlling the matrix elasticity in polyacrylamide gels through the cross-linker (bis-acrylamide) concentration, and by providing adhesion by coating the gels with collagen I, known to support myogenic and osteogenic differentiation, Engler and co-workers (2006) demonstrated that matrix stiffness, in spite of the same culturing conditions and medium supplements, can specify MSC lineage differentiation.

Among the cell's cytoskeletal motors, the non-muscle isoforms of myosin II (NMM II) (Kim, Kovacs et al. 2005) are suitable candidate mechanotransducers, able to generate signals proportional to the matrix deformation. The three existing isoforms (non-myosin II A, B or C) are involved in tensioning cortical actin structures linked to focal adhesion sites. These actin bridges transmit the force from the cell inside to the elastic matrix (Tamada, Sheetz et al. 2004) and are associated with signalling molecules (Bershadsky, Balaban et al. 2003). A counterproof of the involvement of non-myosin II in sensing matrix stiffness in MSC derives from the use of a specific NMM II-inhibitor, blebbistatin, which does not exerts its function on any other form of MSC myosin, other than myosin VI (Limouze, Straight et al. 2004). The administration of this molecule during MSC plating on different matrices blocks cell branching, elongation and spreading, but it has no relevant effect if exposure is carried out after cells have already spread and adopted a specific morphology, or 24 hrs post-seeding. Within this time frame, in fact, blebbistatin, was shown to inhibit the actin-dependent activation of the NNM II ATPase activity; these results were also confirmed by the use of an additional inhibitor specific for the myosin light chain kinase (MLCK), known to activate NNM II (Dhawan and Helfman 2004). Results lead to the conclusion that indeed NNM II appears to be necessary for matrix-elasticity driven lineage specification in MSC.

Traction stresses also modify the surrounding matrix around the cells. Although larger tractions are exerted on stiffer surfaces, typical tractions ($\tau \sim 1$KPa) exceed by far the viscous traction exerted by culture fluid on the cells. Moreover if matrix strain is relatively constant, cells need to be less contractile on soft gels than on stiff ones. Thus their adhesion will not be as strong, as it has consistently been measured by reduced forces needed to peel off cells from gels versus glass (Engler, Griffin et al. 2004).

Local sensing of force and/or geometry are therefore transduced into biochemical signals that regulate cell growth, differentiation shape and even cell death (Vogel and Sheetz 2006). Stiffness sensitivity and consequent cytoskeletal reorganization, however, not only interest the membrane/cytoplasmic compartment: nuclear deformations also take place in response to cytoskeletal modifications, cell cycle and division. Chromatin and laminin B contribute to the viscoelastic properties of the somatic cell nucleus, with single contributions prevailing according to the swelling condition of the nucleus. The nucleus is stiff and resists distortion at short times, whereas it undergoes deformation at longer times, providing essentially an infinite spectrum of timescales for structural reorganization and genome expression kinetics (Dahl, Engler et al. 2005). Nuclear deformation was also reported in response to culture conditions of MSC; static or perfusion cultures on 3D poly(ethylene terephthalate) scaffolds affected the ability of MSC to synthesize and deposit and organized ECM network, but, at the same time, affected nuclear shape: only cells in perfusion cultures displayed uniform spherical nuclei. Interestingly, cells in perfusion systems down-regulated Rex-1 and Oct-4 stemness-related genes, implying that a less primitive stem cell phenotype was retained in the perfusion cultures (Zhao, Grayson et al. 2009). This is in accordance with a possible stiffness sensing-dependent lineage commitment previously described, which would couple a stiff scaffold with a loss of the pluripotency and with an osteogenic-oriented MSC differentiation.

Once naive MSC are exposed to specific matrix stiffness, then, their gene transcription machinery up- or down-regulates specific gene subsets; immunostaining of cytoskeletal markers and transcription factors across the range of the tested matrix stiffnesses proved consistent with the lineage profiling in the experiments carried out by Engler and collaborators (2006): gene expression of markers for neurons, muscle or bone was induced 4- to 6-fold on the corresponding substrate with high specificity. Only stem cells grown on soft substrates with brain-like compliances expressed a phosphorylated form of neurofilament heavy chain. By converse myoD, a marker of muscle differentiation, and CBFα1, a transcription factor required for osteogenic differentiation, were expressed by cells grown on intermediate or rigid substrates, respectively. However it should be noted that expression levels for other markers was limited to only 50% of the standard level in terminally differentiated culture cells; muscle lineage-specific integrins expression was furthermore absent, evidencing that matrix stiffness compliance can drive the progenitor cells toward a developmental route but may not be sufficient *per se* to ignite terminal differentiation. Interestingly a bioinformatic approach also revealed that in mesenchymal stem cells genes regulated by high ECM stiffness included those indicative of the activation of two transcription factors downstream the Hippo signalling pathway, a highly conserved pathway involved in restraining cell proliferation and promoting apoptosis: YAP (Yes-associated protein) and TAZ (a transcriptional co-activator with a PDZ-binding motif, also known as WWTR1). Stiff surfaces induce nuclear translocation and activation of both factors, whereas their location was predominantly cytoplasmic and inactive in cells grown on soft matrices (Dupont, Morsut et al. 2011). In spite of the participation to the Hippo pathway, nuclear localization of the transcription factors was mainly due to the activity of the already cited Rho GTPase activity. Both factors are known to bind to Runx2, a transcription factor essential for osteogenic differentiation of mesenchymal cells. In a relevant set of experiments Dupont and co-workers (2011) demonstrated that depletion of YAP and TAZ prevented osteogenic differentiation of MSC cultured on stiff matrices and conversely promoted their adipogenic differentiation, thus elucidating the primary role of these factor in translating the cell mechanosensitivity from the cytoplasmic machinery to the nuclear/gene expression level.

3.3 Topography and surface chemistry: links to the mechanosensitivity of the cell

The optimization of the interactions between a scaffold matrix and the cell counterparts of the constructs can also be pursued by a specific biomimetic functionalization and/or nanostructuration of the interface. For prosthetic applications in orthopaedics, for example, cell attachment to grooved materials (Eisenbarth, Velten et al. 2007) and to nanocristalline coatings (Nicula, Luthen et al. 2007) has been documented since long. Indeed the interaction of the cells with the surrounding materials is within the nanometer scale. Thus nanoscaled topography of synthetic materials has attracted raising consideration because of its resemblance to in vivo surroundings, and mammalian cells were demonstrated to response to topographical surface variations (Silva, Czeisler et al. 2004; Dalby, McCloy et al. 2006; Dalby, McCloy et al. 2006). For articular chondrocytes, cell motility was increased when cells were cultured on 8 μm-deep grooved plastic-ware. Cells spread and oriented along the long axis of the groove. F-actin condensation was evident along the groove/ridge boundaries, correlated with a doubled velocity at which cells moved and was associated with a loss of the cell chondrogenic potential. Conversely 750 nm-deep grooves induced a reduced migratory capacity (Hamilton, Riehle et al. 2005; Hamilton, Riehle et al. 2005).

Although transdifferentiation of mesenchymal stem cells to neuronal lineages can be forced through specific induction media (Woodbury, Schwarz et al. 2000; Deng, Obrocka et al. 2001; Qian and Saltzman 2004) the mechanisms are not well understood. Nonetheless cultures of human MSC on nano-patterned plastic-ware, with gratings of 350 nm linewidth, an order of magnitude lower than the cells size, showed morphological changes in cell bodies and nuclei. A substantial confirmation of a new phenotype came from gene expression and microarray studies, in which microtubule associated protein 2 (MAP2) and β–tubulin III (Tuj1), both neuronal markers, were detected (Yim, Pang et al. 2007).
In the light of these results, it is reasonable to suppose that specific nano-patterning(s) may be compliant to or guide specific distribution(s) of the cell adhesion molecules within the cell surface. This distribution mirrors the one that cells would adopt in response to specific stiffness and elasticity of an underlying contact surface. The overall result is that nano-patterning may anticipate the cell response to a specific substratum and induce the consequences of cells adhesion onto it.
Clearly, once a cell has somewhat "decoded" its substrate and has ignited a new gene expression program in response to exogenous/endogenous stimuli, the secreted extracellular matrix protein will contribute to modify the microenvironment and to further drive the cell along a specific differentiation pathway. For example passive adsorption of two matrix protein like vitronectin (VN) and type collagen I (Col I) onto polymeric substrates were shown to mediate MSC adhesion and differently induced activation of mitogen-activated protein kinase (MAPK) and phosphatidylinositol-3-kinase (PI3K) signal transduction pathways (Kundu and Putnam 2006). Recent findings reveal that the *de novo* synthesis and deposition of ECM proteins by MSC alters the chemical identity of the polymeric substrate, stimulating changes in the integrin expression profiles. In turn these changes promote modifications in the MAPK and PI3K signalling pathways, therefore influencing the osteogenic differentiation of the seeded cells. Increasing amounts of fibronectin and Col I and decreased amounts of VN are in fact being deposited on poly(lactic) glycolic acid scaffolds over a 28-day period. The cell receptors pattern changed accordingly, providing higher levels for α5 β1 and α2 β1 integrins, (receptors for fibronectin and Col 1, respectively) and reduced levels for αV β3 integrin (VN receptor). Mechanistically, cell adhesion to Col I and fibronectin has been shown to induce the MAPK cascade, in particular the activation of the ERK1/2 system, critical for the activation of the osteogenic transcriptional factor Runx2 (Xiao, Jiang et al. 2000; Franceschi and Xiao 2003). Specific integrins then seem to be preferred or even required for the osteogenic differentiation of MSC. It should be remembered, though, that multiple integrins can bind a single ECM protein and that multiple ECM proteins can bind a single integrin (Miranti and Brugge 2002). Therefore a biofunctionalization of a scaffold surface should not focus on the presentation of a uniform coating to engage a single receptor, but rather identify the properties that control the presentation of integrin-specific epitopes within the coatings (Keselowsky, Collard et al. 2005).
Clearly several additional chemical modifications can be introduced and applied to almost any specific substrata, provided that the proper chemistry is used; indeed many strategies and approaches are currently being tested (Fu, Wang et al. 2011), ranging from simple coatings onto specific substrates (Uygun, Stojsih et al. 2009), to the contemporary use of genetic engineering and structural approaches (Benoit, Schwartz et al. 2008; Gorsline, Tangkawattana et al. 2010), to combinations of matrix-mimicking ligands and engineered structured nanomatrices (Anderson, Kushwaha et al. 2009). The same natural extracellular

matrix is *per se* able to induce specific cell commitment (Chen, Dusevich et al. 2007). Thus the combination of topographical and chemical cues may result in a synergistic effect, in some cases useful enough even to direct cell differentiation of adult MSC stem cells to non-canonical pathways, such as neuronal differentiation. Interestingly the effects of growth conditions onto a nano-patterned surface were stronger than the single biochemical induction on controls grown on un-patterned surfaces (Yim, Pang et al. 2007).

4. Promising bone substitutes in the tissue-engineering scenario

Ideal skeletal reconstruction depends on regeneration of normal tissues that result from initiation of progenitor cell activity. In this context, cells are considered as a key element to achieve the regeneration of the target tissue, since very few biomaterials are osteoinductive by themselves (Goshima, Goldberg et al. 1991; Ohgushi, Dohi et al. 1993; Boo, Yamada et al. 2002; Cancedda, Bianchi et al. 2003; Endres, Hutmacher et al. 2003; Livingston, Gordon et al. 2003; Derubeis and Cancedda 2004; Warren, Nacamuli et al. 2004; Arinzeh, Tran et al. 2005; Kimelman, Pelled et al. 2006; Bernhardt, Lode et al. 2009; Matsushima, Kotobuki et al. 2009).
The most intriguing concept in modern biomaterials is thus obtaining materials able to mimic a specific eventually pre-existing microenvironment and, therefore, inducing stem/progenitor cells to differentiate in a predetermined manner and to regenerate by themselves the bone tissue according to physiological pathways.
Several researches have been conducted using autologous bone marrow-derived osteo-progenitors to repair critical size segmental defects (Bianco and Robey 2001; Cancedda, Bianchi et al. 2003). The results of all these studies were in good agreement suggesting an important advantage in bone formation and, therefore, in the healing of the defect when cells were delivered together with a proper biomaterial scaffold. It is surprising that after the initial enthusiasm demonstrated by the flourishing of very encouraging large animal studies, only two pilot clinical studies have been performed (Quarto, Mastrogiacomo et al. 2001; Vacanti, Bonassar et al. 2001). Although material science technology has resulted in clear improvements in the field of regenerative medicine, no ideal bone substitute has been developed yet and hence large bone defects still represent a major challenge for orthopaedic and reconstructive surgeons. We are now aware, though, that the intended clinical use defines the desired properties of engineered bone substitutes. Anatomical defects in load bearing long bones, for instance, require devices with high mechanic stability whereas for craniofacial applications, initially injectable or moldable constructs are favorable. Therefore, the most intriguing concept is obtaining materials able to mimic a specific eventually pre-existing microenvironment, thus priming the natural processes of bone regeneration driven by cells.

5. Conclusion

In summary, a suitable scaffold for tissue engineering applications must have a structure correctly designed at different spatial scales to mimic the complex SC niche (Dellatore, Garcia et al. 2008). While it will probably not be necessary to mimic all aspects of the niche to enhance stem cells self-renewal and differentiation, it will almost certainly be necessary to simultaneously mimic multiple components of the niche (chemical and multi-scale architectural cues) to induce a specific cell differentiation and tissue ingrowth.
Proper surface sensing, then, has raised as a new requirement for progenitor cells lineage differentiation. Indeed precommitment of MSC grown on a specific matrix cannot be

overcome by the addition of soluble factors to the growth medium. The osteogenic differentiation of MSC seeded onto electrospun poly(ε-caprolactone)/ECM scaffolds is maintained even in the absence of dexamethasone in the culture medium, a molecule normally required in standard osteogenic induction of plastic-adherent MSC cultures (Thibault, Scott Baggett et al. 2010). This observation is therefore of paramount relevance for tissue engineering applications of MSC, considering that specific tissue repair applications, such as bone reconstruction, often lead cell-based applications to relevant rounds of ex-vivo cell duplications, normally performed on standard disposable culture plastic-ware. In this respect, the most recent literature brings new insights onto the sensitivity of stem cells to the mechanical microenvironment, but also raises relevant questions regarding the induction strategies and the physical environments of in vivo and ex vivo microenvironments. Significantly recent findings have also raised the possibility that an injured microenvironment may lose compliance due to insufficient sensitivity and remodelling options of stem cells once in a non-inducing environment such as a fibrotic scar (Berry, Engler et al. 2006). Moreover the current paradigm implies that tissue homeostasis is favoured by a compliant matrix and a relative low integrin-mediated cytoskeletal tension, whereas an elevated integrin-ERK-Rho activity favours a tumorigenic/proliferating behaviour, although an excessively stiff matrix or integrin-dependent activity would promote the generation of stable focal adhesion sites, ultimately antagonizing cell spreading.

Whether all these approaches and specific aspects, (scaffold stiffness compliance, surface topography and tridimensionality, scaffold chemistry) can be integrated into scaffold engineering to properly foster tissue regeneration remains to be seen; these aspects, however, have become even more relevant if the same pluripotent progenitor cells are used within tissue engineered composites proposed for multiple tissue repair, such as in the case of osteochondral defects. Microenvironmental changes may indeed influence the repair outcomes of the different tissues (Djouad, Delorme et al. 2007): the challenge, then, is to provide the proper cell "pre-commitment" in vitro to partially overcome an inappropriate pathological in vivo microenvironment.

6. Acknowledgment

This work was partially funded by the Italian Ministry of Education, Universities and Research (MIUR) (grant no. RBPR05RSM2; RBIP068JL9).

7. References

Alenghat, F. J. and D. E. Ingber (2002). Mechanotransduction: all signals point to cytoskeleton, matrix, and integrins *Sci STKE* 2002(119): pe6.

Anderson, J. M., M. Kushwaha, et al. (2009). Osteogenic differentiation of human mesenchymal stem cells directed by extracellular matrix-mimicking ligands in a biomimetic self-assembled peptide amphiphile nanomatrix *Biomacromolecules* 10(10): 2935-44.

Arinzeh, T. L., T. Tran, et al. (2005). A comparative study of biphasic calcium phosphate ceramics for human mesenchymal stem-cell-induced bone formation *Biomaterials* 26(17): 3631-8.

Badami, A. S., M. R. Kreke, et al. (2006). Effect of fiber diameter on spreading, proliferation, and differentiation of osteoblastic cells on electrospun poly(lactic acid) substrates *Biomaterials* 27(4): 596-606.

Balasundaram, G. and T. J. Webster (2006). Nanotechnology and biomaterials for orthopedic medical applications. *Nanomedicine* (Lond) 1(2): 169-76.

Balasundaram, G. and T. J. Webster (2007). An overview of nano-polymers for orthopedic applications. *Macromol Biosci* 7(5): 635-42.

Barnes, C. P., S. A. Sell, et al. (2007). Nanofiber technology: designing the next generation of tissue engineering scaffolds. *Adv Drug Deliv Rev* 59(14): 1413-33.

Barrilleaux, B., D. G. Phinney, et al. (2006). Review: ex vivo engineering of living tissues with adult stem cells. *Tissue Eng* 12(11): 3007-19.

Bashur, C. A., R. D. Shaffer, et al. (2009). Effect of fiber diameter and alignment of electrospun polyurethane meshes on mesenchymal progenitor cells. *Tissue Eng Part A* 15(9): 2435-45.

Beil, M., A. Micoulet, et al. (2003). Sphingosylphosphorylcholine regulates keratin network architecture and visco-elastic properties of human cancer cells. *Nat Cell Biol* 5(9): 803-11.

Benoit, D. S., M. P. Schwartz, et al. (2008). Small functional groups for controlled differentiation of hydrogel-encapsulated human mesenchymal stem cells. *Nat Mater* 7(10): 816-23.

Bercoff, J., S. Chaffai, et al. (2003). In vivo breast tumor detection using transient elastography. *Ultrasound Med Biol* 29(10): 1387-96.

Bernhardt, A., A. Lode, et al. (2009). In vitro osteogenic potential of human bone marrow stromal cells cultivated in porous scaffolds from mineralized collagen. *J Biomed Mater Res A* 90(3): 852-62.

Berry, M. F., A. J. Engler, et al. (2006). Mesenchymal stem cell injection after myocardial infarction improves myocardial compliance. *Am J Physiol Heart Circ Physiol* 290(6): H2196-203.

Bershadsky, A. D., N. Q. Balaban, et al. (2003). Adhesion-dependent cell mechanosensitivity. *Annu Rev Cell Dev Biol* 19: 677-95.

Bianco, P. and P. G. Robey (2001). Stem cells in tissue engineering. *Nature* 414(6859): 118-21.

Bonzani, I. C., R. Adhikari, et al. (2007). Synthesis of two-component injectable polyurethanes for bone tissue engineering. *Biomaterials* 28(3): 423-33.

Boo, J. S., Y. Yamada, et al. (2002). Tissue-engineered bone using mesenchymal stem cells and a biodegradable scaffold. *J Craniofac Surg* 13(2): 231-9; discussion 240-3.

Boskey, A. L., S. B. Doty, et al. (2008). Modulation of extracellular matrix protein phosphorylation alters mineralization in differentiating chick limb-bud mesenchymal cell micromass cultures. *Bone* 42(6): 1061-71.

Boyde, A., A. Corsi, et al. (1999). Osteoconduction in large macroporous hydroxyapatite ceramic implants: evidence for a complementary integration and disintegration mechanism. *Bone* 24(6): 579-89.

Bruder, S. P., K. H. Kraus, et al. (1998). The effect of implants loaded with autologous mesenchymal stem cells on the healing of canine segmental bone defects. *J Bone Joint Surg Am* 80(7): 985-96.

Cancedda, R., G. Bianchi, et al. (2003). Cell therapy for bone disease: a review of current status. *Stem Cells* 21(5): 610-9.

Chang, B. S., C. K. Lee, et al. (2000). Osteoconduction at porous hydroxyapatite with various pore configurations. *Biomaterials* 21(12): 1291-8.

Chastain, S. R., A. K. Kundu, et al. (2006). Adhesion of mesenchymal stem cells to polymer scaffolds occurs via distinct ECM ligands and controls their osteogenic differentiation. *J Biomed Mater Res A* 78(1): 73-85.

Chen, X. D., V. Dusevich, et al. (2007). Extracellular matrix made by bone marrow cells facilitates expansion of marrow-derived mesenchymal progenitor cells and prevents their differentiation into osteoblasts. *J Bone Miner Res* 22(12): 1943-56.

Cheng, L., F. Ye, et al. (2010). Osteoinduction of hydroxyapatite/beta-tricalcium phosphate bioceramics in mice with a fractured fibula. *Acta Biomater* 6(4):1569-74.

Christley, S., M. S. Alber, et al. (2007). Patterns of mesenchymal condensation in a multiscale, discrete stochastic model. *PLoS Comput Biol* 3(4): e76.

Connelly, J. T., A. J. Garcia, et al. (2008). Interactions between integrin ligand density and cytoskeletal integrity regulate BMSC chondrogenesis. *J Cell Physiol* 217(1): 145-54.

Critchley, D. R. (2000). Focal adhesions - the cytoskeletal connection. *Curr Opin Cell Biol* 12(1): 133-9.

Cukierman, E., R. Pankov, et al. (2001). Taking cell-matrix adhesions to the third dimension. *Science* 294(5547): 1708-12.

Dahl, K. N., A. J. Engler, et al. (2005). Power-law rheology of isolated nuclei with deformation mapping of nuclear substructures. *Biophys J* 89(4): 2855-64.

Dalby, M. J., N. Gadegaard, et al. (2007). The control of human mesenchymal cell differentiation using nanoscale symmetry and disorder. *Nat Mater* 6(12): 997-1003.

Dalby, M. J., D. McCloy, et al. (2006). Osteoprogenitor response to semi-ordered and random nanotopographies. *Biomaterials* 27(15): 2980-7.

Dalby, M. J., D. McCloy, et al. (2006). Osteoprogenitor response to defined topographies with nanoscale depths. *Biomaterials* 27(8): 1306-15.

De Oliveira, J. F., P. F. De Aguiar, et al. (2003). Effect of process parameters on the characteristics of porous calcium phosphate ceramics for bone tissue scaffolds. *Artif Organs* 27(5): 406-11.

Delanoe-Ayari, H., R. Al Kurdi, et al. (2004). Membrane and acto-myosin tension promote clustering of adhesion proteins. *Proc Natl Acad Sci U S A* 101(8): 2229-34.

Dellatore, S. M., A. S. Garcia, et al. (2008). Mimicking stem cell niches to increase stem cell expansion. *Curr Opin Biotechnol* 19(5): 534-40.

Deng, W., M. Obrocka, et al. (2001). In vitro differentiation of human marrow stromal cells into early progenitors of neural cells by conditions that increase intracellular cyclic AMP. *Biochem Biophys Res Commun* 282(1): 148-52.

Derubeis, A. R. and R. Cancedda (2004). Bone marrow stromal cells (BMSCs) in bone engineering: limitations and recent advances. *Ann Biomed Eng* 32(1): 160-5.

Dhawan, J. and D. M. Helfman (2004). Modulation of acto-myosin contractility in skeletal muscle myoblasts uncouples growth arrest from differentiation. *J Cell Sci* 117(Pt 17): 3735-48.

Ding, D. C., W. C. Shyu, et al. (2011). Mesenchymal stem cells. *Cell Transplant* 20(1): 5-14.

Discher, D. E., P. Janmey, et al. (2005). Tissue cells feel and respond to the stiffness of their substrate. *Science* 310(5751): 1139-43.

Djouad, F., B. Delorme, et al. (2007). Microenvironmental changes during differentiation of mesenchymal stem cells towards chondrocytes. *Arthritis Res Ther* 9(2): R33.

Dong, J., H. Kojima, et al. (2001). In vivo evaluation of a novel porous hydroxyapatite to sustain osteogenesis of transplanted bone marrow-derived osteoblastic cells. *J Biomed Mater Res* 57(2): 208-16.

Dong, J., T. Uemura, et al. (2002). Promotion of bone formation using highly pure porous beta-TCP combined with bone marrow-derived osteoprogenitor cells. *Biomaterials* 23(23): 4493-502.

Dupont, S., L. Morsut, et al. (2011). Role of YAP/TAZ in mechanotransduction. *Nature* 474(7350): 179-83.

Eisenbarth, E., D. Velten, et al. (2007). Biomimetic implant coatings. *Biomol Eng* 24(1): 27-32.

El-Amin, S. F., E. Botchwey, et al. (2006). Human osteoblast cells: isolation, characterization, and growth on polymers for musculoskeletal tissue engineering. *J Biomed Mater Res A* 76(3): 439-49.

El-Ghannam, A. (2005). Bone reconstruction: from bioceramics to tissue engineering. *Expert Rev Med Devices* 2(1): 87-101.

Endres, M., D. W. Hutmacher, et al. (2003). Osteogenic induction of human bone marrow-derived mesenchymal progenitor cells in novel synthetic polymer-hydrogel matrices. *Tissue Eng* 9(4): 689-702.

Engler, A. J., M. A. Griffin, et al. (2004). Myotubes differentiate optimally on substrates with tissue-like stiffness: pathological implications for soft or stiff microenvironments. *J Cell Biol* 166(6): 877-87.

Engler, A. J., S. Sen, et al. (2006). Matrix elasticity directs stem cell lineage specification. *Cell* 126(4): 677-89.

Erbe, E. M., J. G. Marx, et al. (2001). Potential of an ultraporous beta-tricalcium phosphate synthetic cancellous bone void filler and bone marrow aspirate composite graft. *Eur Spine J* 10 Suppl 2: S141-6.

Erisken, C., D. M. Kalyon, et al. (2008). Functionally graded electrospun polycaprolactone and beta-tricalcium phosphate nanocomposites for tissue engineering applications. *Biomaterials* 29(30): 4065-73.

Fabbri, M., G. C. Celotti, et al. (1995). Hydroxyapatite-based porous aggregates: physico-chemical nature, structure, texture and architecture. *Biomaterials* 16(3): 225-8.

Fan, H., T. Ikoma, et al. (2007). Surface structural biomimetics and the osteoinduction of calcium phosphate biomaterials. *J Nanosci Nanotechnol* 7(3): 808-13.

Flautre, B., K. Anselme, et al. (1999). Histological aspects in bone regeneration of an association with porous hydroxyapatite and bone marrow cells. *J Mater Sci Mater Med* 10(12): 811-4.

Franceschi, R. T. and G. Xiao (2003). Regulation of the osteoblast-specific transcription factor, Runx2: responsiveness to multiple signal transduction pathways. *J Cell Biochem* 88(3): 446-54.

Fu, R. H., Y. C. Wang, et al. (2011). Differentiation of stem cells: strategies for modifying surface biomaterials. *Cell Transplant* 20(1): 37-47.

Gao, L., R. McBeath, et al. (2010). Stem cell shape regulates a chondrogenic versus myogenic fate through Rac1 and N-cadherin. *Stem Cells* 28(3): 564-72.

Gauthier, O., J. M. Bouler, et al. (1998). Macroporous biphasic calcium phosphate ceramics: influence of macropore diameter and macroporosity percentage on bone ingrowth. *Biomaterials* 19(1-3): 133-9.

Gauthier, O., R. Muller, et al. (2005). In vivo bone regeneration with injectable calcium phosphate biomaterial: a three-dimensional micro-computed tomographic, biomechanical and SEM study. *Biomaterials* 26(27): 5444-53.

Gigante, A., S. Manzotti, et al. (2008). Adult mesenchymal stem cells for bone and cartilage engineering: effect of scaffold materials. *Eur J Histochem* 52(3): 169-74.

Goessler, U. R., K. Bieback, et al. (2006). In vitro analysis of integrin expression during chondrogenic differentiation of mesenchymal stem cells and chondrocytes upon dedifferentiation in cell culture. *Int J Mol Med* 17(2): 301-7.

Gorsline, R. T., P. Tangkawattana, et al. (2010). Accelerated chondrogenesis in nanofiber polimeric scaffolds embedded with BMP-2 genetically engineeered chondrocytes. *Journal Biomedical Science and Engineering* 3: 908-916.

Goshima, J., V. M. Goldberg, et al. (1991). The osteogenic potential of culture-expanded rat marrow mesenchymal cells assayed in vivo in calcium phosphate ceramic blocks. *Clin Orthop Relat Res*(262): 298-311.

Grayson, W. L., P. H. Chao, et al. (2008). Engineering custom-designed osteochondral tissue grafts. *Trends Biotechnol* 26(4): 181-9.

Guarino, V., F. Causa, et al. (2008). The role of hydroxyapatite as solid signal on performance of PCL porous scaffolds for bone tissue regeneration. *J Biomed Mater Res B Appl Biomater* 86B(2): 548-57.

Guo, H., Y. Ma, et al. (2009). Pivotal Advance: PKCzeta is required for migration of macrophages. J *Leukoc Biol* 85(6): 911-8.

Hamidouche, Z., E. Hay, et al. (2008). FHL2 mediates dexamethasone-induced mesenchymal cell differentiation into osteoblasts by activating Wnt/beta-catenin signaling-dependent Runx2 expression. *FASEB J* 22(11): 3813-22.

Hamilton, D. W., M. O. Riehle, et al. (2005). Articular chondrocyte passage number: influence on adhesion, migration, cytoskeletal organisation and phenotype in response to nano- and micro-metric topography. *Cell Biol Int* 29(6): 408-21.

Hamilton, D. W., M. O. Riehle, et al. (2005). The response of primary articular chondrocytes to micrometric surface topography and sulphated hyaluronic acid-based matrices. *Cell Biol Int* 29(8): 605-15.

Harley, B. A., A. K. Lynn, et al. (2010). Design of a multiphase osteochondral scaffold III: Fabrication of layered scaffolds with continuous interfaces. *J Biomed Mater Res A* 92(3): 1078-93.

Harris, L. D., B. S. Kim, et al. (1998). Open pore biodegradable matrices formed with gas foaming. *J Biomed Mater Res* 42(3): 396-402.

He, W., P. Cowin, et al. (2003). Untangling desmosomal knots with electron tomography. *Science* 302(5642): 109-13.

Hench, L. L. and J. M. Polak (2002). Third-generation biomedical materials. *Science* 295(5557): 1014-7.

Hollister, S. J., R. D. Maddox, et al. (2002). Optimal design and fabrication of scaffolds to mimic tissue properties and satisfy biological constraints. *Biomaterials* 23(20): 4095-103.

Hoshiba, T., N. Kawazoe, et al. (2009). Development of stepwise osteogenesis-mimicking matrices for the regulation of mesenchymal stem cell functions. *J Biol Chem* 284(45): 31164-73.

Huang, H., R. D. Kamm, et al. (2004). Cell mechanics and mechanotransduction: pathways, probes, and physiology. *Am J Physiol Cell Physiol* 287(1): C1-11.

Huang, J., Y. W. Lin, et al. (2007). Development of nano-sized hydroxyapatite reinforced composites for tissue engineering scaffolds. *J Mater Sci Mater Med* 18(11): 2151-7.

Hunt, J. A. (2008). Regenerative medicine: Materials in a cellular world. *Nat Mater* 7(8): 617-8.

Hutmacher, D. W. (2000). Scaffolds in tissue engineering bone and cartilage. *Biomaterials* 21(24): 2529-43.

Hutmacher, D. W., J. T. Schantz, et al. (2007). State of the art and future directions of scaffold-based bone engineering from a biomaterials perspective. *J Tissue Eng Regen Med* 1(4): 245-60.

James, R., U. S. Toti, et al. (2011). Electrospun nanofibrous scaffolds for engineering soft connective tissues. *Methods Mol Biol* 726: 243-58.

Jayaraman, K., M. Kotaki, et al. (2004). Recent advances in polymer nanofibers. *J Nanosci Nanotechnol* 4(1-2): 52-65.

Jiang, T., W. I. Abdel-Fattah, et al. (2006). In vitro evaluation of chitosan/poly(lactic acid-glycolic acid) sintered microsphere scaffolds for bone tissue engineering. *Biomaterials* 27(28): 4894-903.

Karageorgiou, V. and D. Kaplan (2005). Porosity of 3D biomaterial scaffolds and osteogenesis. *Biomaterials* 26(27): 5474-91.

Kasten, P., R. Luginbuhl, et al. (2003). Comparison of human bone marrow stromal cells seeded on calcium-deficient hydroxyapatite, beta-tricalcium phosphate and demineralized bone matrix. *Biomaterials* 24(15): 2593-603.

Keselowsky, B. G., D. M. Collard, et al. (2005). Integrin binding specificity regulates biomaterial surface chemistry effects on cell differentiation. *Proc Natl Acad Sci USA* 102(17): 5953-7.

Kim, K. Y., M. Kovacs, et al. (2005). Disease-associated mutations and alternative splicing alter the enzymatic and motile activity of nonmuscle myosins II-B and II-C. *J Biol Chem* 280(24): 22769-75.

Kim, S. S., M. S. Park, et al. (2006). Accelerated bonelike apatite growth on porous polymer/ceramic composite scaffolds in vitro. *Tissue Eng* 12(10): 2997-3006.

Kimelman, N., G. Pelled, et al. (2006). Applications of gene therapy and adult stem cells in bone bioengineering. *Regen Med* 1(4): 549-61.

Kon, E., M. Delcogliano, et al. (2010). Orderly osteochondral regeneration in a sheep model using a novel nano-composite multilayered biomaterial. *J Orthop Res* 28(1): 116-24.

Kon, E., A. Muraglia, et al. (2000). Autologous bone marrow stromal cells loaded onto porous hydroxyapatite ceramic accelerate bone repair in critical-size defects of sheep long bones. *J Biomed Mater Res* 49(3): 328-37.

Kondo, N., A. Ogose, et al. (2006). Osteoinduction with highly purified beta-tricalcium phosphate in dog dorsal muscles and the proliferation of osteoclasts before heterotopic bone formation. *Biomaterials* 27(25): 4419-27.

Kotobuki, N., K. Ioku, et al. (2005). Observation of osteogenic differentiation cascade of living mesenchymal stem cells on transparent hydroxyapatite ceramics. *Biomaterials* 26(7): 779-85.

Kretlow, J. D. and A. G. Mikos (2007). Review: mineralization of synthetic polymer scaffolds for bone tissue engineering. *Tissue Eng* 13(5): 927-38.

Kundu, A. K. and A. J. Putnam (2006). Vitronectin and collagen I differentially regulate osteogenesis in mesenchymal stem cells. *Biochem Biophys Res Commun* 347(1): 347-57.
Lee, J. W. and R. Juliano (2004). Mitogenic signal transduction by integrin- and growth factor receptor-mediated pathways. *Mol Cells* 17(2): 188-202.
Lee, M. Y., J. P. Huang, et al. (2009). Angiogenesis in differentiated placental multipotent mesenchymal stromal cells is dependent on integrin alpha5beta1. *PLoS One* 4(10): e6913.
Lenza, R. F., W. L. Vasconcelos, et al. (2002). Surface-modified 3D scaffolds for tissue engineering. *J Mater Sci Mater Med* 13(9): 837-42.
Leung, V. Y., D. Chan, et al. (2006). Regeneration of intervertebral disc by mesenchymal stem cells: potentials, limitations, and future direction. *Eur Spine J* 15 Suppl 3: S406-13.
Limouze, J., A. F. Straight, et al. (2004). Specificity of blebbistatin, an inhibitor of myosin II. *J Muscle Res Cell Motil* 25(4-5): 337-41.
Liu, H. and T. J. Webster (2006). Nanomedicine for implants: A review of studies and necessary experimental tools. *Biomaterials* 28(2): 354-369.
Livingston, T., P. Ducheyne, et al. (2002). In vivo evaluation of a bioactive scaffold for bone tissue engineering. *J Biomed Mater Res* 62(1): 1-13.
Livingston, T. L., S. Gordon, et al. (2003). Mesenchymal stem cells combined with biphasic calcium phosphate ceramics promote bone regeneration. *J Mater Sci Mater Med* 14(3): 211-8.
Lu, J. X., B. Flautre, et al. (1999). Role of interconnections in porous bioceramics on bone recolonization in vitro and in vivo. *J Mater Sci Mater Med* 10(2): 111-20.
Lynn, A. K., S. M. Best, et al. (2010). Design of a multiphase osteochondral scaffold. I. Control of chemical composition. *J Biomed Mater Res A* 92(3): 1057-65.
Mackay, A. M., S. C. Beck, et al. (1998). Chondrogenic differentiation of cultured human mesenchymal stem cells from marrow. *Tissue Eng* 4(4): 415-28.
Marcacci, M., E. Kon, et al. (1999). Reconstruction of extensive long-bone defects in sheep using porous hydroxyapatite sponges. *Calcif Tissue Int* 64(1): 83-90.
Martin, I., S. Miot, et al. (2007). Osteochondral tissue engineering. *J Biomech* 40(4): 750-65.
Mastrogiacomo, M., S. Scaglione, et al. (2006). Role of scaffold internal structure on in vivo bone formation in macroporous calcium phosphate bioceramics. *Biomaterials* 27(17): 3230-7.
Matsushima, A., N. Kotobuki, et al. (2009). In vivo osteogenic capability of human mesenchymal cells cultured on hydroxyapatite and on beta-tricalcium phosphate. *Artif Organs* 33(6): 474-81.
Mikos, A. G., G. Sarakinos, et al. (1993). Laminated three-dimensional biodegradable foams for use in tissue engineering. *Biomaterials* 14(5): 323-30.
Miranti, C. K. and J. S. Brugge (2002). Sensing the environment: a historical perspective on integrin signal transduction. *Nat Cell Biol* 4(4): E83-90.
Mooney, D. J., C. L. Mazzoni, et al. (1996). Stabilized polyglycolic acid fibre-based tubes for tissue engineering. *Biomaterials* 17(2): 115-24.
Murphy, W. L., S. Hsiong, et al. (2005). Effects of a bone-like mineral film on phenotype of adult human mesenchymal stem cells in vitro. *Biomaterials* 26(3): 303-10.

Mygind, T., M. Stiehler, et al. (2007). Mesenchymal stem cell ingrowth and differentiation on coralline hydroxyapatite scaffolds. *Biomaterials* 28(6): 1036-47.

Nakagawa, S., P. Pawelek, et al. (1989). Extracellular matrix organization modulates fibroblast growth and growth factor responsiveness. *Exp Cell Res* 182(2): 572-82.

Nakamura, T. (2007). Biomaterial osteoinduction. *J Orthop Sci* 12(2): 111-2.

Nam, Y. S. and T. G. Park (1999). Porous biodegradable polymeric scaffolds prepared by thermally induced phase separation. *J Biomed Mater Res* 47(1): 8-17.

Ng, A. M., K. K. Tan, et al. (2008). Differential osteogenic activity of osteoprogenitor cells on HA and TCP/HA scaffold of tissue engineered bone. *J Biomed Mater Res A* 85(2): 301-12.

Nicolas, A., B. Geiger, et al. (2004). Cell mechanosensitivity controls the anisotropy of focal adhesions. *Proc Natl Acad Sci U S A* 101(34): 12520-5.

Nicula, R., F. Luthen, et al. (2007). Spark plasma sintering synthesis of porous nanocrystalline titanium alloys for biomedical applications. *Biomol Eng* 24(5): 564-7.

Niemeyer, P., U. Krause, et al. (2003). Mesenchymal stem cells for tissue engineering of bone: 3D-cultivation and osteogenic differentiation on mineralized collagen. *Z Orthop Ihre Grenzgeb* 141(6): 712-7.

Oberlender, S. A. and R. S. Tuan (1994). Spatiotemporal profile of N-cadherin expression in the developing limb mesenchyme. *Cell Adhes Commun* 2(6): 521-37.

Ohgushi, H., Y. Dohi, et al. (1993). Osteogenic differentiation of marrow stromal stem cells in porous hydroxyapatite ceramics. *J Biomed Mater Res* 27(11): 1401-7.

Paszek, M. J., N. Zahir, et al. (2005). Tensional homeostasis and the malignant phenotype. *Cancer Cell* 8(3): 241-54.

Peters, A., H. Schell, et al. (2010). Standard bone healing stages occur during delayed bone healing, albeit with a different temporal onset and spatial distribution of callus tissues. *Histol Histopathol* 25(9): 1149-62.

Place, E. S., N. D. Evans, et al. (2009). Complexity in biomaterials for tissue engineering. *Nat Mater* 8(6): 457-70.

Qian, L. and W. M. Saltzman (2004). Improving the expansion and neuronal differentiation of mesenchymal stem cells through culture surface modification. *Biomaterials* 25(7-8): 1331-7.

Quarto, R., M. Mastrogiacomo, et al. (2001). Repair of large bone defects with the use of autologous bone marrow stromal cells. *N Engl J Med* 344(5): 385-6.

Ren, J., P. Zhao, et al. (2008). Poly (D: ,L: -lactide)/nano-hydroxyapatite composite scaffolds for bone tissue engineering and biocompatibility evaluation. *J Mater Sci Mater Med* 19(3):1075-82.

Ren, T., J. Ren, et al. (2005). The bone formation in vitro and mandibular defect repair using PLGA porous scaffolds. *J Biomed Mater Res A* 74(4): 562-9.

Saldana, L., S. Sanchez-Salcedo, et al. (2009). Calcium phosphate-based particles influence osteogenic maturation of human mesenchymal stem cells. *Acta Biomater* 5(4): 1294-305.

Santiago, J. A., R. Pogemiller, et al. (2009). Heterogeneous differentiation of human mesenchymal stem cells in response to extended culture in extracellular matrices. *Tissue Eng Part A* 15(12): 3911-22.

Sato, M. and T. J. Webster (2004). Nanobiotechnology: implications for the future of nanotechnology in orthopedic applications. *Expert Rev Med Devices* 1(1): 105-14.

Scaglione, S., P. Giannoni, et al. (2011). Order versus Disorder: in vivo bone formation within osteoconductive scaffolds. *Science* submitted.

Schugens, C., V. Maquet, et al. (1996). Polylactide macroporous biodegradable implants for cell transplantation. II. Preparation of polylactide foams by liquid-liquid phase separation. *J Biomed Mater Res* 30(4): 449-61.

Sherwood, J. K., S. L. Riley, et al. (2002). A three-dimensional osteochondral composite scaffold for articular cartilage repair. *Biomaterials* 23(24): 4739-51.

Silva, G. A., C. Czeisler, et al. (2004). Selective differentiation of neural progenitor cells by high-epitope density nanofibers. *Science* 303(5662): 1352-5.

Smith, L. A., X. Liu, et al. (2008). Tissue Engineering with Nano-Fibrous Scaffolds. *Soft Matter* 4(11): 2144-2149.

Smith, L. A. and P. X. Ma (2004). Nano-fibrous scaffolds for tissue engineering. *Colloids Surf B Biointerfaces* 39(3): 125-31.

Tamada, M., M. P. Sheetz, et al. (2004). Activation of a signaling cascade by cytoskeleton stretch. *Dev Cell* 7(5): 709-18.

Tampieri, A., M. Sandri, et al. (2008). Design of graded biomimetic osteochondral composite scaffolds. *Biomaterials* 29(26): 3539-46.

Thibault, R. A., L. Scott Baggett, et al. (2010). Osteogenic differentiation of mesenchymal stem cells on pregenerated extracellular matrix scaffolds in the absence of osteogenic cell culture supplements. *Tissue Eng Part A* 16(2): 431-40.

Tonti, G. A. and F. Mannello (2008). From bone marrow to therapeutic applications: different behaviour and genetic/epigenetic stability during mesenchymal stem cell expansion in autologous and foetal bovine sera *Int J Dev Biol* 52(8): 1023-32.

Turner, C. E. (2000). Paxillin and focal adhesion signalling. *Nat Cell Biol* 2(12): E231-6.

Uygun, B. E., S. E. Stojsih, et al. (2009). Effects of immobilized glycosaminoglycans on the proliferation and differentiation of mesenchymal stem cells. *Tissue Eng Part A* 15(11): 3499-512.

Vacanti, C. A., L. J. Bonassar, et al. (2001). Replacement of an avulsed phalanx with tissue-engineered bone. *N Engl J Med* 344(20): 1511-4.

Vasita, R. and D. S. Katti (2006). Nanofibers and their applications in tissue engineering. *Int J Nanomedicine* 1(1): 15-30.

Vega, F. M., G. Fruhwirth, et al. (2011). RhoA and RhoC have distinct roles in migration and invasion by acting through different targets. *J Cell Biol* 193(4): 655-65.

Vogel, V. and M. Sheetz (2006). Local force and geometry sensing regulate cell functions. *Nat Rev Mol Cell Biol* 7(4): 265-75.

Voog, J. and D. L. Jones (2010). Stem cells and the niche: a dynamic duo. *Cell Stem Cell* 6(2): 103-15.

Wang, F., V. M. Weaver, et al. (1998). Reciprocal interactions between beta1-integrin and epidermal growth factor receptor in three-dimensional basement membrane breast cultures: a different perspective in epithelial biology. *Proc Natl Acad Sci U S A* 95(25): 14821-6.

Warren, S. M., R. K. Nacamuli, et al. (2004). Tissue-engineered bone using mesenchymal stem cells and a biodegradable scaffold. *J Craniofac Surg* 15(1): 34-7.

Wei, G. and P. X. Ma (2008). Nanostructured Biomaterials for Regeneration. *Adv Funct Mater* 18(22): 3566-3582.

Williams, J. M., A. Adewunmi, et al. (2005). Bone tissue engineering using polycaprolactone scaffolds fabricated via selective laser sintering. *Biomaterials* 26(23): 4817-27.

Wise, J. K., A. L. Yarin, et al. (2009). Chondrogenic differentiation of human mesenchymal stem cells on oriented nanofibrous scaffolds: engineering the superficial zone of articular cartilage. *Tissue Eng Part A* 15(4): 913-21.

Woo, K. M., V. J. Chen, et al. (2003). Nano-fibrous scaffolding architecture selectively enhances protein adsorption contributing to cell attachment. *J Biomed Mater Res A* 67(2): 531-7.

Woo, K. M., J. H. Jun, et al. (2007). Nano-fibrous scaffolding promotes osteoblast differentiation and biomineralization. *Biomaterials* 28(2): 335-43.

Woodbury, D., E. J. Schwarz, et al. (2000). Adult rat and human bone marrow stromal cells differentiate into neurons. *J Neurosci Res* 61(4): 364-70.

Wu, Y. C., S. Y. Shaw, et al. (2006). Bone tissue engineering evaluation based on rat calvaria stromal cells cultured on modified PLGA scaffolds. *Biomaterials* 27(6): 896-904.

Xiao, G., D. Jiang, et al. (2000). MAPK pathways activate and phosphorylate the osteoblast-specific transcription factor, Cbfa1. *J Biol Chem* 275(6): 4453-9.

Yang, F., R. Murugan, et al. (2005). Electrospinning of nano/micro scale poly(L-lactic acid) aligned fibers and their potential in neural tissue engineering. *Biomaterials* 26(15): 2603-10.

Yeung, T., P. C. Georges, et al. (2005). Effects of substrate stiffness on cell morphology, cytoskeletal structure, and adhesion. *Cell Motil Cytoskeleton* 60(1): 24-34.

Yim, E. K., S. W. Pang, et al. (2007). Synthetic nanostructures inducing differentiation of human mesenchymal stem cells into neuronal lineage. *Exp Cell Res* 313(9): 1820-9.

Yuan, H., K. Kurashina, et al. (1999). A preliminary study on osteoinduction of two kinds of calcium phosphate ceramics. *Biomaterials* 20(19): 1799-806.

Yuan, H., Z. Yang, et al. (1998). Osteoinduction by calcium phosphate biomaterials. *J Mater Sci Mater Med* 9(12): 723-6.

Zhang, L. and T. J. Webster (2009). Nanotechnology and nanomaterials: Promises for improved tissue regeneration. *Nano Today* 4: 66-80.

Zhao, F., W. L. Grayson, et al. (2009). Perfusion affects the tissue developmental patterns of human mesenchymal stem cells in 3D scaffolds. *J Cell Physiol* 219(2): 421-9.

18

Influence of Angiogenesis on Osteogenesis

Susanne Jung and Johannes Kleinheinz
Department of Cranio-Maxillofacial Surgery, University Hospital Muenster
Research Unit Vascular Biology of Oral Structures (VABOS)
Germany

1. Introduction

No bone without vasculature. Current research reveals the molecular biology behind this simple statement. Bone formation and regeneration and the associated physiological and pathologically altered angiogenesis is not only of vital importance in the omnipresent fracture repair but also plays a constantly growing role in the diagnosis and treatment of chronically infected bone or osteonecrosis. The jaw and especially the mandible with its exceptional vascularisation in the human body are prone to these diseases.
The understanding of the molecular mechanisms of osteogenic-angiogenic coupling is the fundament of a sufficient diagnosis and therapy.
From the description of the physiological vascularisation of the jaw bones to pure autologous tissue engineering in the context of regenerative medicine we follow the process of bone formation under the influence of a sufficient angiogenesis with special regard to the key protagonist VEGF, the osteogenic-angiogenic interface and their interaction.

2. Vascular anatomy of the jaw bones

Head and neck are supplied with blood by branches of the external carotid arteries (ECA). The anterior and posterior superior alveolar arteries originate from the maxillary artery and provide the nutrient supply for the upper jaw and teeth; the inferior alveolar artery supplies the mandible and lower teeth. The jaw bones are drained of blood by the correspondent veins and eventually by the subclavian and the jugular vein.
The outline of the bony vasculature is closely related to the mechanism of osteogenesis. The comparison between upper and lower jaw is the most striking example for this correlation in the human body: the ramified vessel system of the maxilla is a direct consequence of its intramembraneous ossification. In contrast, the mandible is supplied by a central artery and vein in the wake of a partly enchondral ossification as observed in long bones. This difference is of vital importance for physiological or iatrogenic bone regeneration.
Coupling interface of bone and vasculature is the periosteum. It provides osteoprogenitor cells and keeps the capillary vessel system.
Age dependent changes in the vascular and bony anatomy are the vertical and sagittal bone loss in the edentulous jaw, circulatory disorders, osteoporosis or bisphosphonate associated osteonecrosis. These alterations pose a challenge for successful regenerative intervention.

2.1 Development

The base of any form of growth and development is the continuous and sufficient supply with (progenitor) cells, oxygen and nutrients.

The first appearance of a primitive vessel formation happens shortly after gastrulation, between the 19. and 21. day of embryonal growth. At this moment the first cells that express a VEGF receptor can be visualized. These cells proliferate in the yolk sac, form so called blood islands with endothelial and hemapoetic cells and expand to the first vessel network, the capillary plexus (Ferkowicz et al., 2003).

In a parallel development, first angioblasts form the pairs of dorsal aortas following the midline of the embryo (Lawson et al., 2002).

With the onset of heartbeat and the permanent presence of erthyroblasts in the newly formed vessel structures, the first vascular network is established.

From now on a permanent remodelling and adaptation process set in, that lasts during the whole lifetime of the individual.

Mechanical signals from the blood flow are pivotal factors in the differentiation of the vessel system, of the development of veins and arteries, their walls and the diameters of the growing vasculature.

The head and neck vasculature derives from the first three aortal arches.

2.2 Anatomy

The carotid artery develops as a self-contained vessel from the ventral aorta.

The external carotid artery (ECA) provides the blood supply for the face and the jaws. The maxillary artery as a terminal branch of the ECA supplies the profound regions of the face. It rises behind the mandibular neck to the pterygopalatine fossa; its course is divided into three portions: the mandibular, the pterygoid and the pterygopalatine portion.

The facial bones are supplied by the superior and inferior alveolar artery, branches of the maxillary artery.

The posterior superior alveolar artery divides into numerous branches to supply the upper teeth and the jawbone. The anterior superior artery originates from the infraorbital artery and supplies the upper incisor and canine teeth. Between these branches ramify many anastomoses.

The inferior alveolar artery runs with the inferior alveolar nerve through the mandibular canal, divides into incisor and mental branch and provides the nutrient supply for the bone, the pulp of the lower teeth and the lower lip.

In striking contrast to the ramified vessel system of the upper jaw, the mandible is perfused by the inferior alveolar artery alone.

The venous drainage occurs via concomitant homonymous veins and anastomoses in the jugular and subclavian veins finally (Bouthillier et al., 1996).

Figure 1 illustrates the arterial vascularisation of the upper and lower jaw originating from the internal carotid artery.

2.3 Remodelling

During foetal development the circulation is not only essential to provide the vital oxygen; the mechanical caused by the blood stream exert fundamental impact on the maturating vessel network.

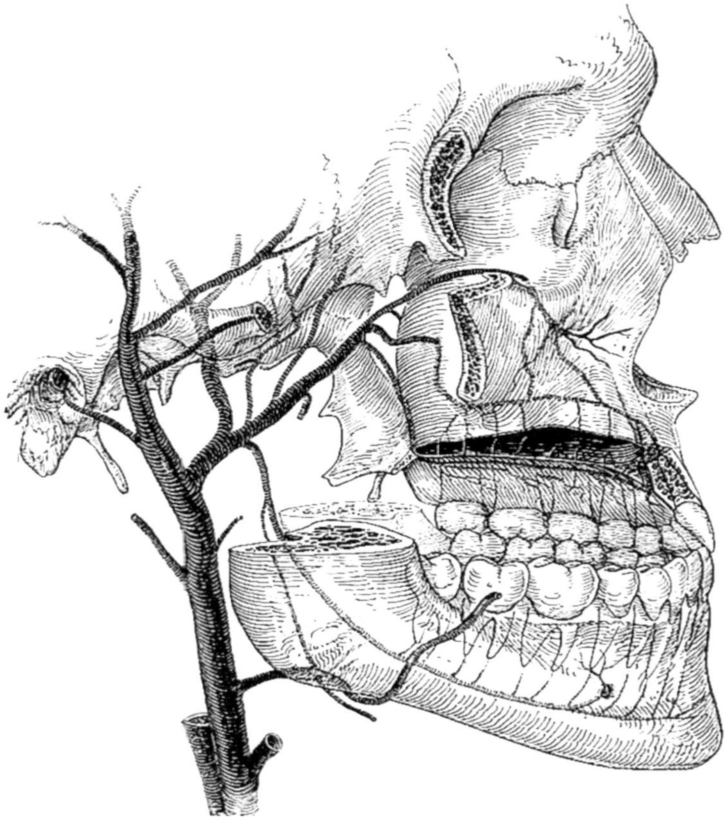

Fig. 1. Vascularisation of the jaw

The development and maturation of the vessel system varies significantly form embryos to adults. One important and often neglected factor is the mechanical influence of an intact blood circulation on vessel anatomy and remodelling.
Vascular remodelling in adults in a physiological surrounding is basically influenced by shear stress and circumferential stretch. These mechanical forces are relevant factors in vessel development and maturation. The molecular mechanisms of transforming a mechanical requirement to a histological response of the involved cells are so far not fully understood (Jones et al., 2011).

3. Basic course of osteogenesis

Bone tissue formation rests upon the balanced interaction between the secretory function of osteoblasts and the degrading activity of osteoclasts. There is intramembraneous and enchondral ossification: both processes are dependent on vascular ingrowth. Most of the facial bones, flat cranial bones, parts of the mandible and the clavicle are formed by enchondral ossification. It combines the growth potential of cartilage and bone to promote skeletal formation and development.
Intramembraneous ossification is the direct ossification of the mesenchym, the primitive connective tissue. It starts with the invasion of capillaries into the mesenchymal zone, where

the resident cells transform into mature osteoblasts and begin to deposit bone matrix extracellularly. The growing bone formations build woven bone, which is eventually organized in stable lamellar bone.

Osteogenesis in general takes place in the vicinity of neo-vessels that mediate the delivery of osteoprogenitors, secrete mitogen for osteoblasts, and transport nutrient and oxygen.

Blood vessels provide a conduit for the recruitment of cells involved in resorption and bone deposition and are therefore a crucial condition for any bone formation or regeneration. Decisive factor of the rate of bone increase is the level of vascularisation of the growth plate.

Bone is the rigid form of connective tissue. It is a dynamic construct, composed of secretory cells, the osteocytes, degrading cells, the osteoclasts and the surrounding matrix, the osteoid. The interaction of its major constituents define its mechanical characteristics: collagen type I guarantees its elasticity; calcium/ phosphate as structural components stand for stability and resistance; dermatan- and chondroitinsulfate represent together with collagen the main part of the extracellular matrix and determine the structural integrity of the tissue.

Skeletal bone is the reservoir of calcium and phosphate that can be activated and released under endocrinological control. Its dynamic (re)modelling is a result of a permanent adaptation to changing physiological and pathological requirements. A fine-tuned balance of (physiological) stress and phases of regeneration leads to a continuous coexistence of osteogenesis and bone degradation and simultaneously to a corresponding vasculature (Martin et al., 2008).

Osteogenesis takes place not only during foetal development but also in the course of facture healing or during infection. One differentiates chondral and intramembranous ossification.

Essential osteoblastic markers are collagen type I, alpha 1 (Col1a1) and alkaline phosphatase (ALP); osteoclasts are characterized by the presence of tartrate-resistant acid phosphatase (TRAP), cathepsin K and dendritic cell-specific transmembrane protein (DC-STAMP) (Suzuki et al., 2007).

3.1 Chondral ossification

The bone formation on the basis of a cartilaginous scaffold is the basic principle of chondral ossification.

One has to distinguish between an enchondral and a perichondral process.

The course of enchondral ossification starts with the ingrowth of nutritive blood vessels in the growth centre of the epiphyseal growth plate. The blood supplies the tissue not only with oxygen and nutrients but also provides a new cell population: Chondroclasts to disintegrate the chondral gantry and osteoblasts and –clasts to construct the bone. Histologically, the growth area of enchondral ossification is characterized by a pillar-like alignment of the chondrocytes in an ascending order from reposing chondral cells to secreting ripe chondrocytes. These active cells produce the initial osteoid that is subsequently mineralized by the locally proliferating osteoblasts. The apoptosis of hypertrophic chondrocytes induces their replacement by osteoblasts in combination with endothelial progenitor cell invasion.

The proliferative chondrocytes produce among other growth and transcription factors VEGF that regulates the vascular ingrowth.

Parts of the os petrosum maturate via chondral ossification.
Perichondral ossification concerns the shafts of the long bones in the extremities: osteoblasts deriving from the well-perfused perichondrium build a three-dimensional cell collar around the cartilage and start to form bone appositionally (Berkowitz et al., 2002).

3.2 Intramembranous ossification

Intramembranous ossification describes the direct bony transformation of the primitive connective tissue and affects mainly the flat bones.
Starting point is the compaction of mesenchymal cells into ossification centres. The cells differentiate to osteoblasts and start to produce the bony matrix, the osteoid. With the emplacement of calcium and phosphate the developing bone maturates; the osteoblasts become resting osteocytes.
Intramembranous ossification takes place in the flat bones of the skull, for example the os frontale (Berkowitz et al., 2002).

3.3 Ossification of the mandible

The skull structures develop from the first and second branchial arch. These mesenchymal tissue areas appear during the fourth week of embryonic growth and determine the base of the individual facial features.
The central vessel/ nerve bundle characterizes the anatomy of the mandible. This special anatomical feature determines the course of the bony development of the lower jaw.
First ossification centres in the mandible can be detected in the area of the branching mandibular nerve in close relation to the developing mental foramen. From here the Meckel´s cartilage indicates course and direction. The Meckel´s cartilage is a transient chondral check rail derived from the first brachial arch to guide the developing bony mandible. It degrades until the 24. week of embryonal development. From its posterior remnants the auditory ossicles malleus and incus are formed.
From the mental foramen to the base of the condylar processus, the course of the mandibular nerve with its corresponding vessels defines the guardrail for the maturating bone. The growing nerval structure with its concomitant vessel system and the tide of growth factors seems to influence and guide the bone maturation.
In an investigation of 25 fetuses the prenatal development and remodelling af the mandible was studied with regard to the morphometry, the histological evaluation of the remodelling processes and a 3D analysis.
The results suggest a strong influence of Meckel´s cartilage configuration and dental primordial on the later anatomy, the signal cascading and specific role of the involved growth and transcription factors in the 3D moulding of the human mandible are not clear (Radlanski et al., 2003).
Another region of interest as far as the ossification of the mandible is concerned, is the mental foramen, the area where nerve, vessels and surrounding extracellular matrix pass the bone. The mental foramen is a centre of bone development and remodelling of the mandible. The pattern of bone configuration and dynamics around the mental foramen differs from the ossification in the bony gutter.
The presence and proximity of the nerve and concomitant vessels does have significant impact on mandibular ossification (Radlanski et al., 2004).

3.4 Role of VEGF in the growth plate

The vital importance of a sufficient VEGF induced vessel formation in the physiological morphological progression of the growth plate was underlined impressively by the work of Gerber et el. 1999: in an mouse model the inactivation of VEGF via soluble antibodies resulted in a significant reduction of vascularisation and consequently ossification. Their data suggests that VEGF-dependent angiogenesis is essential for the processing of he growth plate, cell differentiation and the coupling of chondral degradation and bone formation.

The lack of trabecular bone and the hypertrophy of the chondrocytes were reversible after cessation of VEGF suppression (Gerber et al. 1999).

4. Angiogenic cascade

VEGF and its receptors are the key regulators of angiogenesis.

Angiogenesis is the growth of new blood vessels from pre-existing vasculature that occurs in physiological and pathological contexts. This process is often triggered by a signalling cascade that occurs upon ligand-receptor binding between vascular endothelial growth factor (VEGF) and its receptors (VEGFR1/Flt-1, VEGFR2/KDR). These receptors are expressed by endothelial cells that line the blood vessels. Most potent trigger for angiogenesis is hypoxia.

Angiogenesis, the complex physiological sequence of vasodilatation, degradation of basement membrane, endothelial cell migration, chemotaxis, increasing vascular permeability and eventually endothelial cell proliferation and vessel formation is regulated by VEGF(R). Osteogenic and angiogenic cascade are inseparably interconnected by the functional alliance of endothelial cells, osteoblasts and growth factors.

Angiogenesis is defined as the growth of new blood vessels on the base of existing vascular structures in contrast to vasculogenesis, the development of a new vessel system by sprouting, differentiating endothelial cells. These pivotal processes mark the starting point of any form of vascular development or regeneration processes. In malignant growth, the appearance of new vessels can be a determinant of progression and poor prognosis, whereas a lack of vasculature inevitably leads to loss of functional efficiency and tissue necrosis from the brain to the limbs.

The fine-tuned balance of vasculo- and angiogenesis is controlled by many growth and transcription factors (Pandya, Dhalla et al., 2006).

4.1 Angiogenic cascade – in vitro

The angiogenic cascade in vitro follows a characteristic course of events: Initiation, proliferation or invasion and maturation.

The starting point - the initiation - is the release of angiogenic factors, VEGF, basic fibroblast growth factor (bFGF) or tumour necrosis factor alpha (TGFα). These mitogens and stimulators are released by or inflammatory cells.

Proliferation and invasion includes the degradation and remodelling of the extracellular matrix, the chemotaxis of endothelial (progenitor) cells and the altered expression of adhesion molecules. During the maturation phase, the newly formed vessel ripens, a lumen develops and finally the basal lamina envelops the structure.

These processes can be retraced in vitro:

After the sterile preparation of endothelial cells from an umbilical cord vein according to standard procedure follows the adhesion and proliferation of endothelial cells in their culture well under defined conditions.
After 5 to 7 days a (sub-) confluence of the proliferating cells and formation of a monolayer can be observed.
The formation of circular ring-structures with characteristic cell-cell contacts marks the angiogenic potency of the cultured cells. Special marker of the cell-cell contact is VE-Cadherin, CD 144 antigen.
The adjustment of the culture conditions influences the tissue architecture: in a three dimensional scaffold the formation of a network of capillaries can be observed.
Immunohistochemical markers to characterize the proliferating cell population are CD 31 or von Willebrand factor, for example.
Growth and proliferation of endothelial cells and their surrounding matrix in the course of angiogenesis are governed by a multitude of regulating factors (Fraisl et al., 2009).

4.2 Angiogenic growth factors

Stimulators and inhibitors govern Angiogenesis. The most important pro-angiogenic factors are VEGF, Angiopoietin-1, β-Estradiol, bFGF, IL-8, Leptin, MMPs, NOS, PDGF-BB, TNF-α, Angiogenin and TGFα. They exert their influence on different stages during the angiogenic cascade.
Table 1 subsumes the respective function of the most prominent angiogenic factors.

Factor	Function
VEGF	Cell mitogen, vascular permeability
Angiopoietin-1	Maturation, stability
Angiotropin	Cell migration
bFGF	Angiogenesis in wound healing
IL-8	Chemotaxis, inflammation
NOS	Vasodilatation
PDGF-BB	Chemotaxis, proliferation
Angiogenin	Indirect stimulation, release of angiogenic factors
TGFα	Mitogen, differentiation of endothelial cells

Table 1. Pro-angiogenic factor

The variety of different angiogenic factors with similar functions implies the idea of redundancy: the quick and undisturbed succession of events of angiogenesis is too important for the function of the whole organism to take the risk of relying on unique regulators or promoters.

The interconnection of angio- and osteogenesis is reflected in the expression of angiopoietin by osteoblasts. The role of angiopoietin in bone formation is not yet clear. The results of in vivo studies suggest that one essential osteogenic effect is the induction of angiogenesis. An over expression of angiopoietin led to an augmentation of bone mass (Suzuki et al., 2007).

The anti-angiogenic antagonists are among others angiostatin, anti-angiogenic anti-thrombin III, canstatin, endostatin (collagen XIII fragment), fibronectin fragment, heparinases, IFN-α,β,χ IL4, IL12, IL18, plasminogen activator inhibitor or pigment epithelium derived factor, PEDF.

Their function is to balance angiogenesis and to interfere with excessive vessel formation. The most important inhibiting factors with their functions are summarized in table 2.

Angiogenesis in vivo is a complex, multilayer process involving the degradation of the vessel basement lamina, the migration and proliferation of endothelial cells and the formation of luminal structures. It includes the cell interaction - activation and silencing by a variety of different growth and transcription factors - and depends on the information and provided by the extracellular matrix. Excessive growth has the same devastating pathological impact as a deficient vasculature (Klagsbrun et al., 1991).

Factor	Function
TGFß	Inhibition of endothelial cell motility
Canstatin	Collagen IV, inhibitor
Endostatin	Collagen XVIII, inhibitor
Platelet factor IV	Inhibitor of growth factor-dependent endothelial stimulation
TNFα	Inhibitor of endothelial cell proliferation
Protamin	Block of heparin-binding growth factors
PEDF	Inductor of endothelial cell apoptosis

Table 2. Anti-angiogenic factors

Another pivotal factor in the regulation of angiogenesis is the microRNA (miRNA) expression. MiRNAs are noncoding small RNAs the modulate postransscriptional gene expression. Their up- and down-regulation is partly regulated by hypoxia and is involved in the stimulation or inhibition of VEGF or other (anti-)angiogenic factors. The most prominent angiogenic miRNAs are miR-126 and miR-27b, anti-angiogenic miRNAs are miR-221 or miR-20a (Fraisl et al., 2009).

4.3 Oxygen sensing

How is a lack of oxygen and therefore an urgent need for further vascularisation detected?

VEGF is release is controlled by hypoxia, oxidative stress, pH or glucose concentration. VEGF binds to membrane tyrosine kinase receptors and initiates chemotaxis, cell migration and vascular permeability (Bikfalvi et al., 2002).

The key of oxygen sensing lies in the HIF transcription factor complex. Oxygen-dependent enzymes, the prolyl-hydroxylases (PHDs) bind oxygen and couple it to HIF1-α. Von Hippel Lindau (VHL) protein attacks this complex and initiates its degradation. Decreasing levels of oxygen lead to a rising concentration of HIF1α (Riddle et al., 2009).
HIF initiates pronounced tissue oxygenation, among many other factors by the stimulation of VEGF expression.
In addition, a tight osteogenic-angiogenic coupling was demonstrated by an increase of bone mass and volume up to 70% after HIF1α activation.
The pharmacological activation of the HIF pathway resulted in an accelerated bone healing (Wang et al., 2008, Wan et al., 2007, Towler, 2008).
Under hypoxic conditions the expression of a variety of genes is altered, adapted to the reduced oxygen supply. Among the hypoxia-induced genes there are not genes involved in metabolism, angiogenesis, erythropoiesis, chondrocyte differentiation and the development and maturation of the collagen matrix (Araldi et al., 2010).

5. Interaction between endothelial cells and osteoblasts on different matrix structures

The potency of the matrix has an enormous impact on the development of the new tissue. It influences cell adhesion, spreading and signalling, cell growth and migration as well as the differentiation of the extracellular matrix and the tissue morphogenesis.
One decisive prerequisite of a matrix structure – especially in the context of regenerative medicine - is the ability to support the growth of blood vessels.
As biological surfaces, collagen membranes are accepted as reliable surfaces to support the growth and adherence of endothelial cells.
Titanium emerged as an artificial surface, which clearly supports the growth and adhesion of endothelial cells and therefore complies with the basic requirements for osseointegration of titanium implants.
The differentiation, growth and function of cellular systems not only relies on their interaction and the interference by synthesized soluble factors but also to a considerable extent on the alloplastic or autologous surfaces of the scaffold on which they are supposed to proliferate. This undisturbed or even conducive environment is the precondition for any healing process where the application of any form of implant is involved.
The biological properties of autologous and alloplastic surfaces have been under scrutiny.

5.1 Collagen

The successful clinical application of materials should involve detailed investigations on interaction between them and tissue with which they will contact.
One relevant approach to gain information about the influence of the scaffold on the proliferating cells is to observe the behaviour of endothelial cells on a collagen material, using histological and immunohistochemical methods.
In an in-vitro essay, isolated human umbilical cord vein cells (HUVECs) identified by means of endothelial-specific antibodies were cultivated to perform the trials.
Cells were seeded in a standard density on a collagen membrane (Lycoll, Resorba, Nuernberg, Germany) and on gelatin-coated control plastic surfaces, after two passages. These were maintained for periods of 1, 7, or 14 days. The cells adhered, spread and

proliferated. Within 24 hours the HUVECs started forming a subconfluent monolayer. It could be observed that the cultured cells expressed integrins and synthesized fibronectin.
The results suggested that the collagen material supported growth and attachment of endothelial cells. In addition, the attachment seemed to be related to the fibronectin synthesized by the cells and to its highly expressed receptor, the α5β1 integrin. (Breithaupt-Faloppa et al., 2006).
The undisturbed growth and proliferation on a collagen scaffold is the basis of standard tissue engineering protocols.

5.2 Titanium

Osteogenesis on titanium surfaces is the prerogative of any form of osseointegration of alloplastic titanium implants from the knee to the teeth.
The mechanisms of cell-cell interaction and dependence meet the prerequisites and requirements of an alloplastic interface with its special mechanical and biochemical material properties of the surface, such as roughness or hydrophilicity.
The coculture of osteoblast-like and endothelial cells on titanium surfaces represents a reliable experimental model to evaluate their biological behaviour. It is a simple and reduced model of osteo- and angiogenesis without the complexity of an in vivo investigation.
The cellular integrity and their function ability are measured by the synthesis of characteristic cell markers via immunohistochemistry.
Gene expression of osteo- and angiogenic factors is used to consolidate and quantify the histological results.
In an in vitro trial the endothelial cell behaviour on a titanium surface was investigated by the immunhistochemical detection of characteristic endothelial markers CD 31 and von Willebrand factor, fibronectin and its receptor, vitronectin and its receptor and VE cadherin.
The results indicated that the attachment of ECs on titanium could be related to cellular-derived fibronectin and the binding to its specific receptor, the α5β1 integrin. It was observed that titanium effectively serves as a suitable substrate for endothelial cell attachment, growth and proliferation in the initial phase. It is suggested to describe this feature of titanium if not angiogenic then at least angio-conductive.
Nevertheless the studied cells did not show the physiological cytological development after titanium contact of 7 days. These findings do not correlate with good clinical results.
The high degree of biocompatibility of titanium and its alloys is intimately related to the passively formed oxide film on the metallic surface (Breithaupt-Falloppa et al., 2008).

6. Influence of angiogenesis during osteogenesis

The level of angiogenesis is a pivotal element of osteogenesis. The vasculature provides the transport of oxygen, nutrients, growth regulating factors and metabolites to all tissues in the body.
Many factors act as key protagonists of bone angiogenesis: VEGF, especially its isoforms VEGF120, 164 and 188, bFGF, TGFβ, HIF are the most potential ones. VEGF in its isoforms with the corresponding receptors have emerged as the decisive coupling factors between epi- and metaphyseal vascularisation and cartilage development and therefore enchondral ossification.

The angiopoietins Ang-1 and Ang-2, hepatocyte growth factor HGF, platelet-derived growth factor PDGF, the IGF family and the neurotrophins NGFs also have angiogenic properties.
The course of osteogenesis paves the way for the vascular anatomy: anastomosing vessel net or central artery and vein.

6.1 VEGF in angiogenesis

The development of a new vessel network provides multipotent mesenchymal cells, endothelial progenitor cells and the substrates for any form of regeneration. The sprouting of vessels from surrounding local bone tissue into a defect or regeneration area is considered to be a decisive factor for undisturbed osteogenesis. Quality of bone in traumatology is assessed mostly by its structure and mechanical stability. In the treatment of many other diseases of the skeleton the clinical evaluation of the bony vascularisation is the benchmark for its quality.
The latter is especially disturbed in most patients needing solid and reliable bone regeneration: in clinical routine it is the problem of patients with osteomyelitis, tumour patients after radiation, or patients after trauma in whom local vascularisation is considerably disturbed and not predictable.
For this reason, extensive evaluation of this part of the regeneration process is of great significance as a means of gaining basic information for specific therapy.
Not only the physiological environment of an intact sprouting vessel system, but also its special stimulation by decisive growth and differentiation factors has an enormous impact on the development and maturation of the surrounding bone. The outstanding and field-tested in this context is VEGF. Many publications underline its efficacy in angiogenesis.
The stimulation of bone regeneration by rhVEGF165 was investigated, among many other trials, in a rabbit animal model.
Mandibular critical size defects in 56 animals were treated with VEGF-laden collagen carriers. The significant effects of the stimulation became most obvious during the phase of physiological regression during the regeneration process:
This physiologic process, activated by hyperoxygenation caused by hyperperfusion led to delamination of endothelial cells from the basal membrane and endothelial apoptosis.
The effect of rhVEGF165 as an endothelial cell survival factor led to persisting angiogenesis in the study group, extending beyond day 14. Morphologic and structural effects of angiogenesis on osteogenesis could not be observed until after at least 28 days, because maximum bone regeneration occurs at that time.
In addition to the indirect influence of the increased supply of osteogenic progenitor cells by the developing vascular system, the increased expression of osteogenic growth factors expressed by endothelial cells seemed to play a crucial role in this experimental setting (Kleinheinz et al., 2005).

6.2 Impact of osteoid

In addition to established angiogenic growth stimulators the microenvironment of secretory osteoblasts shows a significant pro-angiogenic effect. The produced extracellular matrix influences demonstrably migration, proliferation and differentiation of endothelial progenitor cells.

Cell culture investigation proved the significant stimulation of endothelial cells to form vessels-like structures.

Post-translationally glycosylated collagen I was unmasked as decisive factor concerning the effective activation of endothelial stimulation via p38/MAPK signalling pathway.

It was demonstrated that the adhesion of endothelial cells to collagen I, one of the major constituents of osteoid, not only supports their growth but represents an essential condition to perform the angiogenic switch.

In cell culture observations the role glycosylated collagen type I became clear:

It induces the expression of molecules like MMP-2 and IL-12. These factors were connected with an increased vessel networking, at the same time an increasing amount of mature bone tissue. The close connection of these processes seems to be linked by the many involved growth factors as well as the variety of progenitors and regenerative cells (Palmieri et al., 2010).

7. Autologous bone tissue engineering

Today approaches to restore the osseous integrity of such primarily non-regenerating lesions rely either on the integration of autologous bone or on the implantation of substitute materials. Autologous bone chips or bars reducing the defect size below the critical threshold enable primary healing processes, but their supply is limited.

Bone tissue engineering subsumes an abundance of approaches to the artificial engineering of osseous structures based on the principle of an extracorporeal linking of cells, matrix and bioactive factors.

As the supply of nutrition and oxygen via diffusion in three-dimensional tissue formations is restricted to an area of 150 μm around the nutritive capillary, resorption and devitalisation in the center of the implant lead to a loss of mechanical stability. A successful approach to bone tissue engineering therefore has to incorporate both osteogenic and vascular development as a combination of osteogenic and angiogenic processes will lead to an improved and accelerated regeneration with simultaneous proliferation of the new complex tissue of these cell types.

A pure autologous combination of all components with an autologous scaffold seems to be of elementary importance for further improved cell behaviour.

Promising approaches concentrate mainly on fibrin as the best possible basis for tissue engineering as fibrin plays an important role in haemostasis and also represents a physiological temporary scaffold for the cell invasion in the following healing process.

Large bone defects or osseous deficiency caused by congenital malformation, trauma, tumour surgery or osseous infection exceeding a critical threshold and showing no signs of spontaneous healing are still a great challenge to reconstructive surgery. These critical size defects (CSDs) are individually determined for every species and vary by age, size and sex (Cima et al., 1991, Glowacki et al., 2004).

Today approaches to restore the osseous integrity of such primarily non-regenerating lesions rely either on the integration of autologous bone or on the implantation of substitute materials. Autologous bone chips or bars reducing the defect size below the critical threshold enable primary healing processes, but their supply is limited. Xenogenous or alloplastic materials seem to be a simple and gentle method with bone regeneration from the

surrounding bone, but malresorption and decreased osteogenic potency of substitute material affect the stability of the defect area.
Besides the repair and reconstruction the in vivo - regeneration of lost tissue using the activation of local regeneration cascades therefore seems to be the most promising method.
Bone tissue engineering subsumes an abundance of approaches to the artificial engineering of osseous structures based on the principle of an extracorporal linking of cells, matrix and bioactive factors. Many approaches to promote the adherence and proliferation of transplanted, osteopotent cells expressing of extracellular matrix proteins (ECM) used xenogenous or alloplastic scaffolds such as corals (Putnam et al., 1996), collagen (Saadeh et al., 2001), PLA/PGA or fibrin glue, but there is no record for the use of autologous scaffolds. Despite the success in some parts of the substitute materials such as the avoidance of donor site morbidity, optimization of the three-dimensional configuration of the transplant and healing processes, infection and non-integration of the implanted materials occur (Frerich et al., 2000).
The potency of the matrix has an enormous impact on the development of the new tissue. This potency should include: a) an optimal scaffold for specific tissue engineering, b) matrix biodegradation with contemporaneous new tissue formation, c) a guaranteed interaction of newly formed tissue with new specific cells forming coherent structures with mechanical function and stability.
As the supply of nutrition and oxygen via diffusion in three-dimensional tissue formations is restricted to an area of 150 µm around the nutritive capillar, resorption and devitalisation in the center of the implant lead to a loss of mechanical stability. A successful approach to bone tissue engineering therefore has to incorporate both osteogenic and vascular development as a combination of osteogenic and angiogenic processes will lead to an improved and accelerated regeneration with simultaneous proliferation of the new complex tissue of these cell types (Pellisier et al., 2003).
A pure autologous combination of all components with an autologous scaffold seems to be of elementary importance for further improved cell behaviour.
Promising approaches concentrate mainly on fibrin as the best possible basis for tissue engineering as fibrin plays an important role in haemostasis and also represents a physiological temporary scaffold for the cell invasion in the following healing process (Nolting et al., not published yet).

8. Conclusion

Chronic infections of the bone, necrosis and drug associated osteomyelitis are diseases that will become of growing dominance in the coming decades. These pathologies result not only from a deficient bone formation or regeneration but are closely related to an impaired vasculature.
To understand and to better evaluate the co-dependence of developing or regenerating bone and its accompanying vasculature it is of vital importance to retrace their serial interface from fetal development to adulthood, from macroscopic anatomy to biomolecular processes.
The future of regenerative approaches to bone healing and regeneration will inevitably combine the field-tested strategies of tissue engineering with modern biomolecular techniques in the scientific environment of stem cells and gene therapy.

Our work showed conclusively the enormous reciprocal impact of angio- and osteogenesis. To stimulate the osteogenesis one clearly has to optimize - or at least: incorporate - the vascular regeneration in the context of regenerative medicine.
Promising approaches for the near future will investigate the potential of stem and regenerative cells: mesenchymal stem cells or cells from the dental pulp have been examined particularly with regard to their ability to differentiate in an angio- or osteogenic direction. The experimental results encourage further experimental work in this field of regenerative medicine to develop an applicable therapeutic strategy of personalized regenerative medicine.

9. References

Araldi E, Schipani E. (2010). *Hypoxia, HIFs and bone development,* Bone 2010, 47:190-196.

Berkovitz BKB, Holland GR, Moxham BJ. (2002). *Oral anatomy, histology and embryology,* Mosby, Edinburgh.

Bikfalvi A, Bicknell R. (2002). *Recent advances in angiogenesis, anti-angiogenesis and vascular targeting,* Trends Pharmacol Sci. 2002 Dec;23(12):576-82.

Bouthillier A, Van Loveren H, Keller J. (1996). *Segments of the internal carotid artery: a new classification,* Neurosurgery 38 (3): 425–432.

Breithaupt-Faloppa AC, de Lima WT, Oliveira-Filho RM, Kleinheinz J. (2008) *In vitro behaviour of endothelial cells on a titanium surface,* Head Face Med. 2008 Jul 23;4:14.

Breithaupt-Faloppa AC, Kleinheinz J, Crivello O Jr. (2006). *Endothelial cell reaction on a biological material,* J Biomed Mater Res B Appl Biomater. 2006 Jan;76(1):49-55.

Cima LG, Vacanti JP, Vacanti C, Ingber D, Mooney D, Langer R. (1991). *Tissue engineering by cell transplantation using degradable polymer substrates,* J Biomech Eng 113:143-51.

Ferkowicz MJ, Starr M, Xie X, Li W, Johnson SA, Shelley WC, Morrison PR, Yoder MC. (2003). *CD41 expression defines the onset of primitive and definitive hematopoiesis in the murine embryo,* Development. 2003 Sep;130(18):4393-403.

Fraisl P, Mazzone M, Schmidt T, Carmeliet P. (2009). *Regulation of angiogenesis by oxygen and metabolism,* Develop Cell 2009, 16(17): 167-179

Frerich B, Kurtz-Hoffmann J, Lindemann N, Müller S. (2000). *Untersuchungen zum Tissue engineering vaskularisierter knöcherner und weichgewebiger Transplantate,* 4 (Suppl 2):490-95.

Gerber HP, Vu TH, Ryan AM, Kowalski J, Werb Z, Ferrara N. (1999). *VEGF couples hypertrophic cartilage remodeling, ossification and angiogenesis during endochondral bone formation,* Nat Med. 1999 Jun;5(6):623-8.

Glowacki J, Shustermann M, Troulis M, Holmes R, Perrott D, Kaban LB. (2004). *Distraction Osteogenesis of the Porcine Mandible: Histomorphometric Evaluation of Bone,* Plast Reconstr Surg 113: 566.

Klagsbrun M, D'Amore PA. (1991). *Regulators of angiogenesis.* Annu Rev Physiol. 1991;53:217-39.

Kleinheinz J, Rottke S, Blasius S, Wiesmann HP, Joos U. (1997). *Immunohistochemical investigation of the microcirculation of mandibular osteomyelitis using endothelial-specific antibodies,* Int J Oral Maxillofac Surg 26, Suppl 1: 245.

Kleinheinz J, Wiesmann HP, Stratmann U, Joos U. (2002). *Beurteilung der Angiogenese und Osteogenese unter dem Einfluss von Vascular endothelial growth factor (VEGF)*, Mund Kiefer GesichtsChir 6: 175-182.

Kleinheinz J, Breithaupt-Faloppa A, Jayaraman M, Wiesmann HP, Joos U. (2004). *Proliferation and attachment of endothelial cells on a collagen surface*, Int Poster J Dent Oral Med 2004, Vol 6 No 3, 230.

Kleinheinz J, Nolting T, Büchter A, Wiesmann HP, Joos U. (2004). *Bone tissue engineering using autologous endothelial cells, osteoblasts and fibrin matrix, J Cranio Maxillofac Surg* 32, Suppl 1: S168.

Kleinheinz J, Fillies T, Brandt B, Joos U. (2005). *Gene expression profile of human osteoblasts and endothelial cells as basis for tissue engineering*, Int J Oral Maxillofac Surg 34, Suppl.1: 44.

Kleinheinz J, Nolting T, Wiesmann HP, Joos U. (2005). *Reinforcement of angiogenesis and osteogenesis in bone tissue engineering using autologous endothelial cells, osteoblasts and fibrinmatrix*, Int J Artif Organs 28: 313.

Kleinheinz J, Stratmann U, Joos U, Wiesmann HP. (2005). *VEGF-activated angiogenesis during bone regeneration*, J Oral Maxillofac Surg 63: 1310-1316.

Kleinheinz J, Jung S, Wermker K, Fischer C, Joos U. (2010). *Release kinetics of VEGF165 from a collagen matrix and structural matrix changes in a circulation model*, Head Face Med 6: 17.

Martin TJ, Seeman E. (2008). *Bone remodelling: its local regulation and the emergence of bone fragility*, Best Pract Res Clin Endocrinol Metab. 2008 Oct;22(5):701-22.

Nolting TC, Wermker K, Jung S, Joos U, Kleinheinz J. *Pure autologeous bone tissue engineering*, Not published yet.

Palmieri D, Valli M, Viglio S, Ferrari N, Ledda B, Volta C, Manduca P. (2010). *Osteoblasts extracellular matrix induces vessel like structures through glycosylated collagen I*, Exp Cell Res. 2010 Mar 10;316(5):789-99. Epub 2009 Dec 16.

Pandya NM, Dhalla NS, Santani DD.(2006). *Angiogenesis--a new target for future therapy*, Vascul Pharmacol. 2006 May;44(5):265-74. Epub 2006 Mar 20.

Pelissier P, Villars F, Mathouli-Pelissier S, Bareille R, Lafage-Proust MH, Vilamitjana-Amedee J. (2003). *Influence of vascularisation and osteogenic cells on ectopic bone formation within a madreporic ceramic in rats*, Plast Reconstr Surg 111(6); 1932-41.

Putnam AJ, Mooney DJ. (1996). *Tissue engineering using synthetic extracellular matrizes*, Nat Med 2:824-26.

Riddle RC, Khatri R, Schipani E, Clemens TL. (2009). *Role of hypoxia-inducible factor-1alpha in angiogenic-osteogenic coupling*, J Mol Med (Berl). 2009 Jun;87(6):583-90.

Ritman E, Bolander M, Fitzpatrick L. (1998). *Micro-CT imaging of structure-to-function relationship of bone microstructure and associated vascular involvement*, Tech Health Care 6:403.

Saadeh PB, Khoslah RK, Mehrara BJ, Steinbrech DS, McCormick SA, DeVore DP, Longaker MT. (2001). *Repair of a Critical Size Defect in the Rat Mandible Using allogenic Type I Collagen*, J Craniof Surg Vol 12,6:573-79.

Suzuki T, Miyamoto T, Fujita N, Ninomiya K, Iwasaki R, Toyama Y, Suda T. (2007) *Osteoblast-specific Angiopoietin 1 overexpression increases bone mass*, Biochem Biophys Res Commun. 2007 Nov 3;362(4):1019-25. Epub 2007 Aug 28.

Towler DA. (2008). *The osteogenic-angiogenic interface: novel insights into the biology of bone formation and fracture repair*, Curr Osteoporos Rep 2008, 6:67-71

Trueta J. (1963). *The role of vessels in osteogenesis,* J Bone Joint Surg 45B:402.
Wang Y, Wan C, Deng L. (2007). *The hypoxia-inducible factor alpha pathway couples angiogenesis to osteogenesis during skeletal development,* J Clin Invest 2007, 117: 1616– 1626.
Wan C, Gilbert SR, Wang Y. (2008). *Activation of the hypoxia-inducible factor-1alpha pathway accelerates bone regeneration,* Proc Natl Acad Sci U S A 2008, 105: 686– 691.